AF790074

Proteins and Steroids in Early Pregnancy

Edited by
H. M. Beier and P. Karlson

With the collaboration of
R.J. Aitken, M. Beato, B.M. Bhatt, D.W. Bullock, H.-W. Denker
W. Elger, F. Fridlansky, J.E. Gadsby, S.R. Glasser, Xi-gen Gu
R.B. Heap, P.W. Jungblut, C. Kirchner, C. Lambadarios
Shu-xiang Li, N.K. Love, S.A. McCormack, E. Milgrom
B.O. Nilsson, Y. Nishino, J.S. Perry, J. Quatacker, C. Rice
M. Sherman, J. de Visser, Yi-chien Yang, K. Yoshinaga

With 153 Figures

Springer-Verlag
Berlin Heidelberg New York 1982

Prof. Dr. med. Dr. rer. nat. Henning M. Beier
Abteilung Anatomie und Reproduktionsbiologie
Medizinische Fakultät der RWTH Aachen
D-5100 Aachen, Federal Republic of Germany

Prof. Dr. rer. nat. Dr. h. c. Peter Karlson
Institut für Physiologische Chemie I
Medizinische Fakultät der Philipps-Universität
D-3550 Marburg/Lahn, Federal Republic of Germany

ISBN-13: 978-3-642-67892-9 e-ISBN-13: 978-3-642-67890-5
DOI: 10.1007/978-3-642-67890-5

Library of Congress Cataloging in Publication Data. Main entry under title: Proteins and steroids in early pregnancy. Bibliography: p. Includes index. 1. Pregnancy. 2. Proteins. 3. Steroids hormones. I. Beier, H. M. II. Karlson, P. (Peter), 1918-. III. Aitken, R. J. [DNLM: 1. Pregnancy proteins. 2. Steroids. 3. Pregnancy. WQ 205 P967] QP281.P76 599.01'6 81-9080

This work is subject to copyright. All rights are reserved, whether the whole or part of materials is concerned, specifically those of translation, reprinting, re-use of illustrations, broadcasting, reproduction by photocopying machine or similar means, and storage in data banks. Under § 54 of the German Copyright Law where copies are made for other than private use, a fee is payable to "Verwertungsgesellschaft Wort", Munich.

© Springer-Verlag Berlin Heidelberg 1982

Softcover reprint of the hardcover 1st edition 1982

The use of registered names, trademarks, etc. in this publication does not imply, even in the absence of a specific statement, that such names are exempt from the relevant protective laws and regulations and therefore free for general use.

Preface

It is about 15 years since the first presentation on *uteroglobin* was given to a group of developmental biologists, reproductive physiologists, and geneticists who had gathered in November 1966 at Konstanz (Germany). In the following decade so much knowledge was accumulated that a special symposium seemed appropriate. This was organized as a satellite symposium to the International Congress of Endocrinology at Hamburg and brought together 50 scientists at Aachen. These scientists, working in the field of proteins and steroids in early pregnancy, recognized the impact of what had been reported, and many of them later agreed to contribute to this book and thus to present their research data available until December 1980.

The present volume covers a relatively broad spectrum of data and observations which shed some light on *preimplantational embryonic life* and on the supports and obstacles provided by the maternal organism with respect to final accomplishment of normal implantation and establishment of pregnancy. The book will serve both as a *textbook* and as a *scientific dictionary* for Ph.D. students, postdoctoral fellows and advanced scientists working in this area.

The course of *early pregnancy* depends very much on a proper balance of steroid hormones, and the induction of *protein synthesis* by *steroid hormones* is one of the well-known fundamental processes in cellular differentiation and embryonic development. This process involves the machinery of transcription and its control by steroid hormone receptor complexes, details of which are presented and discussed in this volume. We hope that the present compilation will stimulate designs for new experiments aimed at providing insight into the molecular biology of protein-steroid interactions as well as the control of gene expression.

It is not unlikely that from a practical point of view there will be an explosion of interest in steroid-controlled proteins which support early pregnancy and enable the mother-to-be to provide an *appropriate pregnancy milieu* for the conceptus. Clinicians will desire more detailed knowledge about the significance of steroids and proteins during the establishment of pregnancy. The conditions for egg transfer and blastocyst development will be elucidated to the benefit of those patients who can only have an own child if fertilization is accomplished extracorporally and if the conceptus is retransferred to its mother within a few days. The data on our basic research on proteins and steroids controlling uterine physiology and embryonic development presented in this book could serve as a really solid basis for further clinical work.

Any success this publication achieves will be due to the contributors who wrote the manuscripts, to the patience of our families, who sacrificed so many weeks and months for the editing work, and finally to the Springer Verlag, who took over the publication of this book.

Aachen and Marburg/Lahn
May 1981

Henning Beier
Peter Karlson

Contents

List of Contributors

R.J. Aitken

Medical Research Council
Dept. of Reproductive Biology
37, Chalmers Street

GB-Edinburgh EH3 9EW, U.K.

M. Beato

Dept. of Physiological Chemistry
Philipps-Universität
Deutschhausstrasse 1-2

D-3550 Marburg/Lahn, W. Germany

H.M. Beier

Dept. of Anatomy and Reproductive Biology, Rheinisch-Westfälische Technische Hochschule, Melatener Strasse 211

D-5100 Aachen, W. Germany

B.M. Bhatt

Dept. of Cell Biology
Baylor College of Medicine

Houston, Texas 77030, USA

D.W. Bullock

Dept. of Cell Biology
Baylor College of Medicine

Houston, Texas 77030, USA

H.-W. Denker

Dept. of Anatomy
Rheinisch-Westfälische Technische
Hochschule, Melatener Strasse 211

D-5100 Aachen, W. Germany

W. Elger

Dept. Endocrine Pharmacology
Schering Research Laboratories
Müllerstrasse 170-178

D-1000 Berlin 65, W. Germany

Francoise Fridlansky

Groupe de Recherches sur la Biochimie Endocrinienne et la Reproduction (INSERM U 135) Faculté de Médecine Paris Sud

F-94270 Bicêtre, France

J.E. Gadsby

Reproductive Endocrinology Program
Dept. of Pathology
University of Michigan

Ann Arbor, Michigan 48109, USA

S.R. Glasser

Dept. of Cell Biology
Baylor College of Medicine

Houston, Texas 77030, USA

Xi-gen Gu

Dept. of Reproductive Physiology
Shanghai Institute of Planned Parenthood
Research, 929 Heng Shan Road

Shanghai, Peoples Republic of China

R.B. Heap

Dept. of Physiology
A.R.C. Institute of Animal Physiology
Babraham
GB-Cambridge CB2, 4AT, U.K.

P.W. Jungblut

Max-Planck-Institut für experimentelle
Endokrinologie
P.O. Box 610309

D-3000 Hannover 61, W. Germany

P. Karlson

Dept. of Physiological Chemistry
Philipps-Universität
Deutschhausstrasse 1-2

D-3550 Marburg/Lahn, W. Germany

C. Kirchner

Dept. of Biology
Philipps-Universität
Lahnberge

D-3550 Marburg/Lahn, W. Germany

Carol Lambadarios

Dept. of Obstetrics & Gynaecology
University of Sheffield
Jessop Hospital for Women
Leavygreave Road

GB-Sheffield 3, U.K.

Shu-xiang Li

Dept. of Reproductive Physiology
Shanghai Institute of Planned Parenthood
Research, 929 Heng Shan Road

Shanghai, Peoples Republic of China

Nancy K. Love

Dept. of Zoology
University of Maryland

College Park, Maryland 20742, USA

Shirley A. McCormack

Dept. of Cell Biology
Baylor College of Medicine

Houston, Texas 77030, USA

E. Milgrom

Groupe de Recherches sur la Biochimie
Endocrinienne et la Reproduction (INSERM
U 135) Faculté de Médecine Paris Sud

F-94270 Bicêtre, France

B.O. Nilsson

Reproduction Research Unit
Dept. of Anatomy, Uppsala Universitets
Biomedicinska Centrum, Box 571

S-751 23 Uppsala, Sweden

Y. Nishino

Dept. of Endocrine Pharmacology
Schering Research Laboratories
Müllerstrasse 170-178

D-1000 Berlin 65, W. Germany

J.S. Perry

Dept. of Physiology
A.R.C. Institute of Animal Physiology,
Babraham

GB-Cambridge CB2 4AT, U.K.

J.R. Quatacker

Laboratorium voor Pathologische
Ontleedkunde, Akademisch Ziekenhuis
Rijksuniversiteit Gent, De Pintelaan 135

B-9000 Gent, Belgium

Catherine Rice

Dept. of Zoology
Arizona State University

Tempe, Arizona 85281, USA

M.I. Sherman

Dept. of Cell Biology
Roche Institute of Molecular Biology

Nutley, New Jersey 07110, USA

J. de Visser

Organon Scientific Development Group
Organon International B.V.

NL-4202 Oss, The Netherlands

Yi-chien Yang

Dept. of Reproductive Physiology
Shanghai Institute of Planned Parenthood
Research, 929 Heng Shan Road

Shanghai, Peoples Republic of China

K. Yoshinaga

National Institutes of Health
Reproductive Science Branch
Center for Population Research NICHD

Bethesda, MD 20205, USA

Proteins and Steroids in Early Pregnancy: General Considerations

Peter Karlson

Physiologisch-Chemisches Institut der Philipps-Universität, Marburg, Germany

There are many relations between steroids, more specifically steroid hormones, and proteins. To evaluate the relevance of these interactions tor early pregnancy, it may be useful to list at first these most important phenomena:

1. Adenotropic Hormones. The production of steroid hormones is under the control of the pituitary. The pituitary hormones are either peptides (e.g. corticotropin) or proteins. Of relevance in our context are the gonadotropins, glycoproteins consisting of two subunits with different functions – one binding to the membrane receptor, the other entering the cell and being responsible for the physiological action. Pituitary control of steroid hormone synthesis and release is important for the cooperation of various hormones in the sexual cycle, resulting in ovulation, as well as for the development of early pregnancy.

2. Enzymes of Biosynthesis. Steroid hormones are the end-products of a biosynthetic pathway involving many enzymes. Cholesterol may be regarded as the common starting material, since cholesterol is present abundantly in biological membranes of all cells (it is, of course, produced from acetyl-CoA in a long sequence of reactions). For the conversion of cholesterol to progesterone, four enzymes are needed; even more are necessary for the biosynthesis of estrone. It must be kept in mind that steroid hormones are not stored in the hormone-producing tissue; thus the control of biosynthesis is the way of controlling hormone production.

3. Transport of Steroid Hormones. Many steroid hormones are only sparingly soluble in water. In blood as well as in other body fluids, they are kept in solution by special binding proteins. This is not only of importance in the blood, where the steroid-binding proteins have been extensively studied, but also in other fluids including the uterine fluid. Is binding and transport of progesterone one of the main functions of uteroglobin? Perhaps the reader will find a tentative answer to this question in this volume.

4. Metabolic "Activation" in the Target Tissue. Some of the steroid hormones are converted in the target tissues to molecules that exert the physiological action. Specific enzymes (proteins) are involved in this "activation". A well-known example is testosterone: It is the main circulating androgen, but in many target tissues (prostate, seminal vesicles, skin) it is reduced by a steroid-5α-reductase to 5α-dihydrotestosterone, the "active androgen". It is a question of definition if we wish to call the secreted and circulating agent "the hormone", or if this term should be used for the substance arising in the target cell and exerting the action. According to the original definition by

Starling (1906), the first alternative would be the appropriate usage. But even in the latter case, it is confusing to term testosterone a "prohormone": This term should be restricted to proteins that serve as precursors of peptide hormones (e.g. proinsulin).

Ecdysone is another example of a steroid hormone converted in most target tissues into a second substance, ecdysterone. Even more complicated is the situation with 1α, 25-dihydroxycholecalciferol where different tissues are involved that are not, *per se*, endocrine glands. A detailed discussion is beyond the scope of this article.

5. Enzymes of Inactivation. For a hormonal control of any physiological function, the active hormone must be either removed from the site of action and/or the circulation, or it must be inactivated. The latter process is exerted mainly by steroid dehydrogenases; they reduce the unsaturated ketone group in ring A with NADH (or NADPH) as hydrogen donor. Though the liver is the main site of steroid hormone metabolism, some inactivation in the target tissue should not be excluded. On the other hand, binding of steroid hormones to proteins may delay their inactivation.

6. Hormone Receptors. It is now generally accepted that the physiological action of hormones is mediated by hormone receptors. In the case of steroid hormones, these receptors are not bound to the membrane; they are soluble proteins occuring within the cell. The generalized picture is that the steroid first binds to the cytosolic receptor. As a result of this interaction, the conformation of the receptor protein is changed, and the receptor-hormone-complex enters the nucleus, as outlined in the left part of Fig. 1. The binding of hormone by receptors is characterized by high affinity, in contrast to the binding by transport proteins. This is discussed in detail in a special chapter in this volume.

7. Induction of Proteins by Steroid Hormones. In the foregoing section, the steroid plays a rather passive part in protein-steroid-interactions. This is reversed in the process to be discussed now, the essential part of steroid hormone action, summarized schematically in Fig. 1.

Evidence for the induction of proteins by steroids came mainly from two lines of research. One was the study of enzymes of amino acid metabolism and gluconeogenesis after cortisol treatment, showing an increase in enzyme activity. The second was the observation that the steroid hormone ecdysone can induce puffs in salivary glands of insects (Chironomus larvae) (Clever and Karlson, 1960). This led us to the hypothesis that steroid hormones may act by retrieval of genetic information in a process comparable to enzyme induction in bacteria as studied by Jacob and Monod in E. coli. This hypothesis, first published in 1961 (Karlson 1961) and later elaborated (Karlson 1963), led to numerous studies on the influence of steroid hormones on mRNA synthesis, corroborating the original concept. Indeed, for a small number of steroid hormones, the mRNA for the specifically induced proteins has been measured and shown to increase after hormone treatment. One of these systems is the induction of uteroglobin by progesterone, which is of special relevance in our context.

This is not the place to go into the details of these studies on steroid hormone action, induction of mRNA and of specific proteins; the reader is referred to recent reviews (Karlson et al. 1975; Karlson 1979). The concept had to be modified in that not the

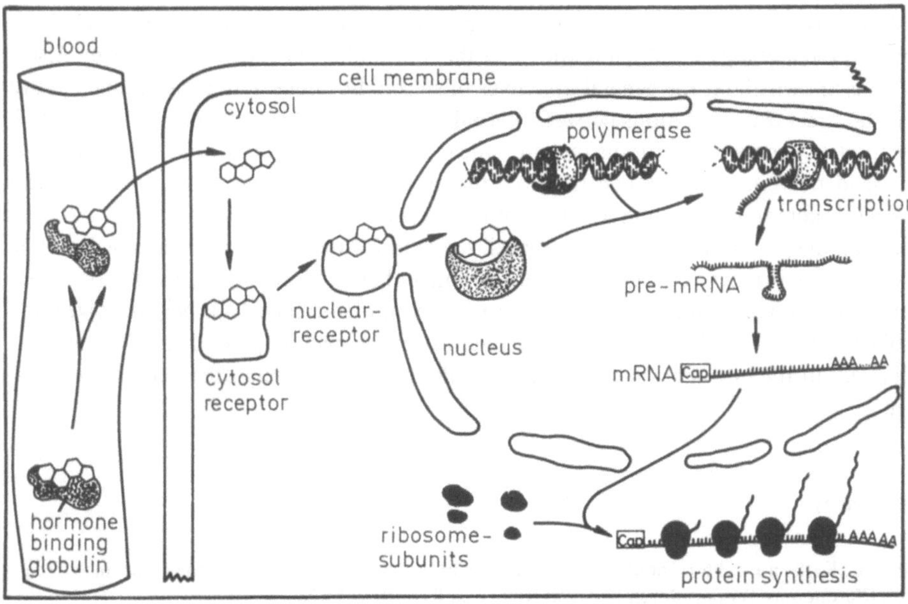

Fig. 1. Schematic representation of the interaction of proteins and steroids in steroid hormone action. In the blood, the steroid hormone is carried by a transport protein *(left)*. It enters the cell by diffusion (or perhaps by a carrier transport) as free steroid and is bound, within the cytosol, to the cytosolic receptor. This interaction results in an allosteric conformational change of the protein; the complex is now able to enter the nucleus where it interacts with the chromatin. The nature of this interaction is still unknown, but it results in transcription of the specific gene to be activated by this hormone. Pre-mRNA is produced, it is then processed to yield mRNA that is exported into the cytosol and forms polysomes. The next step is translation of the message into protein. (From Karlson 1979)

"naked steroid", but the steroid-receptor-complex is involved in control of transcription. How exactly this control is exerted, how the special genes to be activated are recognized and how the process of transcription is started is still largely unknown and is under active study in many laboratories.

As result of transcription, pre-mRNA is formed and processed to mRNA which is released into the cytosol where it is translated into protein. There are good indications that some of the many steps in processing and translation are regulated; however, there is no solid evidence, to my knowledge, that steroid hormones participate in the control of posttranscriptional and translational events.

All these protein-steroid-interactions have to be kept in mind when discussing the role of steroid hormones in early pregnancy. The steroid characteristic for early pregnancy is progesterone, one of the prominent proteins is uteroglobin, which is induced by progesterone. Details of their action and interaction are discussed elsewhere in this volume.

It should be pointed out that early pregnancy is characterized by many developmental processes resulting in differentiation. As we understand differentiation, it is the result of differential gene expression. Since steroid-protein-complexes can modify gene expression, I presume that they have much to do with the differentiation of the uterine tissue, and moreover possibly also the trophoblast and of embryoblast. This system is a challenge to the investigator. Much is known at the morphological and structural level. Also much is known about the general principles of steroid hormone action, as outlined above. But the exact causal relations between steroid (progesterone) and the morphological and functional changes in the uterus remain to be elucidated. A beginning has been made, as is evident from the many contributions presented in this volume. But more information and a lot more understanding of our information is needed.

Among the present problems of mankind, the world-wide control of human population is of vital importance for the survival of the human being. More knowledge about all the details of reproduction in mammals and especially in man can perhaps help to solve our problems. It will not be an easy way; no scientist will expect a "patentrezept" in such a volume. But the information collected here will undoubtedly help others to do the next step.

References

Clever U, Karlson P (1960) Induktion von Puff-Veränderungen in den Speicheldrüsenchromosomen von Chironomus Tentans durch Ecdyson. Expl Cell Res 20:623-626

Karlson P (1961) Biochemische Wirkungsweise der Hormone. Dtsch Med Wschr 86:668-674

Karlson P (1963) New Concepts on the Mode of Action of Hormones, Perpect Biol Med 6:203-214

Karlson P, Doenecke D, Sekeris CE (1975) Intracellular Mechanisms of Hormone Action. In: Florkin M, Stotz EH (eds) Comprehensive Biochemistry, Vol. 25, pp 1-63

Karlson P (1979) Hormonrezeptoren und Hormonwirkung. Klin Wschr 57:607-612

Starling EH (1906) Die chemische Koordination der Körpertätigkeiten. Verhandlg. d. Gesellschaft Deutscher Naturforscher und Ärzte, 78. Versammlung. Stuttgart, p 246-260

Ultrastructure of Trophoblast-Epithelium Relations During Implantation

B. Ove Nilsson

Reproduction Research Unit, Biomedical Centre, University of Uppsala, Sweden

In the uterine cavity, the blastocyst usually waits for about a day before it attaches onto the uterine surface and starts invading the endometrium. The preattachment period, however, can be prolonged – this is called delayed implantation. It appears normally among some wild animals but can also be produced experimentally in mice and rats (Bergström 1978).

An experimental delay of implantation is obtained by ovariectomy of a newly mated mouse or rat. While the animal is given progesterone, the blastocyst will slowly attain a low metabolic activity (Bergström 1972a; Menke 1972; Weitlauf 1974), and it will remain in this inactive state until estrogen is given to the animal. A few hours after this injection, the amount of secretion in the uterus increases (Nilsson 1974b; Bergström and Nilsson 1975) and the blastocyst gets metabolically active (Torbit and Weitlauf 1974; Weitlauf and Kiessling 1980). This is named activation of the blastocyst (McLaren 1973; Webb and Surani 1975). Later follows a change in the antigenic properties of the trophoblast (Hakansson et al. 1975; Searle et al. 1976), a decrease of the number of negative surface sites of the blastocyst (Nilsson et al. 1975), and about 24 h after the injection of estrogen, the trophoblast cells have attached onto the uterine surface and begin invading the endometrium (Bergström and Nilsson 1976). Sometimes during this period the blastocyst also initiates early decidual changes in the endometrium (Finn and McLaren 1967; Lundkvist et al. 1979).

1 Blastocysts in Delay

A steady state of delay is reached by the mouse blastocyst about 4 days after the ovariectomy (Bergström 1972a; Naeslund and Lundkvist 1978). Then the blastocyst lies closely surrounded by the endometrium and also the endometrial surfaces of the implantation sites are closely apposed. Practically no luminal secretion is present (Fig. 1).

Ultrastructurally, the most prominent features of the uterine epithelium are apical protrusions and apical vesicles (Warren and Enders 1964; Psychoyos and Mandon 1971; Nilsson 1972). The apical protrusions contain a homogeneous ground cytoplasm and bulge a few micrometer above the microvillous surface (Fig. 2). They are mushroomlike, and since the trophoblast cells lie so close to the uterine surface, the apical protrusions cause craterlike imprints in the trophoblast surface (Bergström 1972a). The apical vesicles are present in a cytoplasmic zone which is separated from the surface membrane by a thin rim of homogeneous ground cytoplasm and from the nuclear level by the Golgi apparatus and a layer of mitochondria (Fig. 3).

Functionally, the blastocyst has a low metabolic activity. This has been shown by various metabolic tests (Menke 1972; Weitlauf 1974; Nilsson et al. 1980) and it is also

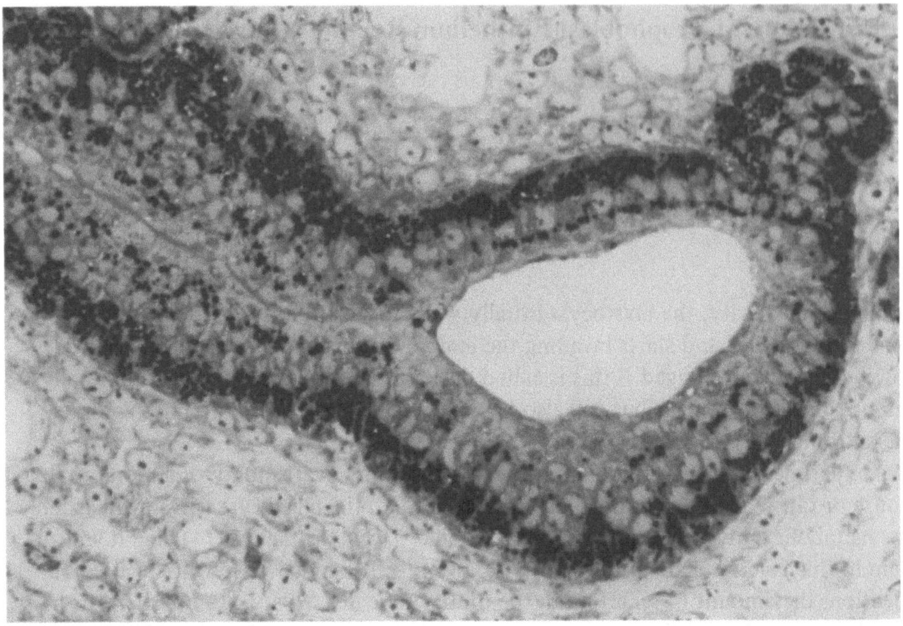

Fig. 1. Blastocyst in uterus from a mouse in delay of implantation. The blastocyst is observed to the *right* and the closed uterine lumen runs from the site of the blastocyst to the *left*. The uterine epithelium contains many dark lipid granules basally and a few lipid granules apically. Light microscopy of Epon-embedded specimen. Mag.400X

visualized in the trophoblast ultrastructure: the ribosomes are scattered randomly and the granular reticulum occurs sparsely (Fig. 4). But although the blastocyst is dormant it does take up nutrients, as judged by the occurrence of endocytotic vesicles (Fig. 5). The endocytotic vesicles are scattered along the borderline and occur also where apical protrusions indent the blastocyst surface. There is thus some form of transport from mother to embryo at these sites.

The apical protrusions themselves are capable of endocytotic activity and have a high capacity for taking up horseradish peroxidase and some stain particles (Enders and Nelson 1973; Parr 1980). But these substances are nonphysiological ones, which might stimulate an endocytosis which is not necessarily relevant for normal conditions.

2 Blastocyst Activation

When estrogen is given, implantation is initiated. Morphologically, the blastocyst gets separated from the uterine mucosa, that is, a uterine secretion appears (Nilsson 1974b; Bergström and Nilsson 1975). One source of the secretion seems to be the apical vesicles of the epithelium. During delay, these vesicles are crowded in an apical region of the cell but are separated from the surface by a layer of ground cytoplasm. A few hours after the estrogen injection, however, they leave their location and empty their – still unknown – contents into the lumen (Nilsson and Lundkvist 1979). At this stage, one

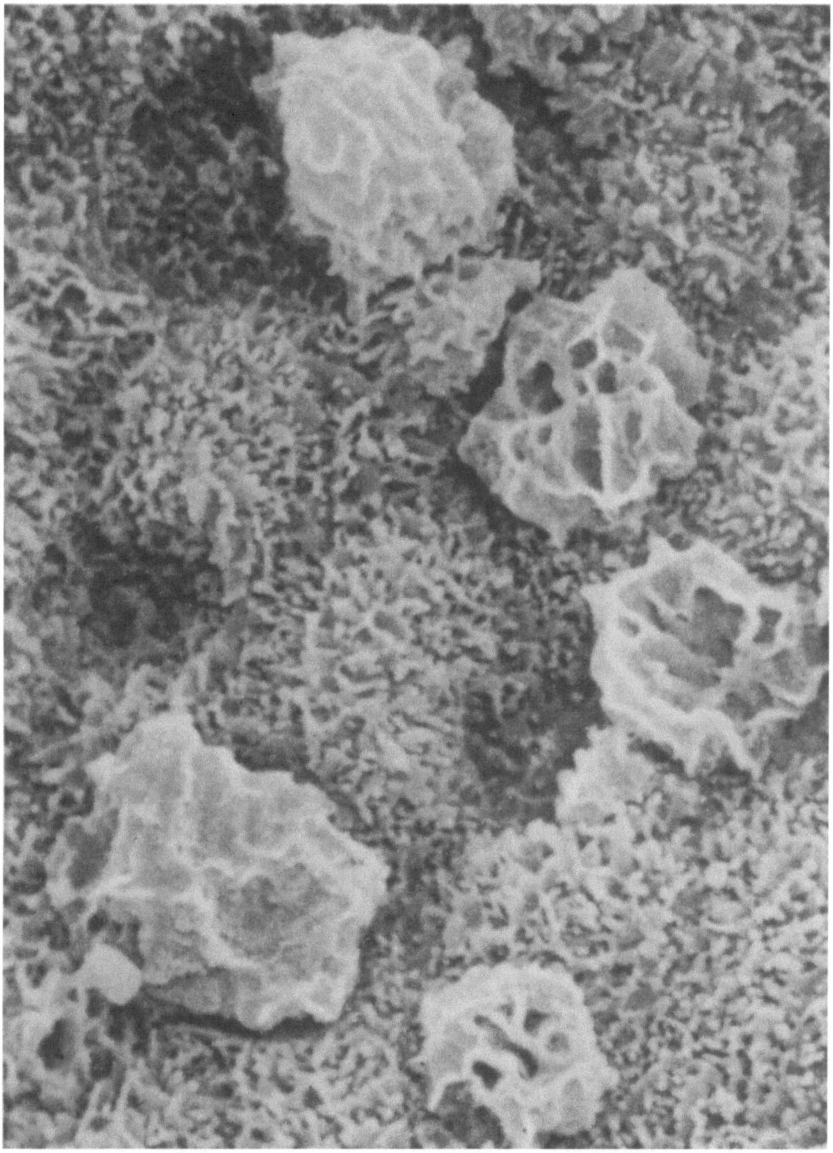

Fig. 2. Luminal surface of uterus from a rat in delay of implantation. Several apical protrusions are noticed among the microvilli of the uterine epithelium. Scanning electron microscopy. Mag. 10,000X

important component of the uterine secretion seems to be some type of carbohydrate. Accordingly, as shown by Aitken (Aitken 1976), the amount of fructose increases in the uterine secretion of the roe-deer when the implantatory delay of that animal comes to an end. In the mouse, in-vitro tests have demonstrated that a newly activated blastocyst prefers glucose to fructose and that it also increases its utilization of glucose (Nilsson et al. 1980). Furthermore, it is essential to delete glucose from the culture

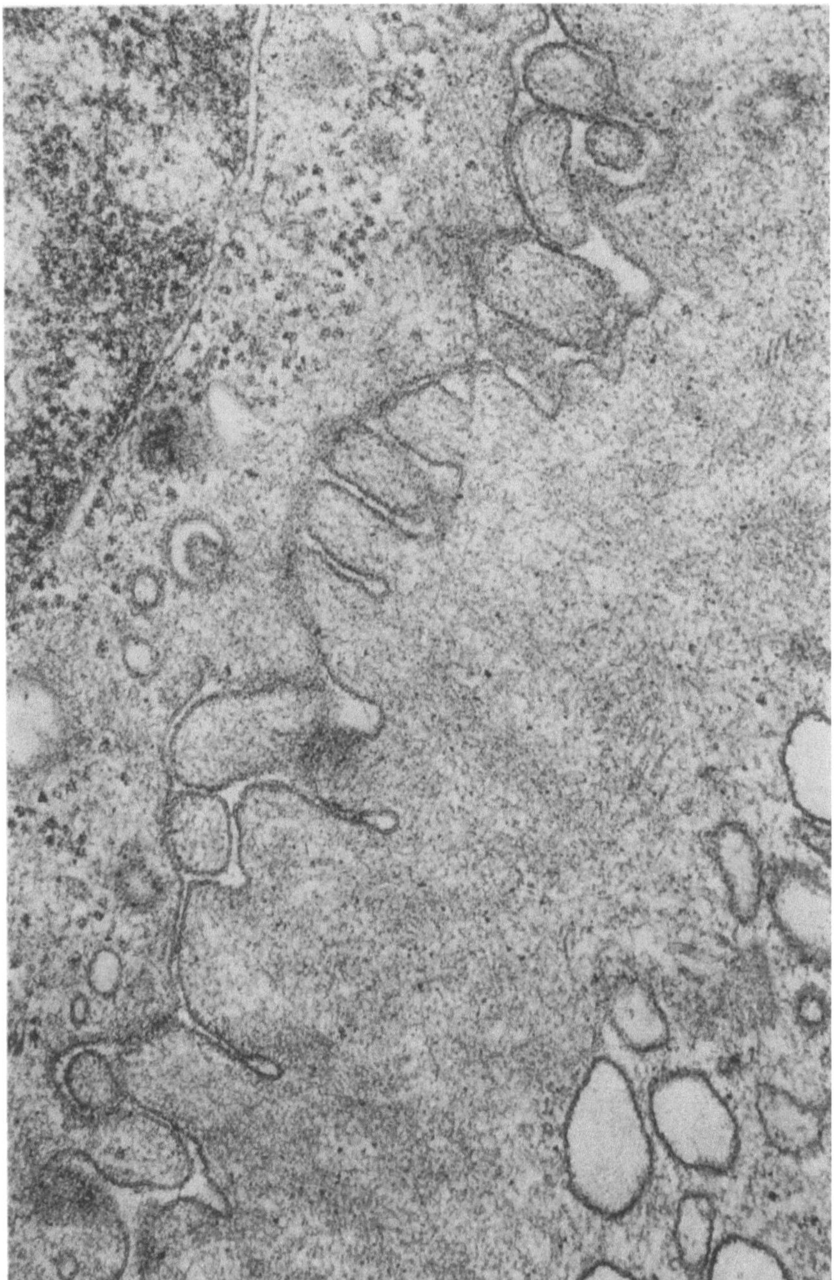

Fig. 3. Blastocyst-epithelium border of uterus from a mouse in delay of implantation. The uterine epithelium (*right*) has a rim of a dense substance separating the apical vesicles from the luminal surface. The tropholblast (*left*) contains some endocytotic vesicles and scattered ribosomes. Mag.70,000X

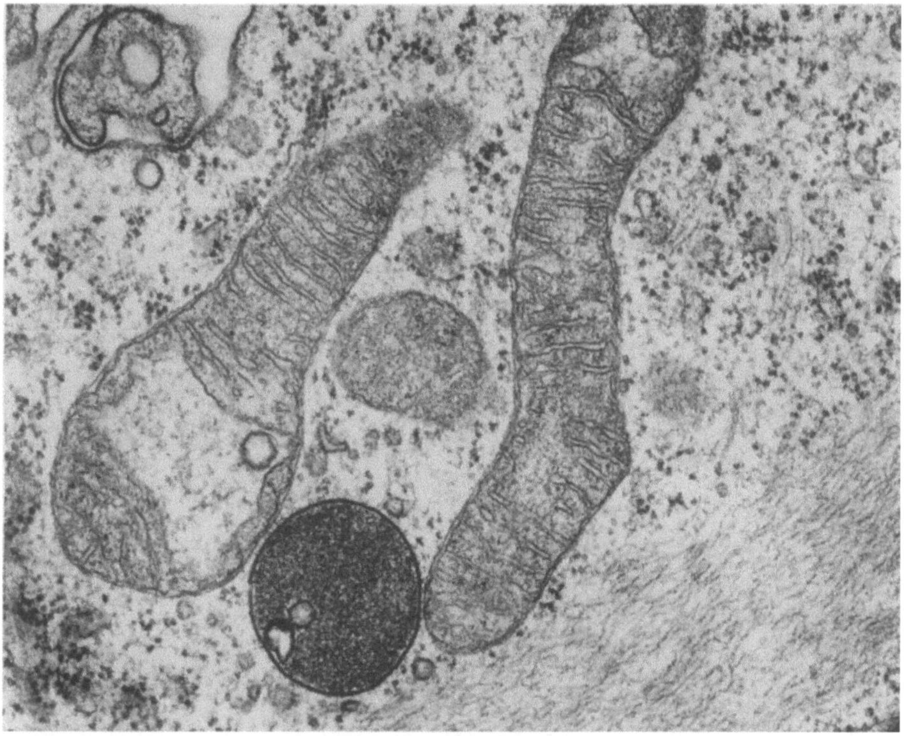

Fig. 4. Trophoblast from a mouse in delay of implantation. The vesiculation of the mitochondria, a feature representative for the mouse trophoblast, is noticed. A peroxisome is observed in the *lower part* of the picture. To the *right*, strands of microfibrils surrounding the nucleus are seen. Mag.50,000X

medium to obtain a delayed outgrowth of the blastocysts in tissue-culture (Naeslund 1979; van Blerkom et al. 1979).

The composition of the mouse uterine fluid, secreted to activate the blastocysts is rather unknown. However, we have demonstrated that glucose is present by analysing uterine flushings (Nilsson et al. 1980). Also the electrolytes are under study, using a new technique for micro-X-ray analysis (Nilsson and Ljung 1979). In involves a careful transfer of Sephadex G-200 particles (Pharmacia, Uppsala, Sweden), sized about 100 μm, into the uterine cavity. The particles are transferred in a small amount of water and left in the uterus for 2 h to obtain a ionic equilibrium. They are then flushed out with a volatile hydrophob silicon oil (MS 200, Dow Corning), mounted on nylon grids with a Formvar membrane, carbon coated and examined in a Philips 400 microscope with an Edax-Edith equipment, using 20 kV and 0.2 nA. As yet we know that the amount of calcium increases in the uterine secretion 8 h after the injection of oestrogen.

When the blastocyst by some mechanism gets activated, this is manifested by ultrastructural changes. Since the endometrial grasp of the blastocyst loosens, the surface imprints disappear (Bergström 1972b). In the trophoblast cytoplasm, polyribosomes and granular reticulum indicate the increased metabolic activity. Also an appearance

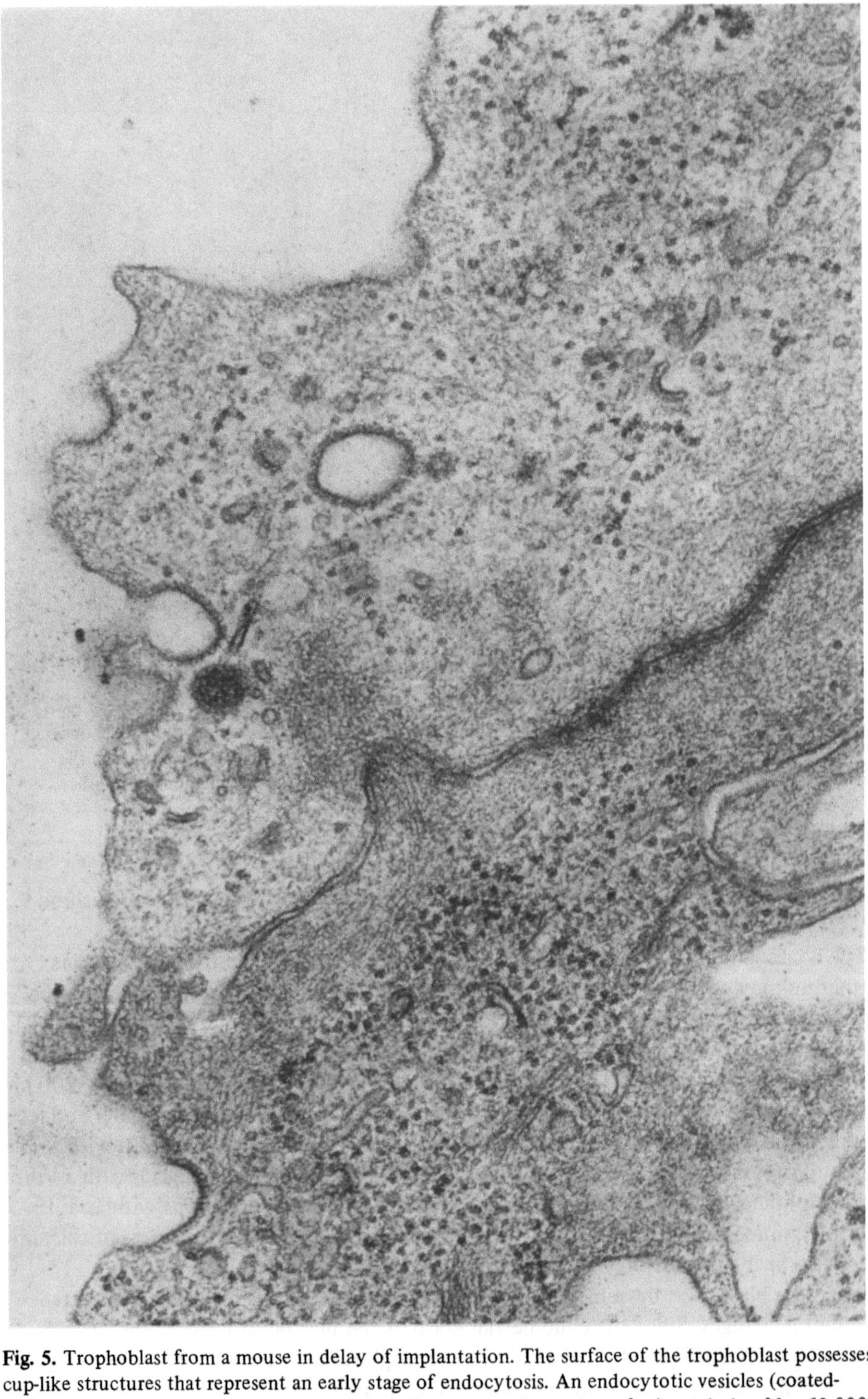

Fig. 5. Trophoblast from a mouse in delay of implantation. The surface of the trophoblast possesses cup-like structures that represent an early stage of endocytosis. An endocytotic vesicles (coated-vesicle) lies in the cytoplasm among scattered ribosomes and a system of microtubules. Mag.60,000X

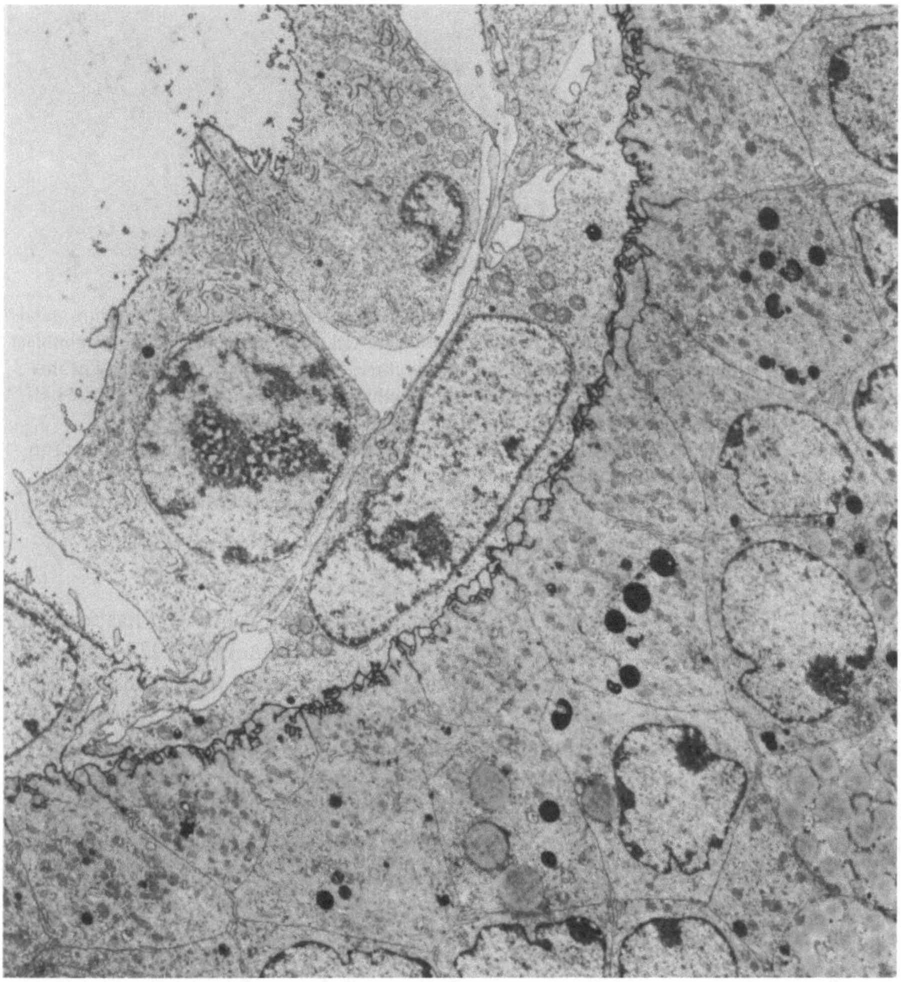

Fig. 6. Blastocyst-epithelium border in uterus from a mouse 24 h after an injection of estrogen. The border between the two tissues is observed as a dark irregularily running line. Inside the trophoblast layer (*left*) lie a few embryoblasts. Mag. 4,000X

of glycogen is associated with the blastocyst activation (Ozias and Stern 1973). About 8 h after the estrogen injection, several structural changes thus indicate the new state of the blastocyst.

Interestingly, the activation of the blastocyst is reversible (Psychoyos 1961; Weitlauf 1969). Accordingly, when activated blastocysts are transferred to the uterus of mice in delay, they revert to the ultrastructural appearance of a delayed blastocyst (Naeslund et al. 1980). The metabolic machinery of the mouse blastocyst thus seems to have a capacity to switch on and off, probably governed by the properties of the blastocyst environment.

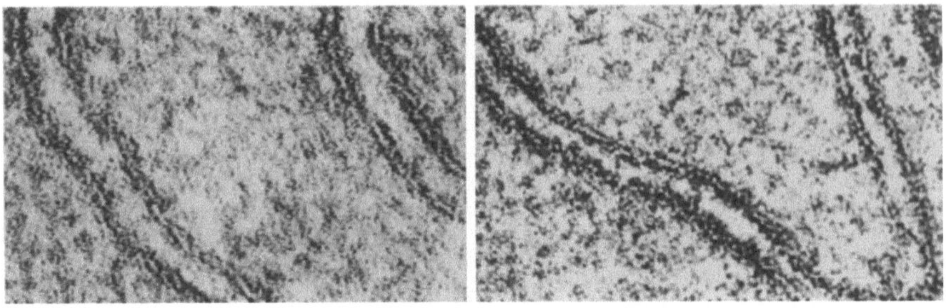

Fig. 7 a,b. Blastocyst-epithelium border. Some areas where the surface of the trophoblast runs parallel to the surface of the epithelial cells are demonstrated. The cell membrane of both the trophoblast and the epithelial cell has the conventional bilaminar structure. Mag. 130,000X. **a** At delay of implantation, the gap between the trophoblast and the epithelium is lightly stained. The distance between the trophoblast cell membrane and the epithelial cell membrane is about 27 nm.
b At blastocyst attachment, the gap between the trophoblast and the epithelium has staining properties similar to those observed at delay of implantation. However, the distance between the trophoblast cell membrane and the epithelial cell membrane now has decreased to about 20 nm

By about 12 h after the injection of estrogen, a signal from the blastocyst has reached the stromal cells, triggering the decidual cell transformation (Lundkvist 1978). The next feature observed during the implantation process is a change in the trophoblast membrane. About 14 h after the injection of estrogen, the number of negative sites of the blastocyst surface decreases (Nilsson et al. 1975; Jenkinson and Searle 1977; Nilsson and Hjerten, to be published). Simultaneously, the paternal H-2 antigens disappear (Hakansson et al.; Searle et al. 1976). Those two changes are probably associated and imply a structural alteration of the trophoblast membrane. Functionally, the changes could be expected to aid the attachment of the blastocyst and to lessen the risk of an immunological rejection of the blastocyst.

3 Blastocyst Attachment

A close contact between the membranes of the trophoblast and the epithelium begins to be established 16-18 h after the injection of estrogen (Fig. 6) (Nilsson 1974b; Bergström and Nilsson 1976), and by 24 h, the blastocyst can be flushed out of the uterine cavity only with great difficulty. Now, the abembryonic pole has a much more ruffled appearance than the embryonic one. This appearance probably is a sign of a growing invasive capacity.

The sites of trophoblast-epithelium contacts at blastocyst attachment look rather similar to these contact sites during blastocyst delay, but a rough calculation indicates that the distance between blastocyst and epithelium is about 27 mm during blastocyst delay while it has decreased to about 20 mm at blastocyst attachment (Fig. 7). There are of course many technical uncertainties in this kind of estimation, but a change in distance between the two tissues might well reflect some of the other changes observed in their surfaces. For instance, the surface morphology of both the blastocyst and the uterine epithelium changes (Bergström and Nilsson 1970; Bergström 1972b; Nilsson 1974a; Enders 1975). One can also include the changes in the extracellular coat of both trophoblast (Enders and Schlafke 1974; Nilsson et al., to be published) and uterine epithelium (Enders and Schlafke 1974; Nilsson 1974a). These changes contribute in various, albeit yet unknown ways, to the attachment of the blastocyst.

References

Aitken RJ (1976) Uterine secretion of fructose in the roe deer. J Reprod Fertil 46:439-440

Bergström S (1972a) Delay of blastocyst implantation in the mouse by ovariectomy or lactation. A scanning electron microscope study. Fertil Steril 23:548-561

Bergström S (1972b) Scanning electron microscopy of ovoimplantation. Arch Gynaekol 212:285-307

Bergström S (1978) Experimentally delayed implantation. In: Daniels JE (ed) Methods in mammalian embryology. Academic Press, New York, pp 419-435

Bergström S, Nilsson O (1970) Morphological changes of the trophoblast surface at implantation in the mouse. J Reprod Fertil 23:339-340

Bergström S, Nilsson O (1975) Embryo-endometrial relationship in the mouse during activation of the blastocyst by oestradiol. J Reprod Fertil 44:117-120

Bergström S, Nilsson O (1976) Blastocyst attachment and early invasion during oestradiol-induced implantation in the mouse. Anat Embryol 149:149-154

Enders AC (1975) The implantation chamber, blastocyst and blastocyst imprint of the rat: a scanning electron microscope study. Anat Rec 182:137-148

Enders AC, Nelson DM (1973) Pinocytotic activity of the uterus of the rat. Am J Anat 138:227-298

Enders AC, Schlafke S (1974) Surface coats of the mouse blastocyst and uterus during the preimplantation period. Anat Rec 180:31-44

Finn CA, McLaren A (1967) A study of the early stages of implantation in mice. J Reprod Fertil 13:259-267

Hakansson S, Heyner S, Sundquist K-G, Bergström S (1975) The presence of paternal H-2 antigens on hybrid mouse blastocysts during experimental delay of implantation and the disappearance of these antigens after onset of implantation. Int J Fertil 20:137-140

Jenkinson EJ, Searle RF (1977) Cell surface changes on the mouse blastocyst at implantation. Exp Cell Res 106:386-390

Lundkvist, Ö (1978) Ultrastructural studies of the endometrial stromal cells in rats during estradiol-induced implantation after an experimental delay Biol Reprod 18:306-316

Lundkvist Ö, Nilsson O, Bergström S (1979) Studies on the trophoblast-epithelial complex during decidual induction in rats. Am J Anat 154:211-230

McLaren A (1973) Blastocyst activation. In: Segal SJ, Crozier R, Corfman PA, Condliffe PG (eds) The regulation of mammalian reproduction. Thomas, Springfield, pp 321-328

Menke TM (1972) Changes in mouse blastocyst carbon dioxide production as a function of time postcoitum in delay of implantation during lactation or following ovariectomy. Biol Reprod 7:414-416

Naeslund G (1979) The effect of glucose-, arginine- and leucine-deprivation on mouse blastocyst outgrowth *in vitro.* Uppsala J Med Sci 84:9-20

Naeslund G, Lundkvist, Ö (1978) Effect of the endocrine state of blastocyst donors on the time required for initiation of trophoblast outgrowth. Uppsala J Med Sci 83:135-139

Naeslund G, Lundkvist Ö, Nilsson O (1980) Transmission electron microscopy of mouse blastocysts activated and growth-arrested *in vivo* and *in vitro.* Anat Embryol 159:33-48

Nilsson O (1972) Ultrastructure of the process of secretion in the rat uterine epithelium at preimplantation. J Ultrastruct Res 40:572-580

Nilsson O (1974a) Changes of the luminal surface of the rat uterus at blastocyst implantation. Scanning electron microscopy and ruthenium red staining. Z Anat Entwicklungsgesch 144:337-342

Nilsson O (1974b) The morphology of blastocyst implantation. J Reprod Fertil 39:187-194

Nilsson O, Hjerten S (to be published) Electrophoretic quantification of the changes in the net negative surface charge of mouse blastocysts implanting *in vivo* and *in vitro*
Nilsson O, Ljung L (1979) Electron probe micro-X-ray analyses of electrolyte composition of fluid microsamples by use of a Sephadex bead. Upsala J Med Sci 84:1-2
Nilsson O, Lundkvist Ö (1979) Ultrastructural and histochemical changes of the mouse uterine epithelium on blastocyst activation for implantation. Anat Embryol 155:311-321
Nilsson O, Lindqvist I, Ronquist G (1975) Blastocyst surface charge and implantation in the mouse. Contraception 11:441-450
Nilsson O, Naeslund G, Curman B (1980) Polar differences of delayed and implanting mouse blastocysts in binding of Alcian Blue and Concanavalin A. J Exp Zool 214:177-180
Nilsson O, Östensson CG, Eide S, Hellerström C (1980) Utilization of glucose by the implanting mouse blastocyst activated by oestrogen. Endokrin 76:82-93
Ozias CB, Stern S (1973) Glycogen levels of preimplantation mouse embryos developing *in vitro*. Biol Reprod 8:467-472
Parr MB (1980) Endocytosis in the uterine epithelium during early pregnancy. In: Hubinont PO (ed) Progress in reproductive biology. Vol 7. S Karger, Basel, pp 81-91
Psychoyos A (1961) Nouvelles recherches sur l'ovoimplantation. C R Acad Sci (Paris) 252:2306-2311
Psychoyos A, Mandon P (1971) Etude de la surface de l'epithelium uterin au microscope electronique a balayage. Observations chez la ratte au 4^e aet au 5^e jour de la gestation. C R Acad Sci (Paris) 272: 2723-2725
Searle RF, Sellens MH, Elson J, Jenkinson EJ, Billington WD (1976) Detection of alloantigens during preimplantation development and early trophoblast differentiation in the mouse immunoperoxidase labelling. J Exp Med 143:348-359
Torbit CA, Weitlauf HM (1974) Effect of oestrogen and progesterone on CO_2 production by "delayed implanting" mouse embryos. J Reprod Fertil 39:379-382
Van Blerkom J, Chavez DJ, Bell H (1979) Molecular and cellular aspects of facultative delayed implantion in the mouse; in Maternal recognition of pregnancy. Ciba Foundation Symposium. Excerpta Medica, Amsterdam, pp 141-172
Warren RH, Enders AC (1964) An electron microscope study of the rat endometrium during delayed implantation. Anat Rec 148:177-194
Webb FTG, Surani MAH (1975) Influence of environment on blastocyst proliferation, differentiation, and implantation. In: Talwar GP (ed) Regulation of growth and differentiated function in eukaryote cells. Raven, New York, pp 519-522
Weitlauf HM (1969) Temporal changes in protein synthesis by mouse blastocysts transferred to ovariectomised recipients. J Exp Zool 171:481-486
Weitlauf HM (1974) Metabolic changes in the blastocysts of mice and rats during delayed implantation. J. Reprod Fertil 39:213-224
Weitlauf HM, Kiessling AA (1980) Comparison of overall rates of RNA synthesis in implanting and delayed implanting mouse blastocysts *in vitro*. Develop Biol 77:116-129

Hormonal Control of Decidualization

Koji Yoshinaga*

Laboratory of Human Reproduction and Reproductive Biology and Department of Anatomy, Harvard Medical School, Boston, Massachusetts, USA

1 Ovarian Hormones and Decidualization

1.1 Ovarian Hormone Secretion Pattern During Early Pregnancy

The secretory pattern of ovarian hormones during the estrous cycle varies in mammals. In the human, estrogen secretion peaks twice during a cycle, once shortly before ovulation and again at the mid-luteal phase. The progesterone level, on the other hand, is very low before ovulation. After ovulation it increases to peak at the mid-luteal phase, then declines to a low level prior to the next menstruation (Ross et al. 1970, Edwards 1980). Since ovum implantation in the human occurs about 8 days after ovulation, the increased secretion of estrogen and progesterone during the luteal phase appears to be necessary to prepare the endometrium for the fertilized ovum and subsequent decidualization.

A second estrogen peak at the mid-luteal phase is not detectable in the rhesus monkey despite its similarity to the human in the length of the menstrual cycle and in other anatomical and physiological features of the reproductive system (Hotchkiss et al. 1971; Atkinson et al. 1975). When rhesus monkeys ovariectomized soon after fertilization were treated with progesterone alone, successful ovum implantation occurred (Meyer et al. 1970). It appears, therefore, that estrogen is not required for ovum implantation and that progesterone alone is essential to induce implantation in the rhesus monkey.

Implantation of the ovum can be induced with progesterone alone in ovariectomized guinea pigs (Deanesly 1960) and in ovariectomized golden hamsters (Orsini and Meyer 1962). In golden hamsters, estrogen secretion is increased prior to ovum implantation (Joshi and Labhsetwar 1972). The significance of the preimplantation increase in estrogen in this species is not clear. In the rabbit a synergistic action of progesterone and estrogen on the uterus appears to facilitate the subsequent acceptance of implanting ova (Chambon 1949; Keyes and Nalbandov 1967; Lee, Keyes and Jacobson 1971; Browning, Keyes and Wolf 1980). In contrast to other species of animals, rats, mice and Mongolian gerbils require both estrogen and progesterone for successful implantation of the ovum. When the ovaries are removed after fertilization and the females treated with progesterone, the fertilized ova develop into blastocysts but remain unattached in the uterine lumen until estrogen becomes available (rat: Canivenc et al. 1956; Cochrane and Meyer 1957; mouse: Yoshinaga and Adams 1966; Mongolian gerbil: Norris and Adams 1971).

*Present address: Reproductive Sciences Branch, Center for Population Research, NICHD, NIH, Bethesda, MD 20205 (USA)

In pregnant rats progesterone secretion increases rapidly during the first 2 weeks of pregnancy (Hashimoto et al. 1968). In pseudopregnant rats the first sign of functional corpora lutea (expressed by sustained increases in peripheral level of progesterone) is seen 2 days after ovulation (Smith et al. 1975). A similar pattern of progesterone secretion is seen in pregnant rats (Yoshinaga, unpublished data). Thus, the uterus is exposed to an increase of progesterone for at least 2 days before implantation occurs in the late afternoon on day 5. (The morning on which a sperm-positive smear is found is designated as day 1 of pregnancy.)

In earlier work using a bioassay method, we found that the ovarian venous level of estrogen increased significantly in the afternoon of day 4 (Yoshinaga et al. 1969). This result is in agreement with the estrogen surge theory proposed by Shelesnyak (1960). Several groups of workers measured estrogen levels in ovarian venous effluent or in peripheral blood during the preimplantation period in the rat. Some groups found a clear peak in the afternoon or evening of day 4; thus the secretion was considered a "surge". Other groups failed to detect a peak, but found a "gradual increase" (Nimrod et al. 1972; Shaikh and Abraham 1969; Shaikh 1971; Pope and Waynforth 1970; Watson et al. 1975). Regardless of the secretory pattern of estrogen, the ovaries on the afternoon of day 4 secrete the amount of estrogen necessary for ovum implantation.

Zeilmaker (1963) showed that the ovaries could be removed after 0100 h on day 4 without causing a delay in implantation if progesterone only was administered after ovariectomy. However, the same operation and treatment delayed implantation when ovariectomy was performed before 1700 h on day 4. Day 4 of pregnancy or pseudopregnancy corresponds to the day of proestrus, if the rat's uterine cervix had not been stimulated by mating or its equivalent. During the estrous cycle estrogen secretion reaches its peak on the day of proestrus (Yoshinaga et al. 1969). The preimplantation increase in estrogen observed in the afternoon of day 4, however, is considerably smaller than at proestrus, presumably reflecting the alteration of the cyclic pattern in the hypothalamo-pituitary-ovarian axis.

1.2 Phases of Uterine Sensitivity for Ovum Implantation and Decidualization in Relation to Ovarian Hormone Secretion

The uteri of pseudopregnant and pregnant rats are maximally receptive to nonspecific stimuli on day 5 [day 1 is the day of ovulation (Yochim and DeFeo 1963)]. In response to a stimulus, the endometrial stromal cells are transformed into decidual cells. Decidual cells proliferate rapidly and form a nidus by surrounding the implanting blastocyst, or in the absence of a blastocyst, they form a mass of tissue, deciduoma.

Various aspects of decidualization were reviewed by DeFeo (1967). We will discuss in this section the three phases of uterine sensitivity in relation to endogenous and exogenous ovarian hormones in the rat. As described in the previous section peripheral serum con concentration of progesterone increases on day 3 of (pseudo-)pregnancy and continues to rise beyond day 5, the day of maximal uterine receptivity. Since traumatization of the endometrium of the pseudopregnant rat uterus does not result in deciduoma formation, the uterus on day 3 is evidently not ready to respond to stimulation (Yoshinaga and Greep 1970). It has been shown that the uterus must be exposed to progesterone for at least 2 days and also to a sensitizing dose of estrogen at the end of the 2-day progesterone exposure in order to become receptive to blastocysts (Psychoyos 1963).

Under the influence of progesterone alone the uterus does not respond to blastocysts and implantation does not take place. However, the same uterus can respond to mechanical trauma and deciduoma can be produced. The magnitude of the response to traumatization, however, is not as large as the decidual response that obtained on day 5 of (pseudo-)pregnancy. The progesterone primed uterus, termed "neutral" by Psychoyos, can become "receptive" to blastocysts when estrogen becomes available. In normal (pseudo-)pregnancy the estrogen rise on the afternoon of day 4 renders the progesterone-primed uterus receptive on day 5. The receptive period, which lasts for less than 1 day, is followed by a "refractory" period after day 6 of (pseudo-)pregnancy. Thus, when blastocysts are collected from donor rats and transferred into the uterus of pseudopregnant rats on day 5, they implant and continue to develop. However, when blastocysts are transferred on day 6 of pseudopregnancy, the transferred blastocysts do not implant and are expelled from the uterus.

Thus, the uterus exhibits three different phases of sensitivy during early (pseudo-)pregnancy – neutral, receptive, and refractory. When progesterone reaches a certain level for 2 days, the uterus becomes neutral. Addition of estrogen makes the uterus receptive and then the uterus becomes refractory (cf. Yoshinaga 1980).

It is well known that implantation of ova is delayed in pregnant rats concurrently nursing a large litter. The delay is not due to lack of progesterone secretion because progesterone secretion rate by the ovary is increased in proportion to the number of young suckled (Yoshinaga et al. 1971; Yoshinaga 1974). Rather, the delay in ovo-implantation is attributable to inadequate estrogen secretion caused by a low gonadotrophin secretion, and this low gonadotrophin secretion is the result of intensive suckling stimulus (Krehbiel 1941; Weichert 1941; Ford and Melampy 1973). Under these circumstances the uterus stays at the neutral phase for a prolonged period of time and implantation is delayed until adequate estrogen becomes available. For this reason estrogen is said to have a "triggering" action in inducing ovum implantation.

Since the neutral phase can be obtained by exposing the uterus to progesterone for 2 days, daily injection of progesterone from day 1 of pseudopregnancy renders the uterus neutral on day 3. If, in addition to daily progesterone injections, a single injection of estrogen is given on day 3 and the uterine endometrium traumatized on day 3, deciduoma is formed (Yoshinaga and Greep 1970). It is, therefore, possible to advance the receptive phase by exogenous hormone treatment.

Once the uterus is made receptive, blastocyst implantation is induced and decidual reaction takes place. Continuous secretions of progesterone and estrogen are required to maintain the continuous growth of decidual tissue. However, if the uterus was not traumatized during the receptive phase and automatically became refractory, this prolonged secretion of progesterone and estrogen maintains the refractory phase of the uterine sensitivity. The refractory phase can be maintained as long as one wishes by continuous treatment with the progesterone and estrogen (Yoshinaga and Greep 1974). The refractoriness can be abolished by estrogen alone, and the uterus can be made neutral again with progesterone alone.

1.3 Local Hormone Action on the Uterus

The uterus which has been exposed to progesterone for at least 2 days will be made receptive for ovum implantation by a synergistic action of estrogen. When estrogen is administered into the mesometrial adipose tissue locally instead of systemically, a limited area of the uterine horn close to the injection site will be influenced by the estrogen (Yoshinaga 1961). The localized stimulatory effect of estrogen is reflected by topographical biochemical changes in the uterine horn in relation to the site of estrogen injection (Yoshinaga and Pincus 1963). This local estrogen stimulation results in asynchrony in the sensitivity of the uterus: only a part of the uterus is made receptive while the rest of the uterus remains neutral. The following experiment illustrates the asynchronous stages of the uterine sensitivity in different portions of the same uterus.

Female rats were mated at postpartum estrus (day 1 of pregnancy) and allowed to nurse a litter of more than eight suckling young in order to cause a delay in implantation. On day 3 of pregnancy, the uterotubal junction was ligated on one side (sterile horn). On day 5, 5 ng estradiol was injected locally in one spot in the endometrial adipose tissue of both the fertile and the sterile horns. In the fertile horn a blastocyst which happened to be close to the site of the estrogen injection implanted in response to the synergistic action of administered estrogen and the endogenous progesterone secreted at a high level due to intensive lactation. Other blastocysts which were beyond the estrogen-sensitized portion of the horn remained unimplanted. In the sterile horn the estrogen-sensitized portion became receptive and then refractory on the following day. The refractoriness of that portion of the uterus was maintained by endogenous progesterone. On day 9 lactation was terminated by removing the suckling young from the mother. The cessation of lactation altered the function of the pituitary and increased gonadotrophin secretion which, in turn, resulted in an increase in estrogen. Thus, the blastocysts which remained unattached implanted soon after the cessation of lactation. On day 9, laparotomy revealed an implantation site close to the estrogen injection site. The sterile horn was traumatized on this day by scratching the endometrium with a curved needle along the entire horn. Since a limited portion of the sterile horn was made sensitive and then refractory by the locally administered estrogen, the traumatization resulted in decidualization of the entire uterine horn except for the refractory portion of the uterus. On day 14 of pregnancy the animals were sacrificed and the uterus examined. As can be seen in Fig. 1, the fertile horn (*left*) had two sets of embryos at different stages of development (superimplantation). The large swelling seen in the left horn contained an embryo corresponding to day 14 of age of normal pregnancy which implanted at the site of the locally injected estrogen (*black arrow*). The small implantation sites (three above and three below the large implantation site) contained embryos that corresponded to those of day 8-9 of normal pregnancy and had implanted after cessation of lactation. In the sterile horn (*right*) deciduomas were observed along the entire horn except in the portion which had been made refractory before the time of traumatization (*near the arrow*).

A similar localized asynchrony in the uterine sensitivity for decidualization was demonstrated in ovariectomized pregnant rats treated with progesterone (Yoshinaga 1977).

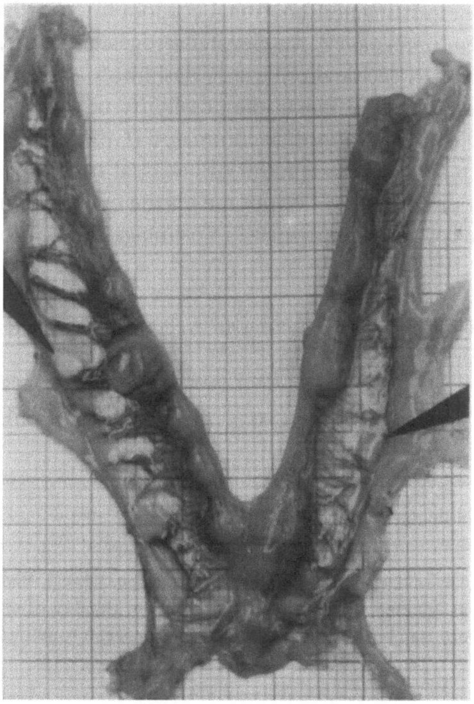

Fig. 1. Rat uterus on day 14 of pregnancy showing asynchrony of uterine sensitivity: the fertile horn (*left*) and the sterile horn (*right*)

1.4 Hormonal Requirements for Maintenance of Deciduoma and Implanted Embryos

Once a properly sensitized uterus receives a stimulation (nonspecific stimulation such as mechanical trauma, injection of air, oil, or saline, or introduction of foreign body into the uterus), transformation of endometrial stromal cells into decidual cells and proliferation of the latter take place very rapidly. As a result of this propagation of decidual cells the uterine horn increases in weight and may attain approximately a tenfold increase in 5 days. The degree of decidual tissue growth depends on the amount of progesterone and because of this dose-response relationship decidual reaction was once used as a bioassay method.

When the uterus is decidualized, the length of pseudopregnancy is prolonged to a time corresponding to the normal duration of pregnancy (Ershoff and Deuel 1943). This prolongation appears to be due to an inhibitory action of the decidualized uterus on the pituitary function. When large doses of progesterone are injected into "cycling" rats pseudopregnancy is induced (Alloiteau and Vignal 1958). However, daily subcutaneous injection of 2 mg progesterone to cycling rats does not result in pseudopregnancy, although this amount of progesterone is sufficient to suppress the cyclic change in the vaginal smear and to allow the uterus to respond to trauma or transferred blastocysts with decidualization (Yoshinaga and Adams 1967).

The ovaries of progesterone-treated cycling rats with decidualized uteri secrete little progesterone compared to the ovaries of pseudopregnant rats (Yoshinaga 1976). Therefore, the decidualized uterus does not appear to have a luteotropic effect in the sense of converting nonfunctional corpora lutea of the cycle into functional corpora lutea of pseudopregnancy. An elevated progesterone secretion by the ovary of pseudopregnant rats with decidualized uteri (Hashimoto et al. 1968) appears to involve the pituitary-ovarian axis, although the possibility of a luteotrophic hormone production by the decidual tissue cannot be eliminated (Gibori et al. 1974). There appears to be a relationship between the amount of decidual tissue present in the uterus and the degree of suppressive effect of deciduoma on the cyclic activity of the pituitary. This was shown by the following experiment (Yoshinaga 1972, Table 1).

Table 1. Suppression of ovarian cyclicity by deciduoma in the rat

Uterine condition	Daily dose of progesterone (mg)			Pseudopregnancy
	2.0	5.0	10.0	
Intact	10.8 ± 1.3[a]	11.3 ± 1.4	11.3 ± 1.6	13.0 ± 0.6
Decidualized	11.7 ± 2.8	14.2 ± 2.1	17.2 ± 3.5	21.2 ± 1.0

[a] Mean$\pm$s.d. duration between two estruses (days). Each group consisted of 6 rats. Progesterone was injected subcutaneously daily from the day of estrus. To produce deciduoma the uterine horns were traumatized on the 4th day of progesterone treatment

Regularly cycling rats were given daily subcutaneous injections of 2, 5, or 10 mg progesterone for 7 days, starting on the day of estrus. The vaginal cyclicity was suppressed for 11 days on the average, indicating that the suppression of the cycle was not due to induction of pseudopregnancy (13-day suppression for pseudopregnancy). When the uterus was traumatized on the 4th day of the progesterone treatment, deciduomas were formed and the amount of decidual tissue produced was in proportion to the dose of progesterone. The estrous cycle resumed 11.7, 14.2, and 17.2 days after the first injection of progesterone in these rats treated daily with 2, 5, and 10 mg progesterone, respectively. In progesterone-treated rats with deciduoma, the length of the diestrous period was proportional to the dose of progesterone, whereas in progesterone-treated rats with intact uteri the length of the diestrous period was 11 days on the average and was not proportional to the progesterone dose. It appears that the suppressive effect of the decidualized uterus on the hypothalamo-pituitary axis involves degeneration and absorption of decidual tissue in the uterus. Thus, in rats, the larger the amount of decidual tissue, the longer it takes to resume the ovarian cyclicity.

In the rat the quantity and quality of ovarian hormones required for maintenance of embryos varies depending on the stage of pregnancy. Embryos in early pregnancy require not only larger amounts of progesterone than those at later stages but also concomitant estrogen. When day-14 pregnant rats were bilaterally ovariectomized and 1 mg progesterone was injected subcutaneously daily thereafter, 85% of embryos were found alive 5 days later (Table 2). However, only 23% of the embryos were maintained when pregnant rats were ovariectomized on day 9 and 2 mg progesterone was injected daily. The maintenance rate was not improved with larger doses of progesterone. A high proportion of embryos was maintained when ovariectomized rats were treated with a com-

Table 2. Hormonal requirements for maintenance of embryos at different stages of pregnancy

Progesterone (mg/day)	No. of rats	Day of ovariectomy	No. of embryos at ovariectomy	No. of embryos surviving 5 days later	Survival %
0.5	5	14	65	13	20
1.0	3	14	48	40	85
1.0	5	9	73	0	0
2.0	8	9	130	30	23
4.0	4	9	51	2	4
8.0	5	9	68	0	0
0.5 + $E_2$0.1μg	4	9	53	0	0
2.0 + $E_2$0.1μg	6	9	77	60	78

bination of 2 mg progesterone and 0.1 μg estradiol. As described earlier in this chapter, we can experimentally produce superimplantation, i.e., two sets of embryos at different developmental stages in the same uterus, using asynchronous implantation (Yoshinaga 1961). Superimplantation rats with two sets of embryos 5 days apart in their development were ovariectomized on day 14 of pregnancy (old embryos at 14 days and young embryos at 9 days of pregnancy). They were injected subcutaneously with various daily doses of progesterone for 5 days and survival of the embryos was examined on the 6th day. As can be seen in Table 3, over two-thirds of the day-14 embryos were maintained with progesterone at daily doses larger than 1 mg. On the other hand, no more than 30% of the day-9 embryos in the same uterus were maintained with the highest dose of progesterone (4 mg daily). These results clearly demonstrate that older embryos require less progesterone than young embryos.

Table 3. Maintenance of superimplantation embryos with progesterone after bilateral ovariectomy in the rat

Daily dose of progesterone (mg)	No. of rats	No. of embryos at ovariectomy		No. of embryos survived 5 days		Survival %	
		Young	*Old*	*Young*	*Old*	*Young*	*Old*
0.5	6	39	61	0	19	0	31
1.0	4	48	11	13	8	27	72
2.0	9	80	24	0	16	0	67
4.0	7	88	19	26	16	27	84

2 Control of Decidualization through the Hypothalamo-pituitary Axis

2.1 Pituitary Control

The ovarian hormone secretion which is necessary to prepare the endometrium for decidualization at the time of ovum implantation is controlled by the pituitary gland. The ovum is prevented from implanting if the pituitary is removed from rats between days 1 and 4 of pregnancy (Pencharz and Long 1931). Luteinizing hormone (LH) has been shown to play an important role in stimulating the ovary to produce the steroids necessary for ovum implantation by the following studies: Macdonald et al. (1967) induced implantation of the ovum with LH in rats hypophysectomized after mating and treated with progesterone, prolactin, or pituitary graft under the kidney capsule. Antiserum to LH prevented ovum implantation in the rat (Hayashida and Young 1963; Madhwa Raj et al. 1968) and in the mouse (Bindon 1971). Plasma LH, but not follicle-stimulating hormone (FSH), was elevated prior to ovum implantation in the mouse (Murr et al. 1974).

Other studies indicate that FSH is important for ovum implantation. In the rat and mouse pituitary content of FSH fell significantly prior to ovum implantation (Bindon 1969). A single injection of FSH or PMS abolished lactational delay in implantation in the mouse (Bindon 1971) and in the rat (Raud 1974). Raud (1974) found that the peripheral serum level of FSH was significantly lower in pregnant lactating rats than in normal pregnant rats and that the effect of FSH in initiating implantation was reduced by prolactin. A supplement of FSH to lactating pregnant rats abolished the delay in implantation but FSH was ineffective in rats whose ovaries were irradiated to destroy follicles. These experiments show that preimplantation increase in estrogen secretion appears to be produced by the follicles.

Lactation influences ovum implantation in rats and mice through the hypothalamo-pituitary-ovarian axis. Intensive lactation is necessary to cause a delay in implantation, and there is no delay if the suckling litter is small (Yoshinaga 1976). The intensive suckling stimulus results in an increase in prolactin secretion and a decrease in gonadotrophins (Ford and Melampy 1973), presumably due to an inhibition of prolactin inhibiting factor (PIF) and gonadotrophin releasing hormone (Gn-RH) production and a stimulation of prolactin releasing hormone (PRH) in the hypothalamus. Injection of prolactin to pregnant rats nursing two pups caused a delay in implantation (Moudgal et al. 1974). It appears, therefore, that a proper combination of prolactin and gonadotrophin is necessary for progesterone secretion by the newly formed corpora lutea and for a preimplantation rise in estrogen secretion by the follicles. An excess of prolactin appears to prevent the follicular response to gonadotrophins perhaps by inhibiting or reducing the number of receptors for gonadotrophins in the follicles.

2.2 Hypothalamic Control

Mating (or equivalent mechanical stimulation) changes the function of the hypothalamo-pituitary-ovarian axis from the cyclic type to the pregnant type in the incomplete cyclers such as rats, mice, and hamsters. The stimulus transmitted from the uterine cervix to the hypothalamus is thought to initiate prolactin secretion by reducing PIF and perhaps by increasing PRH. Thus initiated, prolactin secretion is characterized by a pattern of two daily surges, one diurnal and the other nocturnal (Freeman et al. 1974).

The hypothalamic content of Gn-RH increases rapidly immediately after cervical stimulation. This hormone reaches a high level 2 h after cervical stimulation, comparable to the Gn-RH levels on day 5 of pseudopregnanty (Takahashi et al. 1975). The increase in the hypothalamic content of Gn-RH may be considered an indication of reduction in gonadotrophic hormone secretion. However, the reduced level of gonadotrophin secretion does not imply a reduction in the steroidogenic roles of gonadotrophins. It has been shown that "tonic" secretion of LH is required for progesterone secretion under various physiological conditions (Yoshinaga et al. 1971; Moudgal et al. 1972). Moreover, secretion of gonadotrophin essential for preimplantation increase of estrogen has been shown to be under the influence of the hypothalamic Gn-RH secretion. Arimura et al. (1976) injected a highly specific Gn-RH antiserum into pregnant rats and prevented ovum implantation. This inhibitory effect of Gn-RH antiserum was nullified by concomitant injection of Gn-RH or estrogen. Thus, it is obvious that preimplantation increase in estrogen secretion is exerted through the hypothalamo-pituitary-ovarian axis. This function can be interrupted physiologically by intensive lactation or pharmacologically by administration of reserpine and tranquilizers. It appears that the Bruce effect (a phenomenon where pregnant mice abort if they are exposed to the odor of strange male mice) works through the same route of the hypothalamo-pituitary-ovarian axis, because Bloch was able to abolish a delay in ovum implantation during lactation by placing males near the pregnant lactating mice (Bloch 1973).

The finding that Gn-RH is involved in the preimplantation increase in estrogen led us to examine the effect of Gn-RH on early pregnancy in the rat (Lin and Yoshinaga 1976a). In the following study, 50 μg Gn-RH was injected subcutaneously twice daily to pregnant rats. When Gn-RH was injected from day 1 to day 7 or 9, implantation of the ovum was delayed by 4 or 6 days respectively. When the daily injection was continued to day 14, implantation of the ovum was completely inhibited. Peripheral serum concentration of progesterone, 20α-hydroxypregn-4-en-3-one, and estradiol were measured by radioimmunoassay, and all these steroids were significantly lower in the Gn-RH-treated rats than in the control rats (Table 4). Serum levels of LH and FSH were, on the other hand, significantly higher in the Gn-RH-treated rats than in the control.

The same treatment to pseudopregnant rats also had a suppressive effect on deciduoma formation after uterine traumatization (Lin and Yoshinaga 1976b). In these animals FSH and LH levels were much higher (several times the control values) in the Gn-RH-treated rats than in controls. Steroid concentrations in peripheral serum were also markedly lowered by Gn-RH. These results suggest that an abnormal ration of prolactin to LH or FSH may have an adverse effect on ovarian hormone secretion. This assumption is based on the finding that a large dose of LH injected into the periovarian bursa of rats nursing a large litter of pups resulted in a decrease in progesterone output (Yosinaga 1974). Therefore, an abnormal ratio of prolactin/LH may be one of the causes of lowered steroidogenic fuction. Recent reports indicate that Gn-RH and its analogues can also act directly on the gonads and reduce gonadotrophin receptors (cf Yoshinaga, 1981). The loss of gonadotrophin receptors may be the major cause of lowered steroidogenic function of the ovary.

Table 4. Effect of Gn-RH treatment on peripheral serum concentration of ovarian steroids on day 4 of pregnancy

Treatment	No. of rats	Steroid concentration Progesterone (ng/ml)	20α-OH-P(ng/ml)	Estradiol (pg/ml)
Control	7	51.6±2.2 [a]	43.0±2.4	934±42
Gn-RH	5	28.6±1.6	39.5±3.1	376±26

[a] Mean±s.d.

Acknowledgments. The author wishes to express his gratitude to the National Science Foundation for the grant PCM76-22376, and to Mrs. Stella Nieland for her assistance in preparation of the manuscript.

References

Alloiteau J-J, Vignal A (1958) Pseudogestation apres injection de progesterone chez la ratte. C R Acad Sci (Paris) 246:2804-2810

Arimura A, Nishi N, Schally AV (1976) Delayed implantation caused by administration of sheep immunogammaglobulin against LHRH in the rat. Proc Soc Exp Biol Med 152:71-75

Atkinson LE, Hotchkiss J, Fritz GR, Surve AH, Neill JD, Knobil E (1975) Circulating levels of steroids and chorionic gonadotropin during pregnancy in the rhesus monkey, with special attention to the rescue of the corpus luteum in early pregnancy. Biol Reprod 12:335-345

Bindon BM (1969) Follicle-stimulating hormone content of the pituitary gland before implantation in the mouse and rat. J Endocrinol 44:349-356

Bindon BM (1971) Gonadotrophin requirements for implantation in the mouse. J Endocrinol 50: 19-27

Bloch S (1973) Nidation induced in mice during the lactational delay by the presence of strange males. J Endocrinol 57:185-186

Browning JY, Keyes PL, Wolf RC (1980) Comparison of serum progesterone, 20α-dihydroprogesterone, and estradiol-17ß in pregnant and pseudopregnant rabbits: evidence for postimplantation recognition of pregnancy. Biol Reprod 23:1014-1019

Canivenc R, Laffargue M, Mayer G (1956) Nidations retardees chez la ratte castree et injectee de progesterone: influence du moment de la castration sur la chronologie de l'ovo-implantation. C R Soc Biol (Paris) 150:2208-2212

Chambon Y (1949) Besoins endocriniens qualitatifs et quantitatifs de l'ovoimplantation chez la lapine. C R Soc Biol 143:1172-1175

Cochrane RL, Meyer RK (1957) Delayed nidation in the rat induced by progesterone. Proc Soc Exp Biol Med 96:155-159

Deanesly R (1960) Implantation and early pregnancy in ovariectomized guinea-pigs. J Reprod Fertil 1:242-248

DeFeo VJ (1967) Decidualization. In: Wynn RM (ed) Cellular biology of the uterus. Appleton-Century-Crofts, New York, pp 191-290

Edwards RG (1980) Conception in the human female. 1st Edition, Academic Press, London, New York, Toronto, Sydney, San Francisco

Ershoff BH, Deuel HJ Jr (1943) Prolongation of pseudopregnancy in induction of deciduomata in the rat. Proc Soc Exp Biol Med 54:167-168

Ford JJ, Melampy RM (1973) Gonadotropin levels in lactating rats: Effect of ovariectomy. Endocrinology 93:540-547

Freeman ME, Smith MS, Nazian SJ, Neill JD (1974) Ovarian and hypothalamic control of the daily surges of prolactin secretion during pseudopregnancy in the rat. Endocrinology 94:875-882
Gibori G, Rothchild I, Pepe GJ, Morishige WK, Lam P (1974) Luteotrophic action of decidual tissue. Endocrinology 95:1113-1118
Hashimoto I, Henricks DM, Anderson LL, Melampy RM (1968) Progesterone and pregn-4-en-20α-ol-3-one in ovarian venous blood during various reproductive states in the rat. Endocrinology 82:333-341
Hayashida T, Young WP (1963) Interruption of pregnancy in rats with antiserum. Anat Rec 145: 323-324
Hotchkiss J, Atkinson LE, Knobil E (1971) Time course of serum estrogen and luteinizing hormone (LH) concentrations during the menstrual cycle of the rhesus monkey. Endocrinology 89:177-183
Joshi HS, Labhsetwar AP (1972) The pattern of ovarian secretion of oestradiol and oestrone during pregnancy and the post partum period in the hamster. J Reprod Fertil 31:299-302
Krehbiel RH (1941) The effects of Theelin on delayed implantation in the pregnant lactating rat. Anat Rec 81:381-392
Keyes PL, Nalbandov AV (1967) Maintenance and function of corpora lutea in rabbits depend on estrogen. Endocrinology 80:938-946
Lee C, Keyes PL, Jacobson HI (1971) Estrogen receptor in rabbit corpus luteum. Science 173:1032-1033
Lin YC, Yoshinaga K (1976a) Inhibitory effect of Gn-RH on ovum implantation in the rat. 58th Annual Meeting of the Endocrine Society, June 23-25 1976, San Francisco, Abstract
Lin YC, Yoshinaga K (1976b) Chronic effects of Gn-RH on the pituitary ovarian axis in pseudo-pregnant rats. 9th Annual Meeting of the Society for the Study of Reproduction; August 10-13, Philadelphia, Abstract
Macdonald GJ, Armstrong DT, Greep RO (1967) Initiation of blastocyst implantation by luteinizing hormone. Endocrinology 80:172-176
Madhwa Raj HG, Sairam MR, Moudgal NR (1968) Involvement of luteinizing hormone in the implantation process of the rat. J Reprod Fertil 17:335-341
Meyer RK, Wolf RC, Arslan M (1970) Implantation and maintenance of pregnancy in progesterone-treated ovariectomized monkeys (*Macaca mulatta*) In: Experimental medicine and surgery in primates (Goldsmith EI and Moore-Jankowski J (eds) Proc 2 nd Conference New York 1969, Karger, Basel, pp 30-35
Mougdal NR, Behman HR, Greep RO (1972) Effect of luteinizing hormone antiserum on progesterone and 20α-dihydroprogesterone secretion in the pregnant rat. J Endocrinol 52:413-418
Mougdal NR, Rao AJ, Maneckjee R, Muralidhar K, Mukku V, Rani CSS (1974) Gonadotropins and their antibodies. Recent Prog Horm Res 30:47-71
Murr SM, Bradford GE, Geschwind II (1974) Plasma luteinizing hormone, follicle stimulating hormone and prolactin during pregnancy in the mouse. Endocrinology 94:112-116
Nimrod A, Ladany S, Lindner HR (1972) Perinidatory ovarian oestrogen secretion in the pregnant rat, determined by gas chromatograph with electron capture detection. J Endocrinol 53:249-260
Norris ML, Adams CE (1971) Delayed implantation in the Mongolian gerbil, *Meriones unguiculatus.* J Reprod Fertil 27:486-487
Orsini MW, Meyer RK (1962) Effect of varying doses of progesterone on implantation in the ovariectomized hamster. Proc Soc Exp Biol Med 110:713-715
Pencharz RI, Long JA (1931) The effect of hypophysectomy on gestation in the rat. Science 74:206
Pope GS, Waynforth HB (1970) Secretion of oestrogens into the ovarian venous blood of pregnant rats. J Endocrinol 48:i-ii
Psychoyos A (1963) Precisions sur l'etat de "non-receptivite" de l'uterus. C R Acad Sci (Paris) 257: 1153-1156
Raud HR (1974) The regulation of ovum implantation in the rat by endogenous and exogenous FSH and prolactin: posibble role of ovarian follicles. Biol Reprod 10:327-334
Ross GT, Cargill CM, Lipsett MB, Rayford PL, Marshall JR, Strott CA, Rodbard D (1970) Pituitary and gonadal hormones in women during spontaneous and induced ovulatory cycles. Rècent Prog Horm Res 26:1-26
Shaikh AA (1971) Estrone and estradiol levels in the ovarian venous blood from rats during the estrous cycle and pregnancy. Biol Reprod 5:297-307

Shaikh AA, Abraham GE (1969) Measurement of estrogen surge during pseudopregnancy in rats by radioimmunoassay. Biol Reprod 1:378-380
Shelesnyak MC (1960) Nidation of the fertilized ovum. Endeavour 19:81-86
Smith MS, Freeman ME, Neill JD (1975) The control of progesterone secretion during the estrous cycle and early pseudopregnancy in the rat: prolactin, gonadotropin and steroid levels associated with rescue of the corpus luteum of pseudopregnancy. Endocrinology 96:219-226
Takahashi M, Ford JJ, Yoshinaga K, Greep RO (1975) Effects of cervical stimulation and anti-LH releasing hormone serum on LH releasing hormone content in the hypothalamus. Endocrinology 96:453-457
Watson J, Anderson FB, Alam M, O'Grady JE, Heald PJ (1975) Plasma hormones and pituitary luteinizing hormone in the rat during the early stages of pregnancy and after post-coital treatment with Tamoxifen (ICI 46,474). J Endocrinol 65:7-17
Weichert CK (1941) The effectiveness of estrogen in shortening delayed pregnancy in the rat. Anat Rec 81:106
Yochim JM, DeFeo VJ (1963) Hormonal control of the onset, magnitude and duration of uterine sensitivity in the rat by steroid hormones of the ovary. Endocrinology 72:317-326
Yoshinaga K (1961) Effect of local application of ovarian hormones on the delay in implantation in lactating rats. J Reprod Fertil 2:35-41
Yoshinaga K (1972) Suppression of ovarian cyclic activity by deciduoma in the rat. Anat Rec 172: 432 (Abstract)
Yoshinaga K (1974) Ovarian progestin secretion in lactating rats: Effect of intrabursal injection of prolactin-antiserum, prolactin and LH. Endocrinology 94:829-834
Yoshinaga K (1976) Ovarian hormone secretion and ovum implantation. In: Yoshinaga K, Meyer RK, Greep RO (eds) Implantation of the ovum. Harvard University Press, Cambridge, Massachusetts, pp 3-17
Yoshinaga K (1977) Hormonal interplay in the establishment of pregnancy. In: Greep RO (ed) International review of physiology. Reproductive physiology II. University Park Press, Baltimore, pp 201-223
Yoshinaga K (1979) Suppression of ovarian function by LHRH and its analogues in pregnant rats. In: Ovarian follicular and corpus luteum function (Channing CP, Marsh JM, Sadler WA eds) Plenum Publish Corp, New York pp 729-734
Yoshinaga K (1980) Inhibition of implantation by advancement of uterine sensitivity and refractoriness. In: Blastocyst-endometrium relationships (Leroy F, Finn CA, Psychoyos A, Hubinont PO, eds) Progr reprod biol Vol 7, Karger, Basel, pp 189-199
Yoshinaga K (1981) Induction of delayed implantation in rats by LH-RH. J Reprod Fertil Suppl 29: 131-133
Yoshinaga K, Adams CE (1966) Delayed implantation in the spayed, progesterone treated adult mouse. J Reprod Fertil 12:593-595
Yoshinaga K, Adams CE (1967) Luteotrophic activity of the young conceptus in the rat. J Reprod Fertil 13:505-509
Yoshinaga K, Greep RO (1970) Precocious sensitization of the uterus in pseudopregnant rats. Proc Soc Exp Biol Med 134:725-727
Yoshinaga K, Greep RO (1974) Uterine sensitivity with regard to ovo-implantation. In: Husain SM, Guttmacher AF (eds) Progress in reproduction research and population control. Publications International, Quebec, pp 137-146
Yoshinaga K, Pincus G (1963) Local effect of estrogen on cholesterol synthesis in the uterus of ovariectomized rats. Steroids 1:656-663
Yoshinaga K, Hawkins RA, Stocker JF (1969) Estrogen secretion by the rat ovary *in vivo* during the estrous cycle and pregnancy. Endocrinology 85:103-112
Yoshinaga K, Mougdal NR, Greep RO (1971) Progestin secretion by the ovary in lactating rats: Effects of LH-antiserum, LH and prolactin. Endocrinology 88:1126-1130
Zeilmaker GH (1963) Experimental studies on the effects of ovariectomy and hypophysectomy on blastocyst implantation in the rat. Acta Endocrinol (Kbh) 44:355-366

Ultrastructural Observations on Rabbit Luteal Cells and Interstitial Gland Cells During Pseudopregnancy

Jozef R. Quatacker

Department of Pathology, Akademisch Ziekenhuis, State University, Gent, Belgium

The rabbit ovary has been the subject of extensive studies in reproductive endocrinology and numerous papers deal with the function and regulation of luteal cells and interstitial gland cells. It is however surprising to note that few morphological studies have been devoted to luteal cells (Blanchette 1966) and interstitial gland cells (Davies and Broadus 1968). Moreover there are no systematic and quantitative descriptions at all.

This contribution describes qualitative and quantitative changes in corpora lutea (CL) and interstitial gland tissue during pseudopregnancy (PSP) and, where relevant, results from corpora lutea during pregnancy are included. All our material was fixed by vascular perfusion. The quantitative evaluation is based on a morphometric analysis, which allows an objective determination of the different cytoplasmic structures. These parameters can be correlated with biochemical findings, as for example the progesterone and 20α-OH-progesterone concentration in the venous blood (Hilliard et al. 1968; Hilliard and Eaton 1971; Horrell et al. 1972; Challis et al. 1973; Browning et al. 1980). Interesting observations can also be made by comparing the morphology of luteal cells to that of interstitial gland cells. Special attention is given to the septatelike cell contacts which occur between steroid-producing cells (Enders 1973; Friend and Gilula 1972), and our recent findings (Quatacker 1975b; 1979) are reported. In target cells, endocytic activity has been implicated in hormone-receptor transport to the lysosomes (Szego 1975). In rabbit luteal cells estradiol binding to cytosol and nuclear fraction has been demonstrated (Miller and Keyes 1978). In this respect we studied the changes in endocytic activity during pseudopregnancy with horseradish peroxidase as a tracer.

1 Development of Luteal Cells During Pseudopregnancy

1.1 Morphology

Human chorion gonadotrophin (hCG) injection is followed approximately 10 h later by ovulation and subsequent luteinization of granulosa cells. Four days later the luteal cells have the distinct appearance of steroid-producing cells. In perfusion-fixed material the corpora lutea have large intercellular spaces, which seem inherent to stimulation. Villi are already present at day 4, but at days 8 and 12 the number of villi is much greater. Structures resembling bile canaliculi (Adams and Hertig 1969; Enders 1973) can be observed in some places. The membrane however is enzyme-cytochemically no different from plasma membrane in other areas (Quatacker 1973). Enzyme-cytochemically it can be shown that the plasma membrane contains an adenyl cyclase, an ATP-ase, an ADP-ase, an AMP-ase and an alkaline phosphatase (Quatacker 1976, 1979). The properties of the ATP-ase and ADP-ase being closely related. Biochemically the activity and characteristics of the adenyl cyclase on membrane fractions of rabbit CL have been determined, and nucleotides may have a regulatory function at the plasma membrane level (Hunzicker-Dunn and Birnbaumer 1976; Hunzicker-Dunn and Jungmann 1978). In bovine corpus luteum a mechanism for nucleotide inhibition of gonadotrophin binding to cell membranes is proposed (Rao 1975). The types of membrane specialization present are septatelike cell contacts (Enders 1973; Friend and Gilula 1972; Quatacker 1975b; and below) and gap junctions. At day 4 the rough endoplasmic reticulum (RER) cisternae are tortuous and the smooth endoplasmic reticulum (SER) is composed of random tubules (Christensen and Gillim 1969). The larger areas of SER (Fig. 1) at days 8 and 12 are formed by concentric fenestrated lamellae (Christensen and Gillim 1969) and at that time the RER shows parallel cisternae. Functional differences have been attributed to the tubular and whorled SER (Mc Clellan et al. 1975). Enzyme-cytochemically glucose-6-phosphatase, unspecific esterase (Quatacker 1973) and locally also arylsulphatase (Quatacker 1979) can be found in the SER. At the moment of maximum RER proliferation many free ribosomes and polysomes are found in the cytoplasm. Already at day 12 some signs of degradation in the SER may appear. At day 16 the SER has shrunk and the concentric membrane areas have collapsed completely by day 20. Small peroxisomes are associated with the SER. They are a common feature of luteal cells and of other steroid-producing cells (Gulyas and Yuan 1975).

Soon after ovulation the mitochondria acquire their typical aspect with wide open lamellar cristae and a dense matrix (Fig. 1). In porcine CL this aspect seems to coincide with the capacity to form progesterone (Dimino 1977). In young corpora lutea the mitochondria are elongated and may even be branching. The lipid droplets remain osmiophilic (Fig. 1) until involution (Adams and Hertig 1969; Blanchette 1966) and then become electron-lucent, as the result of cholesterol ester storage (see below). Biochemical findings also indicate that in pseudopregnant rabbits the activity of the cholesterol ester synthetase (Flint et al. 1974) increases during gonadotrophin induced luteolysis. The Golgi complex forms a cap over the nucleus with coated vesicles, dense bodies and multivesicular bodies (MvB's) filling up the space beneath the cap. Mostly a Golgi stack is composed of three or four continuous rigid lamellae with a few additional curved saccules on the outside and the so-called GERL elements on the inside. The development and regression of the Golgi apparatus runs parallel with the growth (Adams and Hertig 1969; Enders 1973) and involution (Adams and Hertig 1969) of luteal cells. Also in pregnant animals our results seem to indicate, that the volume of the Golgi apparatus largerly follows the curve of cell development. There are practically no autophagic vacuoles at day 4, some are present at day 8, but they are obvious in the cytoplasm of day 12 luteal cells. Apart from its normal role the Golgi apparatus has also been implicated in the formation of relaxine (Belt et al. 1970) and the conjugation of steroids (Adams and Hertig 1969). The way in which the steroids are secreted is still not clear. The special junctions (Friend and Gilula 1972), the "bile canaliculi" (Adams and Hertig 1969) and small dense granules in the sheep and goat (Gemmell et al. 1977b) have been proposed as possible pathways for steroid secretion. Evidence for a steroid-containing secretory organelle has also been presented in the adrenal (Gemmell et al. 1977a). In rabbit luteal cells the granular structures, resembling dense bodies, are located centrally and there is no direct evidence for a directional flow out of the cell. Neither our morphometric results provide any argument for a correlation with steroid production in rabbit luteal cells (cf. Reviews in: Channing et al. 1979).

1.2 Morphometric Analysis

Corpora lutea from days 4, 8, 12 and 16 of pseudopregnancy have been studied by morphometric methods. An early manifestation of stimulation and growth is the appearance of RER and polysomes in the cytoplasm. The highest relative concentration of RER ist noted at day 4, decreasing subsequently at days 8 and 12 (-50%), and with a further reduction at day 16 (-80%). The ribosomal density on the membrane is positively correlated with the proliferative activity. Similar observations have been made in the developing rat hepatocyte (Dallner et al. 1966). On a per cell basis (Table 1) and after correction for ribosomal density a linear increase in the amount of RER is observed up to day 12, but a clear-cut decrease is noted at day 16. From the amount of RER it appears that the development progresses up to day 12 of pseudopregnancy. Preliminary results from pregnant animals show that the highest relative concentration of RER is obtained in early pregnancy and that the amount of RER per cell decreases as the cell reaches maximal cell volume. Taken together it thus appears that the amount of RER mainly reflects the development of SER and is therefore also closely related to the increase in cell volume. It remains to be cleared which hormonal stimulus regulates the described cellular events and is responsible for the induction and maintenance of the steroidogenic capacity. The effect of growth hormone (Gospodarowicz 1977) on bovine granulosa cells have been determined. In rabbit CL the first three days the cells seem to be preprogrammed by a preovulatory luteinizing hormone (LH) surge (Miller and Keyes 1978; Suzuki et al. 1977); as they are independent of LH during that period. Thereafter the cells can again be stimulated with LH (Suzuki et al. 1977). Up to day five they are independent of estrogen (Miller and Keyes 1978). After implantation an integrated regulatory system based on an estrogen-LH luteotrophic complex has been proposed, in which LH regulates cAMP formation, and in which estrogen is responsible for induction of steroidogenic enzymes (Hunzicker-Dunn and Birnbaumer 1976). However, this does not explain the high enzym synthesizing activity at a moment (day 4) that the luteal cells are independent of estrogen (Mills and Osteen 1977). At that time LH may be the most likely candidate for this function. The greatest increase in SER per cell (Table 1) is noted between 4 and 8 days, as a consequence of the high RER concentration at day 4. A less pronounced increase in SER between 8 and 12 days is noted. In pregnancy the maximal amount of SER is reached at day 16 and only slightly decreases at day 28. In PSP from day 12 on regression phenomena (see above) have to be taken into account and become important at day 16. The SER which represents the bulk of the microsomal fraction contains enzymes which provide reducing equivalents needed for lipid and cholesterol synthesis (Gemmell et al. 1977b; McClellan et al. 1975) and for reduction and hydroxylation reactions in steroid production (McKerns 1969).

The relationship between lipid droplets and the SER can be deduced from morphometric determinations. The high increase in SER per cell between 4 and 8 days is followed by a similar increase in the amount of lipid droplets per cell (Table 1) in the next

Table 1. Volume and composition in relative units (means$\pm$s.e.m.) of luteal cells during pseudopregnancy

Days after hCG injection	4	8	12	16
No. of observations	4	4	4	4
Cell volume	88$\pm$2	222$\pm$8	349$\pm$24	338$\pm$22
Smooth ER (vol./cell)	9.8$\pm$1.0	39.5$\pm$2.7	49.1$\pm$6.6	29.9$\pm$2.3
Mitochondria (vol./cell)	7.1$\pm$0.2	16.1$\pm$1.3	23.1$\pm$1.4	23.7$\pm$2.7
Lipid droplets (vol./cell)	9.2$\pm$1.2	36.0$\pm$1.6	74.7$\pm$10.7	81.1$\pm$7.3
Golgi apparatus (vol./cell)	1.4$\pm$0.2	2.9$\pm$0.5	4.8$\pm$0.6	3.0$\pm$0.5
Lysosomes (vol./cell)	0.26$\pm$0.01	0.38$\pm$0.02	0.58$\pm$0.09	0.60$\pm$0.06
Rough ER (area/cell)	8.2$\pm$0.8	13.6$\pm$1.7	19.5$\pm$1.8	5.2$\pm$0.7

period (8-12 days). This is also obvious in the SER between 8 and 12 days and the fat droplets in the following period. During pseudopregnancy the proportional increase of lipid droplets doubles the proportional increase in cell volume. The mitochondrial volume per cell grows less rapidly than the cell volume (Table 1) and this is more pronounced at day 12 than at day 8. At day 16 the mitochondrial volume has slightly increased, but a significantly lower size of the mitochondrial profiles is noted. In rat luteal cells a correlation between progesterone production and mitochondrial size has also been observed (Breinl 1970). Comparison of SER with mitochondria reveals an outburst of SER until day 8, with on the other hand a steady increase of the mitochondria until day 12. From our studies we can conclude that the SER develops more specifically and is more sensitive to impaired conditions in the cell, whereas the mitochondria increase continuously and are more resistant to an involutive environment. The finding that the microsomal cytochome P-450 is more sensitive to hypophysectomy gives the same indication (Purvis 1973). This seems to show that the amount of SER most adequately reflects the steroid production as determined by biochemical means (Browning et al. 1980; Challis et al. 1973; Hilliard et al. 1968; Horrell et al. 1972). Also in the adrenal cortex the amount of SER seems to be the most important parameter for assessing the steroid-producing activity (Nussdorfer and Mazzocchi 1970). During pregnancy progesterone production has also been determined (Challis et al. 1973; Hilliard et al. 1968; Hilliard and Eaton 1971). However in corpora lutea during pregnancy the amount of SER is more or less correlated with the cell volume and it has been shown by Hilliard et al. (1968) that in the last half of pregnancy the cell volume does not run parallel with the progesterone production.

During PSP some authors report a correlation between progesterone production (Hilliard et al. 1968) and corpus luteum weight, whereas others (Horrell et al. 1968) observe maximum progesterone production at day 7 and maximum weight at day 13. Furthermore it remains impossible to determine to what extent the SER and/or mitochondria contribute to the steroid production. If the cholesterol side-chain cleavage enzyme, which has been shown to be present in microsomes from rat and cow, is also present in the SER of rabbit luteal cells, it might provide together with the Δ 5-3 β-OH-steroid dehydrogenase a complete microsomal pathway for progesterone production. For the porcine CL it has been demonstrated (Dimino 1977) that highly purified mitochondria can produce progesterone and it is suggested that the hormone may be produced in different ways by mitochondria and endoplasmic reticulum. As indicated some regressive changes are present at day 12 of PSP and relatively important deviations in the amount of lipid droplets between animals become obvious. The vascular space is also somewhat reduced. This indicates an impairment to the luteal function, but on the other hand the cells are still stimulated, as is indicated by the relatively high amount of RER present. The significantly lower value for RER together with the minimal decrease in cell volume at day 16, seems to indicate that a functional luteolysis precedes a structural one (Flint et al. 1974). The amount of Golgi apparatus per cell (Table 1) increases until day 12, with at day 16 a reduction of the same order of magnitude as the reduction of SER. Similar changes in the volume of the Golgi apparatus in the adrenal cortex have also been observed (Magalhaes and Magalhaes 1969). Both the stacks and the Golgi associated vesicles are diminished, probably as the result of the low RER concentration. From the results in PSP it is impossible to decide whether the observed reduction at day 16 is related to reduced secretion or to the absence of cell growth, but in pregnancy the volume of the Golgi apparatus is at large correlated with the development of the cell. The higher lysosomal volume per cell in PSP (Table 1) at 12 and 16 days coincides with regressive changes in the cytoplasm.

2 The Interstitial Gland Cells During Pseudopregnancy

2.1 Morphology

Injection of gonadotrophin results in drastic changes in the composition of the cytoplasm (Davies and Broadus 1968). In the first 10 h after injection the interstitial gland cells are heterogeneous and the tissue is loosely organized, showing interstitial edema and hyperemia. A marked depletion of electron-lucent lipid droplets occurs in the cells and after 4 h large concentric areas of SER already appear in conjunction with osmiophilic lipid droplets. At 10 h the SER has further increased. The mitochondria are also heterogeneous with sometimes round, sometimes elongated forms prevailing. During a 10-h period after coitus 20α-OH-progesterone can be detected in the ovarian venous effluent (Hilliard et al. 1968; Hilliard and Eaton 1971). By 20 h after hCG injection the cell population has become more homogeneous, but the tissue is still loosely organized. Ultrastructurally the interstitial gland cells now have the closest resemblance to luteal cells. However the Golgi apparatus is very tortuous and does not show rigid saccules as in luteal cells. The mitochondria are small, round to elongated with open lamellar cristae (Fig. 2) and the cytoplasm is mostly occupied by extremely large areas of

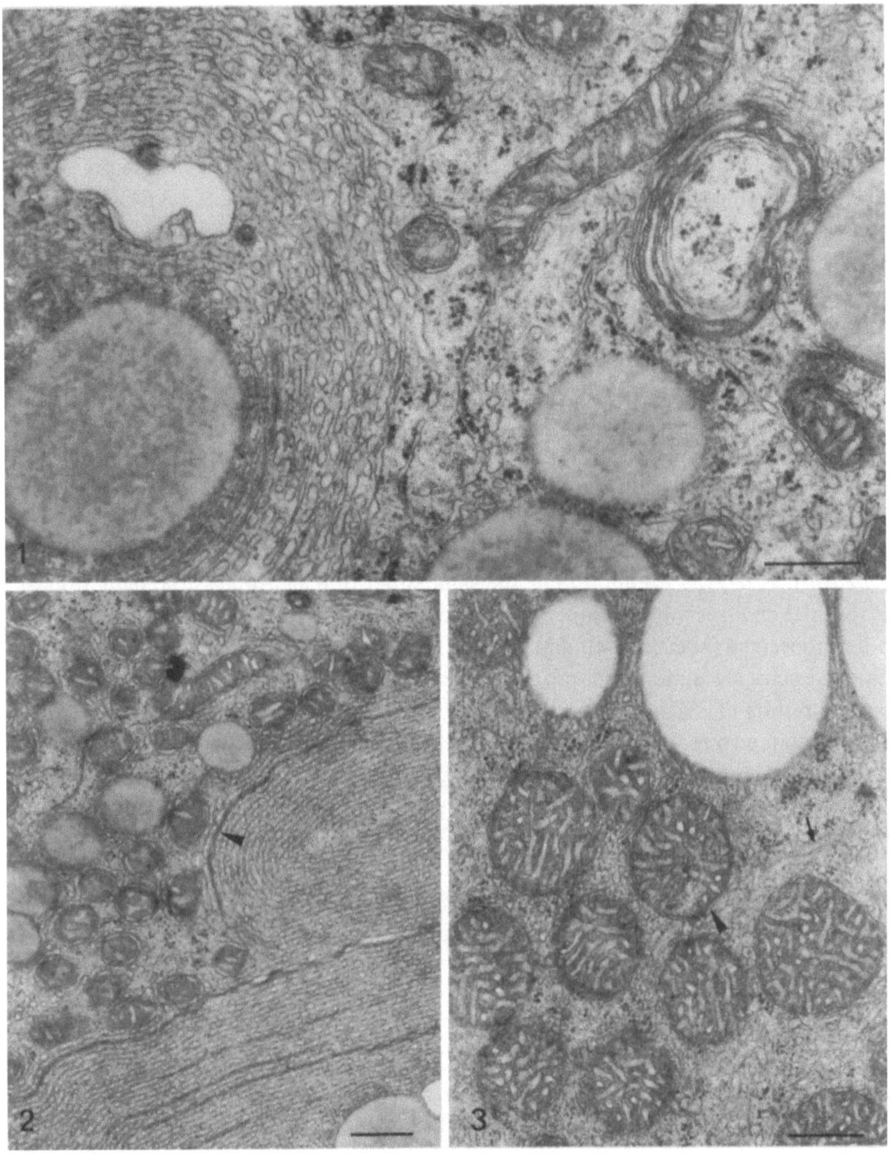

Fig. 1. Besides osmiophilic lipid droplets and typical mitochondria the luteal cells contain concentrically arranged smooth ER lamellae with rough ER elements on the outside. x 26 000 (bar=0.5μm)

Fig. 2. Interstitial gland cell 20 h after hCG injection showing the appearance of luteal cells. Collapsed membranes (*arrowhead*) are present in the large areas of smooth ER. x 16 000

Fig. 3. Interstitial gland cell 12 days after hCG injection showing highly differentiated mitochondria with sometimes tubular cristae (*arrowhead*). The lipid droplets are electron-lucent and between the mitochondria anastomosing smooth ER tubules (*arrow*) are present. x 21 000

SER associated with osmiophilic lipid droplets (Fig. 2). But regression phenomena in the form of ring-shaped collapsed membranes are present (Fig. 2). This seems to indicate that stimulation has ended and may explain why after LH stimulation only low levels of 20α-OH-progesterone are detected at that time (Browning et al. 1980; Hilliard et al. 1968; Horrell et al. 1972). However determination of 20α-OH-progesterone in our animals has not been carried out and stimulation with hCG, which has a longer half-life than LH may prolong secretion. However a similar pattern, with a fall to basal levels at 8 h is observed in the cyclic AMP production in rabbits made pseudopregnant with hCG (Goff and Major 1975).

At day 5 the cells are polygonal and grouped together. The intercellular space is reduced and the vascular bed is less dense. At this point the SER is less prominent, although in some animals large areas of SER are still present. The lipid droplets are no longer osmiophilic, as the result of an increased sterol content (Claesson 1954; Guraya and Greenwald 1964). These changes in the composition of the lipid droplets occuring under gonadotrophic stimulation have already been described biochemically by Claesson (1954) and histochemically by Guraya and Greenwald (1964). The mitochondria are heterogeneous, and also between animals they may vary considerably. They may be round with few cristae or more elongated with transverse lamellar ones. As they grow and differentiate they tend to become spherical and sometimes show an increasing number of cristae (Davies and Broadus 1968). In some animals many cristae, sometimes tubular, are found filling up large spherical mitochondria (Fig. 3). A similar differentiating process has been reported in regenerating adrenal cortex (Yago et al. 1972). On day 9 and following days the SER is further reduced and the number of electron-lucent lipid droplets increases. On day 20 the cells contain large numbers of electron-lucent lipid droplets.

2.2 *Morphometric Analysis*

In order to estimate the changes in the interstitial gland cells a morphometric analysis was performed at 20 h, 5, 9, 14 and 20 days after hCG injection. The percentage of the cytoplasm occupied by mitochondria, SER and lipid droplets was determined. The relative volume of mitochondria does not change significantly during the period studied and varies from 12% to 17% of the cytoplasm depending on the animal. The SER occupies 17% at 20 h, 15% at day 5, approx. 11% at days 9 and 14 and 7% at day 20. Taking into account its importance in steroidogenesis, we have calculated for each animal the ratio of the percentage of SER to mitochondria in the cells (Table 2). This ratio indicates an increasing predominance of the mitochondria over the SER during a period of rising steroid production (Browning et al. 1980; Hilliard et al. 1968). This tendency proceeds up to day 20 and is even accelerated at the end of pseudopregnancy. The number of lipid droplets is lowest at 20 h (Table 2), somewhat higher at day 5, and later increases steadily until day 20. This fits in with the pattern of biochemically measured cholesterol ester concentration (Hilliard et al. 1968) in the cells during pseudopregnancy, except at day 20. At that time Hillard et al. (1968) observed a decrease, which was correlated with a high level of 20α-OH-progesterone in the venous effluent. Horrell et al. (1972) detected the highest secretion at day 13 with a high concentration of 20α-OH-progesterone in the ovarian interstitial tissue during the following days.

Table 2. Composition of the interstitial gland cells during pseudopregnancy (means$\pm$s.e.m.)

Time after hCG injection	20 h	5 days	9 days	14 days	20 days
No. of observations	2	5	4	4	5
% smooth ER/ % mitochondria	1.35$\pm$0.15	1.07$\pm$0.10	0.69$\pm$0.06	0.67$\pm$0.07	0.47$\pm$0.03
% lipid droplets	6.5$\pm$0.9	7.5$\pm$0.7	12.1$\pm$0.6	18.1$\pm$1.7	25.8$\pm$1.8

3 Comparison Between Luteal Cells and Interstitial Gland Cells

The luteal cell in rabbit shows large amounts of SER, osmiophilic lipid droplets and elongated mitochondria. In the interstitial gland cell a similar picture is found shortly after stimulation. In both cases this pattern is accompanied by a dense vascular bed, allowing each cell to be in contact with a capillary. All this seems to indicate that the cells function at a maximal level of stimulation (Holt et al. 1975). Indeed rabbit luteal cells cannot be further stimulated by additional estrogen (Holt et al. 1975). After the acute stimulation the interstitial gland cells are gradually transormed. The relative predominance of the SER decreases (see above) and the impact of the mitochondria on the steroid synthesis may increase. As a Δ5-3β-OH-steroid dehydrogenase (Jackanicz and Armstrong 1968) has been reported to be present also in mitochondria of the interstitial gland cells of rabbit, they may in preference carry out the steroid synthesis under these submaximal stimulatory conditions. In the light of the hypothesis that two pathways for steroid production may exist (Dimino 1977), the cells may thus use the microsomal pathway under high stimulatory conditions and the mitochondrial pathway under low stimulatory conditions (Flint and Armstrong 1971). Each pathway may have its own regulatory mechanism, but further biochemical evidence is needed to substantiate this hypothesis.

4 Septatelike Cell Contacts

Septatelike cell contacts have been extensively studied by Friend and Gilula (1972) in a number of steroid-producing cells and similar structures are also described in other papers (Enders 1973). Between rabbit (Quatacker 1975b), rat, and human (Quatacker 1973) luteal cells we observed electron-lucent septa bridging the intercellular space. However, circular particles (Friend and Gilula 1972) were not detected. The septa are only clearly seen occasionally, unless colloidal lanthanum is applied. In that case repeating structures appear more regularly. At high magnification the repeating units after colloidal lanthanum application seem to be composed of a larger central particle (65-70 Å), accompanied on both sides by a smaller globule (25-30 Å) in contact with the plasma membrane (Fig. 4). In tangential view clear granules are observed against a dense background. These granules are not as regularly spaced as the intercellular particles seen in

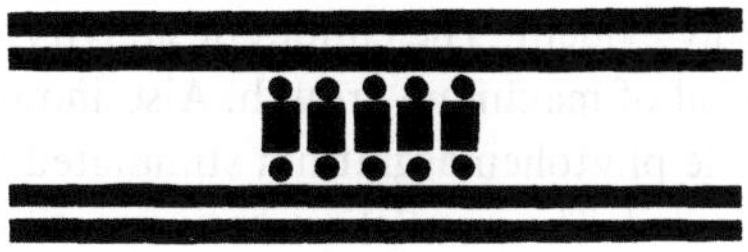

Fig. 4. This schematic drawing represents our observations on septatelike cell contacts after colloidal lanthanum application. The repeating units are composed of a large central particle connected with the plasma membrane by symmetrically arranged smaller globular structures

gap junctions. Our findings disagree with earlier observations (Friend and Gilula 1972) that alveolar structures are present in tangentially cut septatelike cell contacts.

The symmetric nature of the repeating units observed in cross-section makes it reasonable to suppose that particles or projections from both plasma membranes are involved in the formation of the cell contact. Although these cell contacts have only been described between steroid-producing cells, they do not seem to be always present, as they are almost certainly missing between interstitial gland cells. It can be presumed that these contacts provide support and adhesion with minimal blocking of the diffusion pathway, although their precise role remains to be determined.

5 Endocytic Activity in Luteal Cells During Pseudopregnancy

5.1 Endocytic Process

Endocytosis has been studied in rabbit corpus luteum (Quatacker 1975a) by a single injection of horseradish peroxidase directly into the corpora lutea on days 5, 9, 14, and 20 of pseudopregnancy. The peroxidase is taken up rapidly and is transported to the inner side of the Golgi region close to the nucleus. Fifteen to twenty minutes after injection comma-shaped structures, tailed or round bodies, clear vacuoles and also multivesicular bodies (MvBs) are found to be positive. The transformation of endocytic vacuoles into the different types of MvB can be traced. Peroxidase-positive vesiculo-tubular structures and small positive vesicles are found near the MvBs and may bring more material to the already formed MvBs. Vesicles derived from the inner side of the Golgi apparatus (Quatacker 1975a), as can be deduced from their heavy staining with phosphotungstic acid at low pH, are also found close to the MvBs. They probably represent the link between the endocytic vacuole and the lysosome system. In order to exclude that endocytosis was induced by the horseradish peroxidase injection, counts of the MvBs were made in control animals. They revealed that more endocytic vacuoles and early MvBs are present in the cytoplasm at day 4, justifying the results of our tracer experiments.

5.2 Relationship Between Age of Luteal Cells and Uptake of Horseradish Peroxidase

The horseradish peroxidase-positive structures in 5-day-old luteal cells are mainly concentrated in the Golgi area, but this distribution is less clear in 9-day-old luteal cells. However the estimated number of positive structures in the cells at days 5 and 9 is not different. At 14 days the cells show fewer cytoplasmic extensions and the uptake of horseradish peroxidase is drastically diminished. Uptake in 20 day old luteal cells is hardly evident. The endocytic activity in luteal cells seems thus to be related to the period of maximum growth. Also in rapidly developing cells of other origins, as for example phytohemagglutinin stimulated lymphocytes, a high endocytic activity has been observed. The parallelism between stimulation and lysosomal activity has already prompted Szego (1975) to collect arguments in favour of an endocytic uptake and a lysosomal transport to the nucleus of an agonist. This may be a hormone bound to its receptor site at the plasma membrane and delivered afterwards to the nucleus in order to switch on the transscription process (Miller and Keyes 1978).

Our findings indicate that transition of plasma membrane areas into the interior of the cell and the formation of MvBs occurs rapidly and is correlated to the activity of the cell. But some findings with iodinated gonadotrophins seem to indicate that the hormone uptake does not progress so rapidly. Han et al. (1974) demonstrated that in rat ovarian tissue after 1 h, 60% of the label was present over the plasma membrane and only 10% over subplasmalemmal vesicles,with no significant activity over the other organelles. Recently (Chen et al. 1977) an initial labeling of the periphery of ovine luteal cells with iodinated hCG was reported. After 4 1/2 h the silver grains were mainly distributed over the MvBs and dense bodies. This is an obvious sign of endocytosis, but does not seem to indicate a rapid transfer of gonadotrophin to the nuclear compartment. Thus the exact significance of this phenomenon remains unclear. In fibroblasts it seems (Carpenter and Cohen 1976) that the iodinated growth factor is degraded in the lysosomes. However it cannot be decided, whether internalization with or without degradation is required in order to increase DNA synthesis or whether it is simply a mechanism for inactivation of the hormone (Carpenter and Cohen 1976; Chen et al. 1977).

References

Adams EC, Hertig AT (1969) Studies on the human corpus luteum. I. Observations on the ultrastructure of development and regression of the luteal cells during the menstrual cycle. J Cell Biol 41:696-715

Belt WE, Cavazos LF, Anderson LL, Kraeling RR (1970) Fine structure and progesterone levels in the corpus luteum of the pig during pregnancy and after hysterectomy. Biol Reprod 2:98-113

Blanchette EJ (1966) Ovarian steroid cells. II. The lutein cell. J Cell Biol 31:517-542

Breinl H (1970) Biometrische Objektivierung funktions-abhängiger Grössenänderungen der Mitochondrien in Luteinzellen des Rattenovars. Arch Gynäk 208:453-470

Browning JY et al. (1980) Comparison of serum progesterone, 20α-dihydroprogesterone, and estradiol-17ß in pregnant and pseudopregnant rabbits: evidence for postimplantation recognition of pregnancy. Biol Reprod 23:1014-1019

Carpenter G, Cohen S (1976) ^{125}I-labeled human epidermal growth factors. Binding, internalization and degradation in human fibroblasts. J Cell Biol 71:159-171

Challis JRG et al. (1973) The concentration of progesterone, estrone and estriol-17ß in the plasma of pregnant rabbits. Endocrinology 93:971-976

Channing CP et al. (1979) Ovarian follicular and corpus luteum function. Plenum Press, New York 1979 (1st Edition)

Chen TT et al. (1977) Localization of gonadotropic hormones in lysosomes of ovine luteal cells. Cytobiologie 14:412-420

Christensen AK, Gillim SW (1969) The correlation of the fine structure and function in steroid-secreting cells with emphasis on those of the gonads. In: McKerns KW (ed) The gonads. Amsterdam, North-Holland, pp 415-490

Claesson L (1954) The intracellular localization of the esterfied cholesterol in the living interstitial gland cell of the rabbit ovary. Acta Physiol Scand Suppl 113, 31:53-78

Dallner G et al. (1966) Biogenesis of endoplasmic reticulum membranes. Structural and chemical differentiation in developing rat hepatocyte. J Cell Biol 30:73-96

Davies J, Broadus CD (1968) Studies on the fine structure of ovarian steroid-secreting cells in the rabbit. I. Normal interstitial cells. Am J Anat 123:441-474

Dimino MJ (1977) Differences in mitochondrial steroidogenesis between follicular and luteal tissues of porcine ovaries. Endocrinology 101:1844-1849

Enders AC (1973) Cytology of the corpus luteum. Biol Reprod 8:158-182

Flint APF, Armstrong DT (1971) Intracellular localization of cholesterol side-chain cleavage enzyme in corpora lutea of cow and rat. Nature (New Biol) 231:60-61

Flint APF et al. (1974) Metabolism of the corpus luteum during luteolysis in the pseudopregnant rabbit. Endocrinology 94:509-517

Friend DS, Gilula NB (1972) A distinctive cell contact in the rat adrenal cortex. J Cell Biol 53:148-163

Gemmell RT et al. (1977a) Ultrastructural and biochemical evidence for a steroid-containing secretory organelle in the perfused cat adrenal gland. J Cell Biol 72:209-215

Gemmell RT et al. (1977b) Secretion of granules by the luteal cells of the sheep and the goat during the estrous cycle and pregnancy. Anat Rec 189:161-167

Goff AK, Major PW (1975) Concentration of cyclic AMP in rabbit ovarian tissue during the preovulatory period and pseudopregnancy after induction of ovulation by administration of human chorionic gonadotrophin. J Endocr 65:73-82

Gospodarowicz D (1977) Effects of fibroblast and epidermal growth factors on ovarian cell proliferation in vitro. II. Proliferative response of luteal cells to FGF but not EGF. Endocrinology 100: 1121-1128

Gulyas BJ, Yuan LC (1977) Association of microperoxisomes with the endoplasmic reticulum in the granulosa lutein cells of the rhesus monkey (Macaca mulatta). Cell Tis Res 179:357-366

Guraya SS, Greenwald GS (1964) Histochemical studies on the interstitial gland in the rabbit ovary. Am J Anat 114:495-519

Han SS et al. (1974) Gonadotropin receptor in rat ovarian tissue. II. Subcellular localization of LH binding sites by electron microscopic radioautography. Endocrinology 95:589-598

Hilliard J et al. (1968) Cholesterol storage and progestin secretion during pregnancy and pseudopregnancy in the rabbit. Endocrinology 82:157-165

Hilliard J, Eaton LW Jr. (1971) Estradiol-17ß, progesterone and 20α-hydroxyprogesterone – 4-en-3-one in rabbit ovarian venous plasma. II From mating through implantation. Endocrinology 89: 522-527

Holt JA et al. (1975) Premature regression of corpora lutea in pseudopregnant rabbits following removal of polydimethylsiloxane capsules containing 17ß-estradiol. Endocrinology 97:76-82

Holt JA et al. (1976) Lack of gonadotropic activity in the rabbit blastocyst prior to implantation. Endocrinology 98:904-909

Horrell et al. (1972) Progestational steroids during pseudopregnancy in the rabbit. J Endocr 55:89-96

Hunzicker-Dunn M, Birnbaumer L (1976) Adenylyl cyclase activities in ovarian tissues. II. Regulation of responsiveness to LH, FSH and PGE_1 in the rabbit. Endocrinology 99:185-197

Hunzicker-Dunn M, Jungmann RA (1978) Rabbit ovarian protein kinases. III. Gonadotrophin-induced activation of soluble adenosine 3',5'-monophosphate dependant protein kinases. Endocrinology 103:441-451

Jackanicz TM, Armstrong MT (1968) Progesterone biosynthesis in rabbit ovarian interstitial tissue mitochondria. Endocrinology 83:769-776

Magalhaes MC, Magalhaes MM (1969) A stereologic study of the effects of metopiron on the rat adrenal. Lab Invest 21:411-496

McClellan MC et al. (1975) Luteinizing hormone, progesterone and the morphological development of normal and superovulated corpora lutea in sheep. Cell Tiss Res 164:291-307

McKerns KW (1969) Studies on the regulation of ovarian function by gonadotrophins. In: McKerns KW (ed) The gonads. North-Holland, Amsterdam, pp 137-174

Miller JB, Keyes PL (1975) Progesterone synthesis in developing rabbit corpora lutea in the absence of follicular estrogens. Endocrinology 97:83-90
Miller JB, Keyes PL (1978) Transition of the rabbit corpus luteum to estrogen dependence during early luteal development. Endocrinology 102:31-38
Mills TM, Osteen KG (1977) 17 beta-Estradiol receptor and progesterone and 20 alpha-hydroxy-4-pregnen-3-one content of the developing corpus luteum of the rabbit. Endocrinology 101:1744-50
Nussdorfer GG, Mazzocchi G (1970) Correlated morphometric and autoradiographic studies on the effects of corticosterone on adrenocortical cells of intact and hypophysectomized ACTH-treated rats. Z Zellforsch 111:92-105
Purvis JL et al. (1973) Lifetime of microsomal cytochrome P-450 and steroidogenic enzymes in rat testis as influenced by human chorion gonadotrophin. Arch Biochem Biophys 159:39-49
Quatacker JR (1973) De luteale cellen van het pseudozwangere konijn. PhD Thesis, Gent
Quatacker JR (1975a) Endocytosis and multivesicular body formation in rabbit luteal cells during pseudopregnancy. Cell Tiss Res 161:541-553
Quatacker JR (1975b) Cell contacts in rabbit corpora lutea. J Ultrastruct Res 50:299-305
Quatacker JR (1976) Adenyl cyclase and adenosinephophatases in the rabbit ovary. Proceedings of the fifth international congress of histochemistry and cytochemistry, pp 287
Quatacker JR (1979) Different aspects of membrane differentiation at the inner side (GERL) of the Golgi apparatus in rabbit luteal cells. Histochem. Journal 11:399-416
Rao ChV (1975) Mechanism of nucleotide inhibition of gonadotropin binding to cell membranes of bovine corpus luteum. Molecular and Cellular Endocrinology 3:255-271
Suzuki A et al. (1977) Formation of steroid hormones in vitro by developing corpora lutea of the rabbit. J Endocrinol 75:355-61
Szego CM (1975) Lysosomal function in nucleocytoplasmic communication. In: Dingle JT and Dean RT (eds) Lysosomes in biology and pathology 4. North-Holland, Amsterdam Oxford, pp 385-461
Yago et al. (1972) Growth and differentiation of mitochondria in the regenerating rat adrenal cortex. J Cell Biol 52:503-513

Uteroglobin and Other Endometrial Proteins: Biochemistry and Biological Significance in Beginning Pregnancy

Henning M. Beier

Department of Anatomy and Reproductive Biology, Rheinisch-Westfälische Technische Hochschule, Aachen, Germany

1 Historical Background of Uterus Protein Research

The fluids of male and female genital tracts were considered by the scientists of the 16th and 17th century to represent the male and female "semen". They believed that the fusion of male and female semen resulted in the formation of an embryo, obviously emerging from the semen pool within the cavity of the uterus. Ham and Leeuwenhoek detected in 1677 (cited by Bargmann 1967) that the male semen fluid contained the spermatozoa, which immediately were referred to as the male germ cells. However, the female gamete was lacking, and consequently von Haller (cited by Bodemer 1969), the leading physiologist of his time, taught that a fluid was delivered from the ovary into the fallopian tube, then flooding the uterine cavity and giving rise to the oocyte by a process of coagulation. It was not until 1827 that von Baer finally detected and described properly the mammalian oocyte and its ovulation from the ovary. This discovery put an end to the theories of "uterine oocytes". At the same time, however, the old question was again open, of whether or not the genital tract fluids, and in particular the uterine secretion, have a certain biological significance. Von Baer (1827) extended his studies on early mammalian development logically to the search for human early embryonic stages. When dissecting the cadaver of a young woman who had committed suicide through fear of being pregnant and not being married, von Baer investigated the uterus at 8 days after the supposed conception. He described the contents of the opened uterus as being a viscous and mucous fluid, most likely containing much protein. From the recent biochemical point of view this was an assumption; however, we know now that von Baer was right, and that with this observation the "modern history" of the scientific analysis of the female genital tract fluids, in particular the uterine fluid, began. Several investigators of the early years of our century have recognized the physiological dependency of the genital tract secretions on the stages of the reproductive or menstrual cycles (Bond 1898, 1899, 1906; Gerlinger 1922; Hartmann 1923); however, it was not until sophisticated physicochemical methods were developed that a detailed analysis was possible (Bishop 1956; Homburger et al. 1955; Junge and Blandau 1958; Lutwak-Mann 1960, 1962). Finally, when high resolution electrophoretic and chromatographic methods became available, particularly use of polyacrylamide gels, the first informative macromolecular component patterns of uterine secretions were demonstrated (Schwick 1963; Beier 1966, 1967, 1968a; Kirchner 1969).

2 Morphology of Secretory Epithelia of the Endometrium

The synthesis and release of uterine secretion proteins is accompanied by characteristic morphological transformations of the endometrial epithelium and connective tissue. These features have been well known to endocrinologists earlier than the biochemical events within the nucleus and the cytoplasm of the secretory cells. Practical implications of these tissue and cell transformations have been used as useful Fools by means of the Clauberg-test and the McPhail-index.

By the recent methodological improvements higher resolutions by scanning and transmission electron microscopy can demonstrate clearly a significant and rapid cell transformation, which is induced by estrogens and progesterone. The histological (Figs. 1, 4), scanning (Figs. 3, 6, 8) and transmission electron micrographs (Figs. 2, 5, 7) document the classical surface and cell transformation of the rabbit endometrial luminal epithelium . These transformations occur in the rabbit, regardless whether the endocrine status of the animal is a normal physiological one (pregnant or pseudopregnant) or whether the endocrine regulation is artificially built up by hormone supplementation on ovariectomized animals (Beier and Mootz 1979).

3 Endometrial Secretion

At the same time as morphological and cytological transformations are evident, preparing the endometrium for implantation of the blastocyst, we find even more clearly physicochemically alterations of the uterine secretions, the characteristics of which can be shown by quantitative and qualitative changes of the secretory components. Volume and viscosity are dependent on the chronological stage of the reproductive cycle and the endocrine background of the maternal system. Homburger et al. (1955), Homburger and Tregier (1957) and Harpel et al. (1968) demonstrated this by data on preovulatory uterine fluid, Lutwak-Mann et al. (1960) were able to show that the viscosity of uterine fluid increases after ovulation. thereby increasing the protein concentration, which in turn is further increased by higher secretion rates, triggered by postovulatory progesterone.

As our knowledge of the biochemistry of uterine secretion components accumulates, new advances are made concerning the various ways in which these components (proteins) are released into the lumen of the genital tract, and about the origin of these molecules. Analytical evidence from qualitative and quantitative studies employing acrylamide gel electrophoresis, Ouchterlony double immunodiffusion, and other immunoelectrophoretic combinations suggests that, whereas many genital tract fluid proteins are identical to those of the blood serum, several are only found in uterine secretion. As shown earlier (Beier 1967), there exists a selective transport of serum proteins into the uterine luminal fluid. Proteins identical to serum proteins do not fall within a certain category of small molecular sizes, but have, on the contrary, rather diverse molecular weights, which rules out any preference for small protein components that would suggest a simple sieve effect by the blood-uterine secretion barrier. The partial disparity in the relative amounts of these few identified serum proteins in uterine secretion, compared to their serum proportions, further emphasizes the selectivity of the transudation

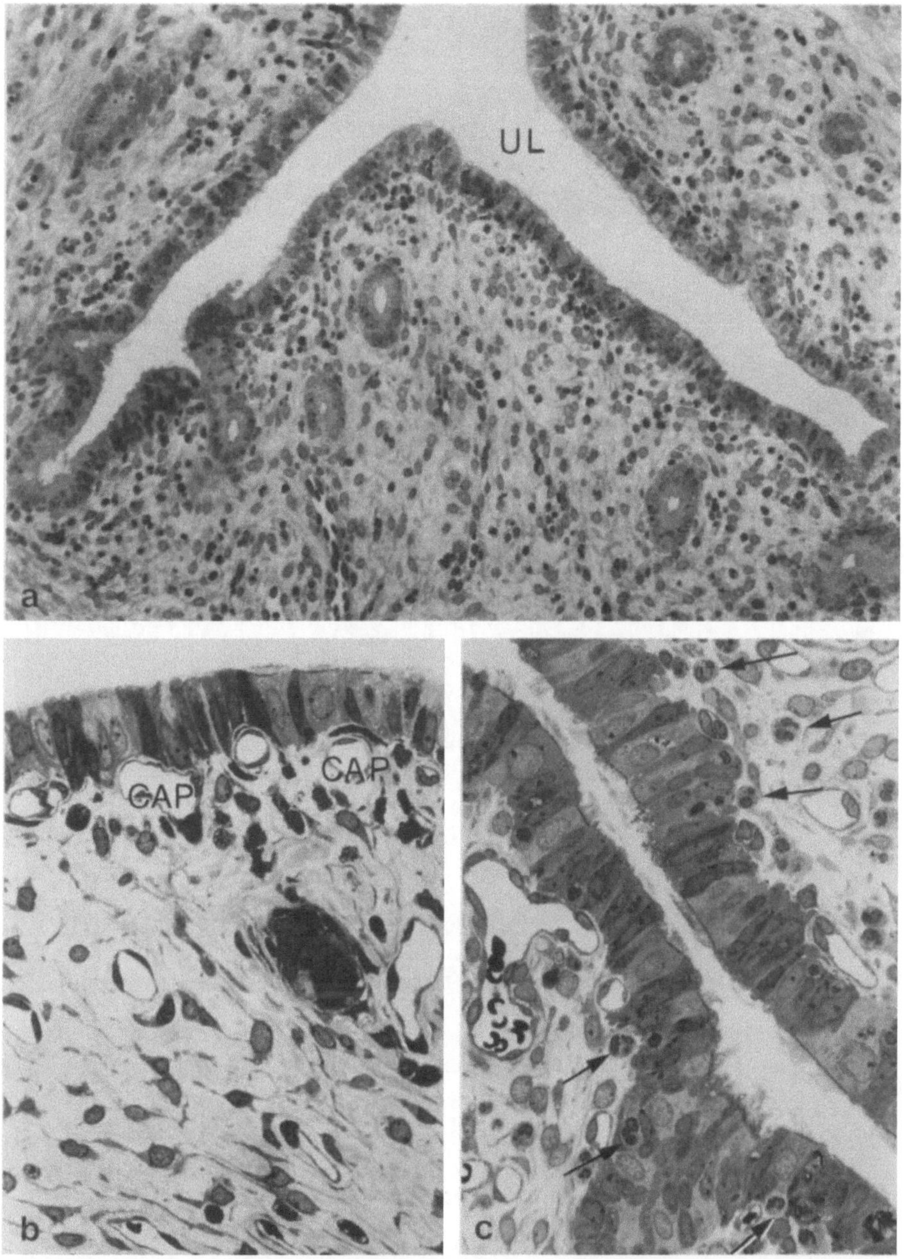

Fig. 1 a-c. Rabbit uterus at estrous. **a** The endometrium is poorly proliferated, the luminal surface appears smooth, and there are only very few glands cut. *UL* uterine lumen. Semi-thin section, toluidine blue staining, x 230. **b** At higher magnification there is a good demonstration of the lumenal epithelium and the connective tissue, including numerous subepithelial capillaries (*CAP*). **c** Beneath these epithelial layers, which contain a relatively large number of ciliated cells, various leucocytes are visible (*arrows*). These leucocytes even seem to penetrate the epithelium to reach the uterine lumen. Semi-thin section, toluidine blue staining, x 460

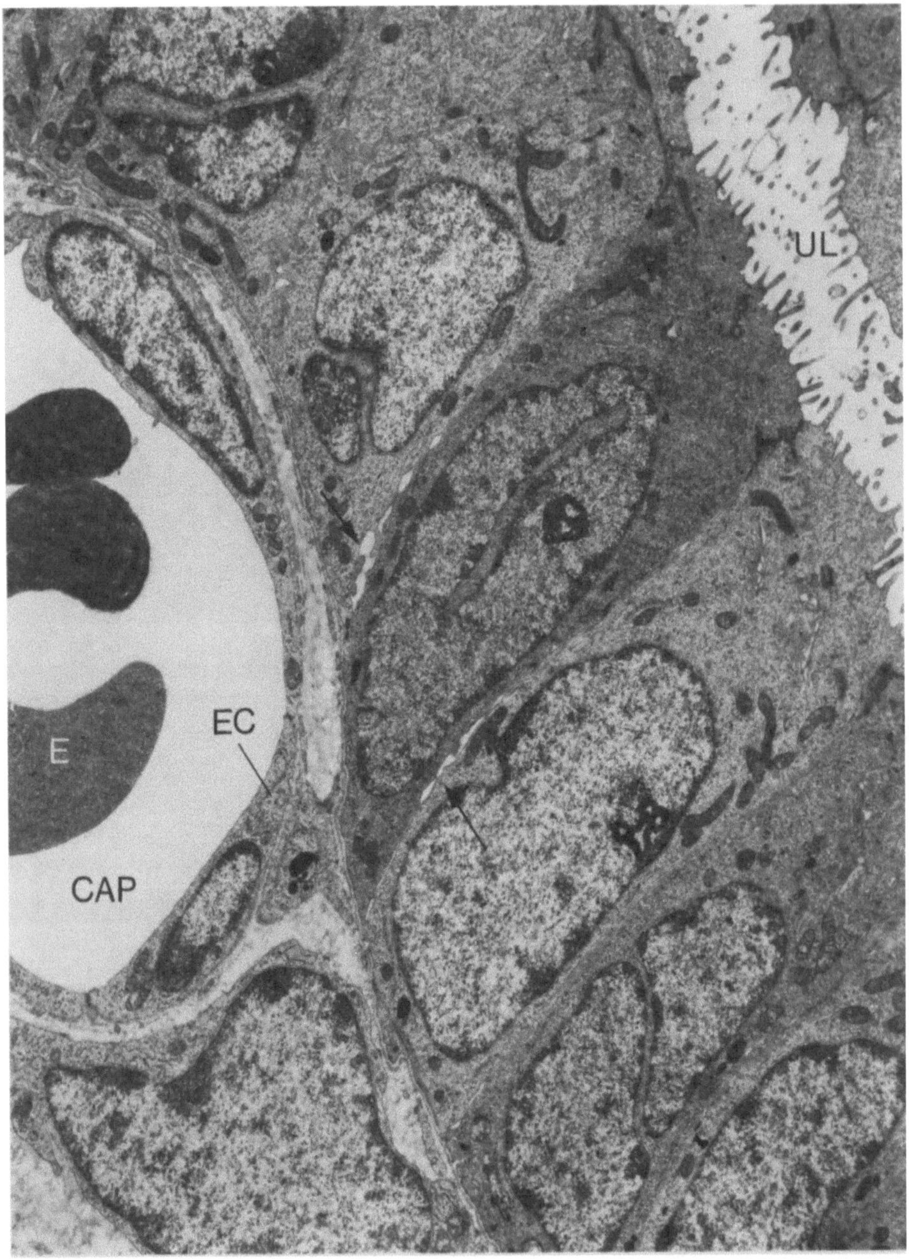

Fig. 2. Endometrial luminal epithelium at estrous. The columnar epithelium cells carry many short microvilli at the lumenal surface (*UL* uterine lumen). The lateral intercellular space is partly expanded (*arrows*). At the left, a typical subepithelial capillary is cut (*CAP*). *E* erythrocyte *EC* endothelial cell. TEM x 7200 (From: Beier 1973)

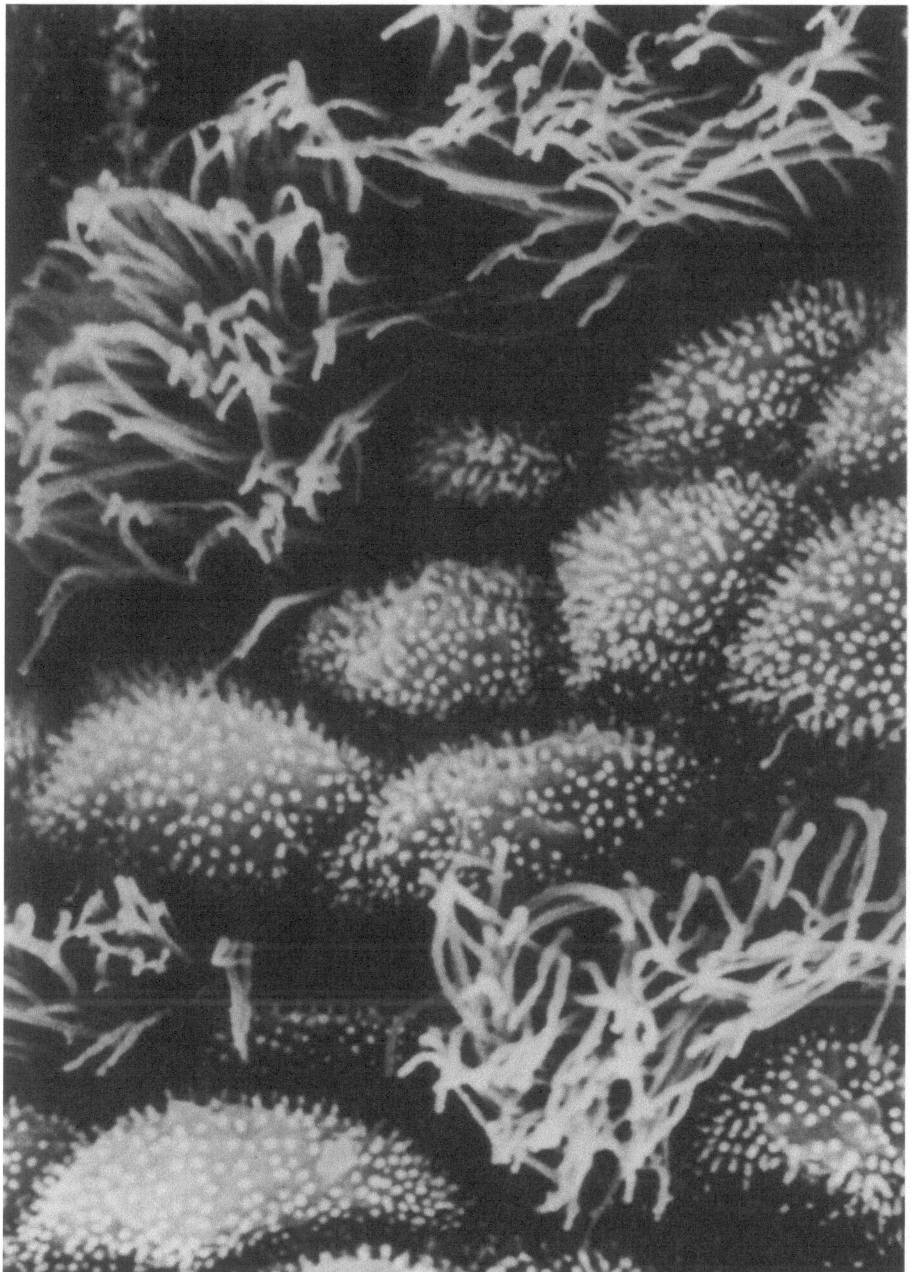

Fig. 3. Endometrial luminal surface. Endometrial surface shows apical regions of epithelium cells at the time before secretory transformation. Many cells are covered with small slender microvilli (see Fig. 2) and other cells are abundantly ciliated. SEM x 8000

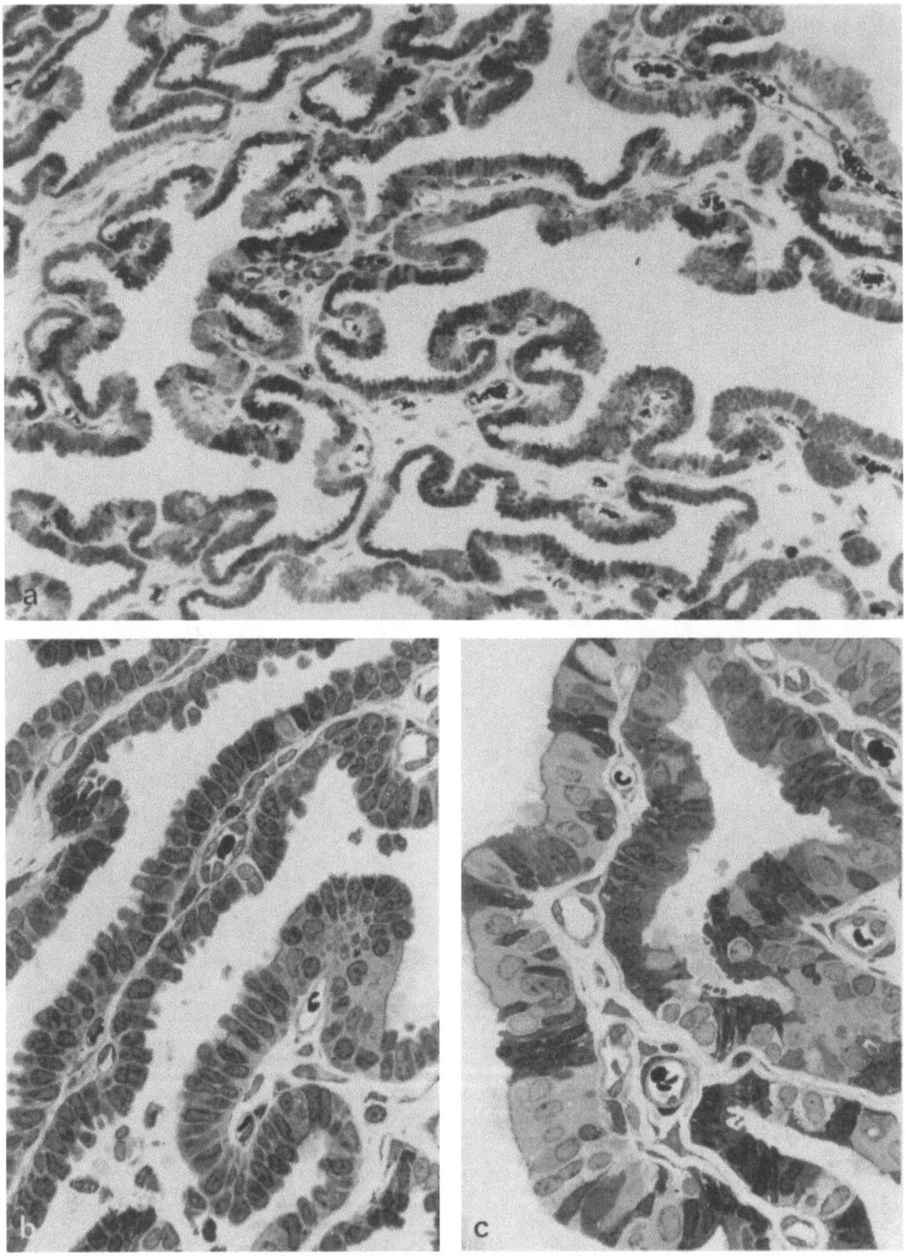

Fig. 4 a-c. Rabbit uterus at preimplantation period. **a** The endometrium is in its highly secretory transformed stage (about day 6 post coitum). Large epithelial tissue areas are covering the branching connective tissue portions. The rel. percentage of epithelial tissue has dramatically increased since estrous. Semi-thin section, toluidine blue staining, x 230. **b** Numerous epithelial cells show apical protrusions, and the connective tissue is partly diminished to extremely slender bridges. Semi-thin section, toluidine blue staining, x 460. **c** At other parts, the lightly staining ciliated cells are dominating. Again we see numerous subepithelial capillaries. Semi-thin section, toluidine blue staining, x 460

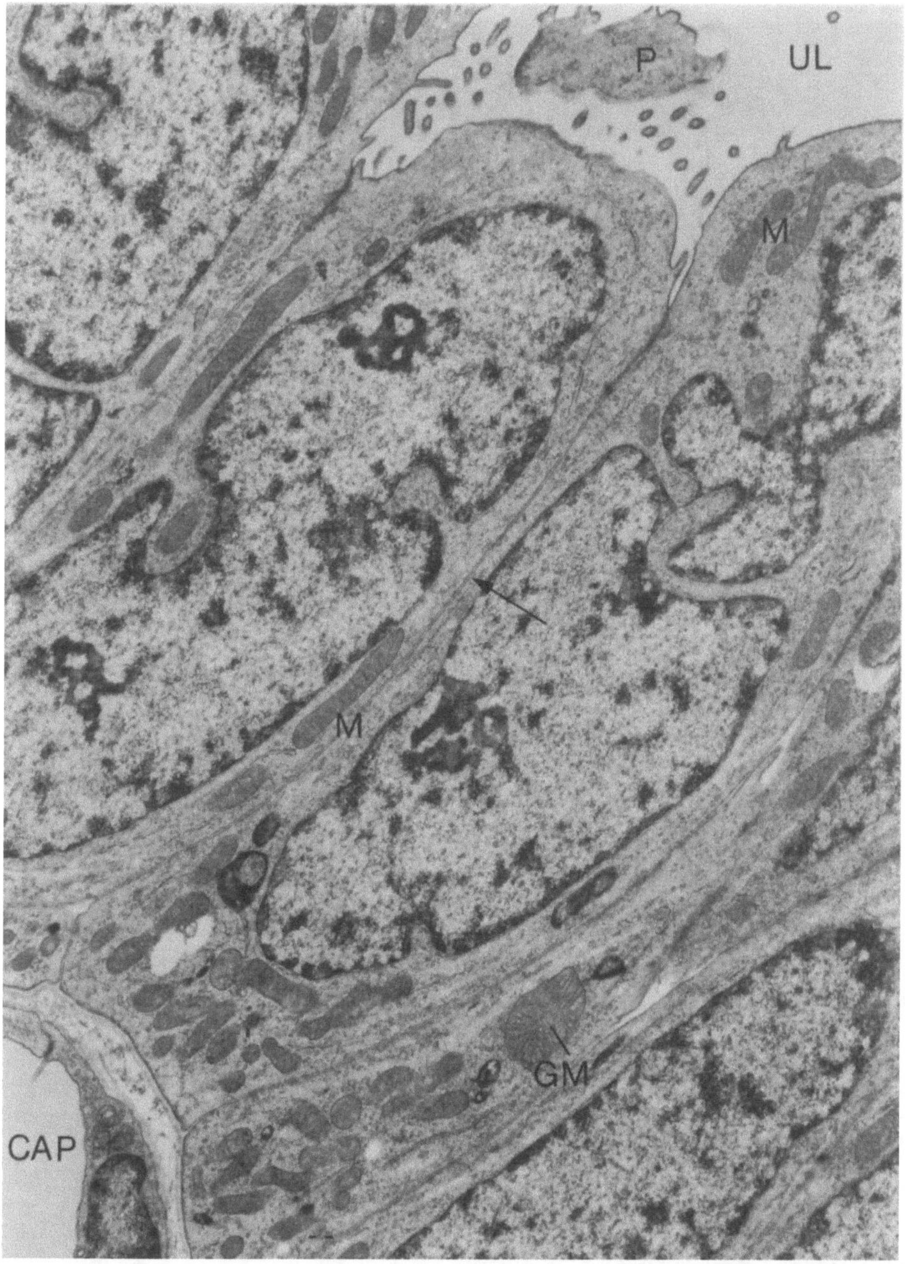

Fig. 5. Endometrial luminal epithelium at the preimplantation period. The columnar epithelial cells develop apical protrusions, which seem to pinch off at some sections (*P*), however there is controversal discussion about the protruding apical portions of these cells and whether they release the secretion by apocrine extrusion into the uterine lumen (*UL*). The intercellular space is closed and the lateral cell membranes are smooth (*arrow*). Progestational transformation results in the formation of giant mitochondria (*GM*) or extremely elongated mitochondria (*M*). (*CAP* subepithelial capillary) TEM x 12000 (From: Beier 1973)

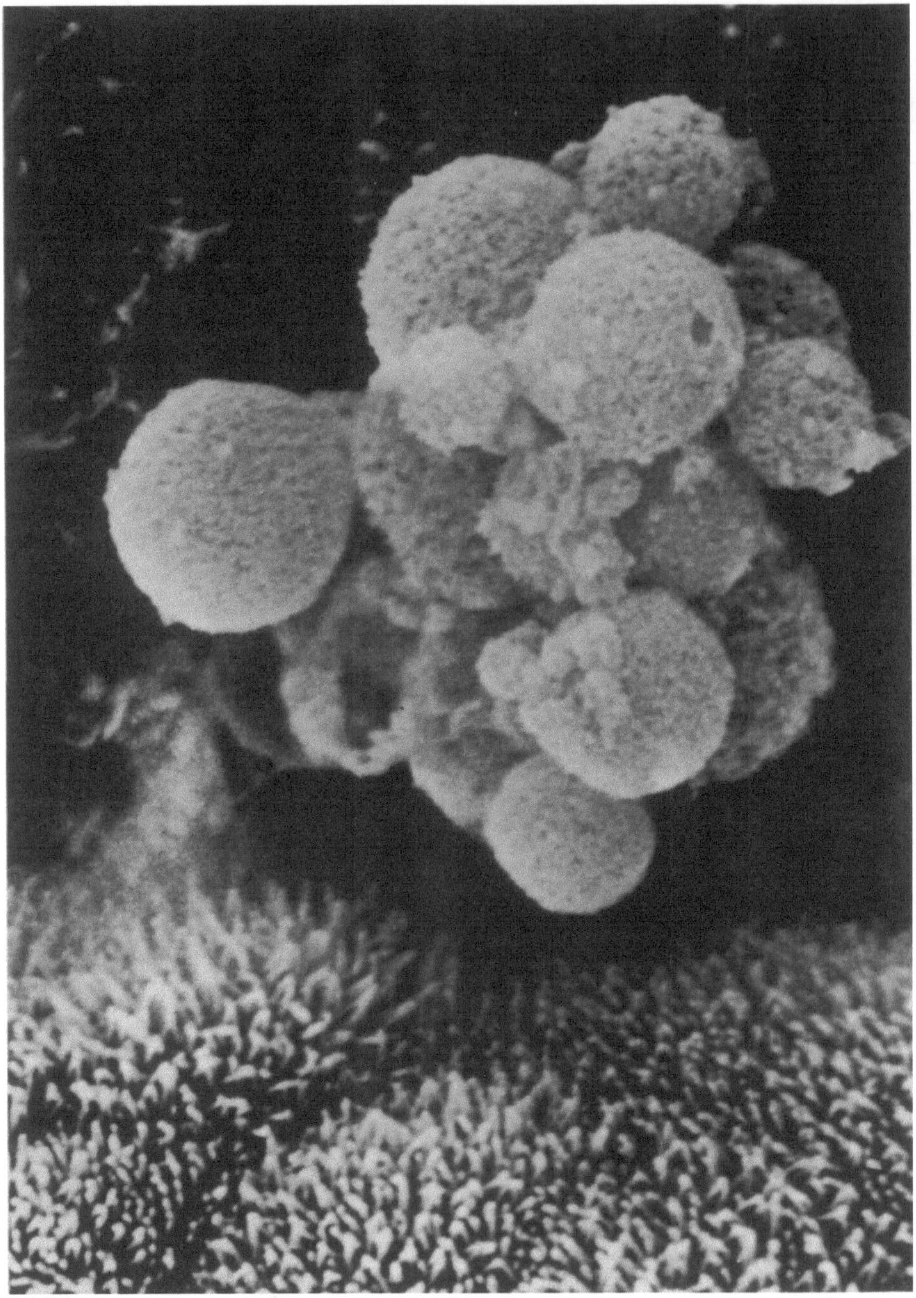

Fig. 6. Luminal secretion in the rabbit uterus. The endometrial glands deliver globular secretion "packets", which differ significantly from apical cell portions. Below are microvilli covered epithelial cells at the gland opening. SEM x 8800

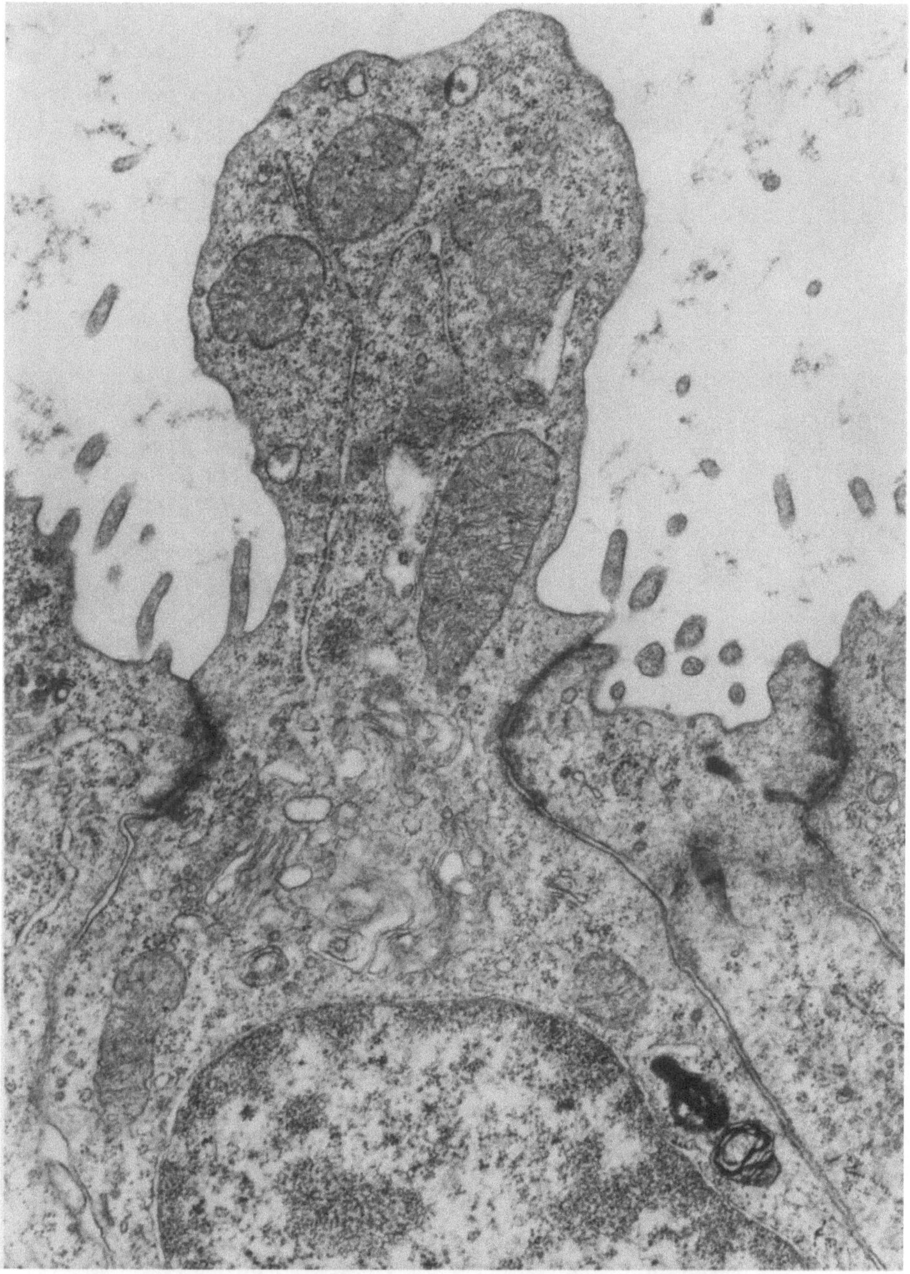

Fig. 7. Endometrial epithelium cell at the preimplantation period. The apical protrusion is filled with giant mitochondria and endoplasmic reticular portions, however, no secretory granules. TEM x 22400 (From: Beier 1973)

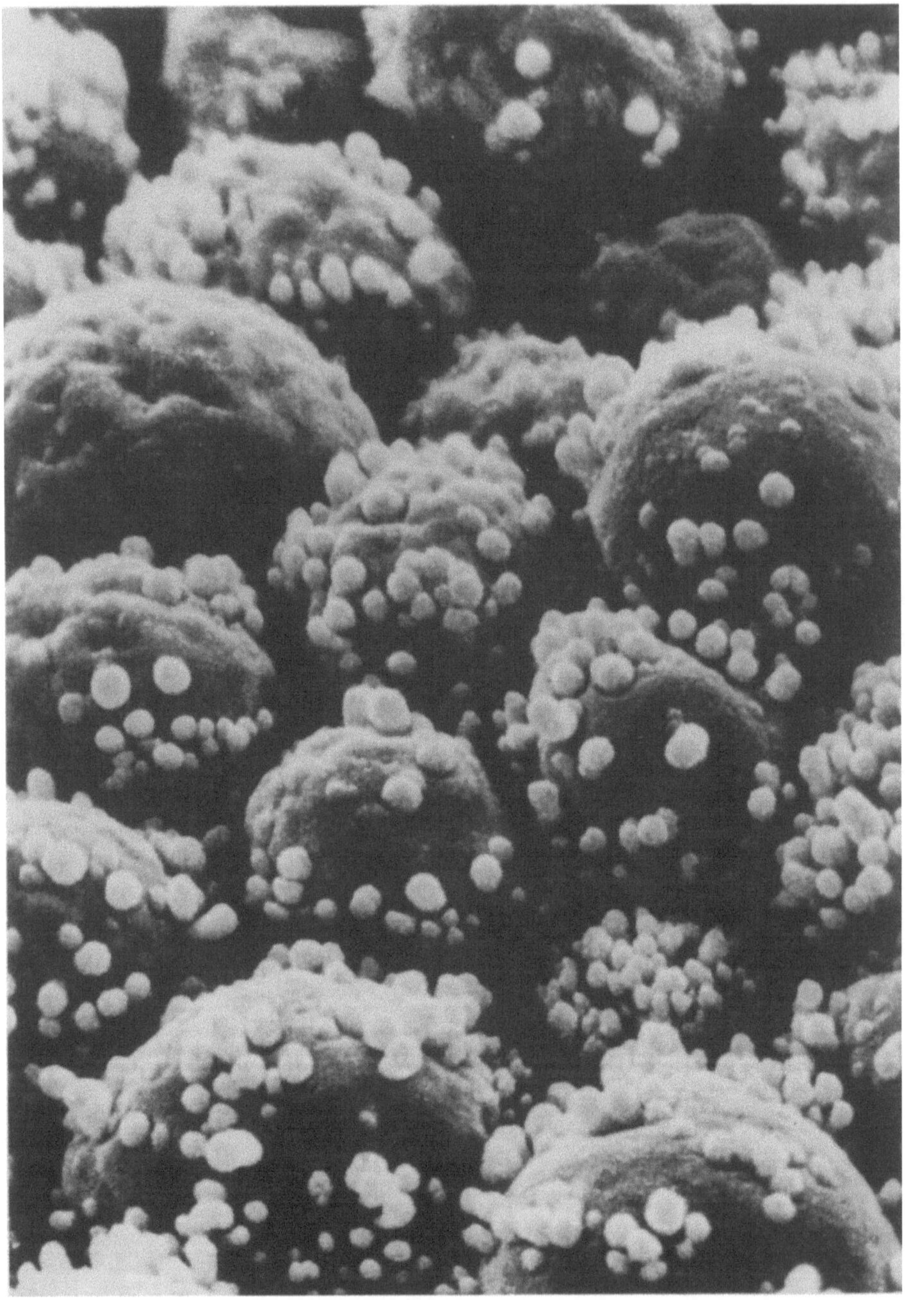

Fig. 8. Endometrial luminal surface at the preimplantation period. The typical apical protrusion as they appear in secretory transformed shape carry only small, short microvilli (more like 'microbubbles'). These epithelial surface resembles that from Fig. 4b very clearly. SEM x 8000

process. The albumin/globulin ratio has been calculated for blood serum as 1.32, but for rabbit day-6 uterine secretion as only 0.15, indicating a significant relative albumin decrease in uterine fluid. This feature is especially obvious in the amounts of immunoglobulins in blood serum and uterine fluid. Whereas serum contains normally 15-17 rel.% of immunoglobulins, uterine preimplantational secretion has no more than 2 rel.%.

4 Comparative Animal Data

Changes in the nature of uterine secretion proteins have been extensively studied recently in several species. Always, it has been found that endometrial secretion patterns are hormone-dependent and develop time-specific characteristics according to the estrogen/progesterone ratio. The following references give basic information:

Rabbit: Schwick (1963), Beier (1966, 1967, 1968a, b, 1973, 1976, 1978), Kirchner (1969), Krishnan and Daniel (1967), Urzua et al. (1970), Shapiro et al. (1971), Johnson (1972), Feigelson et al. (1977), Bullock (1977), Beato (1977)
Mouse: Mintz (1971), Aitken (1977a, b), Fishel (1979)
Rat: Kunitake et al. (1965), Surani (1976)
Roe deer: Aitken (1974)
Cow: Roberts and Parker (1974), Laster (1977), Dixon and Gibbons (1979)
Pig: Squire et al. (1972), Murray et al. (1972)
Sheep: Roberts (1976)
Macaque and baboon: Mastroianni et al. (1969, 1970), Joshi (1974), Peplow et al. (1973)
Man: Beier et al. (1970), Moghissi (1970), Bernstein (1971), Daniel (1971a), Shirai et al. (1972), Beier and Beier-Hellwig (1973), Wolf and Mastroianni (1975) Noske and Feigelson (1976), Voss and Beato (1977), Beier (1978), Shapiro and Forbes (1978), Aitken (1979), Sutcliffe et al. (1978, 1980), Joshi et al. (1980 a, b).

The characteristic protein patterns of some of the various species' uterine secretions are shown in Figs. 9 (rabbit), 10 (mouse), 11 (rat), 12 (sheep). Selective transudation of serum proteins evidently occurs into the uterine cavity, irrespective of the size of the protein molecules. This could be shown by immunoelectrophoretic means using available antibodies. More relevant information, however, on the synthesis and transudation of macromolecules into the uterine secretion would be gained by the use of labeled precursors. Studies of the uterine luminal proteins of the rat have been carried out using ^{3}H-leucine and ^{3}H-fucose (Surani 1977). There is little or no protein synthesized and secreted when progesterone alone dominates the spayed animal; however, when it was administered together with 17β-estradiol, a substantial increase in the amounts of radiolabeled proteins in uterine secretion was found (Surani 1977). The nature of the proteins released into the uterine cavity depends on the correct balance of estradiol and progesterone. Similar findings were made in the mouse (Aitken 1977b; Fishel 1979). By contrast, the most pronounced changes have been found in the endometrial secretion of the rabbit, where uteroglobin represents the major fraction (50%-70%) of all proteins and is produced in remarkable response to progesterone (Beier 1968; Beier et al. 1970). Interestingly, no clear evidence is so far available that uteroglobin is more than a unique rabbit protein. Similar protein fractions have been described in other animals (see Table 1), but no immunological identity has been demonstrated, despite the fact that the methods used have been as many and as varied as those used in research on the human uterine secretions and wash fluids.

Table 1. Uteroglobin analogues in various species

Species	Criterion for similarity with rabbit uteroglobin			References
	Chromatographic	Electrophoretic	Immunological	
Wallaby		+		Renfree (1973)
Opossum	–	–	–	Renfree (1975)
Mouse	–	–		Mintz (1971), Aitken (1977b), Fishel (1979)
Rat	–	–	–	Kunitake et al. (1965), Surani (1976)
Hamster		–	–	Noske and Daniel (1974)
Guinea pig	–	–	–	Bullock (personal communication)
Roe deer		+	–	Aitken (1974)
Cow	–	+(?)		Roberts and Parker (1974), Laster (1977), Dixon and Gibbons (1979)
Pig	+	+	–	Squire et al. (1972)
Mink	+(?)	–	–	Daniel (1968), Daniel and Krishnan (1969)
Ferret	–	–	–	Daniel (1971a)
Black bear	–	–	–	Daniel (1968) Daniel and Krishnan (1969)
Dog	+		–	Daniel and Krishnan (1969)
Fur seal	+(?)	+	–	Daniel (1971b)
Armadillo	–	–		Daniel (1968), Daniel and Krishnan (1969)
Baboon		+(?)		Peplow et al. (1973)
Man		+	–/+(?)	Beier et al. (1970), Shirai et al. (1972), Daniel (1973), Beier and Beier-Hellwig (1973), Wolf and Mastroianni (1975), Noske and Feigelson (1976), Voss and Beato (1977), Maathuis and Aitken (1978), Beier (1978), Aitken (1979)

This table was corrected and extended after a table of Daniel (1976)

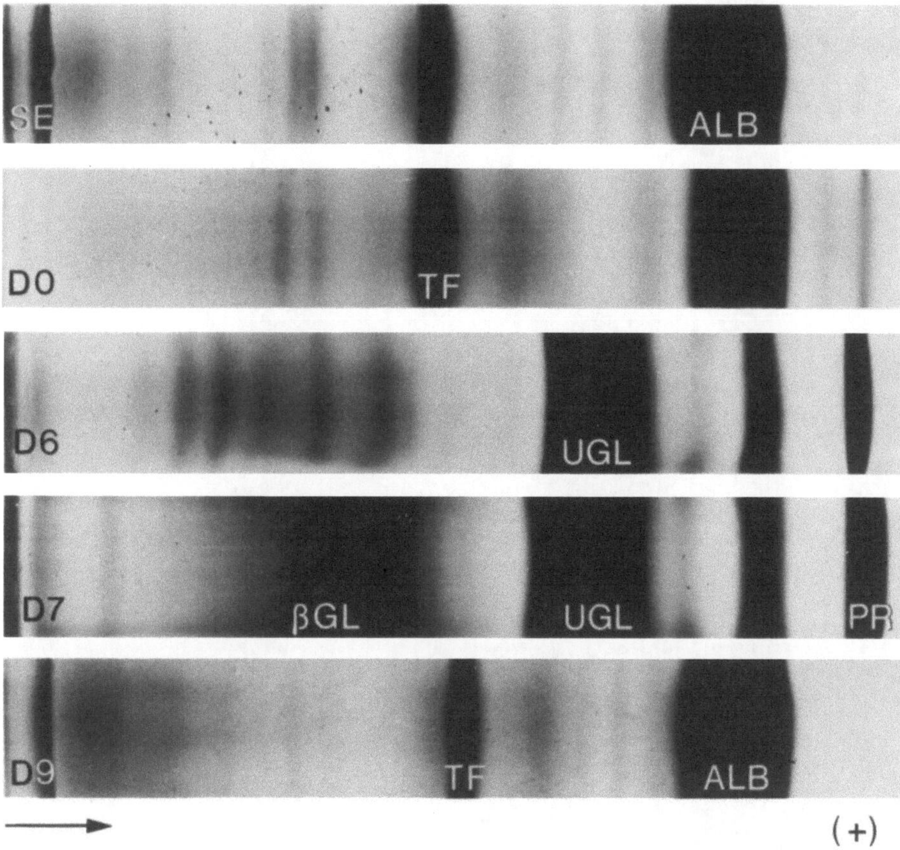

Fig. 9. Uterine secretions of the rabbit. Polyacrylamide gel electrophoretic separation of rabbit uterine flushings during estrus (D0) and early pregnancy, compared to blood serum (SE). The early pregnancy protein patterns characterize dynamic pictures which change daily (D6= day 6, D7= day 7, D9= day 9 post coitum). Note the prominent fractions of prealbumins (PR), albumin (ALB), uteroglobin (UGL), transferrin (TF), and β-glycoproteins (β-GL). Tris-glycine buffer pH 9.0, amido black-10B staining

5 Endocrine Regulation of Uteroglobin Synthesis and Release

It has been an interesting question since the first observations on progesterone effects on uteroglobin appearance in uterine secretion, whether this steroid acts by directly stimulating the synthesis, or only the release, or even both processes together. Evidence has now been found that progesterone, and also several other progestogens, are capable of stimulating synthesis to a large degree, however release totally. Uteroglobin synthesis is mainly activated by progesterone, although it is produced in detectable amounts by the uterus of estrogen-treated castrated does. Under such experimental conditions uteroglobin is only present in the epithelial cells, but not extruded into the cavum uteri in amounts which could be detected by immunohistochemical means, as work by Kirchner (1976) indicates. However, within the uterus we were never able to obtain any uteroglobin release under the influence of estrogens, even if a considerable dosage of 17β-estradiol was used (Beier 1974). The release response to progesterone is to a certain

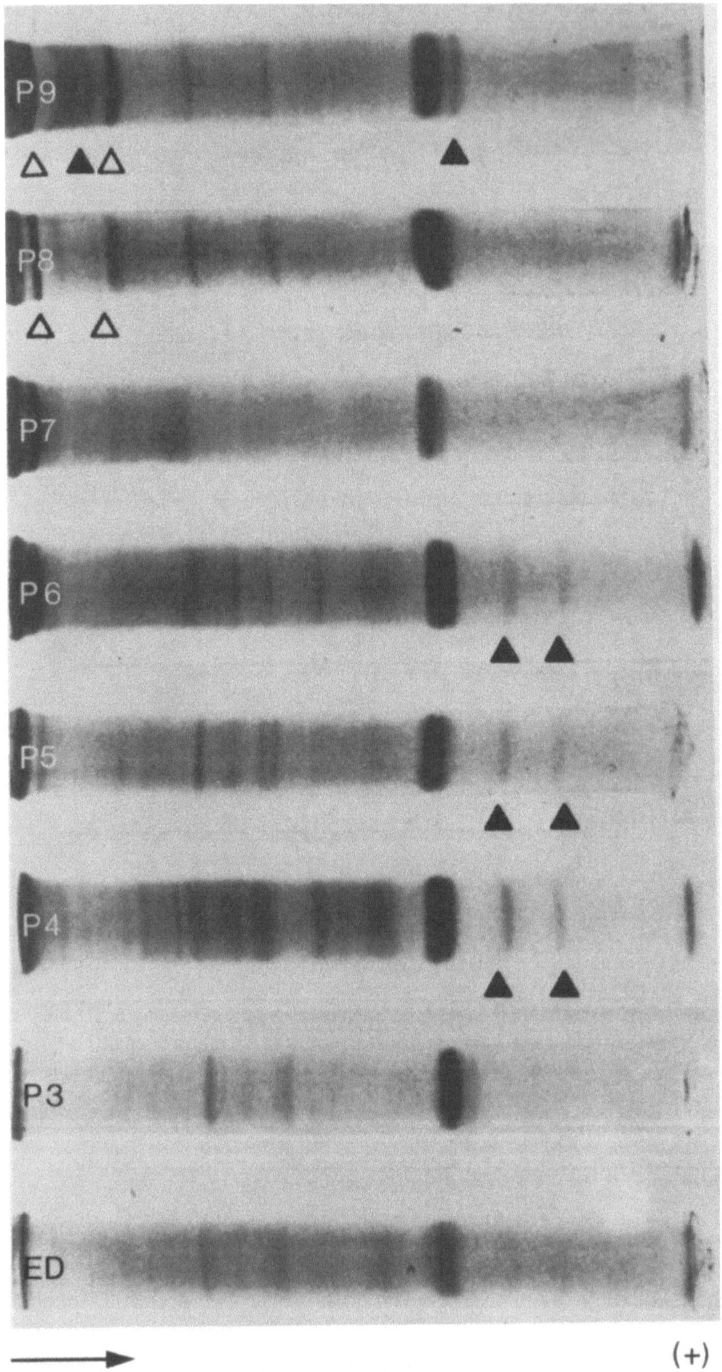

Fig. 10. Uterine secretions of the mouse. Polyacrylamide gel electrophoretic separation of mouse uterine flushings during experimental delayed implantation (*ED*) and pseudopregnancy (*P*), the days of which are indicated (*P3, P4, P5, P6, P7, P8, P9).* Note the increase in the number and the intensity of protein bands on day P4, the presence of prominent prealbumins on days P4-P6 (*closed arrows*), the progressive appearance of dominant bands at the macromolecular regions on days P5-P9 (*open arrows*), and the appearance of other dominant fractions in the prealbumin and macromolecular regions on day P9 (*closed arrows*). (Modified from Aitken 1977b)

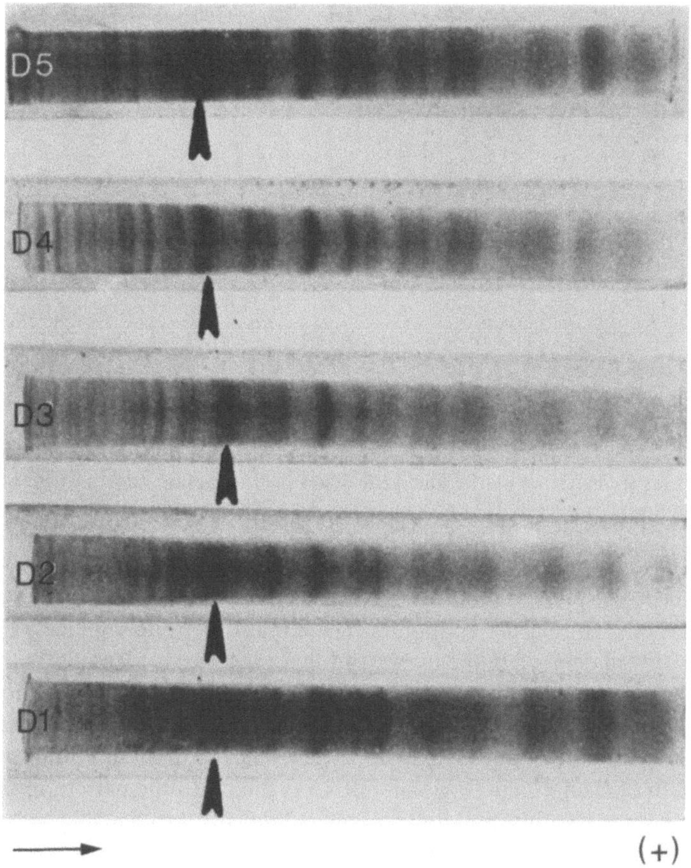

Fig. 11. Uterine secretions of the rat. SDS acrylamide gel electrophoretic separation of rat uterine flushings. The samples are taken from normal pregnant preimplantational animals and during implantation: days 1 to 5 (*D1, D2, D3, D4, D5*). Changes appear in the macromolecular region, where a whole "family" of bands changes from day to day. (Modified from Surani 1976)

extent dose-dependent, the most effective dosage being in the range of 0.6-3.0 mg/day per animal (average weight 2.5-3.0 kg). All higher dosages have had no remarkable effect on increase of uteroglobin concentrations in uterine secretion (Arthur and Daniel 1972; Nishino p. 83).

Doubtless, progesterone is required to establish the characteristic secretion pattern of uteroglobin during preimplantation in vivo. But, Beato and Arnemann (1975) have shown that this secretion is unaffected even when the uterus is removed from the mother, isolated, and perfused under defined experimental conditions.

Production of uteroglobin as the major endometrial secretion protein logically called for identification of its actual synthesis by demonstration of the messenger RNA (mRNA). The endometrium of progesterone-stimulated rabbits has been used as a source for the isolation of uteroglobin mRNA. The translation of poly (A)-containing mRNA for uteroglobin has been demonstrated in different systems (Beato and Rungger 1975; Bullock et al. 1976). Identification of uteroglobin newly synthesized in vitro was made by

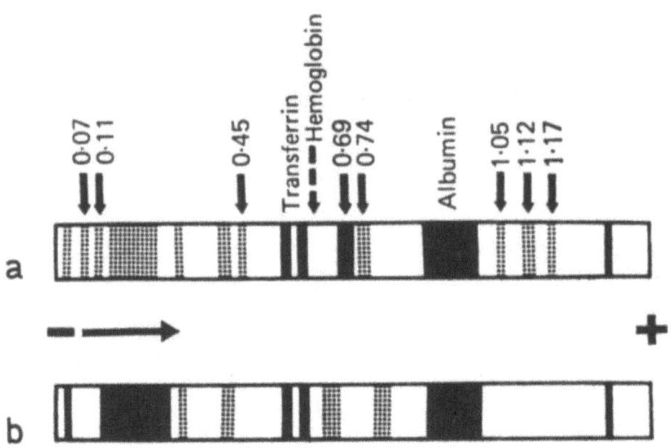

Fig. 12 a, b. Ovine uterine secretion. Comparison of polyacrylamide gel electrophoretic patterns of ovine blood serum (**b**) and uterine wash fluid (**a**) from a 15-day-pregnant sheep. The characteristic bands are indicated by the solid arrows and their migration rates relative for albumin shown by the numbers. Interestingly, a prealbumin group and also pre- and posttransferrins appear which were not detected in blood serum. (Modified from Roberts et al. 1976)

means of monospecific antibodies. These results definitely indicate that uteroglobin is synthesized de novo in the uterus and that progesterone stimulates this synthesis. Bullock et al. (1976) have demonstrated that the specific mRNA for uteroglobin accounts for an increasing proportion of the poly(A)-rich endometrial mRNA during the preimplantation period, reaching a maximum at day 4 post coitum in normal pregnancy. The pattern of change in uteroglobin mRNA is similar to the pattern of secretion of uteroglobin during normal preimplantation and reflects the changing endocrine control mechanisms of the maternal system.

So far, it is unclear what process, in detail, terminates uteroglobin presence in the secretion. It may well be that the progesterone/estradiol ratio is more important than the level of one of the two steroids alone. Since the protein pattern observed during pseudopregnancy suggests that implantation itself terminates uteroglobin release or synthesis, there may be a specific "message" delivered from the blastocyst or from the decidual tissue, as has been proposed by Johnson (1974). Perfusion experiments on isolated uteri and their uteroglobin production (Beato and Arnemann 1975) indicate that the "switch-off" in vivo is a termination of synthesis rather than only of release, since these studies show no long-term accumulation of uteroglobin within the cytosol of the endometrial cells.

6 Approaches to the Study of Uteroglobin Significance by the Comparison of Endometrial and Ectopic Production

Some biochemical and biological parameters of uteroglobin, as presently known, are compiled in Tables 2 and 3. Approaches to the analysis of uteroglobin physiology concentrate on its biochemical properties, including binding and transport functions, as well as enzyme inhibitor activities. Additionally, a differential analysis of uteroglobin

Table 2. Biochemical properties of uteroglobin

a) Separation from other uterine secretion proteins by acrylamide gel electrophoresis and identification as postalbumin fraction (Beier 1966, 1967)

b) Identification of uteroglobin, the postalbumin fraction as identical antigen to the β_1-U-globulin from classical agarimmunoelectrophoresis performed by Schwick (1965), by means of a direct acrylamide-agar immunoelectrophoretic combination (Beier 1966, 1967)

c) Sedimentation in analytical ultracentrifugation (Svedberg units analysis) $S_{c=1.0\%}=1.38$ (Beier 1968a)

d) Molecular weight around 14 000-15 000 (Murray et al. 1972; Bullock and Connell 1973)

e) Composition of two identical subunits of approx. 7000-8000 MW, linked by S-S bonds and composed of 70 amino acids (McGaughey and Murray 1972; Beato and Baier 1975; Beato 1977; Ponstingl 1978)

f) Isoelectric focussing reveals an isoelectric point of 5.4 (McGaughey and Murray 1972)

g) Binding of progesterone with high affinity, in a reaction which does not involve the sulfhydryl groups, requires the opening of the disulfide bridges (Beato 1977). Affinity for progesterone has been reported with different association constants:
5.95×10^8 M^{-1} (Arthur et al. 1972)
3.00×10^6 M^{-1} (Beato and Baier 1975; Beato 1977)
Very much lower than these (Rahman et al. 1975)
Not calculated, but estimated as low (Urzua et al. 1970)

h) Inhibition of trypsin, tested in fibrin-agar electrophoresis (Beier 1970, 1976; Johnson 1974)

i) Uteroglobin is synthesized by endometrial cells, as indicated by isolation of mRNA for uteroglobin from this tissue, and translation of this mRNA in *Xenopus laevis* oocytes (Beato and Rungger 1975), mRNA isolation by Bullock et al. (1976)

j) Uteroglobin amino acid sequence analysis shows the molecule to be composed of two identical polypeptide chains of 70 amino acids linked by 2 disulfide bonds. The sequence is not homologous to any known primary protein structure, except for a small acidic region (residues 22-29) resembling a sequence found in somatotropin (Ponstingl 1978)
The larger size of pre-uteroglobin is accounted for by a 21-amino-acid leader sequence, containing 15 hydrophobic residues, at the NH_2 terminus (Malsky et al. 1979)

k) Cloning the chromosomal gene for uteroglobin (Atger et al. 1980)

and its homologous molecules of the male reproductive tract and the lung is necessary and will be useful.

The uteroglobinlike protein from seminal plasma and from bronchial secretion are immunologically identical antigens, as several independent investigators have been able to confirm (Beier et al. 1975, 1978; Beier 1977; Noske and Feigelson 1976; Bullock and Bhatt, p. 117. The most challenging questions arise from comparative studies of uteroglobin and the uteroglobinlike antigens. Does this "ectopic uteroglobin" show similar or identical properties, particularly with regard to the most intriguing features of steroid binding and trypsin inhibition? Purified proteins have been used in recent studies on progesterone binding to authentic uteroglobin and the uteroglobinlike antigen from the lung (Beato and Beier 1978). Progesterone binding to both proteins is identical. However, synthesis and secretion are controlled by ovarian steroids within the uterus only, since all investigated lungs, even from ovariectomized animals and from male and newborn rabbits, contained the uteroglobinlike protein in fairly similar concentrations.

Table 3. Biological features of uteroglobin

a) Significant and predominant protein in rabbit uterine secretion during preimplantation in normal pregnancy and in equivalent stages of pseudopregnancy (Beier 1966, 1967, 1968b; Krishnan and Daniel 1967; Kirchner 1969; Urzua et al. 1970; Beier et al. 1970; Daniel 1971a)

b) Predominant protein of blastocyst fluid (Beier 1966, 1967; Hamana and Hafez 1970; Petzold 1974)

c) Uteroglobin is not detectable in cultured blastocysts, when development is accomplished from the 2-cell stage up to the large expanded blastocyst stage in BSA-supplemented in vitro culture media (Beier and Maurer 1975)

d) Release from uterine epithelial cells is stimulated and controlled by progesterone (Beier 1968a; Beier et al. 1970; Urzua et al. 1970; Arthur and Daniel 1972)

e) Release can be stimulated by chlormadinone acetate and norgestrel (Beier and Beier-Hellwig 1973)

f) Release from uterine epithelial cells is delayed by postcoital estrogen injections (17β-estradiol benzoate) (Beier et al. 1971)

g) Release into uterine lumen is terminated mainly by estrogen after implantation (Bullock and Willen 1974)

h) Uteroglobin is antigenic in mice (Beier 1966), rats (Schwick 1963) guinea pigs (Beier 1966, 1968b; Kirchner 1972; Johnson 1972), goats (Noske and Feigelson 1976), sheep (Beier et al. 1975)

i) Antigenic cross-reactions can be attained under certain conditions by means of all highly potent antiserum preparations with antigens from seminal vesicle secretion, seminal plasma (Beier et al. 1975), from lung tissue extracts (Noske and Feigelson 1976), and from lung lavage (Beier et al. 1978)

Additional studies on rabbit esophagus tissue and flushings, jejunal flushings, and also human endometrium, human oviduct, and human seminal plasma indicate uteroglobin cross-reacting antigens to be present there (Feigelson et al. 1977)

Rather surprisingly, we have obtained quite different results using fibrin-agar electrophoresis to test the trypsin inhibitory activity of uteroglobin and the homologous antigen from bronchial secretion. Preliminary results, which need further confirmation, show that the uteroglobin fraction of endometrial secretion does inhibit trypsin, whereas the uteroglobinlike protein fraction of lung tissue does not (Beier 1977). It remains to be clarified whether an unknown factor (molecule) is involved in this proteinase inhibition, since the highest activities are always found, in our system, when total uterine flushings are tested and compared to purified or isolated fractions. Another possible difficulty may arise from the chemical purification procedure itself, by altering the biological activity of the uteroglobin fraction. Regardless of further studies and results on the comparison of uteroglobin and the uteroglobinlike antigen, it should be pointed out here that we can no longer accept that immunological identity is proof of the total identity of two antigen molecules. The biological activities of a protein may not be localized or dependent on the immunologically determinant parts of the molecule.

7 Studies of Biological Systems Where Uteroglobin May Act: The Blastocyst Growing in Vivo and in Vitro

Injections of 17β-estradiol benzoate 6 h and 20 h after mating reveal a significant delay in the secretory pattern sequence of the rabbit uterus (Fig. 13). Compared to the normal preimplantation patterns, there is a delay of 2-5 days, dependent on the stages compared. At earlier stages, the delay is less extended (2-3 days) than in later stages (4-5 days). This feature is not only true for the protein patterns, but also for the histology of the endometrium and the enzyme histochemistry of the endometrial epithelia. We have claimed that these asynchronous protein patterns contribute particularly to an unfavorable uterine environment for blastocyst development (Beier 1970, 1974; Beier et al. 1970). Subsequent egg transfer experiments have shown clearly that normally developed blastocysts require a normally developed uterine environment to accomplish implantation and further development (Beier et al. 1972; Adams 1974). We transferred normal day-4 blastocysts into the uteri of day-8-pregnant rabbits treated post coitum with estrogen and showing delayed secretion. From these transferred blastocysts, implanted around day 12 (recipient's reproductive stage), there developed in an experimental series approx. 40% normal fetuses. Several of these fetuses were allowed to develop in to viable young rabbits. In conclusion, we are convinced that the clearly synchronized uterine environment, particularly the protein secretion patterns, plays an essential part in early mammalian embryogenesis. The model of delayed secretion has not been studied merely to demonstrate a questionable growth-inhibition effect on the native blastocysts. One particular item, however, is very important: the asynchronous egg transfer in delayed secretory uteri provides biological evidence for the necessity of a proper uterine environment to blastocyst development. Consequently, the essential protein environment for normal blastocyst development is composed of a considerable number of macromolecules. Within this, uteroglobin obviously is the major component, moreover so characteristic that it represents a specific marker molecule for the ovarian hormonal status of the animal.

Interestingly, experimentally induced delay of uterine secretion is not the only possible way of desynchronizing the maternal and embryonic systems. Prefertilization treatment of estrous rabbits with progesterone (up to 2 mg/day per animal) for 8 days (days -6 to +1) and induction of ovulation (by HCG injection on day 0), with subsequent artificial insemination, results in normal egg development during oviducal passage; however, degeneration of blastocysts appears after their arrival in the uterus, mostly within 2 days of exposure to the uterine secretion environment (De Visser, p. 89). With this treatment the intrauterine milieu is changed in so far that as the protein patterns are advanced compared to the normal preimplantation. The exogenous progesterone induces uteroglobin synthesis and release before ovulation and fertilization. There is a maximum of uteroglobin secretion from day -1 until day +2, whence the designation "advanced secretion". Comparable observations are reported by Kendle and Telford (1970) and by McCarthy et al. (1977). This phenomenon leads as clearly to failure of implantation as delayed secretion does (Fig. 14).

We have paid particular attention to the presence of uterine secretion proteins in blastocyst fluid. In a recent investigation on the protein patterns of rabbit blastocyst fluid and blastocyst homogenates after development in vivo and in vitro, we tried to present evi-

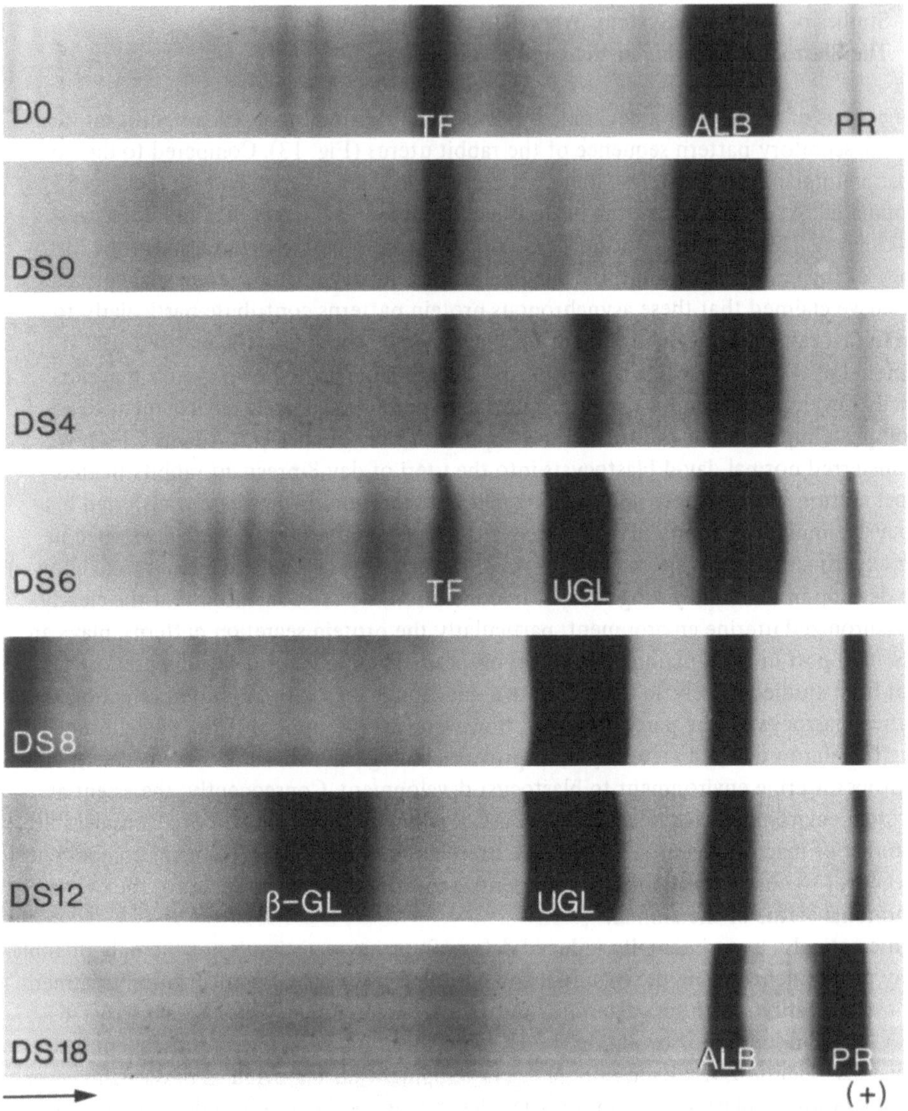

Fig. 13. Uterine fluid protein patterns of the rabbit during delayed secretion. Polyacrylamide electrophoretic separation of protein fractions from estrous rabbits (*DO*), from estrogen-treated controls after two injections of 100 and 150 μg 17β-estradiol respectively, (*DSO*) and from delayed secretion stages at day 4 (*DS4*), day 6 (*DS6*), day 8 (*DS8*), day 12 (*DS12*), and day 16 (*DS16*). Note the delay of the release of prominent protein fractions uteroglobin and β-glycoproteins. *PR*, prealbumin; *ALB*, albumin; *UGL*, uteroglobin; *TF*, transferrin; *β-GL*, glycoprotein. Tris-glycine buffer pH 9.0, amido black-10B staining.

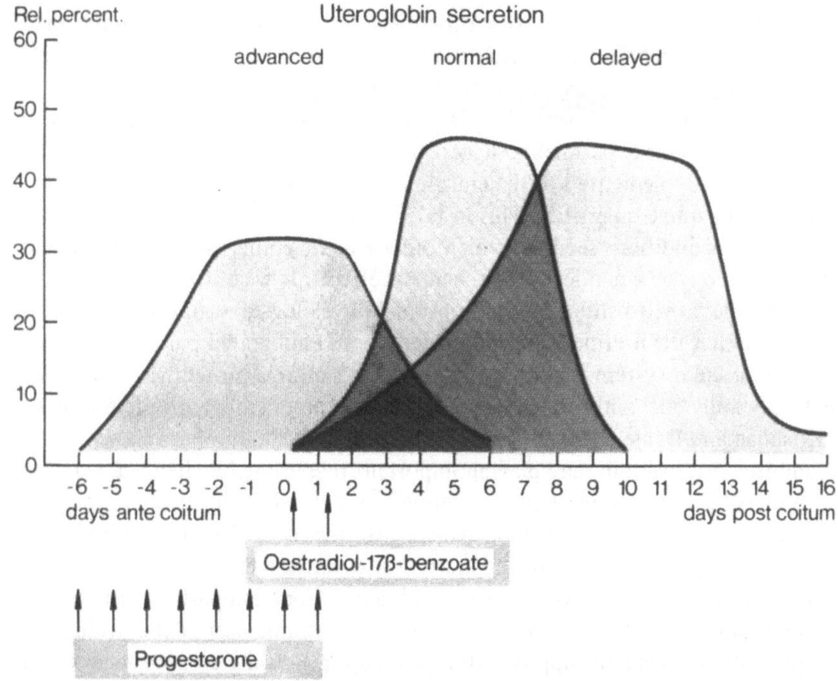

Fig. 14. Desynchronization of uteroglobin secretion. This diagram shows the profiles of relative percentages of uteroglobin in uterine secretion protein during normal and desynchronized endometrial secretion. By two different experimental regimens, either delayed or advanced secretory activity can be triggered. 17β-Estradiol benzoate given in two injections, 6 h and 30 h post coitum, 100 μg and 150 μg per animal, respectively, induces delay of uterine secretion. Progesterone, injected on eight consecutive days (2 mg/animal/day), designated day -6 until day +1 according to artificial ovulation induction by HCG on day 0, stimulates advanced uterine secretion. Desynchronized uteroglobin profiles are consistent with retarded blastocyst development and inhibition of implantation. Experiments on advanced uterine secretion have been performed by de Visser (p. 89), experiments on delayed uterine secretion by Beier (1973) (From: Ludwig and Tauber 1978)

dence for the origin of the blastocyst fluid proteins (Beier and Maurer 1975). Special emphasis was directed to the protein patterns of in vitro grown blastocysts, since these embryos developed from the two- and four-cell stages into expanded blastocysts without any rabbit protein in the culture medium. There, bovine serum albumin was used as the only protein source. Patterns from in vivo and in vitro development differ significantly, as judged by means of acrylamide gel electrophoresis and by several immunochemical test methods. These results demonstrate that blastocysts grown in vitro do not contain uteroglobin or β-glycoprotein in detectable amounts. Compared to the in vivo developed blastocysts, this is a striking difference, because uteroglobin and β-glycoprotein have been demonstrated in fluids of in vivo expanded blastocysts in considerable amounts, at least in such quantities that our routinely immunochemical tests give positive results. However, our study indicates that the in vitro developing blastocyst cannot synthesize uteroglobin. It may well be true that in vivo growing blastocysts too do not synthesize molecules identical to the uterine secretion proteins. All evidence now agrees with the concept that uterine secretion proteins permeate into the blastocysts fluid (Beier 1967; Kulangara and Crutchfield 1973; Schlafke and Enders 1973), into the blastocyst cells, particularly trophoblast cells on the 7th day post coitum (Kirchner 1976), and into the blastocyst coverings (Kirchner 1972). It seems that the blastocyst utilizes these environmental proteins. Our experiments suggest that under the conditions of in vitro culture, other proteins supplemented to the culture medium pass into the blastocyst compartments in comparable quantities. This in turn indicates that the environmental proteins under in vitro conditions seem not to enfold a specific embryotropic activity, but more likely general physicochemical or biochemical effects. If we extrapolate these conclusions from the in vitro situation to the normal physiological situation in utero, it appears conceivable that the uterine secretion proteins act as integrated parts of a linked mother-embryo molecular system, the function of which is more a protective activity for the genetically individual embryo (Beier 1974), than a direct embryotropic role as suggested by Krishnan and Daniel (1967).

In addition, uteroglobin may play an important role in the biochemical reactions and the "metabolism" of the blastocyst coverings, particularly, when the blastocyst enters the uterine cavity on day 4. Striking evidence for the involvement of uteroglobin in the physicochemical conditioning of the blastocyst coverings, the zona pellucida and the mucin coat (mucoprotein layer), has been obtained from comparison of two experiments on rabbit blastocysts. Blastocysts flushed from the isthmic part of the oviduct or from the upper uterine segment (approx. 70 h post coitum) during delayed secretion show herniations of the trophoblast. The same picture of trophoblast herniation can be easily obtained by culturing early cleavage stages up to the blastocyst stage in vitro, using a defined medium (Maurer and Beier 1976) with or without bovine serum albumin as the only protein source. Both sets of blastocysts are faced with an unfavorable environment, the ones with a desynchronized uterine secretion and the others with a nonnatural medium, also "desynchronized" by lack of uteroglobin or any other uterine protein.

Probably in consequence of this disproportion of uterine proteins, in partcular of uteroglobin, the mucin coat does not show the elasticity and flexibility that is usual on day 4, when the normally developing blastocysts start expansion. This expansion cannot take place when the rigid coverings under desynchronized conditions act as a straitjacket for the blastocyst. We have presented evidence for the influence of the uterine proteins on the coverings and their physicochemical alteration by supplementing in vitro protein-free culture media with uterine secretion proteins (Maurer and Beier 1976). Expansion does appear frequently; however, the overall development of the embryonic system occurs more slowly than normal. Since the uteroglobin fraction acts like a protease-inhibitor (Beier 1976, 1977), it may be regulating proteases which control the structural metabolism of the blastocyst coverings. These proteases can be of uterine or embryonic (trophoblast) origin, and may be controlled by uteroglobin and other uterine secretion components (Denker 1977, 1978).

8 Uteroglobin: Immunological Protection of the Implanting Blastocyst?

The recent structural analysis on uteroglobin including its amino acid sequence of the molecular subunit (see Beato, this volume) has stimulated new discussions on its biological function. Uteroglobin is composed of two identical monomers of 70 amino acids, hold together by two disulfide bonds at amino acid residues 3 and 69. Since the enzyme transglutaminase (Factor XIIIa) is increased 4-to 5-fold in the uterus during early pregnancy, Mukherjee et al. (1980) have compared this phenomenon with the synchronous increase of uteroglobin synthesis and release, taking into account, that transglutaminase is known to catalytically crosslink fibrin molecules by the ϵ-(γ-glutamyl)lysine bonds. The crosslink possibilities of human and bovine fibrin are indicated in Fig. 15. Striking similarities were put together by these authors in the case of uteroglobin and β_2-microglobulin (H-2 or HL-A antigen), and it is shown in Fig. 16 that there are 3 areas where GLN-LYS matching occurs when a β_2-microglobulin strand is opposed to an uteroglobin monomer. Thus, the hypothethis is presented that crosslinking of β_2-microglobulin with uteroglobin could render a masking effect on the H-2 antigen of the embryonic and trophoblast cells, preventing the maternal lymphocytes from recognizing the blastocyst as an allograft (Fig. 16). In fact, this formation of a β_2-microglobulin-uteroglobin complex may be a local mechanism for the non-recognition of the alloantigens of the embryo, a local mechanism more specifically even restricted to the pregnant uterus. Consequently, the pregnant animals' ability to recognize foreign (genetically different) antigens is intact but with the only exemption of the uterus.

This immunological protection mechanism could additionally explain the occurrance of uteroglobin in the male genital tract, particularly in seminal plasma (Beier et al. 1975) and the masking of sperm, thus the generally physiological non-immunogenicity of ejaculated sperm in the uterus (Mukherjee et al. 1980).

9 Secretory Proteins of the Human Endometrium

We have subjected human female genital tract secretion to acrylamide disk electrophoresis (Figs. 17 and 18) and several immunochemical tests, expecting an answer to the long-standing question of whether human uterine secretion follows the same principles as our animal model, the rabbit. Follicular and oviducal fluids, uterine and cervical secretions do not contain a predominant postalbumin component comparable to rabbit uteroglobin. However, rather surprising results have been obtained on cross-reactions of human uterine proteins with anti-uteroglobin sera from the goat and sheep, antibodies that had been raised against authentic endometrial uteroglobin from the rabbit. The uterine secretion samples from four of 12 patients did contain a uteroglobinlike antigen which cross-reacted with the antibodies against rabbit uteroglobin, demonstrated by means of the acrylamide agar immunoreaction. The appearance of a human uterine protein that cross-reacts with antibodies to uteroglobin is challenging, since for most of the data from animal (rabbit) experiments it remains to be seen whether the results could be "transferred" to human reproductive physiology. Even samples obtained during the midsecretory phase of the human menstrual cycle, when implantation is normally expected, do not always contain a typical uteroglobin band or a protein with the same electrophoretic mobility that stains well with amido black. To date, we have not been able to isolate by chromatography a predominant postalbumin fraction with the characteristcs of uteroglobin. Such results indicate that the immunological cross-reactions are probably not highly specific, indeed they may even be totally nonspecific. Since there is no immunological cross-reaction of anti-uteroglobin sera with any of the

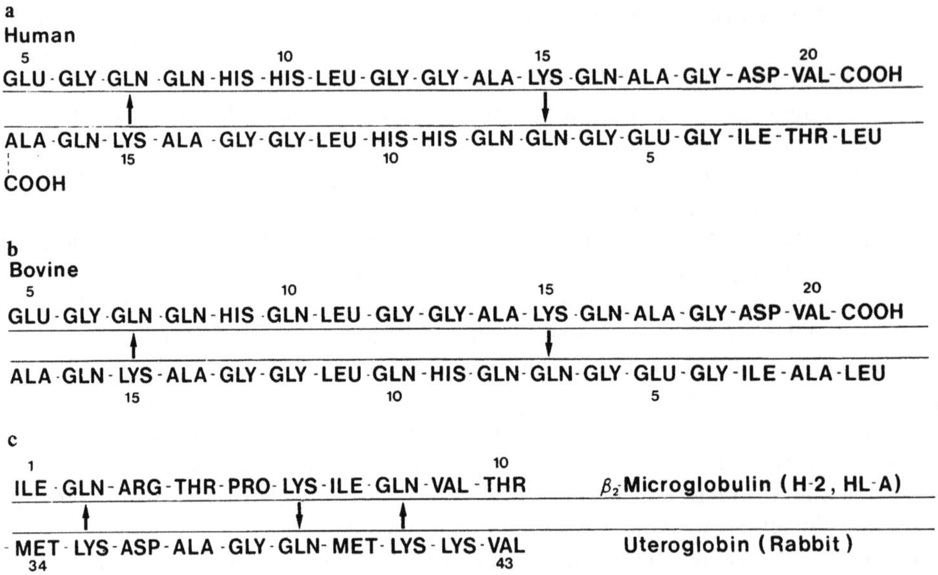

Fig. 15 a-c. Possibilities of Gln-Lys-Crosslinks. Mode of crosslinking between fibrin molecules as well as between uteroglobin and β_2-microglobulin. Note: only one uteroglobin monomer is shown here. **a** *Upper row* Human fibrin molecules (γ-chain). **b** *Middle row* Bovine fibrin molecules (γ-chain). The arrows indicate the reciprocal crosslinks between lysine and glutamine residues in the two molecules. **c** *Bottom row* β_2-microglobulin-uteroglobin complex. (From: Mukherjee, Laki, and Agrawal 1980)

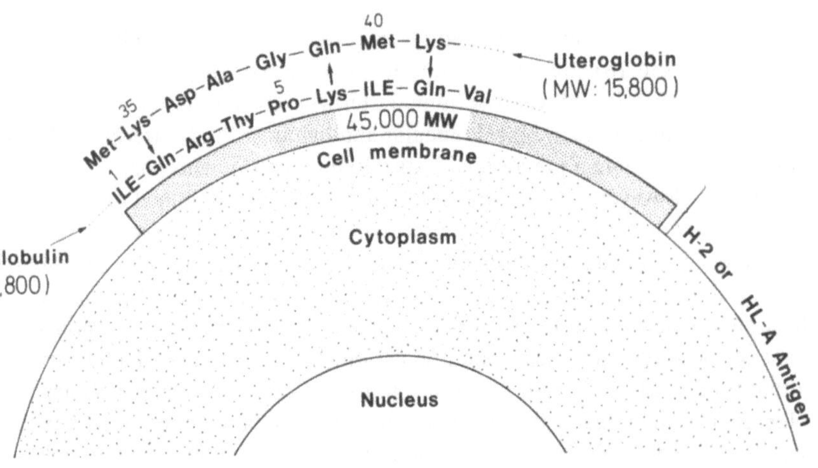

Fig. 16. New Hypothesis on Uteroglobin function. Diagramatic representation of masking of the H-2 antigens on the cell surface. Note the three point crosslinking of the two molecules forming the complex. Only part of one uteroglobin monomer and β_2-microglobulin amino acid sequence are shown here. (From: Mukherjee, Laki, and Agrawal 1980)

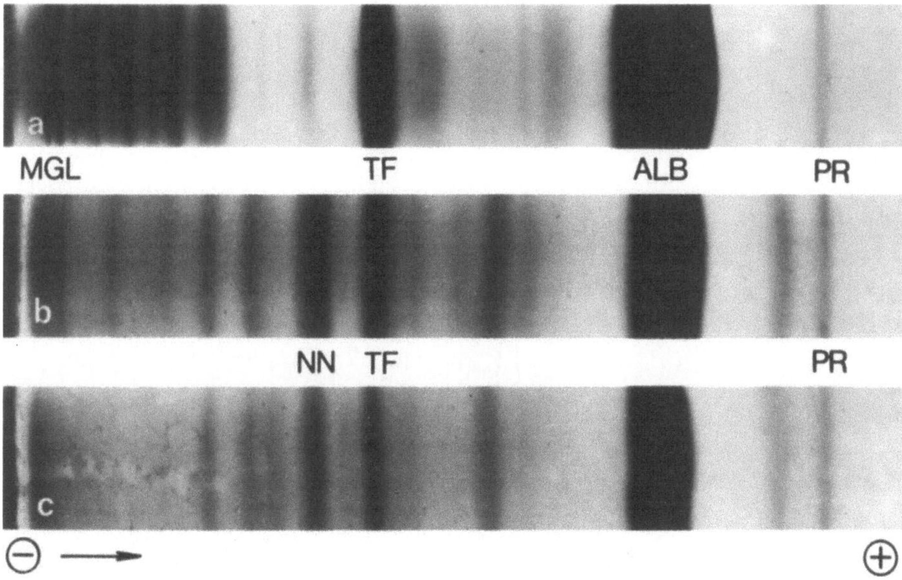

Fig. 17 a-c. Human genital tract secretions. Polyacrylamide electrophoretic separation of uterine flushing (**b**) and oviductal flushing (**c**) from the 14th day of the cycle (patient H.R., 38 years, no pathological reproductive parameters), compared to the paired blood serum (**a**). Note the similarity of protein patterns of uterine and oviductal secretions, *PR*, prealbumin; *ALB*, albumin; *MGL*, macroglobulins; *TF*, transferrin; *NN*, posttransferrin. Tris-glycine buffer pH 9.0, Amido black-10B staining. (Modified from Beier and Beier-Hellwig 1973)

uterine flushings from the well known laboratory animals, such as the mouse, rat, guinea pig, hamster, mink, or from larger animals, e.g., goat, sheep, pig, cow, and even baboon, we may conclude that the interspecies cross-reactions with human uterine wash-fluid or endometrial tissue homogenate are exceptional reactions. For immediate research, it is essential to biochemically identify uteroglobinlike protein or analogous molecules in human uterine secretion. The concentrations of such components may differ totally from the concentrations known in the rabbit uterus, because implantation of the human blastocyst differs remarkably from that in the rabbit (interstitial *vs.* superficial type). However, we have no reason to deny the existence of uteroglobinlike protein in the human uterus so long as this assumption has not been ruled out by convincing research data.

There are characteristic oviducal and uterine proteins in the human being, which we were able to demonstrate immunologically by specific antisera, though not yet directly. We have observed several remarkable fractions, among which one prealbumin and one posttransferrin seem to lack any equivalent fraction in blood plasma from the same patient. Comparable posttransferrin bands have been reported by Mastroianni et al. (1970) in oviducal fluid of the rhesus monkey, by Moghissi (1970) in the human oviduct, and by Wolf and Mastroianni (1975) in human uterine washings. The prealbumin and the posttransferrin fractions both show a clear PAS-positive reaction indicating their glycoprotein nature.

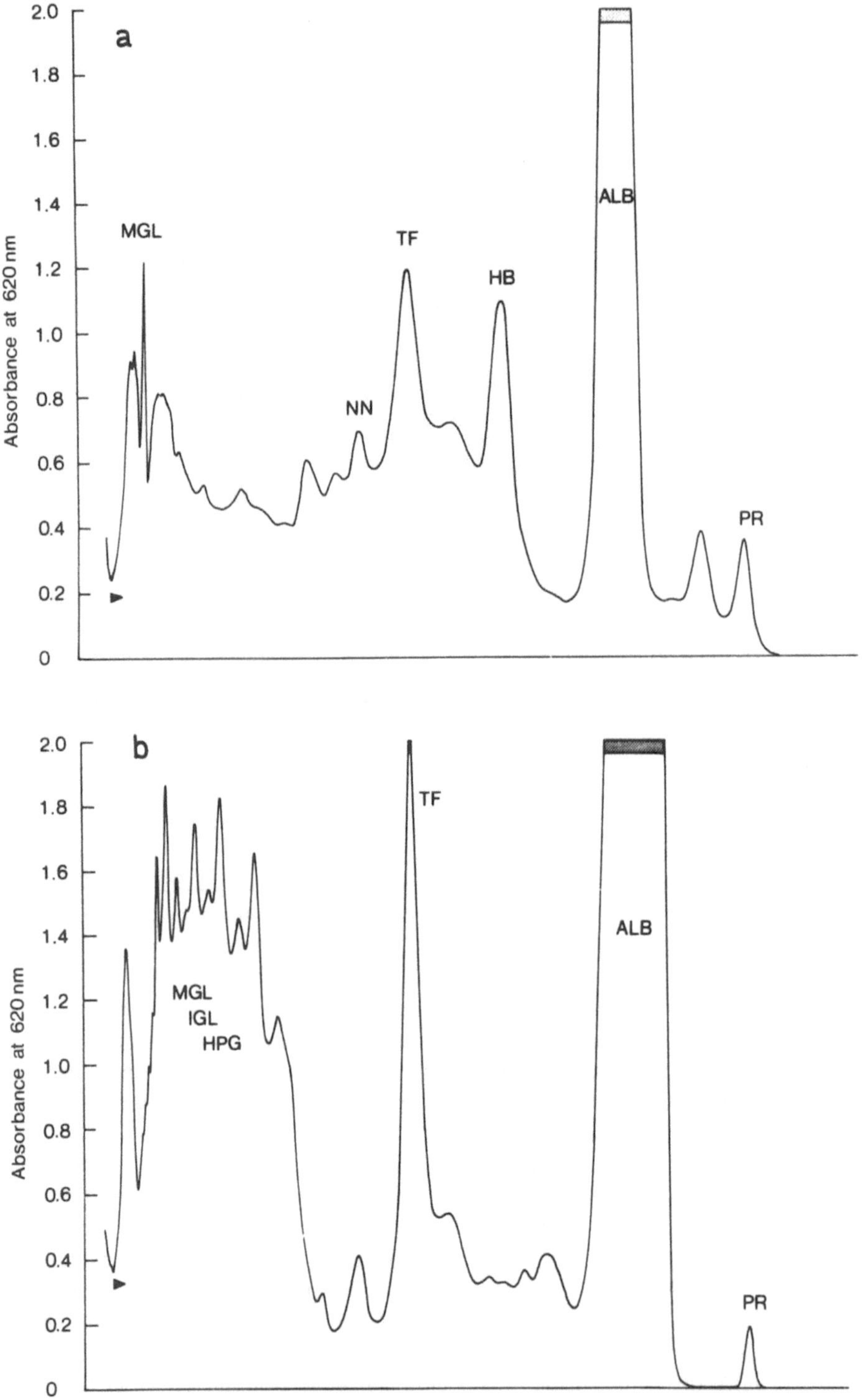
a
Absorbance at 620 nm
2.0
1.8
1.6
1.4
1.2
1.0
0.8
0.6
0.4
0.2
0
MGL
NN
TF
HB
ALB
PR
b
Absorbance at 620 nm
2.0
1.8
1.6
1.4
1.2
1.0
0.8
0.6
0.4
0.2
0
MGL
IGL
HPG
TF
ALB
PR

◁ **Fig. 18 a,b.** Human uterine secretion protein patterns. Densitometric curves of polyacrylamide electrophoretic separation of uterine flushing (**a**) from the 28th day of the cycle (Patient M.J., 47 years, no pathological reproductive parameters, cycle length usually 30 days). The blood serum pattern is shown for comparison (**b**). Note that uterine secretion of this premenstrual stage shows contamination by hemoglobin (*HB*). The possibly uterine-specific prealbumin fraction is migrating next to the fast prealbumin (*PR*) and does not appear in serum. The posttransferrin band characteristic for uterine secretion is present (*NN*). *ALB*, albumin; *TF*, transferrin; *MGL*, macroglobulins; *HPG*, haptoglobins; *IGL*, immunoglobulins. (Modified from Beier and Beier-Hellwig 1973)

Sutcliffe et al. (1978, 1980) have detected a human uterine protein that appears in pregnant uterine tissue, in decidual tissue and in amniotic fluid. The probably tissue-specific protein is characterized as α_2-uterine globulin, composed of two approximately 25,000 molecular weight subunits. This molecule is not found by the immunological means used by Sutcliffe et al. (1978) in male reproductive tissues, and not in blood serum of pregnant or non-pregnant women. It may be justified to expect interesting results from further studies on this protein.

A similar protein, if not the identical molecule, has been described as a progesterone-dependent protein of the human endometrium in a series of recent publication by Joshi et al. (1980 a-c). The specific protein was isolated biochemically from homogenized tissues and recovered in the crude nuclei fraction, the heavy particulate fraction and the cytosol of decidua-rich material from very early pregnancies. Further isolation of the protein was achieved by a combination of Concanavalin A-Sepharose chromatography and polyacrylamide gel electrophoresis. The preliminary determination of the molecular weight was approximately 48,000. The isoelectric point is about 4.9. The protein was also found in amniotic fluid with maximal concentrations during the 15th to the 18th week of pregnancy (Joshi et al. 1980c).

Lactoferrin was initially described as the iron-binding protein in bovine and human milk. We have now found it in uterine and cervical secretions, as reported earlier by Masson et al. (1968). In contrast, we were not able to show its presence in follicular and oviducal fluid. The occurrence of lactoferrin in uterine and cervical secretions in addition to transferrin may indicate that it has a wider physiological role than the binding of iron and copper. Possibly these proteins protect the mucosae against poisoning by heavy metal ions, and also against infection, since bacteriostatic activities, dependent on the chelation of ionized iron and copper from the substrate fluid, could be demonstrated for both lactoferrin and transferrin (Masson and Heremans 1967; Schade 1967).

Among the plasma protein components, special attention has been paid to the immunoglobulins in genital tract secretions. In contrast to other body secretions (milk, saliva, intestinal fluid) where IgA is the major immunoglobulin, the ratio of IgA/IgG in uterine and cervical mucus is approximately the same as in the blood plasma. Both immunoglobulins are present in similar proportions in follicular and oviducal fluid, however in much lower concentrations in uterine and cervical secretions. Interestingly, IgM occurs only in uterine secretion as a permanent component (Beier and Beier-Hellwig 1973; Beier et al. 1976).

10 Conclusions

During the periovulatory time of all species so far investigated, including man, a voluminous, protein-rich uterine-secretion of low viscosity is produced. By contrast, the endometrium delivers a second type of secretion, the luteal phase secretion, which serves to induce and maintain blastocyst growth and development over the implantation period. In species with facultative or obligatory delayed implantation, such as the roe deer, the wallaby, fur seal, mouse, and rat, the reactivation of the dormant blastocyst at the end of the diapause is temporally associated with an increase in uterine protein secretory activity. In the rabbit, we find an unusual situation, since there uteroglobin predominates over all other uterine proteins, when enough progesterone is present. In most species investigated, the progestational endometrial secretions consist of a complex mixture of proteins (Surani 1977; Aitken 1979), many of which possibly have a transport function.

In the mouse, wallaby, rabbit, and pig, evidence suggests that the implanting or expanding blastocyst may exert local stimulatory effects on protein *synthesis.* The evidence further suggests that the implanting blastocyst may have a local stimulatory effect upon the *secretory* activity of the endometrium and on the protein *transudation* from blood plasma (Renfree 1973; Wyatt et al. 1976; Aitken 1977a, 1979). Since the progestational transformations of the secretory patterns are rather species-specific, as evidenced by the diverse reactions of release of one or only a few major proteins on the one hand, and of the release of a whole group (rather a "family") of proteins of higher molecular weight without any predominant fraction on the other hand, it may well be true that protein fractions stimulating (facilitating) implantation or attachment only appear in endometrial secretion in the presence of a blastocyst. Such an activating unit, trophoblast cells and probably also embryoblast cells, may be particularly important in human beings, because all results of our own laboratory and that of Maathuis and Aitken (1978) indicate that during the luteal phase of the menstrual cycle the quantity of uterine protein released is rather low. Two consequences have to be discussed therefore: It could be necessary to have the interaction of a blastocyst to trigger real preimplantational uterine protein secretion in women, or it could well be that the implanting human blastocyst does not need a uteroglobinlike protein in the uterine cavity during the phase before invasion. If the latter is true, one may find such proteins as "tissue-bound" components in the endometrium, probably in the local vicinity of the implantation site. These thoughts underline the importance of studying the nature of the mutual interactions between the blastocyst and the uterus and the need to search for relevant models in available animals.

Acknowledgments. Cordial thanks are due to my wife, Dr. med. Karin Beier-Hellwig, who has always cooperated in my research on reproductive biology and never failed to keep in mind the clinical relevance of reproductive physiology research. Also I am grateful to my colleague Dr. Ursula Mootz for invaluable help. The investigations were supported by the Deutsche Forschungsgemeinschaft in the Special Research Programme "Biologie und Klinik der Reproduktion".

Excellent technical assistance by Mrs. Maria Petuelli, Mrs. Jennie Barbour, Mrs. Barbara Bonn, Mrs. Elisabeth Jung, Mrs. Gaby Bock, and Mrs. Ria Becht is gratefully acknowledged.

References

Adams CE (1974) Asynchronous egg transfer in the rabbit. J Reprod Fertil 35:613-614

Aitken RJ (1974) Delayed implantation in roe deer (*Capreolus capreolus*). J Reprod Fertil 39:225-233

Aitken RJ (1977a) Embryonic diapause. In: Johnson HH (ed) Development in mammals, vol. 1. North-Holland, Amsterdam New York Oxford, pp 307-359

Aitken RJ (1977b) The protein content of mouse uterine flushings during pseudopregnancy. J Reprod Fertil 50:191-192

Aitken RJ (1979) Tubal and uterine secretions; the possibilities for contraceptive attack. J Reprod Fertil 55:247-254

Arthur AT, Daniel JC Jr (1972) Progesterone regulation of blastokinin production and maintenance of rabbit blastocysts transferred into uteri of castrate recipients. Fertil Steril 23:115-122

Arthur AT et al. (1972) Steroid binding to blastokinin. Fertil Steril 23:85-92

Atger M, Mornon JP, Savouret JF, Loosfelt H, Fridlansky F, Milgrom E (1980) Uteroglobin: A model for the study of the mechanism of action of steroid hormones. In: Steroid induced uterine proteins (Beato M, Ed) Elsevier/North-Holland Biomedical Press, Amsterdam New York Oxford, pp 341-350

Baer KE von (1827) On the origin of the ovum of mammals and man: Letter to the St. Petersburg Imperial Academy of Sciences (in Latin). Voss, Leipzig

Bargmann W (1967) Histologie und mikroskopische Anatomie des Menschen, 6th edn. Thieme, Stuttgart

Beato M (1977) Physico-chemical characterization of uteroglobin and its interaction with progesterone. In: Johnson MH (ed) Development in mammals, vol 2. North-Holland, Amsterdam New York Oxford, pp 173-198

Beato M, Arnemann J (1975) Hormone-dependent synthesis and secretion of uteroglobin in isolated rabbit uterus. FEBS Lett 58:126-129

Beato M, Baier R (1975) Binding of progesterone to the proteins of the uterine luminal fluid. Identification of uteroglobin as the binding protein. Biochim Biophys Acta 392:346-356

Beato M, Beier HM (1978) Characteristics of the purified uteroglobin-like protein from rabbit lung. J Reprod Fertil 53:305-314

Beato M, Rungger D (1975) Translation of the messenger RNA for rabbit uteroglobin in *Xenopus* oocytes. FEBS Lett 59:305-309

Beier HM (1966) Das Proteinmilieu in Serum, Uterus und Blastocysten des Kaninchens vor der Nidation. In: Beermann W (ed) Biochemie der Morphogenese. Deutsche Forschungsgemeinschaft, Konstanz, pp 1-10

Beier HM (1967) Veränderungen am Proteinmuster des Uterus bei dessen Ernährungsfunktion für die Blastocyste des Kaninchens. Verh Dtsch Zool Ges 31:139-148

Beier HM (1968a) Uteroglobin: a hormone-sensitive endometrial protein involved in blastocyst development. Biochim Biophys Acta 160:289-291

Beier HM (1968b) Biochemisch-entwicklungsphysiologische Untersuchungen am Proteinmilieu für die Blastocystenentwicklung des Kaninchens (Oryctolagus cuniculus). Zool Jahrb Anat 85:72-190

Beier HM (1970) Hormonal stimulation of protease inhibitor activity in endometrial secretion during early pregnancy. Acta Endocrinol (Kbh) 63:141-149

Beier HM (1973) Die hormonelle Steuerung der Uterussekretion und frühen Embryonalentwicklung des Kaninchens. *Habilitationsschrift*, Kiel

Beier HM (1974) Oviducal and uterine fluids. J Reprod Fertil 37:221-237

Beier HM (1976) Uteroglobin and related biochemical changes in the reproductive tract during early pregnancy in the rabbit. J Reprod Fertil [Suppl] 25:53-69

Beier HM (1977) Immunologische und biochemische Analysen am Uteroglobin und dem Uteroglobin-ähnlichen Antigen der Lunge. Med Welt NF 28:788-792

Beier HM (1978) Control of implantation by interference with uteroglobin synthesis, release and utilization. In: Ludwig H, Tauber PF (eds) Human fertilization. Thieme, Stuttgart, pp 191-203

Beier HM, Beier-Hellwig K (1973) Specific secretory protein of the female genital tract. Acta Endocrinol (Kbh) [Suppl] 180:404-425

Beier HM, Maurer RR (1975) Uteroglobin and other proteins in rabbit blastocyst fluid after development in vivo and in vitro. Cell Tissue Res 159:1-10
Beier HM, Mootz U (1979) Significance of maternal uterine proteins in the establishment of pregnancy. In: Maternal Recognition of Pregnancy (Ciba Found Symp 64 NS), Excerpta Medica, Amsterdam Oxford New York, pp 111-132
Beier HM, Petry G, and Kühnel W (1970) Endometrial secretion and early mammalian development. In: Gibian H, Plotz EJ (eds) Mammalian reproduction. (Colloquium der Gesellschaft für biologische Chemie in Mosbach, vol 21, pp 264-285) Springer, Berlin Heidelberg New York
Beier HM, Mootz U, and Kühnel W (1972) Asynchrone Eitransplantationen während der verzögerten Uterussektretion beim Kaninchen. Proc 7th Int. Congr. Anim. Reprod. Artif. Insem. Munich, *3*, 1891-1896
Beier HM, Bohn H, and Müller W (1975) Uteroglobin-like antigen in the male genital tract secretions. Cell Tissue Res 165:1-11
Beier HM, Mettler L, Beier-Hellwig K, Kühnel W (1976) Antigens of human genital tract secretions. In: Biological and clinical aspects of reproduction (Ebling FJG and Henderson IW, eds), Excerpta Medica, Amsterdam, pp 194-197
Beier HM, Kirchner C, and Mootz M (1978) Uteroglobin-like antigen in the pulmonary epithelium and secretion of the lung. Cell Tissue Res 190:15-25
Bernstein GS, Aladjem F, and Chen S (1971) Proteins in human endometrial washings. A preliminary report. Fertil Steril 22:722-726
Bishop DW (1956) Active secretion in the rabbit oviduct. Am J Physiol 187:347-352
Bodemer CW (1969) History of the mammalian oviduct. In: Hafez ESE, Blandau RJ (eds) The mammalian oviduct. University of Chicago Press, Chicago London, pp 3-26
Bond CJ (1898) Preliminary note on certain undescribed features in the secretory function of the uterus and Fallopian tubes in the human subject and in some of the mammalia. J Physiol (Lond) 22:296
Bond CJ (1899) On the experimental production of hydrosalpinx in the human subject. Lancet 2:200
Bond CJ (1906) An inquiry into some points in uterine and ovarian physiology and pathology in rabbits. Br Med J 2:121-127
Bullock DW (1977) Progesterone induction of messenger RNA and protein synthesis in rabbit uterus. Ann. New York Acad. Scienc. 286:260-272
Bullock DW, Connell KM (1973) Occurrence and molecular weight of rabbit uterine "blastokinin". Biol Reprod 9:125-132
Bullock DW, Willen GF (1974) Regulation of a specific uterine protein by estrogen and progesterone in ovariectomized rabbits. Proc Soc Exp Biol Med 146:294-298
Bullock DW, Woo SLC, and O'Malley BW (1976) Uteroglobin messenger RNA. Translation in vitro. Biol Reprod 15:435-443
Daniel JC Jr (1968) Comparison of electrophoretic patterns of uterine fluid from rabbits and mammals having delayed implantation. Comp Biochem Physiol 24:297-299
Daniel JC Jr (1971a) Uterine proteins and embryonic development. Adv Biosci 6:191-203
Daniel JC Jr (1971b) Growth of the preimplantation embryo of the northern fur seal and its correlation with changes in uterine protein. Dev Biol 26:316-322
Daniel JC Jr (1973) A blastokinin-like component from the human uterus. Fertil Steril 24:326-328
Daniel JC Jr, Krishnan RS (1969) Studies on the relationship between uterine fluid components and the diapausing state of blastocysts from mammals having delayed implantation. J Exp Zool 172:267-281
Denker HW (1977) Implantation. The role of proteinases, and blockage of implantation by proteinase inhibitors. Adv Anat Embryol Cell Biol 53:1-123
Denker HW (1978) The role of trophoblastic factors in implantation. In: Spilman CH, Wilks JW (eds) Novel aspects of reproductive physiology. Spectrum, New York London, pp 181-212
Dixon SN, Gibbons RA (1979) Proteins in the uterine secretions of the cow. J Reprod Fertil 56: 119-127
Feigelson M, Noske IG, Goswami AK, and Kay E (1977) Reproductive tract fluid proteins and their hormonal control. Ann NY Acad Sci 286:273-286

Fishel SB (1979) Analysis of mouse uterine proteins at pro-oestrus, during early pregnancy and after administration of exogenous steroids. J Reprod Fertil 56:91-100
Gerlinger H (1922) The existence of a secretory cycle in the horns of the uterus of mammals during the heat period. C R S Biol (Paris) 87:582
Hamana K, Hafez ESE (1970) Disc electrophoretic patterns of uteroglobin and serum proteins in rabbit blastocoelic fluid. J Reprod Fertil 21:557-560
Harpel PC, Homburger F, Tregier A (1968) Mouse uterine fluid plasminogen activator, acid phosphatase, and contraceptive hormones. Am J Physiol 215:928-931
Hartman CG (1923) The oestrus cycle in the opossum. Am J Anat 32:353
Homburger F, Tregier A (1957) Endocrine factors determining the rate of accumulation of endometrial secretions in experimental hydrometra of mice. Endocrinology 61:634-639
Homburger F, Grossman MS, and Tregier A (1955) Experimental hydrouteri (hydrometra) in rodents and some factors determining their formation. Proc Soc Exp Biol Med 90:719
Johnson MH (1972) The protein composition of secretions from pregnant and pseudopregnant rabbit uteri with and without a copper intrauterine device. Fertil Steril 23:123-130
Johnson MH (1974) Studies using antibodies to the macromolecular secretions of the early pregnant uterus. In: Carretti N, Centaro A (eds) Immunology in obstetrics and gynecology. Excerpta Medica, Amsterdam, pp 123-133
Joshi SG (1974) Regulation of protein synthesis in baboon endometrium by endogenous estrogens and progestins. J Steroid Biochem 5:340 (Abstr)
Joshi SG, Ebert KM, Swartz DP (1980a) Detection and synthesis of a progestagen-dependent protein in the human endometrium. J Reprod Fertil 59:273-285
Joshi SG, Ebert KM, Smith RA (1980b) Properties of the progestagen-dependent protein of the human endometrium. J Reprod Fertil 59:287-296
Joshi SG, Smith RA, Stokes DK (1980c) A progestagen-dependent endometrial protein in human amniotic fluid. J Reprod Fertil 60:317-321
Junge JM, Blandau RJ (1958) Studies on the electrophoretic properties of the cornual fluids of rats in heat. Fertil Steril 9:353-367
Kendle,KE, Telford JM (1970) Investigations into the mechanism of the antifertility action of minimal doses of megestrol acetate in the rabbit. Brit J Pharmacol 40:759-774
Kirchner C (1969) Untersuchungen an uterusspezifischen Glycoproteinen während der frühen Gravidität des Kaninchens (*Oryctolagus cuniculus*). Wilhelm Roux Arch Entwicklungsmech Org 164: 97-133
Kirchner C (1972) Immune histologic studies on the synthesis of a uterine-specific protein in the rabbit and its passage through the blastocyst coverings. Fertil Steril 23:131-136
Kirchner C (1976) Uteroglobin in the rabbit: I. Intracellular localization in the oviduct, uterus, and the preimplantation blastocyst. Cell Tissue Res 170:415-424
Krishnan RS, Daniel JC Jr (1967) "Blastokinin": Inducer and regulator of blastocyst development in the rabbit uterus. Science 158:490-492
Krishnan RS, Daniel JC Jr (1968) Composition of "blastokinin" from rabbit uterus. Biochim Biophys Acta 168:579-582
Kulangara AC, Crutchfield FL (1973) Passage of bovine serum albumin from the mother to rabbit blastocysts. II. Passage from uterine lumen to blastocyst fluid. J Embryol Exp Morphol 30:471-482
Kunitake GM, Nakamura RM, Wells BG, and Moyer DL (1965) Studies on uterine fluid. I. Disc electrophoretic and disc-gel Ouchterlony analysis of rat uterine fluid. Fertil Steril 16:120-124
Laster DB (1977) A pregnancy-specific protein in the bovine uterus. Biol Reprod 16:682-690
Ludwig H, Tauber PF (eds) (1978) Human fertilization. Thieme, Stuttgart
Lutwak-Mann C (1960) On the nature and chemical content of fluids in rabbit blastocysts (in Polish). Acta Biochim Pol 7:331-339
Lutwak-Mann C (1962) Some properties of uterine and cervical fluid in the rabbit. Biochim Biophys Acta 58:637-639
Lutwak-Mann C, Boursnell JC, and Bennett JP (1960) Blastocyst-uterine relationships: uptake of radio-active ions by the early rabbit embryo and its environment. J Reprod Fertil 1:169-185
Maathuis JB, Aitken RJ (1978) Protein patterns of human uterine flushings collected at various stages of the menstrual cycle. J Reprod Fertil 53:343-348

Malsky ML, Bullock DW, Willard JJ, and Ward DN (1979) Progesterone-induced secretory protein. NH_2-terminal sequence of pre-uteroglobin. J Biol Chem 254:1580-1585
Masson PL, Heremans JF (1967) Studies of lactoferrin, the iron-binding protein of secretions. Protides Biol Fluids Proc Colloq 14:115-124
Masson PL, Heremans JF, and Ferin J (1968) Presence of an iron-binding protein (lactoferrin) in the genital tract of the human female. I. Its immunohistochemical changes in the female genital tract. Int J Fertil 14:1-7
Mastroianni L, Urzua M, Avalos M, and Stambaugh R (1969) Some observations on Fallopian tube fluid in the monkey. Am J Obstet Gynecol 103:703-709
Mastroianni L, Urzua M, and Stambaugh R (1970) Protein patterns in monkey oviductal fluid before and after ovulation. Fertil Steril 21:817-820
Maurer RR, Beier HM (1976) Uterine proteins and development in vitro of rabbit preimplantation embryos. J Reprod Fertil 48:33-41
McCarthy SM, Foote RH, and Maurer RR (1977) Embryo mortality and altered uterine luminal proteins in progesterone-treated rabbits. Fertil Steril 28:101-107
McGaughey RW, Murray FA (1972) Properties of blastokinin: amino acid composition, evidence for subunits, and estimation of isoelectric point. Fertil Steril 23:399-404
Mintz B (1971) Control of embryo implantation and survival. Adv Biosci 6:317-342
Moghissi KS (1970) Human Fallopian tube fluid. I. Protein composition. Fertil Steril 21:821-829
Mukherjee AB, Laki K, Agrawal AK (1980) Possible mechanism of success of an allo-transplantation in nature: Mammalian pregnancy. Medical Hypotheses 6:1043-1055
Murray FA, McGaughey RW, and Yarus MJ (1972) Blastokinin: Its size and shape, and an indication of the existence of subunits. Fertil Steril 23:69-77
Noske IG, Daniel JC Jr (1974) Changes in uterine and oviducal fluid proteins during early pregnancy in the golden hamster. J Reprod Fertil 38, 173-176
Noske IG, Feigelson M (1976) Immunological evidence of uteroglobin (blastokinin) in the male reproductive tract and in non-reproductive ductal tissues and their secretions. Biol Reprod 15:704-713
Peplow V, Breed WG, Jones CMJ, and Eckstein P (1973) Studies on uterine flushings in the baboon. Am J Obstet Gynecol 116:771-779
Petzoldt U (1974) Micro-disc electrophoresis of soluble proteins in rabbit blastocysts. J Embryol Exp Morphol 31:479-487
Ponstingl H, Nieto A, and Beato M (1978) Amino acid sequence of progesterone-induced rabbit uteroglobin. Biochemistry 17:3908-3912
Popp RA, Foresman KR, Wise LD, and Daniel JC Jr (1978) Amino acid sequence of a progesterone-binding protein. Proc Natl Acad Sci USA 75:5516-5519
Rahman SSU, Velayo N, Domres P, and Billiar RB (1975) Evaluation of progesterone binding to uteroglobin. Fertil Steril 26:991-995
Renfree MB (1973) Proteins in the uterine secretions of the marsupial, *Macropus eugenii.* Dev Biol 32:41-49
Renfree MB (1975) Uterine proteins in the marsupial, *Didelphis marsupialis virginiana,* during gestation. J Reprod Fertil 42:163-166
Roberts GP, Parker JM (1974) Macromolecular components of the luminal fluid from the bovine uterus. J Reprod Fertil 40:291-303
Roberts GP, Parker JM, and Symonds HW (1976) Macromolecular components of genital tract fluids from the sheep. J Reprod Fertil 48:99-107
Schade AL (1967) Non-heme metalloproteins: their distribution biological functions, and chemical characteristics. Protides Biol Fluids Proc Colloq 14:13-23
Schlafke S, Enders AC (1973) Protein uptake by rat preimplantation stages. Anat Rec 175:539-560
Schwick HG (1963) Untersuchungen über die Zusammensetzung der Uterus- und Blastocystenflüssigkeit des Kaninchens *(Oryctolagus cuniculus).* Medical dissertation, Marburg
Shapiro AA, Jentsch JP, and Yard AS (1971) Protein composition of rabbit oviducal fluid. J Reprod Fertil 24:403-408
Shapiro SS, Forbes SH (1978) Alterations in human endometrial protein synthesis during the menstrual cycle and in progesterone-stimulated organ culture. Fertil Steril 30:175-180

Shirai E, Iizuka R, and Notake Y (1972) Analysis of human uterine fluid protein. Fertil Steril 23: 522-528
Squire GD, Bazer FW, and Murray FA (1972) Electrophoretic patterns of porcine uterine protein secretions during the estrous cycle. Biol Reprod 7:321-325
Surani MAH (1976) Uterine luminal proteins at the time of implantation in rats. J Reprod Fertil 48:141-145
Surani MAH (1977) Cellular and molecular approaches to blastocyst uterine interactions at implantation. In: Johnson MH (ed) Development in mammals, vol 1. North Holland, Amsterdam, pp 245-305
Sutcliffe RG, Brock DJH, Nicholson LVB, Dunn E (1978) Fetal- and uterine-specific antigens in human amniotic fluid. J Reprod Fert 54:85-90
Sutcliffe RG, Bolton AE, Sharp F, Nicholson LVB, MacKinnon R (1980) Purification of human alpha uterine protein. J Reprod Fert 58:345-442
Urzua MA, Stambaugh R, Flickinger G, and Mastroianni I (1970) Uterine and oviduct fluid protein patterns in the rabbit before and after ovulation. Fertil Steril 21:860-865
Voss H-J, Beato M (1977) Human uterine fluid proteins: gel Electrophoretic pattern and progesterone-binding properties. Fertil Steril 28:972-980
Wolf DP, Mastroianni L (1975) Protein composition of human uterine fluid. Fertil Steril 26:240-247
Wyatt C, Heap RB, and Perry JS (1976) Protein synthesis in co-cultures of blastocyst and endometrium explants during blastocyst steroidogenesis in the pig. Proc. 5th Int. Congr. Endocr. Hamburg, Abstract No 266

Localization of Uteroglobin and Related Secretion Proteins in Oviducal, Uterine and Blastocyst Cells

Christoph Kirchner

Department of Biology, Philipps-Universität, Marburg, Germany

Electrophoretic and immunologic determinations of uterine proteins are usually made on uterine secretion. The immune histologic technique, on the other hand, allows the proteins to be identified one step earlier in the reaction chain at the intracellular level, long before their release into the uterine lumen. The drawback that histologic results are difficult to quantify, is compensated by the unique possibility of studying the proteins within a cell and the distribution of the fluorescence positive cells within a tissue. As well as this, since the presence of uterine proteins can be determine in single cells, a degree of sensitivity is reached, which is comparable to that of a radioimmunoassay.

Consequently, we investigated the distribution of uteroglobin and other uterine secretory proteins in the cells of the rabbit oviduct, uterus, and blastocyst during early pregnancy and after experimental hormone injections. The results are described in detail elsewhere by Kirchner (1976a,b; 1979, 1980). In this paper I will restrict myself to the discussion of some cell-topographic aspects of these studies.

1 Immunohistochemically Means

Antiserum against uteroglobin and other protein (antigen) preparations, isolated and purified by Sephadex G75 SF and DEAE-cellulose A50, was obtained from guinea pigs. For immune reaction, we used the sandwich technique with FITC-conjugated rabbit immunoglobulin G anti-guinea-pig immunoglobulin G. Oviducal and uterine material was fixed in Bouin's solution at 4°C and embedded in low-melting-point paraffin (48°-52°C). For washing we used physiological NaCl solution buffered with 0.01 m phosphate buffer at pH 7.2. Each washing and incubation step was limited to 1 h. Following the first incubation with guinea pig serum anti-uteroglobin and the first washing step, the microscope slides were dipped in 0.5% agarose dissolved in phosphate buffered saline, so that the subsequent reaction with the FITG-conjugate took place through the film.

Examinations was carried out with a Leitz Orthoplan microscope with overhead fluorescent light source. The specifity of the uteroglobin and uterine protein reactions was controlled by the following five tests: Treatment with FITC conjugate without previous treatment with the anti-uteroglobin serum (1), after treatment with normal guinea pig serum (2), after treatment with the anti-uteroglobin serum previously neutralized with uteroglobin (3) and uterus homogenate (4), and treatment with FITC-conjugated normal rabbit immunoglobulin after treatment with the anti-uteroglobin serum (5). None of these controls revealed fluorescence comparable to that specific for uteroglobin. On the other hand, absorptions of the anti-uteroglobin serum with rabbit serum and other uterine fractions did not change the uteroglobin-positive reactions.

2 Uteroglobin Distribution in Epithelial Cells

We found three different types of uteroglobin-containing cells in the uterine epithelium (Figs. 1 and 2). Type I cells contained uteroglobin-positive material in small amounts in a single accumulation at the top of the cell (Fig. 2a,b). Type II cells are characterized by the filling of the apical half with homogenous uteroglobin material (Fig. 2b,c), type III by the filling of the whole cell (Fig. 2d). Cells with apical protrusions occur in type II (Fig. 2c) as well as in type III. Our hypothesis is that type I is an early stage of uteroglobin synthesis, type II a later stage, and that the apical protrusions represent release or extrusion stages. This will be sustained later. Type III. however, is difficult to interpret. Some cells might represent a later stage than type II ready for extrusion; others might contain peculiar or possibly foreign resorbed uteroglobin which may not be extruded. As a result, several possible sequences of progressing synthesis with an without type III cells can be postulated.

In contrast to the uterus, in the oviduct (Fig. 3) the uteroglobin-positive material first appears at the base of the cells and forms discret granules (Fig. 3a). Later stages show a filling of the basal half, and further of the whole cell, with these uteroglobin granules. The apical protrusions are more striking (Fig. 3b). Here, the indicated sequence seems easier to follow (Fig. 3c).

3 Onset of Uteroglobin Synthesis

Some years ago it was observed that uterine epithelium during estrus contains uteroglobin-positive cells (Johnson 1972; Kirchner 1972). These cells can be categorized mainly as type II (Fig. 2c). Accordding to our hypothesis we must suppose that an important part can be simulated in ovariectomized does by means of estrogen treatment (Figs. 1b and 4). One, three, five and seven days after two injections each of 100 μg 17β-estradiol, on consecutive days, we were able to show that uteroglobin synthesis starts parallel to endometrial proliferation. The first cells are of type I (Fig. 4a). Seven days after estradiol treatment the deepest epithelial folds adjacent to the myometrium are full of type I cells (Fig. 4b) while the other parts may also contain type II and III cells (Fig. 4c).

4 Uteroglobin in the Endometrium at Implantation

The following zonation of uteroglobin-positive cells is found in the late 6th day of pregnancy (Figs. 1c and 5): The epithelium around the cavum is characterized by an irregular mosaic of type III and uteroglobin-negative cells (Fig. 5a). The epithelium in the folds usually contains no uteroglobin, whereas the lumen reacts positively, as expected.

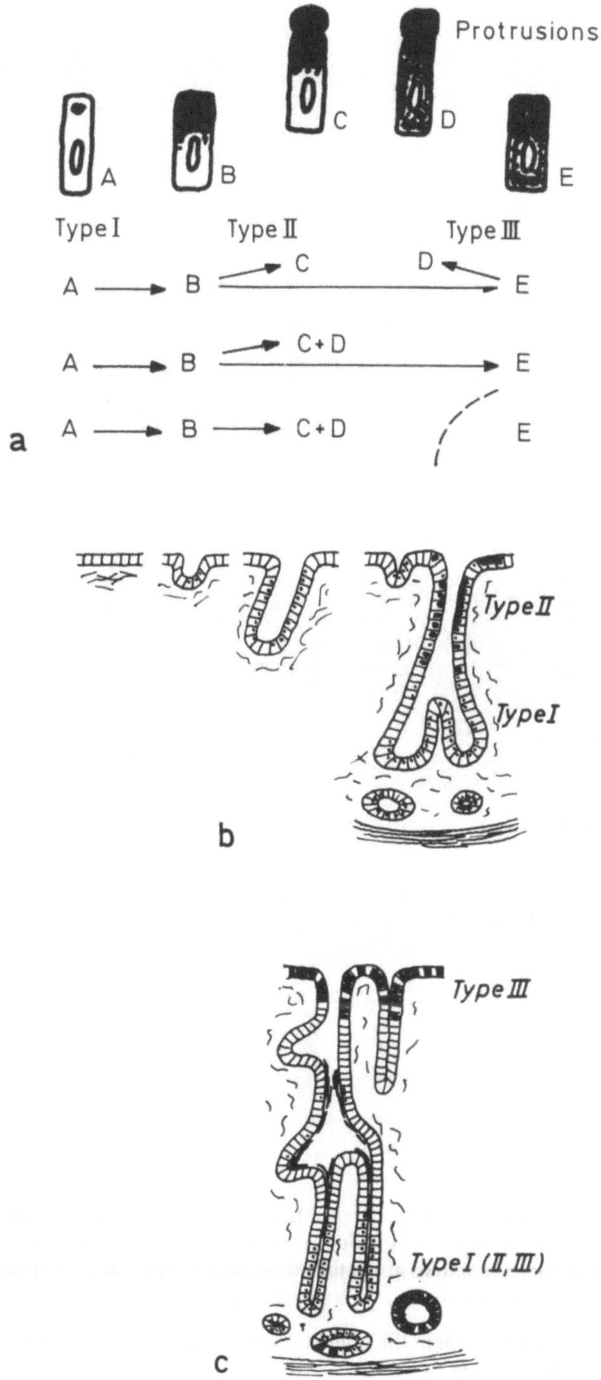

Fig. 1 a-c. Types of Uteroglobin-positive epithelium cells shown in the uterus (**a**). Distribution of Uteroglobin-positive cells after estrogen treatment of ovariectomized rabbits (**b**) and of normal rabbits at the 6th day post coitum (6d 20h) at the time of beginning implantation (**c**)

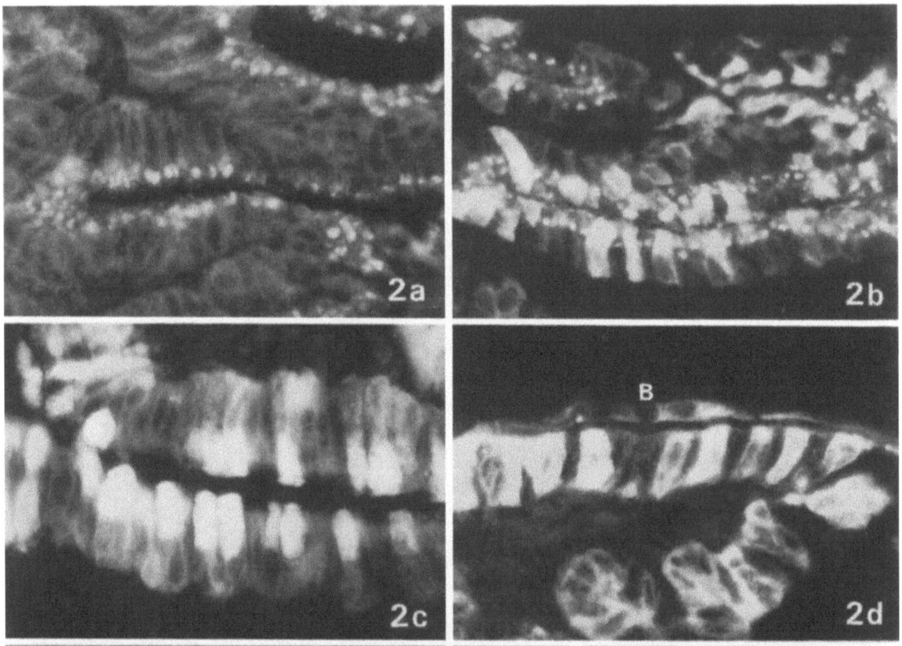

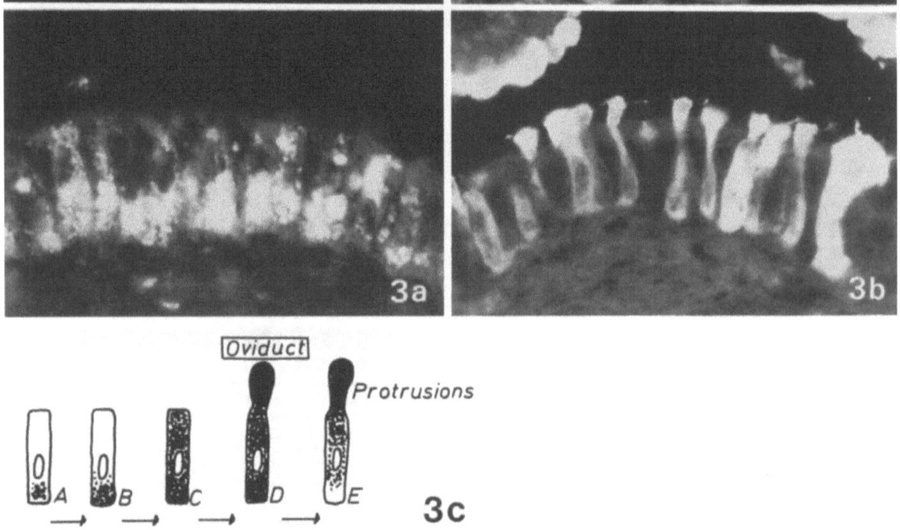

Fig. 2 a-d. Uteroglobin in the uterus. Uteroglobin-positive cells of type I (**a, b**) and type II (**b**) in the crypts adjacent to myometrium at the late 6th day post coitum. Type II cells with protrusions in the endometrium folds at the time of estrus (**c**). Type III cells in the cavum epithelium at the late 6th day post coitum (**d**). *B*, blastocyst. x 350

Fig. 3 a-c. Uteroglobin in the oviduct. x 350 (**a, b**). Diagram of uteroglobin-positive cells in the oviductal epithelium (**c**)

Here the uteroglobin originates possibly also from the uterine secretion. The epithelium of crypts adjacent to the myometrium, however, shows positive uteroglobin reactions again (Fig. 5b). In this area it is the type I cell which is predominant. Fig. 1c summarizes these findings.

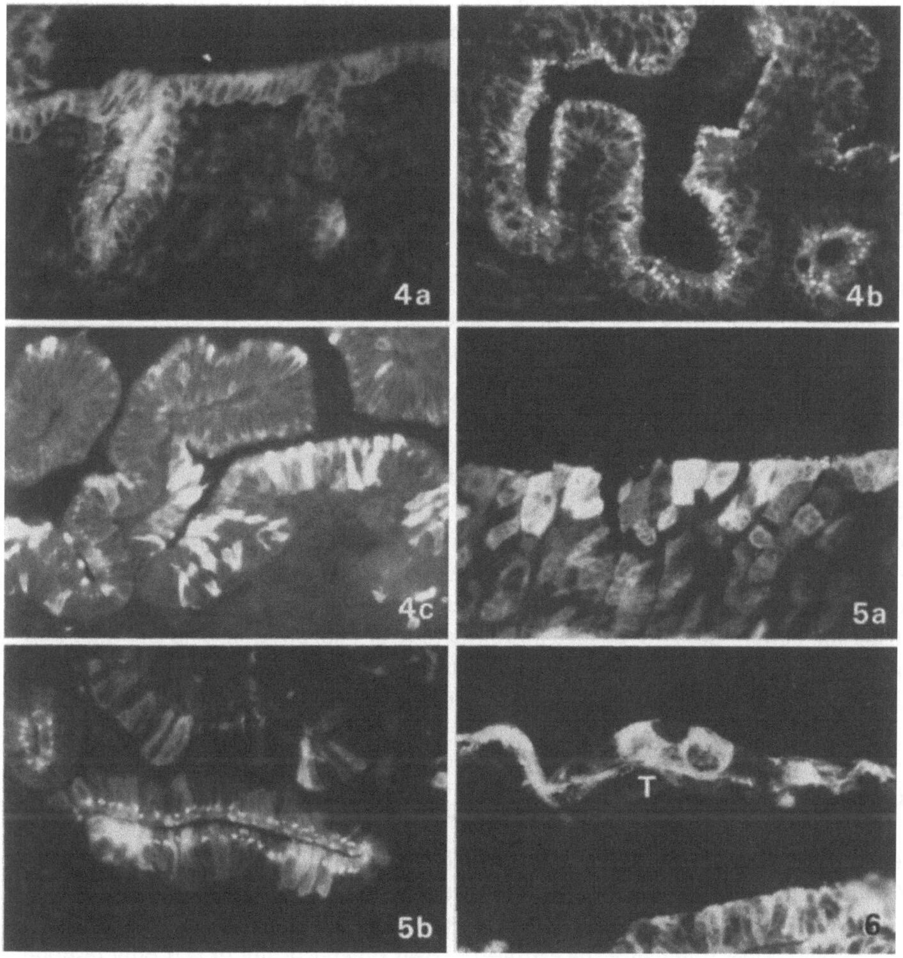

Fig. 4 a-c. Uteroglobin after estrogen treatment: Progressive endometrial proliferation and positive reaction in ovariectomized 3 days (**a**) and 5 days (**b**) after the last estrogen injection. The endometrial folds contain Uteroglobin-positive cells of the type I, whereas the cavum near epithelium contains also type II and III cells (**c**). x 250

Fig. 5 a, b. Uteroglobin in the uterus at the time of implantation. The cavum near epithelium contains a mosaic of type III cells and uteroglobin-negative cells (**a**), whereas the endometrial folds are uteroglobin-negative. The crypts adjacent to the myometrium contain uteroglobin-positive cells again (**b**). Here the type I is predominant. x 250

Fig. 6. Uteroglobin in the blastocyst. Positive trophoblast cells at the 7th day post coitum. x 350

5 Uteroglobin in Blastocyst Cells

Finally it can be shown that uteroglobin also appears in the blastocyst. The positive cells seem to contain large amounts of uteroglobin distributed throughout the entire plasma, comparable to the type III cells of the uterus. The majority of cells, however, respond negatively. The trophoblast (Fig. 6) as well as entoderm or embryonic ectoderm contains cells which react uteroglobin-positive. We could find no systematical pattern in the distribution of these cells. We were also unable to decide whether that uteroglobin originates from the uterus or whether it is a synthetic product of the blastocyst cells.

6 Other Uterine Secretory Proteins

Besides uteroglobin there were three other antigens of the rabbit uterine secretion isolated and antibodies raised: the uterine α_2-globulin, the uterine β_1-globulin and β_2-globulin (Fig. 8). Since uteroglobin was defined as a β_1-globulin (Beier 1968), the β_1-globulin referred to here, is a different antigen at the immunoelectrophoretical β_1-region. Figure 7 shows four subsequent sections of the uterine epithelium at the first day post coitum incubated with antibodies against uteroglobin (Fig. 7b) α_2-(Fig. 7a) β_1-(Fig. 7c) and β_2-(Fig. 7d) uterine globulins. Figs. 7e and f represent horizontal sections incubated with anti-α_2 and anti-uteroglobin. The cells containing α_2, remain uteroglobin-negative. The cells containing uteroglobin are also β_2- and β_1-positive but remain α_2-negative. α_2- and β_2-globulins are distributed throughout the entire cytoplasma, while uteroglobin is localized in the apical half and β_1 only in granules near the top of the cells. This distribution pattern does not change up to the time of implantation. The endometrium shows an irregular mosaic of fluorescence positive cells containing uteroglobin as well as β_1 and β_2 while the other cells remained negative or only α_2-positive.

7 Conclusion

The results raise many new questions, of which a few should be mentioned here:

Why do the oviduct and the uterus synthesize their uteroglobin in different parts of the cells? If estrogen can initiate the synthesis of uteroglobin, what is the function of progesterone? Does it enhance the estrogen answer synergistically or is it only responsible for the release of uteroglobin? Few epithelial cells seem to produce the main amount of uterine secretion proteins. What do the others do? The most interesting question is, what does uteroglobin do in the blastocyst? Are the uteroglobin-positive cells the storing site or even the producting site for this protein in the embryo? And what is the further fate of uteroglobin in the embryo? We must develop new techniques in order to pursue these questions.

With respect to immunohistochemical techniques, there are only a few reports in the literature on the intracellular localisation of uteroglobin in oviductal, uterine or blastocyst cells and in the lung (Johnson 1972; Kirchner 1972, 1976a,b; Sacco and Shivers 1973; Beier et al. 1978). Only recently have comparable studies on other uterine pro-

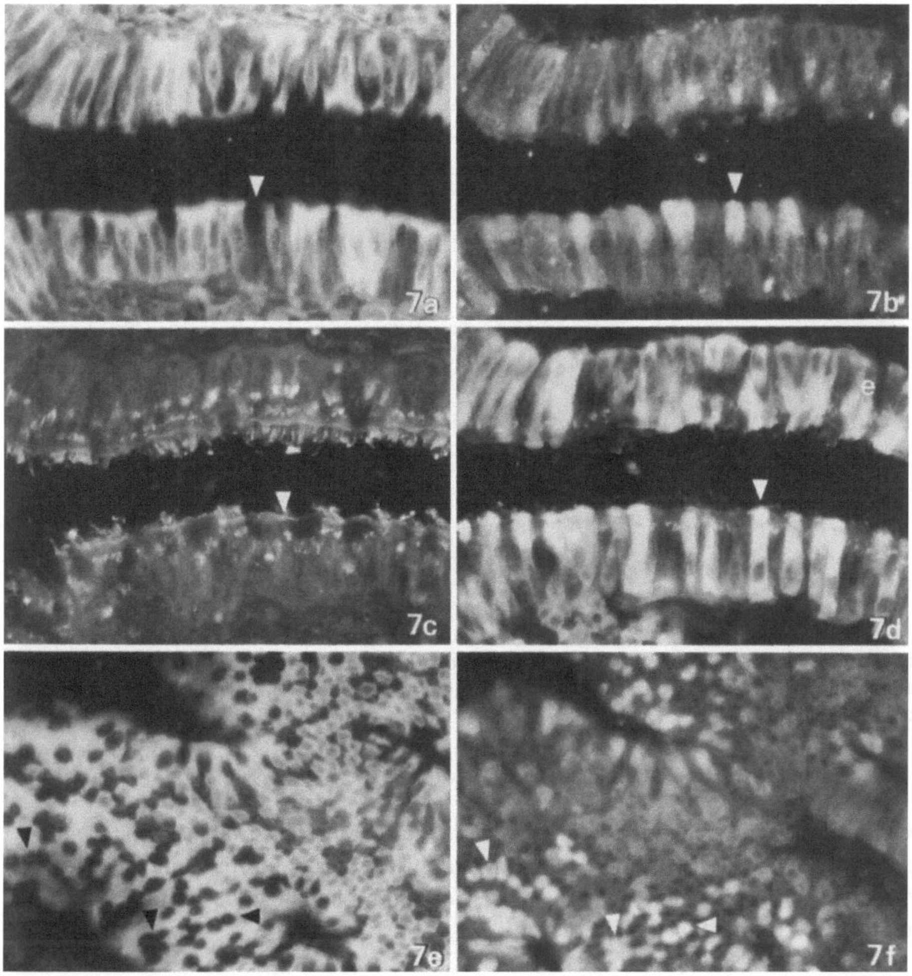

Fig. 7 a-f. Uteroglobin and other secretory proteins in the uterus at the first day post coitum. Serial sections after incubation with Anti-α_2-serum (**a, e**), Anti-uteroglobin (**b, f**), Anti-β_1-(**c**) and Anti-β_2-serum (**d**) showing the distribution of these four secretory proteins within the endometrial cells. The triangles indicate identical cell groups in the series. x 350

teins been performed in our own laboratory (Kirchner 1979). The present investigation again demonstrates the temporal difference of the occurrence of a protein in the endometrial cells and in the secretion. Provided that, at least during the first days, the immunohistologic staining actually indicates the site of synthesis of a protein, the synthesis of β_1 starts long before its release into the uterine lumen. The results obtained from serial sections suggest that some of the proteins, e.g. α_2 and uteroglobin, are producted by different cells, others, e.g. uteroglobin and β_2, are obviously produced by the same cells.

The reactions of the endometrium and of the blastocystic tissue from the seventh day post coitum are much more difficult to interpret: We were unable to distinguish bet-

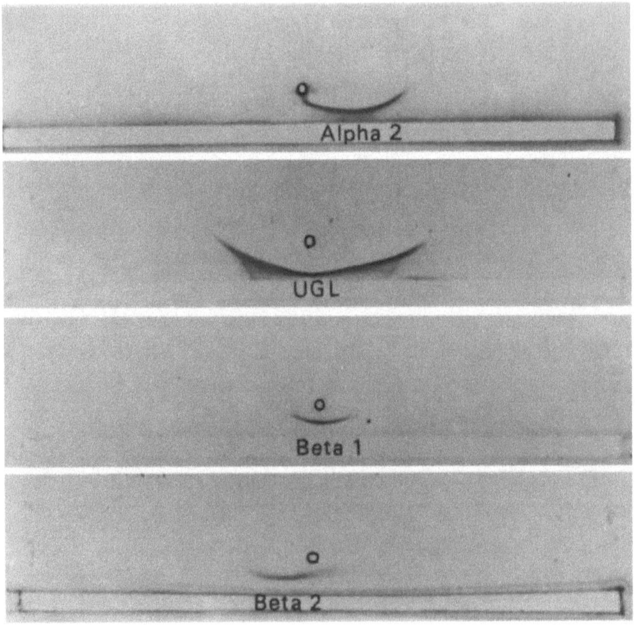

Fig. 8. Immunoelectrophoreses of uterine secretion of the 7th day post coitum after incubations with monospecific antisera against the secretory proteins uteroglobin, α_2-, β_1- and β_2-globulin. Coomassie blue staining

ween intrinsic synthetic products and extrinsic (absorbed or passively penetrated) proteins in the stained cells. The mosaic of immunohistologically positive and negative cells within the epithelium adjacent to the uterine cavum is random and bears no relation to the site of attachment of the blastocyst.

Acknowledgements. Supported by grant Ki 154/5-7 of the Deutsche Forschungsgemeinschaft.

References

Beier HM (1968) Biochemisch-entwicklungsphysiologische Untersuchungen am Proteinmilieu für die Blastocystenentwicklung des Kaninchens (Oryctolagus cuniculus). Zool Jb Anat 85:72-190

Beier HM, Kirchner C, Mootz U (1978) Uteroglobin-like antigen in the pulmonary epithelium and secretion of the lung. Cell Tissue Res 190:15-25

Johnson MH (1972) The distribution of a blastokinin-like uterine protein studied by immunefluorescence. Fertil Steril 23:929-939

Kirchner C (1972) Immune histologic studies on the synthesis of a uterine-specific protein in the rabbit and its passage through the blastocyst coverings. Fertil Steril 23:131-136

Kirchner C (1976a) Uteroglobin in the rabbit. I. Intracellular localization in the oviduct, uterus and preimplantation blastocyst. Cell Tissue Res 170:415-424

Kirchner C (1976b) Uteroglobin in the rabbit. II. Intracellular localization in the uterus after hormone treatment. Cell Tissue Res 170:425-434

Kirchner C (1979) Immunohistochemical localization of secretory proteins in the endometrial epithelium of the rabbit. Cell Tissue Res 199:25-36

Kirchner C (1980) Non-Uteroglobin proteins in the rabbit. In: Beato M (ed) Steroid-induced uterine proteins. Elsevier/North-Holland Biomedical Press, Amsterdam, New York, Oxford, pp 69-86
Sacco AG, Shivers CA (1973) Localisation of tissue-specific antigens in the rabbit ovary, oviduct and uterus by the fluorescent antibody technique. J Reprod Fertil 32:415-420

Uteroglobin as a Sensitive Indicator for the Biological Activity of Progestogens

Y. Nishino

Department of Endocrinepharmacology, Schering Research Laboratories, Berlin (West), Germany

Progesterone has a number of biological actions which form the basis of bioassay for progestational compounds. The glandular development of rabbit endometrium is one of the well-known responses of the uterus to progesterone and is widely used as a parameter for progestational effect in the Clauberg test.

According to recent studies, the glandular development of endometrium appears to be accompanied by a coincident increase in a specific protein, named uteroglobin, in the rabbit uterine flushings. Since the uteroglobin concentration in the uterine flushings seems to be controlled by progesterone (Beier 1968; Urzua et al. 1970; Arthur and Daniel 1972; Bullock and Willen 1974; Rahman et al. 1975), it is suggested that uteroglobin might be a useful indicator for testing the biological activity of progestational compounds.

In the present study, progesterone and some synthetic progestogens were thus tested, in order to see the correlation between uteroglobin content in rabbit uterine fluid and the glandular development of endometrium, using the conventional Clauberg procedure.

In this experimental series, immature New Zealand rabbits weighing about 1 kg (6 weeks old) were used. Animals were first primed by a daily injection with estradiol for 6 days (total dosage=5μg). After priming, a daily treatment with progestogens for 5 days was initiated. On the 11th day after the treatment was initiated, animals were sacrificed, and uteri were immediately excised. The uterine fluid was collected by the procedure described by Beier (1968). A part of the uterine horn was used for the histological examination of the endometrium. The grade of endometrial development was scored by the criteria of McPhail (1934). Total protein in the uterine fluid was measured by the method of Lowry et al. (1951).

The concentration of uteroglobin and albumin in the uterine fluid was determined by scanning at 660 nm after separation of proteins by disk electrophoresis. An integrator was used to calculate the percent concentration of uteroglobin and the uteroglobin/albumin ratio from the peaks registered.

Following steroids dissolved in castor oil containing a small amount of benzylbenzoate were tested: progesterone (4-pregnene-3, 20-dione), chlormadinone acetate (6-chlor-17α-acetoxy-4,6-pregnadiene-3,20-dione), medroxyprogesterone acetate (17α-acetoxy-6α-methyl-4-pregnene-3,20-dione), norethisterone acetate (17-acetoxy-17α-ethinyl-4-estrene-3-one), norethinodrel (17-hydroxy-17α-ethinyl-5(10)-estrne-3-one), *d*-norgestrel (*d*-17-hydroxy-18β-methyl-17α-ethinyl-4-estrene-3-one).

As can be seen from Fig. 1a, progesterone increased not only total protein but also the uteroglobin concentration and the ratio between uteroglobin and albumin in a dose-

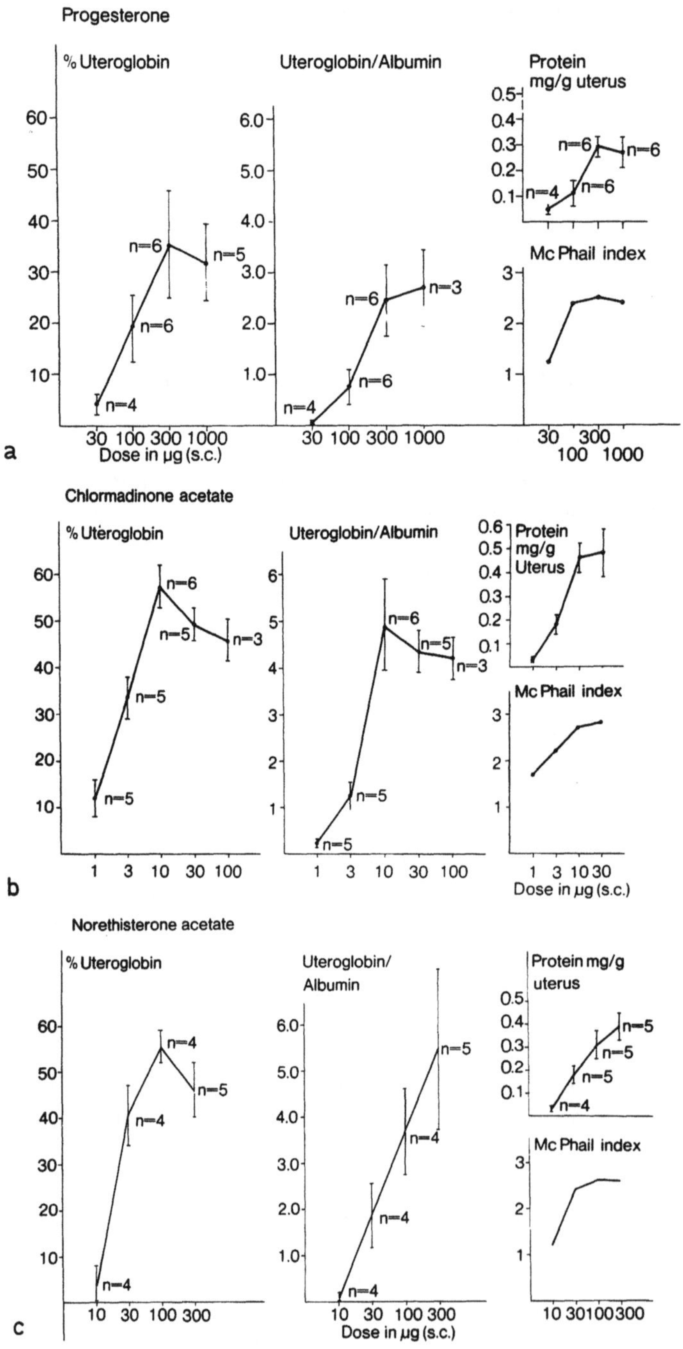

Fig. 1a-c. Total protein, uteroglobin concentration, uteroglobin/albumin ratio and McPhail index in rabbits after estrogen priming and subcutaneous treatment with various doses (total dosage for 6 days) of progesterone (**a**), chlormadinone acetate (**b**), or norethisterone acetate (**c**)

dependent manner. The maximal effect was attained at a dose of 300 μg. The threshold dose of progesterone required for increasing the uteroglobin concentration or uteroglobin/albumin ratio was found to be 30-100 μg. By determining the McPhail index, a dose-response relationship was also found in the dose range of 30-300 μg progesterone.

Figure 1b,c shows the data obtained with chlormadinone acetate and norethisterone acetate. These synthetic progestogens were effective both in increasing the uteroglobin concentration and in stimulating endometrium. The effective dose range was found to be 1-10 μg for chlormadinone acetate and 10-100 μg for norethisterone acetate.

Statistical data can be seen in Table 1. The dose-response curve of the progestogens, with the exception of norethinodrel, showed a linear and significant regression with a positive correlation coefficient. In a six-point assay, all slopes were statistically confirmed to show no significant deviation from parallelism. The biological potencies of synthetic progestogens are also listed in Table 1.

Table 1. Statistical data and relative potencies of progestogens

	Dose range (s.c.) μg/for 6 days	Regression coefficient β	Precision index	Relative potency
Progesterone	30-1000	a) 2.0 b) 31.7	0.59 0.61	1.0 1.0
Chlormadinone acetate	1-10	a) 4.8 b) 45.3	0.32 0.25	89.1 77.7
Medroxyprogesterone acetate	3-30	a) 3.2	0.49	27.3
Norethisterone acetate	10-300	a) 3.6 b) 50.7	0.68 0.19	8.5 7.1
Levonorgestrel	3-30	a) 3.2	0.36	27.8
Norethinodrel	100-3000	---	---	Inactive

Endpoints: a) Uteroglobin/albumin ratio
b) Uteroglobin concentration

The results confirm that progesterone induces the appearance of uteroglobin in rabbit uterine flushings. The uteroglobin concentration increases in proportion to the daily dose of progesterone, and the increase is always well correlated with the McPhail index, indicating the intensity of the glandular development of the endometrium. The minimum effective dose of progesterone needed to increase the uteroglobin concentration or uteroglobin/albumin ratio was found to be between 30 and100 μg/animal (body weight about 1 kg). This agrees well with the results of Rahman et al. (1975), who found the minimum dose of progesterone (s.c.) to be 75 μg/kg in mature castrated rabbits. These authors reported, moreover, that the amount of uteroglobin in the uterine flushings from infantile castrated rabbits treated with 2 mg progesterone/kg plus 2 μg estradiol/kg was 0.62 ± 0.14 mg. In the present study using infantile intact rabbits primed with estradiol, we found a maximal increase in the uteroglobin concentration of 35.5±10.7% of total protein (about 0.32 mg uteroglobin) at the progesterone dose of 300 μg/animal.

Synthetic progestogens tested, excepting norethinodrel, were active in this respect, and the order of the potency compared well with that found in the Clauberg test. The increase in the uteroglobin concentration or the uteroglobin/albumin ratio is always observed in parallel with the glandular development of the endometrium.

In view of the statistical data, it is proposed that the uteroglobin concentration and uteroglobin/albumin ratio may be used as a quantitative parameter in evaluating the biological activity of progestogens. The covariance analysis by a six-point method provided an estimate of the difference in the potency of several synthetic progestogens for increasing the uteroglobin concentration and uteroglobin/albumin ratio.

19-Nortestosterone derivatives were generally less potent than hydroxyprogesterone derivatives in the uteroglobin test and the Clauberg test. This does not correspond well with the clinical efficacy of progestogens: levonorgestrel, one of the most potent progestogens in women, is much less active in the uteroglobin test and the Clauberg test. According to a recent study of Arthur and Chang (1974), some progestogens possessing a methyl group at C-19 (chlormadinone acetate, medroxyprogesterone acetate, etc.) induce the appearance of uteroglobin in rabbit uterine flushings, but 19-nor-steroids (norethisterone acetate and norethinodrel) are inactive in this respect. The oral doses administered by these authors are excessive, especially norethisterone acetate tested in a dose range between 0.5 and 10 mg/kg in mature rabbits weighing 3.4-5.1 kg. The minimum effective dose of norethisterone acetate, administered orally, has been found to be about 30 μg/kg in the Clauberg test (Neumann et al. 1964). It is well known that norethisterone acetate at an excessively high dose shows a predominantly estrogenic effect rather than a progestational effect in several test models (Neumann and Elger 1971). On the other hand, norethinodrel is known to be weakly active in the Clauberg test in rabbits (Elton 1962) and to show only estrogenic effects when tested in mice and rats (Neumann and Elger 1971). Additional aspects of these pharmacological problems may become clear out of the discussion by Elger and Beier (this volume).

References

Arthur AT, Chang MC (1974) Induction of blastokinin by oral contraceptive steroids: Implications for fertility control. Fertil Steril 25:217-221

Arthur AT, Daniel JC (1972) Progesterone regulation of blastokinin production and maintenance of rabbit blastocysts transferred into uteri of castrated recipients. Fertil Steril 23:115-122

Beier HM (1968) Biochemisch-entwicklungsphysiologische Untersuchungen am Proteinmilieu für die Blastocystenentwicklung des Kaninchens (Oryctolagus cuniculus). Zool Jahrb Anat 85:72-190

Bullock DW, Willen GF (1974) Regulation of a specific uterine protein by estrogen and progesterone in ovariectomized rabbits. Proc Soc Exp Biol Med 146:294-298

Elton RL (1962) Morphological changes in the glandular epithelium of rabbit endometrium due to hormonal treatment. Anat Rec 142:469-477

Lowry OH, Rosebrough NJ, Farr AL, Randall RJ (1951) Protein measurement with folin phenol reagent. J Biol Chem 193:265-275

McPhail MK (1934) The assay of progestin. J Physiol (Lond) 83:145-156

Neumann F, Elger W (1971) Kritische Überlegungen zu den biologischen Grundlagen von Toxizitätsstudien mit Steroid- (Sexual-) hormonen. In: Plotz EJ, Haller J (eds) Methodik der Steroidtoxikologie for Forschung und klinische Anwendung der Steroide. Thieme, Stuttgart, pp 6-48

Neumann F, Kramer M, Junkmann K (1964) Tierexperimentelle Untersuchungen mit 17α-Aethinyl-19-nor-testosteron-acetat. Med Exp [Suppl] 11:1-36
Rahman SS, Billiar RB, Little B (1975) Induction of uteroglobin in rabbits by progestogens, estradiol-17β and ACTH. Biol Reprod 12:305-314
Urzua MA, Stambaugh R, Flickinger G, Mastroianni L (1970) Uterine and oviduct fluid protein patterns in the rabbit before and after ovulation. Fertil Steril 21:860-865

Prefertilization Progesterone Treatment: Effect on Endometrium Uteroglobin Secretion, and Embryonic Development

J. de Visser

Organon International B.V., Endocrinological Research and Development Laboratories, Oss, The Netherlands

Biological methods of demonstrating direct effects of steroids on the female genital tract are important for the development of contraceptive or interceptive preparations which do not inhibit hypothalamic-pituitary-ovarian function.

Steroid-induced changes in the cervical and uterine environment during the preovulation and preimplantation phase in the rabbit have been the object of thorough investigations during the last decade.

Based on the observations of Boyarski et al. (1947), who found unfertilized ova in rabbits after HCG (human chorion gonadotrophin) injection and vaginal artificial insemination (AI) after the 5th day of pseudopregnancy, a test method was developed (de Visser 1969). Preovulatory treatment of estrous rabbits with progesterone for 7 days before HCG-induced ovulation and artificial insemination induced a change in cervical mucus resulting in a blockade of migration through the cervix and prevention of fertilization of ova. This procedure is similar to that published by Nutting and Mares (1970).

The effect of progesterone or synthetic progestagens administered for 3 days to estrous rabbits before HCG-induced ovulation and *vaginal* AI, which partly prevented fertilization, has been reported by Chang (1967), Kendle and Telford (1970), and Allen and Foote (1973). These authors found indications of acceleration of ovum transport and degeneration of fertilized ova 3-6 days later and suggested the effect was partly due to advanced endometrial development.

Treatment of estrous rabbits with 0.25-5 mg progesterone for 3 days prior to HCG injection and *intrauterine* AI resulted in a limited number of implantations in the low dose (0.25-0.5 mg) group at laparotomy on day 10, but none went to term (McLean-Morris 1973). This author presumed that under this treatment ova were fertilized and that the failure to implant suggested that either accelerated transport resulted in degeneration of the ovum or pretreatment with progesterone altered the receptiveness of the endometrium.

The mechanism of antifertility actions of progesterone in the rabbit was studied by our group and preliminary results were communicated earlier (de Visser and Turpijn 1974). These results and more recent data on uteroglobin in rabbit uterine fluid are now reported. The technical details are described briefly.

Crystalline progesterone was dissolved in arachis oil, 10 mg/ml. HCG (Pregnyl, Organon) was dissolved in 0.9% NaCl solution (100 IU/ml). Mature 7-month-old Dutch-belted rabbits were housed in air-conditioned quarters. To obtain a comparable estrous period of limited duration before starting a treatment with progesterone the animals were mated with a vasectomized buck to induce pseudopregnancy. Corpora lutea regress after a functional period of 17 days and the estrous stage reappears. Daily progesterone or placebo treatment for 8 days started on the 21st day after sterile mating. This day was designated in the experiments as day -6. Ovulation was induced by intravenous injection of 25 IU HCG on day 0.

Sperm was collected from fertile bucks by means of an artificial vagina according to the method of Adams (1961, 1972). Immediately after HCG injection the AI was performed either deep in the vagina in conscious rabbits or under sodium pentobarbitone anesthesia into the lumen of the exposed uterine horns.

A whole ejaculate was used for vaginal AI and about 10 million spermatozoa (0.05 ml of an ejaculate) for insemination in each uterine horn. In order to avoid pyometra, which occurs in progesterone-treated rabbits, a small dose of an antibiotic (0.01 ml = 3000 IU procaine penicillin G, Mycofarm) was added to 0.05 ml of undiluted semen (Killingbeck 1964; Chang 1967; Haynes 1967). At the time of autopsy, indicated for each experiment, the animals were killed by an overdose of sodium pentobarbitone. The genital tract was removed and trimmed. The number of ruptured follicles (corpora lutea) was determined. The oviducts and the uterine horns were flushed with 0.9% NaCl to collect the ova, which were carefully examined under the microscope. In one experiment the number of implantations on day 13 of pregnancy was recorded.

The effect on sperm migration after vaginal and intrauterine insemination was determined indirectly in experiment 1 in untreated estrous, pseudopregnant and progesterone-treated rabbits by determination of the number of fertilized and unfertilized ova. HCG injection and AI were given on day 7 of pseudopregnancy. Estrous rabbits were treated with progesterone from day -6 to day +1 according to the schedule in Fig. 1. The animals were killed 48h after HCG injection and AI.

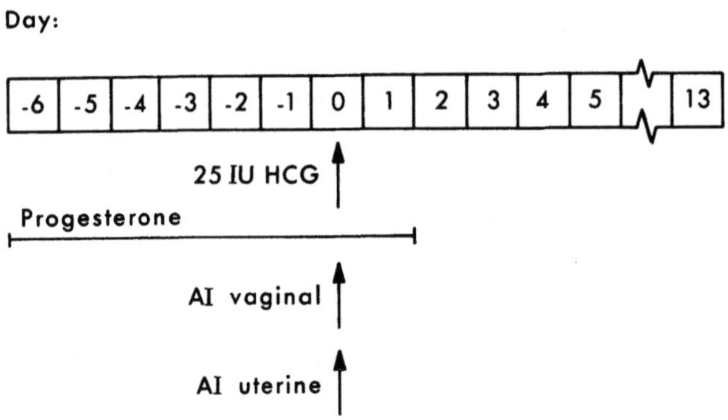

Fig. 1. General experimental schedule

After vaginal insemination and HCG injection unfertilized ova were found in the oviducts of pseudopregnant and of progesterone-treated rabbits. Fertilized ova, however, were observed in the oviducts after intrauterine AI (Table 1). These facts clearly show an increase of the barrier function of the cervix to migrating sperm in progesterone-dominated rabbits. Apparently capacitation of sperm after uterine AI was not inhibited by progesterone treatment of estrous rabbits or endogenous progesterone in pseudopregnant animals.

Table 1. Effect of progesterone dominance on fertilization of ova in mature rabbits treated with HCG to induce ovulation and vaginal or intrauterine AI

	AI HCG i.v. day	No. of rabbits	Vaginal insemination		No. of rabbits	Uterine insemination	
			Recovery of ova	Fertilized ova		Recovery of ova	Fertilized ova
Estrous rabbits	0	4	30/32	29/30	4	27/28	25/27
Pseudopregnant rabbits	8	2	23/32	0/23	2	14/16	14/14
Progesterone-treated rabbits (4 mg/d, s.c., days -6 to +1)	0	4	36/37	0/36	4	31/33	31/31

The fate of the fertilized ova obtained after intrauterine AI and HCG treatment of rabbits given daily subcutaneous doses of from 0.125 mg to 1 mg progesterone was investigated in experiment 2.

The progesterone-treated animals were killed on day +13; the results are shown in Table 2. The pregnancy rate: (No. live fetuses)/(No. corpora lutea) was 64% in the control group and 8% and 0 in the groups treated with 0.5 and 1 mg progesterone respectively. No indications of fetal death were found in the progesterone-treated animals.

Table 2. Effect of preovulatory progesterone treatment on expected pregnancy in mature rabbits treated with HCG to induce ovulation and intrauterine AI

Treatment day -6 to +1 s.c.	AI (intra-uterine) HCG i.v. day	No. of rabbits	No. of corpora lutea	No. of implantations	Implantation rate (%)
Controls	0	8	70	45	64
Progesterone					
0.125 mg/d	0	3	30	24	80
0.25 mg/d	0	2	14	7	50
0.5 mg/d	0	6	53	4	8
1 mg/d	0	6	63	0	0

The fertilized ova were probably affected in some way, resulting in prevention of nidation. Two rabbits given 0.5 mg progesterone each had two normally developing embryos.

In experiment 3 the rabbits were treated according to the schedule (Fig. 1) with daily subcutaneous doses of 0.5, 1, and 2 mg progesterone. Groups of rabbits were sacrificed on days +3, +4, and +5, and the ova were flushed from the tubes and uterine horns. The percentage of normally developing ova and degenerating ova was determined. The results are shown in Table 3.

Table 3. Effect of preovulatory progesterone treatment on development of fertilized ova in rabbits treated with HCG to induce ovulation and intrauterine AI

Treatment day -6 to +1 daily s.c.	No. of rabbits	Autopsy day	No. of ruptured follicles	No. of ova recovered	Tubal ova		Uterine ova	
					Ferti-lized	Unfer-tilized	Ferti-lized	Degenerated
Controls	8	3	58	39	38	0	0	1
Progesterone								
0.5 mg	3	3	30	28	22	0	0	6
1 mg	5	3	48	30	10	6	11	3
2 mg	3	3	23	9	6	0	2	1
Controls	8	4	65	42	0	0	37	5
Progesterone								
0.5 mg	3	4	27	19	0	0	10	9
1 mg	6	4	52	12	0	0	0	12
2 mg	5	4	43	7	0	0	0	7
Controls	6	5	46	22	0	0	19	3
Progesterone								
0.5 mg	3	5	23	0	0	0	0	0
1 mg	7	5	62	9	0	0	3	6
2 mg	3	5	29	1	0	0	0	1

On day +3 most of the ova were found in the tubes of progesterone-treated rabbits; some ova were found in the uterus, a few of them showing an abnormal appearance. On day +4 ova could be flushed only from the uterine horns. The number of ova in the groups treated with 1 or 2 mg progesterone was small in comparison with the control group and the group given 0.5 mg progesterone daily.

Degeneration of ova was apparent in the progesterone-treated groups. On day +5 very few ova could be recovered in the progesterone-treated groups, and those found had degenerated. Because of the advanced endometrial development, the recovery of ova might have been incomplete. Microphotographs (Fig. 2) of ova flushed from the oviducts on days +2 and +3 showed normal development. On days +4 and +5, ova flushed from the uterus of control rabbits showed normal development and expansion. The uterine ova from progesterone-treated rabbits showed definite signs of degeneration.

In experiment 4, using the same treatment schedule as in experiment 3, the concentration of uteroglobin in rabbit uterine fluid during the preovulatory progesterone treatment was determined. Untreated estrous rabbits were killed on days +3 through +6. Groups of rabbits were sacrificed on days -4 through +1 and on days +2 through +5, i.e., from after the third up to the eighth injection of 2 mg progesterone.

At autopsy the uterine horns were flushed with Hank's solution (1 l Hank's solution contains 8 g NaCl; 0.4 g KCL; 0.13 g $CaCl_2$; 0.2 g $MgSO_4$; 0.075 g $Na_2HPO_4.2H_2O$; 0.06 g KH_2PO_4; 0.350 g $NaHCO_3$; 1 g glucose). If ova were present they were removed.

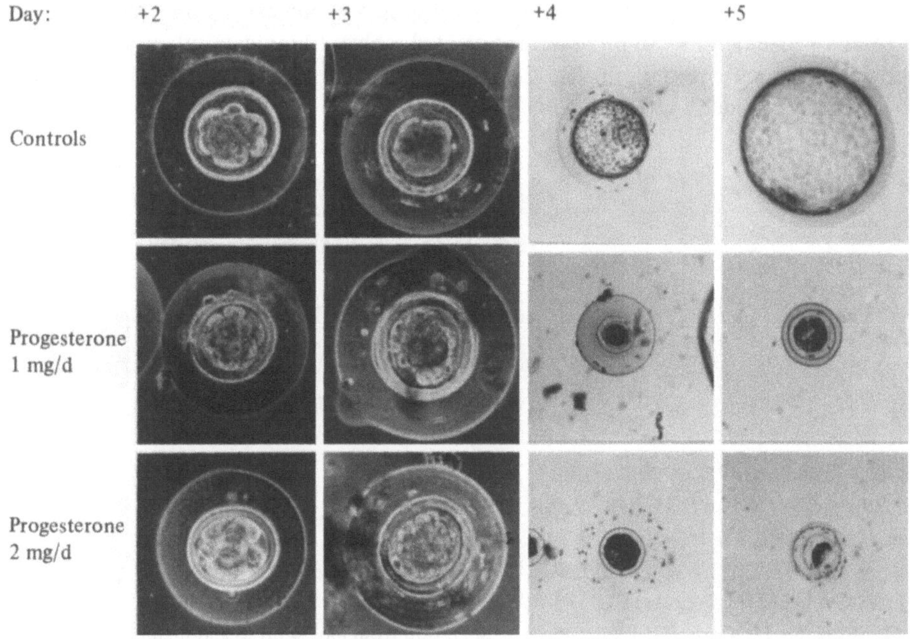

Fig. 2. Microphotographs of embryos from control animals and progesterone-treated animals flushed from the tubes on day +2 (80x) or from the uterine horns on day +3 (80x) +4, or +5 (38x)

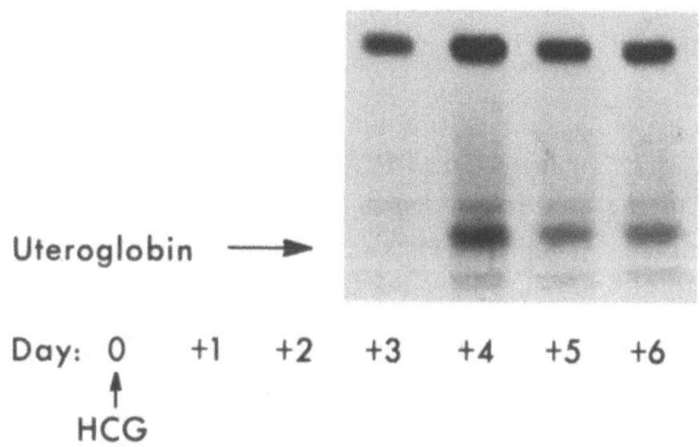

Fig. 3. Electrophoretic protein patterns of rabbit uterine secretion on days +3, +4, +5, and +6 of normal pregnancy

The uterine horns were cut open along the antimesometrial side and were placed in an Erlenmeyer flask together with the previously collected flushing. The flask was shaken for 5 min to ensure maximum collection of released uteroglobin. The fluid was decanted into glass tubes and centrifuged for 10 min at 5000 rpm. The volume of the fluid was adjusted to 6 ml and stored at $-18^{\circ}C$ until determination by agarose-agar gel electrophoresis according to the method of Noble (1968).

Uteroglobin was detected in the uterine fluid of untreated pregnant rabbits on days +4, +5 and +6 (see Fig. 3). In progesterone-treated rabbits (2 mg/day, subcutaneously), uteroglobin was present in uterine fluid on days -2 through +1 and uteroglobin was absent on days +2 through +5 (see Fig. 4).

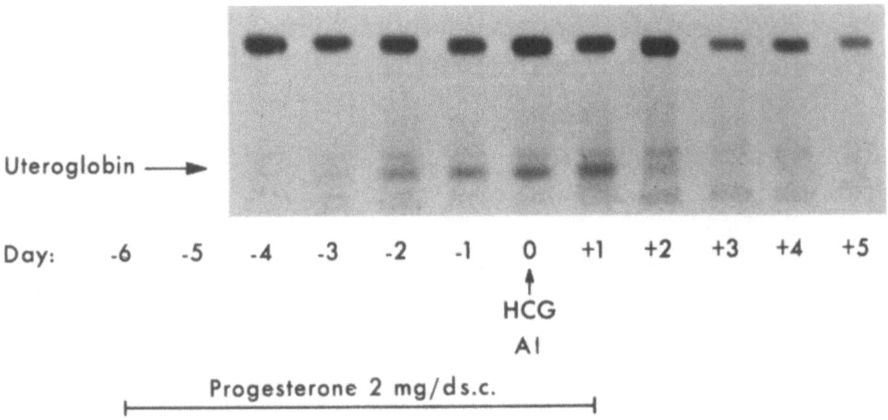

Fig. 4. Electrophoretic patterns of rabbit uterine secretion from day -4 up to and including day +5 of rabbits treated with progesterone before HCG-induced ovulation and intrauterine AI

These results clearly show an increase of the uteroglobin concentration during the preovulatory progesterone treatment and a diminished uteroglobin concentration in the uterus at the time fertilized ova arrived in the uterus of rabbits pretreated with progesterone.

The histological evaluation according to McPhail (1934) of the uterus of these rabbits is shown in Table 4. After HCG-induced ovulation and intrauterine AI of estrous rabbits the endometrium responded to endogenous progesterone from day +3 on. In progesterone-treated rabbits this response was observed after 3 days' treatment and was maintained for 8 days (Table 5).

Considered collectively, these experiments show that progesterone pretreatment of rabbits for 7 days before HCG-induced ovulation and insemination displayed two mechanisms of contraceptive action.

Progesterone treatment of estrous rabbits affected cervical mucus production and brought about a change in the physical properties of the mucus which resulted in inhibition of migration of spermatozoa through the cervix and prevention of fertilization (experiment 1). Even preovulatory treatment for 7 days with a relatively high dose of progesterone did not inhibit capacitation of spermatozoa, for fertilization took place after intrauterine insemination. Normal pregnancy follows after HCG-induced ovulation and intrauterine insemination of estrous rabbits. Notwithstanding the fact that fertilized ova were found in the oviducts of rabbits pretreated with progesterone, none of the animals was pregnant in the 1-mg dose group (experiment 2). It was shown that after arrival in the uterus the fertilized ova did not develop and degenerated within 2 days, and this was accompanied by a considerable loss of ova (experiment 3).

Table 4. Histological evaluation of the uterus of mature rabbits given 25 IU HCG and intrauterine AI and killed on days +3 through +6

Experiment 4

McPhail/score	day +3	day +4	day +5	day +6
	Number of rabbits			
0				
1				
2	2			
3	1	2	3	3
4				

Table 5. Histological evaluation of the uterus of mature rabbits given 25 IU HCG and intrauterine AI on day 0 and pretreated with 2 mg progesterone on days -6 through +1

Experiment 5

McPhail/score	day -4	-3	-2	-1	0	+1	+2	+3	+4	+5
	Number of rabbits									
0										
1	3									
2		2						1		3
3		1	2	2	3	5	3	1	3	
4			1	1				1		

Uteroglobin, a low-molecular-weight protein (MW 15,000), is produced in the uterus of rabbits from the 4th to the 7th day after mating (Arthur and Daniel 1972; Daniel 1976; Beier 1968, 1974, 1976). This protein might be of importance during blastocyst development in the uterus. Maurer and Beier (1976) could not confirm the earlier interpretation that uteroglobin was the "inducer and regulator" of blastocyst development under in-vitro conditions (cf. Krishnan and Daniel 1967). The uteroglobin fraction from pregnant uteri on day +6 did not support development of eight-cell embryos to the blastocyst stage in vitro. Unfractionated uterine proteins appeared to be more efficient (Maurer and Beier 1976). However, uteroglobin may act on the blastocyst coverings and influence their structure and involvement in expansion (Beier, Mootz, Fischer 1980).

It was assumed that the treatment with progesterone before ovulation would induce temporary uteroglobin production which, however, would be low at the time of arrival of fertilized ova in the uterus, notwithstanding the presence of corpora lutea. The presence of uteroglobin in uterine fluid during early pregnancy was confirmed in experiment 4. In the rabbits, pretreated with progesterone, uteroglobin secretion occurred during the treatment, and was absent at the time of arrival of fertilized ova in the uterus. Administration of progesterone to rabbits for about 3 days before ovulation as reported by Chang (1967) and Allen and Foote (1973) did not fully prevent nidation. As shown in our experiment a longer pretreatment, for about 8 days, was more effective. Vickery and Bennett (1969), however, treated rabbits with 0.1 mg chlormadinone acetate orally for 3 days before ovulation and vaginal insemination and pregnancy was prevented. This dose did not affect sperm migration and fertilization of ova. In contrast to the preovulatory treatment, postovulatory treatment with an excessively high dose of 100 mg/kg progesterone given for 3 days did not affect pregnancy, as shown by Mc-Lean-Morris (1973).

McCarthy et al. (1977) undertook a study to elucidate the location, time and nature of embryo mortality induced by preovulatory progesterone administration. Progesterone was injected into rabbits on days -2, -1, and 0 (the day of mating) at doses of 0.5, 1.0, and 1.0 mg respectively. Normal fertilization rates resulted, but embryonic death occurred by day 4. Embryos residing in progesterone-treated does for up to 3 days survived normally when transferred to normal recipients, whereas day -4 embryos from treated does exhibited a reduced ability to implant. Uterine fluid protein pattern was examined on days 1-7 after mating. Total uterine fluid protein levels were significantly greater in treated animals on day 4 than in controls. Uteroglobin secretion was significantly advanced in the treated animals by day 3. As shown in our experiments, a longer period of progesterone pretreatment may have advanced receptivity of the uterus by some days, inducing nonreceptivity at the time of arrival of blastocysts.

Schacht and Foote (1978) showed that normal 3- and 4-days-old rabbit blastocysts exposed for 24h to the uterus of preovulatory progesterone-treated 3 and 4 days post coitum rabbits, respectively, were irreversibly damaged. The interpretation is that an embryotoxic condition is created in the uterus by the progesterone-induced embryo-uterine asynchrony.

The duration of the phase during which blastocysts can implant is rather short (24h or less). Dickmann and Noyes (1960) showed implantation to occur after transfer of 5-day ova into the uterus of 5-day-pseudopregnant rats; there was no implantation, however, after transfer to 6-day-pseudopregnant rats.

The conclusions of Yochim (1975) based on the work of Kirby (1965, 1967, 1971) are that (a) the uterus normally is not a fit place for nidation of the blastocyst unless it is conditioned by ovarian hormones, and (b) the blastocyst will initiate the implantation given the proper tissue substrate. Adams (1971) reported on degeneration of 60-h morulae transferred to 3- or 11-day-pseudopregnant rabbits, whereas a high percentage of 60-h morulae survived and developed to early blastocysts in the ligated oviducts. This author suggested that degeneration of ova might be due to the existence of a deleterious factor rather than to a deficiency of a vital substance. The incubation time of fertilized ova in the oviducts of the rabbit is about 3 days, and further development to the blastocyst stage requires another 3 days in the uterine horns.

The finding by Eiler et al. (1977) of detectable estrogen and progesterone levels in the uterine luminal fluid 1-2 h after injection of luteinizing hormone or mating of rabbits and a steep increase between days 3 and 7 will be of importance in studying the influence of these luminal steroids on the physiology of the synchronous development of the embryo and the endometrium until nidation and the way in which preovulatory progesterone treatment affects these intraluminal steroid levels (see Yang et al., this volume).

With respect to the similarity between the woman and the rabbit in duration of the incubation period for ova in the fallopian tubes and the time of arrival of blastocysts in the uterus (Croxatto 1972; Cheviakoff 1976; Coutinho 1976), preovulatory gestagen treatment might, in addition to affecting the cervical mucus, also induce a hostile uterine environment in women and such enlarge the possibilities of progesterone and progestogen involvement in contraception (Rudel et al. 1965).

Acknowledgments. The author is grateful for the technical assistance of Mr. J. Odeh, Mr. E. Turpijn and Mr. C. v.d. Kerkhof, and for the advice of Drs. A. Coert.

References

Adams CE (1961) Artificial insemination in the rabbit. Commer Rabbit Assoc Techn Bull 1

Adams CE (1971) The fate of fertilized eggs transferred to the uterus or oviduct during advancing pseudopregnancy in the rabbit. J Reprod Fertil 26:99-111

Adams CE (1972) Induction of ovulation and A.I. techniques in the rabbit. Vet Rec 91:194-197

Allen MC, Foote RH (1973) The effect of progesterone on the early development of the rabbit embryo. Fertil Steril 24:220-226

Arthur AT, Daniel JC (1972) Progesterone regulation of blastokinin production and maintenance of rabbit blastocysts transferred into uteri of castrate recipients. Fertil Steril 23:115-122

Beier HM (1968) Uteroglobin: A hormone-sensitive endometrial protein involved in blastocyst development. Biochim Biophys Acta 160:289-291

Beier HM (1974) Oviducal and uterine fluids. J Reprod Fertil 37:221-237

Beier HM (1976) Uteroglobin and related biochemical changes in the reproductive tract during early pregnancy in the rabbit. J Reprod Fertil [Suppl] 25:53-69

Beier HM, Mootz U, Fischer B (1980) New aspects on the physiology of uteroglobin. In: Beato M (ed) Steroid induced uterine proteins. Elsevier/North-Holland, Biomedical Press, Amsterdam, New York, Oxford; pp 47-67

Boyarski LH, Bayliess H, Casida LE, Meyer RK (1947) Influence of progesterone upon the fertility of gonadotrophin treated female rabbit. Endocrinology 41:312-321

Chang MC (1967) Effects of progesterone and related compounds on fertilization, transportation and development of rabbit eggs. Endocrinology 81:1251-1260

Cheviakoff S, Diaz S, Carril M, Patritti N, Croxatto HD, Llados C, Ortiz ME, Croxatto HB (1976) Ovum transport in women. In: Harper MJK, Pauerstein CJ, Adams CE, Coutinho EM, Croxatto HB, Patow DM (eds) Ovum transport and fertility regulation. WHO Symposium, San Antonio, Texas. Scriptor, Copenhagen, pp 416-424

Coutinho EM (1976) Interference with ovum transport: Implications for fertility control. In:Harper MJK, Pauerstein CJ, Adams CE, Coutinho EM, Croxatto HB, Patow DM (eds) Ovum transport and fertility regulation. WHO Symposium, San Antonio, Texas. Scriptor, Copenhagen, pp 544-556

Croxatto HB, Diaz S, Fuentealba B, Croxatto HD, Carillo D, Fabres C (1972) Studies on the duration of egg transport in the human oviduct I. The time interval between ovulation and egg recovery from the uterus in normal women. Fertil Steril 23:447-458

Daniel JC (1976) Blastokinin and analogous proteins. J Reprod Fertil 25:71-83

Dickmann Z, Noyes RW (1960) The fate of ova transferred into the uterus of the rat. J Reprod Fertil 1:197-212

Eiler H, Bahr J, Nalbandov AV (1977) Ovarian steroids in the uterine lumen. I. Effect of LH injection and mating in rabbits. Biol Reprod 17:459-464

Haynes NB (1967) Influence of the uterine environment of the phagocytosis of spermatozoa. In: Laming GE, Amoroso EC (eds) Reproduction in the female mammal. Butterworth, London, p 500-509

Kendle KE, Telford JM (1970) Investigations into the mechanism of the antifertility action of minimal doses of megestrol acetate in the rabbit. Br J Pharmacol 4:759-74

Killingbeck MJ (1964) Influence of ovarian hormones on the inflammatory response of the uterus to infection . PhD dissertation, Nothingham

Kirby DRS (1965) The role of the uterus in early stages of mouse development. In: Wolstenholme GEW, O'Connor M (eds) Preimplantation stages of pregnancy. Little Brown, Boston, pp 325-339

Kirby DRS (1967) Ectopic autografts of blastocysts in mice maintained in delayed implantation. J Reprod Fertil 14:515-517
Kirby DRS (1971) Blastocyst-uterine relationship before and during implantation. In: Blandau RJ (ed) Biology of the blastocyst. University of Chicago Press, Chicago, pp 393-411
Krishnan RS, Daniel JC Jr (1967) "Blastokinin": inducer and regulator of blastocyst development in the rabbit uterus. Science 158:490-492
Maurer RR, Beier HM (1976) Uterine proteins and development *in vitro* of rabbit preimplantation embryos. J Reprod Fertil 48:33-41
McCarthy SM, Foote RH, Maurer RR (1977) Embryo mortality and altered uterine luminal proteins in progesterone-treated rabbits. Fertil Steril 28:101-107
McLean-Morris J (1973) Mechanisms involved in Progesterone Contraception and Estrogen Interception. Am J Obstetrics and Gynecology 117:167-176
McPhail MK (1934) The assay of progestin. J Physiol (Lond.) 83:145-156
Noble RP (1968) Electrophoretic separation of plasma lipo-proteins in agarose gel. J Lipid Res 9: 693-699
Nutting EF, Mares SE (1970) Inhibition of fertilization in rabbits during treatment with progesterone. Biol Reprod 2:230-238
Rudel HW, Martinez-Manautou J, Maqueo-Topete M (1965) The role of progestogens in the hormonal control of fertility. Fertil Steril 16:158-169
Schacht CJ, Foote RH (1978) Progesterone-induced asynchrony and embryo mortality in rabbits. Biol Reprod 19:534-539
Vickery BH, Bennett JP (1969) Mechanisms of antifertility action of chlormadinone acetate in the rabbit. Biol Reprod 1:372-377
Visser J de (1969) Development of an antifertilization test in rabbits. Arch Int Pharmacodyn 182: 407-408
Visser J de, Turpijn E (1974) Mechanisms of antifertility actions of progesterone in the rabbit. (Abstract and Oral presentation) Fifth Joint Meeting of the Dutch and British societies for endocrinology, August 1974
Yochim JM (1975) Development of the progestational uterus: Metabolic aspects. Biol Reprod 12: 106-133

Evidence for the Existence of Human Uteroglobin?

Carol Lambadarios

Department of Obstetrics & Gynaecology, University of Sheffield, Jessop Hospital for Women, Sheffield, U.K.

1 Implantation

Implantation is characterized by the distinction of an area of the uterus adjacent to the blastocyst from other uterine areas. It occurs at a time when the presence of a blastocyst is necessary to annul the luteolytic effect of the uterus in a number of species. Between ovulation and implantation the ovum passes along the oviducts encased in a mucopolysaccharide membrane – the zona pellucida. This layer gives the ovum a neutral charge, whereas after its dissolution the ovum becomes negatively charged. If the ovum has been fertilized it will have cleaved to form a morula and finally develop into a blastocyst. Attachment must occur at the blastocyst stage before further development can proceed. The blastocyst is differentiated into two parts: the embryonic pole, which gives rise to the embryo, and the trophoblast which attaches to the uterine wall and forms part of the placenta. In the human, the blastocyst remains small and passes through the uterine epithelium at one point and comes to lie in the subepithelial connective tissue, with the embryonic pole towards the uterine epithelium. Attachment takes place when the microvilli of the endometrium and the trophoblast interlink, and invasion occurs when the endometrium around the trophoblast disappears due to the breakdown of junctional complexes at the apices of the epithelial cells. This breakdown could be due to the secretion of proteolytic enzymes by the trophoblast, the production of which is under the influence of ovarian steroids (Finn and Porter 1975).

The mechanism for directing the small unexpanded blastocysts to certain areas of the uterus is not fully understood, but there must be some attraction at this point and it is unlikely that the blastocyst plays an active part in this process. The stimulus therefore probably comes from the uterine epithelium or endometrium.

2 Cyclic Changes in the Endometrium

During the menstrual cycle striking changes occur in the epithelium of the uterus. In the early proliferative phase the glands are small, with a low columnar epithelium and a few ciliated cells. Mitosis is common and lipid droplets are numerous. Around the gland openings are numerous ciliated cells. By the later proliferative phase the glands have become tortuous, the microvilli on the cells are well developed, projecting into the lumen, and the Golgi complex is also more developed. Just after ovulation the glands are even more tortuous and the epithelial cells contain irregular, small, subnuclear vacuoles of glycoprotein. By the mid-luteal phase secretions are being expelled into the lumen. There is glycogen throughout the cytoplasm and secretory vesicles accumulate beneath the surface plasma membrane. The mitochondria and Golgi exhibit maximal activity. Using the scanning electron microscope one can see ciliated cells but the microvilli are shorter and less numerous. Some cells have apical protrusions rich in glycogen, enzymes and mucopolysaccharides. By day 24 all the secretions are out of the glands and lying on the surface of the endometrium. The number of secretory vesicles and ribosomes has now decreased and the Golgi has regressed. After menstruation the cycle of product build-up and secretion repeats (Cooke and Lambadarios 1974; Ferenczy and Richard 1973, 1974; Ferenczy 1980; Wynn 1979).

3 Animal Evidence

Returning to the implantation stimulus: data obtained from investigations carried out on small mammals (e.g. rat, rabbit and mouse) indicate the existence of a specific protein present at the time of implantation. This protein, called uteroglobin or blastokinin, has a molecular weight of approximately 14 500 (McGaughey and Murray 1972), it is a globular molecule containing few aromatic amino acids. It has an isoelectric point of 5.4 and exists as two nearly identical subunits (each containing 70 amino acid residues) linked by two disulphide bridges and other covalent interactions (Nieto et al. 1977). Its function may be related to the hormonal interaction between the maternal tissue and the blastocyst because uteroglobin binds ovarian steroids, especially progesterone (Arthur et al. 1972; Beato et al. 1980). Progesterone binding to uteroglobin is enhancd by SH reducing agents but there is no difference in the equilibrium dissociation constants under both conditions ($K_D = 4.1 \times 10^{-7}$M) (Fridlansky and Milgrom 1976). Oestradiol also binds to uteroglobin but at a low affinity and binding is unaffected by SH reducing agents.

At the time when the blastocyst is free, before implantation at 4-6 days post coitum in the rabbit, uteroglobin represents 40%-50% of the total protein content of the uterine fluid. It is produced after the oestrogen peak but this is not a prerequisite as its production is controlled by progesterone (Beier 1967, 1968, 1974). Uteroglobin can be detected in rabbit fallopian tube cells during oestrus (Kirchner 1976), appearing as granules towards the base of cells. The cells become filled and droplike protrusions of apical cytoplasm appear, suggesting apocrine extrusion. All stages of filling and extrusion are visible during the entire preimplantation period. During oestrus, synthesis of uteroglobin in the uterine cells becomes advanced enough for extrusion to have begun or about to begin. Uteroglobin appears in supranuclear regions of the cells then later in the entire cytoplasm. There is an uneven distribution of uteroglobin-producing cells in the endometrium, glands adjacent to the myometrium and cells of cavum epithelium containing uteroglobin. It has also been detected in blastocyst cells 7 days post coitum, in both the endoderm and the ectoderm. Earlier work showed maximal uteroglobin production on the 4-6th day post coitum (Beier 1974; Kirchner 1972, 1974; Johnson 1972) and it was seen to pass through discrete channels in the blastocyst covering, invading the cavity. When this protein was isolated, then added to cultured blastocysts, it promoted growth and incorporation of amino acid and uridine (El-Banna and Daniel 1972). Its production can be induced by steroids containing a C19 methyl group (Arthur and

Chang 1974) and this could be a mode of action of certain contraceptive preparations, causing asynchrony between embryonic development and an altered uterine environment. When antiserum to rabbit uteroglobin was raised in chickens (Garcea et al. 1977) then injected into rabbits at varying itervals post coitum, there was a high perinatal mortality and a reduced weight of the stillborn, especially in rabbits injected 4 days post coitum. Some authors have reported the presence of uteroglobin in places other than the female reproductive tract, e.g. rabbit lung tissue, the male genital tract, (Noske and Feigelson 1976) bronchial and trachial tract (Beier et al. 1975; Beier et al. 1978; Feigelson et al. 1977).

4 Investigations in the Human

Initial analyses of human uterine secretion proteins were reported by Beier et al. (1970) and Beier and Beier-Hellwig (1973). Wolf and Mastroianni (1975) have reported four characteristic proteins in human uterine fluid, but only one was claimed specific to the secretory phase. Daniel (1973) published that protein in human uterine secretions, around day 23, moved electrophoretically the same as rabbit uteroglobin, it should constitute 20%-30% of the total uterine protein and by Ouchterlony double diffusion against a pure, monospecific goat antiserum to rabbit uteroglobin, it produced a precipitation arc without spurs. Human progestational endometrial extracts, uterine and oviducal fluids and seminal plasma all cross-react with an antiserum to rabbit uteroglobin (Noske and Feigelson 1976). Thus there appears to be tentative evidence for the existence of a human equivalent to rabbit uteroglobin, to prove or to disprove this further investigations have been carried out at this laboratory. (Review on uterine proteins see Aitken 1979).

4.1 Choice of Patient

Patients attending the hospital for gynaecological complaints were screened for a clear history, i.e. lack of previous complaint and absence of infertility or disease. Women attending for sterilization were preferred, but in general those with no abnormalities were included. Histological investigations were carried out to eliminate any tissue not falling into the "normal" category, i.e. those showing advanced or retarded development for the stated day of the cycle, presence of carcinoma, etc. Patients whose records indicated they were taking oral contraceptives were immediately eliminated from the group, thus drastically reducing the quantity of available tissue.

4.2 Collection of Material

The endometrium obtained by curettage was placed in sterile saline then set up in organ culture (Trowell 1959). Organ culture was employed in an attempt to exclude traces of blood, often seen when the uterine washing procedure was used; any blood present in or on the tissue would be removed by saline washing. The explants were supported on lens paper on a grid over a central well of culture medium. The medium was supplemented with penicillin, streptomycin, insulin and initially calf serum, although in later experiments the serum was omitted to reduce the albumin contamination. This regime would enable products to pass from the tissue into the medium as they were secreted. The cultures were gassed and incubated at 37°C for 16 h to remove any remaining blood or damaged tissue. This medium was discarded and replenished with fresh. After a further 72 h the medium was removed and stored frozen. When 10-15 samples from each phase of the cycle had been collected, the bulked samples were dialysed against distilled water and then lyophilised or concentrated by ultrafiltration.

4.3 Chromatography on Sephadex

Previous workers on rabbit uteroglobin (Murray et al. 1972) used Sephadex G200 to prepare the material. In the investigations using secreted human products a G100 column proved as effective as a larger G200 column. The void volume was estimated and two marker proteins of known molecular weight (chymotrypsinogen 25 000 and ribonuclease A 13 700) were used to set location limits. Phosphate buffer was used to elute the proteins.

Samples of the unused (fresh) medium and medium in which endometrium from various stages of the cycle had been cultured, were then applied and eluted. This was repeated to ensure reproducibility. It can be seen in Fig. 1 that the fresh and proliferative tissue medium showed no peaks in the region of the marker proteins (indicated by arrows). The medium from the mid-cycle tissue could possibly contain some protein in this region since the bulk of the peak is outside the markers, but a small definite peak can be seen in medium exposed to secretory tissue. The large peak seen in all cases and eluted around the void volume is probably the albumin from the calf serum.

4.4 Affinity Chromatography

Since G100 only estimates molecular size, an affinity gel (Con-A Sepharose 4B) was employed. This gel is a group-specific adsorbent for polysaccharides and glycoproteins. The binding site is specific for α-D-mannosyl, α-D-glycosyl and sterically similar residues, so that elution is by satururation with α-methyl-D-glucoside. After flushing the column with phosphate buffer the samples were applied in turn. A further volume of buffer was used to flush away non-adsorbed material off the column, then a linear gradient of α-methyl-D-glucoside was employed and finally the gel reconstituted with an acetate buffer containing Ca^{2+}, Mn^{2+} and Mg^{2+}. Samples were collected, their optical densities read and protein content determined by the Lowry method (Lowry et al. 1951), using a bovine serum albumin (BSA) standard. The optical densities were used for the profile of the gradient and a standard curve using various concentrations of α-methyl-D-glucoside was constructed to check the gradient. Samples of fresh medium, mid-cycle and secretory tissue medium were applied to the column and eluted as described. Each sample was repeated to ensure reproducibility. Figure 2 shows the gradient profile and chromatograph of the fresh medium. Three major peaks are observed: one is near fractions 10-15 (before the gradient has started), probably albumin, and there are also peaks at fractions 43-46 and 60-72.

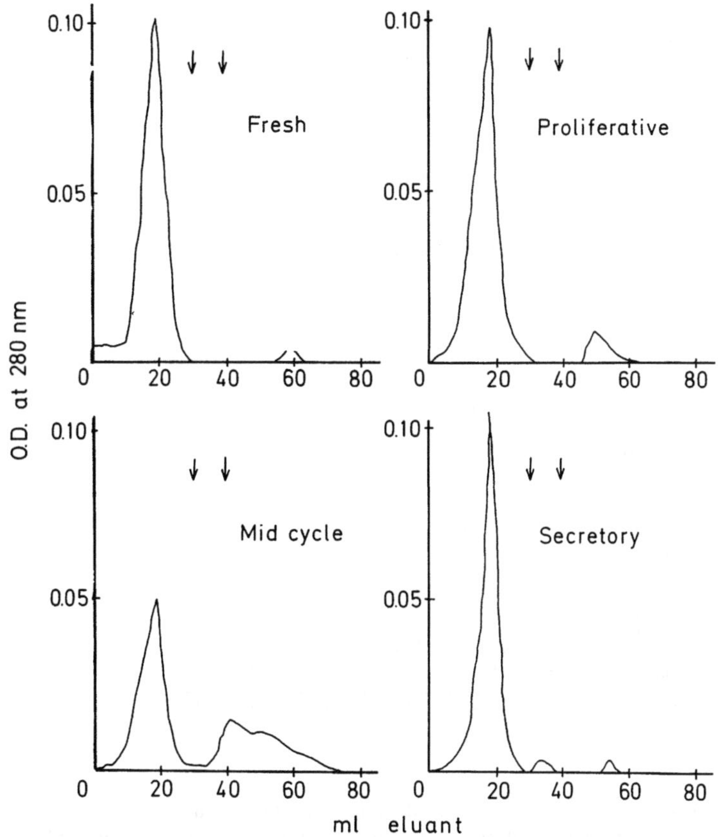

Fig. 1. Chromatography of aliquots from the three phases of the menstrual cycle (compared with unused medium) on Sephadex G100. The *arrows* indicate marker proteins of molecular weights 25 000 and 13 700

Figure 3 shows the chromatographs of the mid-cycle and secretory tissue medium. Again the first peak is very large and found in fractions 10-15. The second peak around fractions 42-50 (I) has increased, especially in the secretory sample. However the third peak in fresh medium (60-72) has disappeared, probably metabolised but in both cases there are peaks at the concentrated end of the gradient (100-120). The small peaks (II) are not seen in fresh medium. Thus it was decided to concentrate on secretory tissue samples only. These peaks were also observed in samples cultured without calf serum.

4.5 Protein Precursor Incorporation

Con-A-Sepharose is also an affinity gel for pure polysaccharides, so protein production by the tissue was examined. After the initial 16-h equilibration period samples of secretory endometrium were fed labelled protein precursors, i.e. ^{3}H-leucine and ^{14}C-glucosamine, then the medium was collected and treated as previously. The aliquots off the

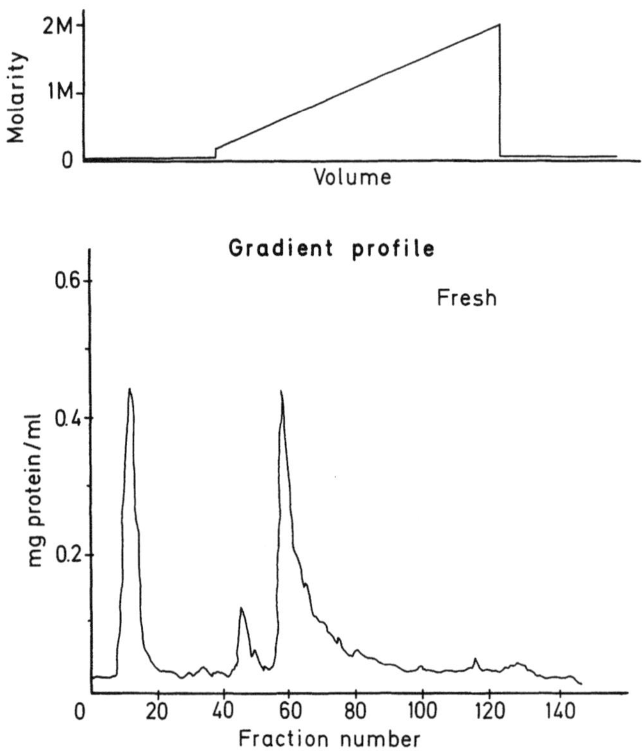

Fig. 2. Affinity chromatography of the unused medium employing a linear 0.05-2.0 M α-methyl-D-glucoside gradient for elution. The *top diagram* illustrates the gradient and buffer profiles

affinity gel were then counted. Figure 4 shows a profile of incorporated protein precursors. Labelled protein precursors appear in the same peaks as shown by unlabelled secretory tissue medium. Therefore, the bound material is glycoprotein, not pure protein or polysaccharide. This was particularly interesting since previous work (Lambadarios et al. 1976) (Table 1) showed that the secretory products of the endometrium were altered quantitatively and possibly qualitatively by synthetic progestagens; this is one mode of action of contraceptive compounds. Now these same proteins are found around the time of implantation.

4.6 Sample Purification

Since the albumin present in the culture medium produced such a large peak in early samples, the collection of adequate amounts of the other peaks was impossible without column overloading. Thus a Sephadex G50 column was employed to collect material of molecular weight less than 40 000. As before the collection limits were determined by marker proteins and elution was by phosphate buffer. The purified samples of secretory medium were concentrated by ultrafiltration to give material in a range of 40 000-

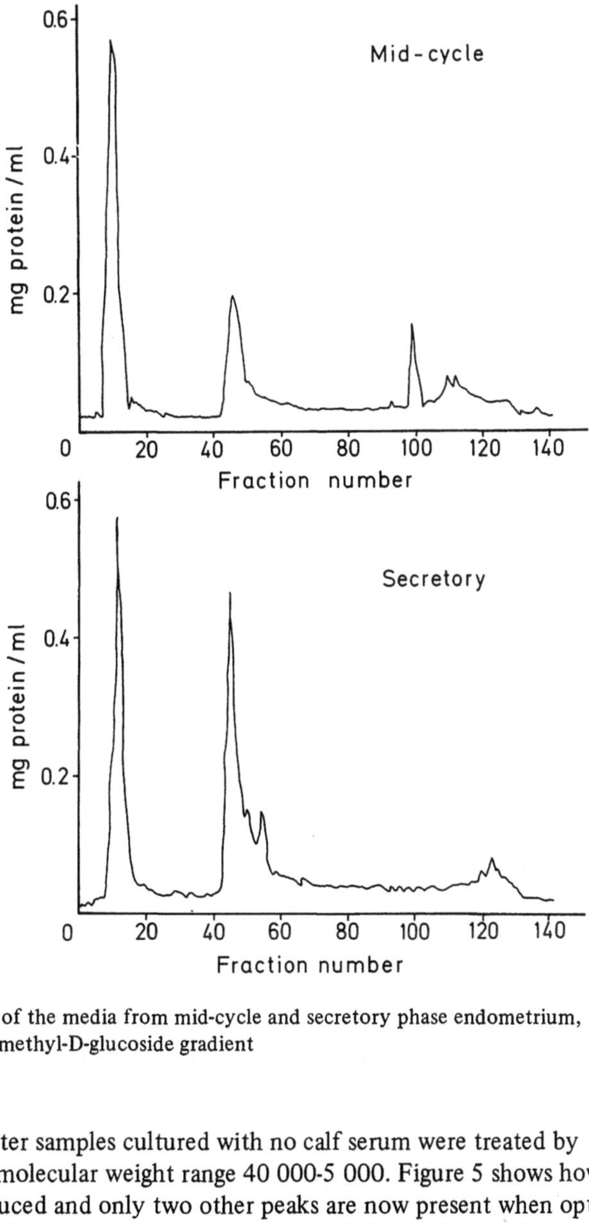

Fig. 3. Affinity chromatography of the media from mid-cycle and secretory phase endometrium, employing a linear 0.05-2.0 M α-methyl-D-glucoside gradient

10 000 molecular weight. Later samples cultured with no calf serum were treated by ultrafiltration to collect the molecular weight range 40 000-5 000. Figure 5 shows how the initial peak is greatly reduced and only two other peaks are now present when optical density (O.D.) is measured. One peak is at the beginning of the gradient and the other at the end, i.e. one is eluted by a weak sugar and the other by a strong sugar solution. When tissue was homogenized after culture, for secretory tissue a similar profile was observed as that for culture medium.

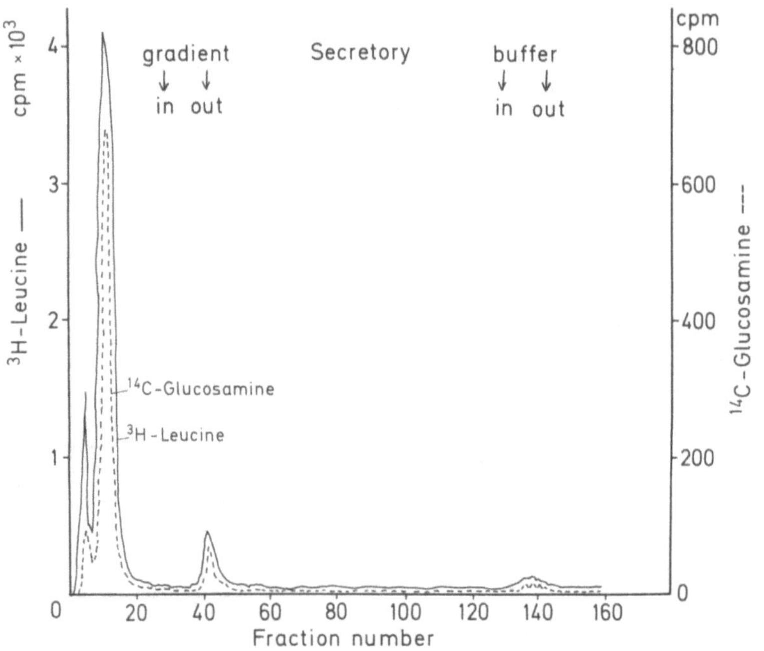

Fig. 4. Affinity chromatography of medium from secretory tissue, employing the same linear sugar gradient. In this instance labelled protein precursors (^{3}H-leucine and ^{14}C-glucosamine) have been added to the culture medium to demonstrate protein/glycoprotein synthesis

4.7 Steroid Binding

Since progesterone and estradiol binding are features of rabbit uteroglobin (Arthur et al. 1972; Beato et al. 1980), aliquots of each peak off the gradient (I and II) were tested for steroid binding. Aliquots of each containing 200 μg protein were incubated with 4 or 44 nmol of either ^{3}H-progesterone or ^{3}H-oestradiol. After equilibration for 16 h at 4°C the "bound" and "free" were separated using dextran-coated charcoal. The "free" was counted and results corrected to 1 mg protein/ml. The first peak off the gradient bound 20% progesterone and 23% oestradiol. whilst the second peak bound 72% progesterone and 50% oestradiol. Due to the minute amounts of material available and the problems associated with continued freezing and thawing, very little material was available after each run so, to date, no Scatchard plots have been attempted, nor have binding inhibition studies.

4.8 Ion-Exchange Chromatography

To date, there is evidence of a specific secretory (possible mid-cycle) phase material which is glycoprotein. After purification it appears to consist of two components by affinity chromatography, one eluted by a weak sugar solution (I) and the other by a strong sugar solution (II). If each of these components is tested further by ion-exchange chromatography, the purity of each peak should be demonstrated. For this a column

Table 1. Incorporation of ^{14}C-glucosamine and ^{3}H-Leucine (d/min/mg protein) onto microsomes and into extracellular glycoprotein of human secretory endometrium in organ culture, under the influence of added steroids (10^{-9} M oestradiol and progesterone, and 10^{-6} M MPA)

	^{14}C-Glucosamine			
	Incubation time (h)			
	1	2	4	8
Microsomes				
Oestradiol + progesterone	1 900	2 500	3 500	7 500
Oestradiol + progesterone + MPA	1 500	1 700	2 000	3 000[a]
Glycoprotein				
Oestradiol + progesterone	500	1 100	1 400	4 500
Oestradiol + progesterone + MPA	450	600	700	1 500[b]
	^{3}H-Leucine			
	Incubation time (h)			
	1	2	4	8
Microsomes				
Oestradiol + progesterone	15 250	38 500	55 000	82 250
Oestradiol + progesterone + MPA	11 750	35 750	57 900	77 750
Glycoprotein				
Oestradiol + progesterone	8 000	9 150	19 000	34 700
Oestradiol + progesterone + MPA	10 000	9 500	16 750	23 000[a]

Significantly different from control $^{a}P<0.01$; $^{b}P<0.05$

of DE52 equilibrated with distilled water was prepared. Using a discontinuous sugar gradient samples of both peaks were prepared by affinity chromatography and then concentrated by ultrafiltration. These were then applied to the DE52 column, flushed with distilled water, eluted off using a linear slaine gradient and reflushed with distilled water. Collected samples had their optical denisty (O.D.) read at 280 nm.

The first fraction, I (eluted off the affinity gel by a weak sugar), was eluted off the ion-exchange column in fractions 130-180, i.e. 0.26 M saline, and consisted of one peak (Fig. 6). The spikes at the end of the gradient could be due to contamination from using a discontinuous gradient for the preparation of peaks I and II. The second fraction, II (eluted off the affinity gel by a strong sugar), appears to consist of two components (Fig. 7): one component in fractions 145-155, i.e. 0.28 M saline, and the other in fractions 164-185, i.e. 0.34 M saline. Here contamination was greatly reduced and both components appear similar.

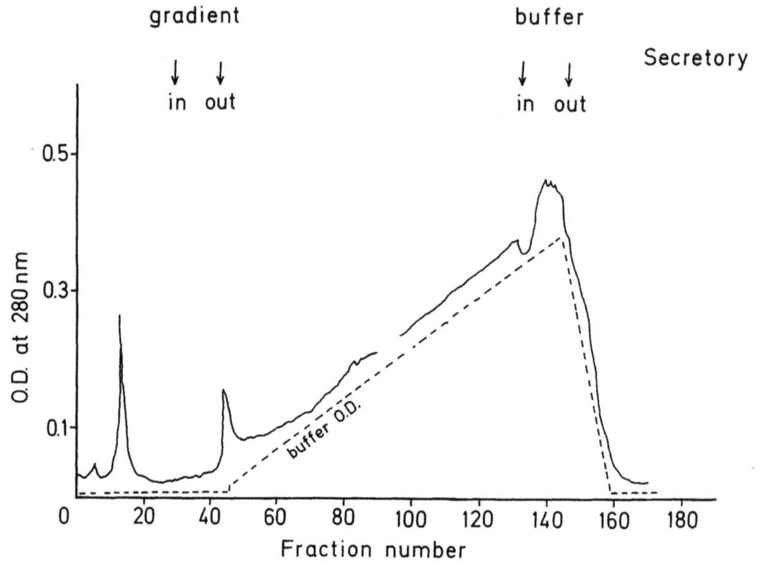

Fig. 5. Affinity chromatography of medium from secretory endometrium, after purification on a Sephadex G50 column. The same linear sugar gradient as previously used was again employed

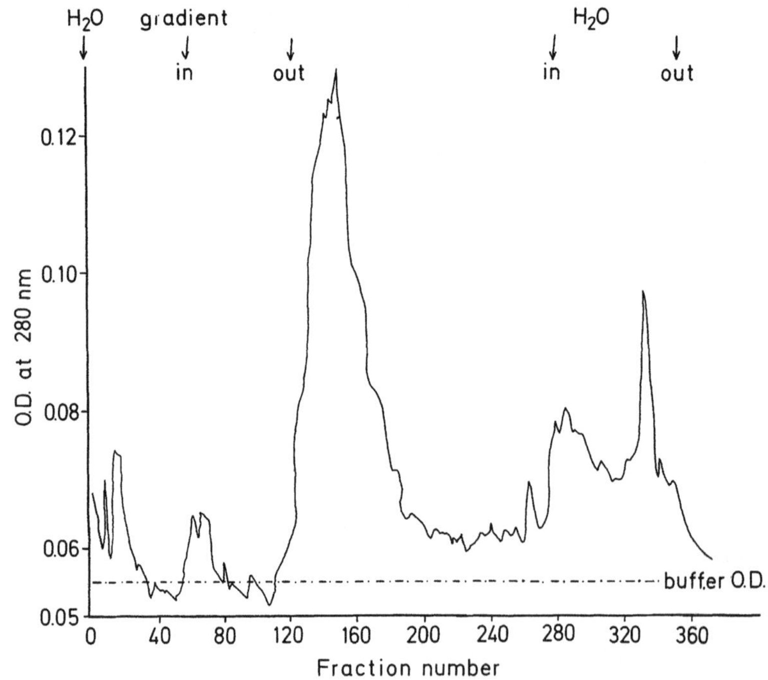

Fig. 6. Ion-exchange chromatography on DE52 of peak I off the affinity gel, employing a 0.05-1.0 M saline gradient

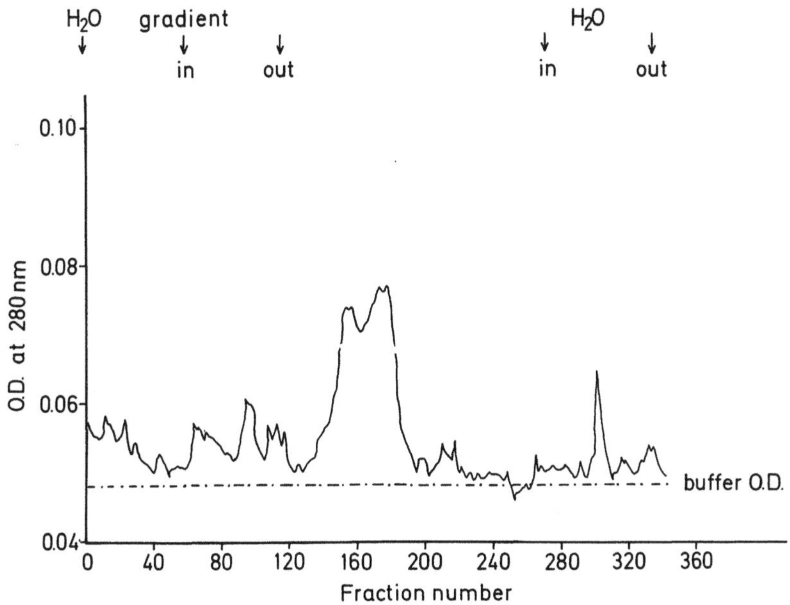

Fig. 7. Ion-exchange chromatography on DE52 of peak II off the affinity gel, employing a 0.05-1.0 M saline gradient

4.9 Electrophoresis

On 7.5% polyacrylamide gel using a Tris-glycine buffer at pH 9.5 and staining for glycoprotein, mid-cycle, early and late secretory tissue medium produced distinct single bands of similar Rf and moving more slowly than the two staining bands produced by fresh medium. Proliferative tissue medium produced no staining bands, but protein concentrations were very low. This was also observed in 15% gels although Rf values were lower and in this case peaks I and II were also run, producing bands in a similar area to those in secretory tissue. Isoelectric focusing of the same samples was carried out on a commercially prepared Ampholine PAG plate pH 3.5-9.5. This failed to produce any precipitation lines on fixing with trichloroacetic acid or staining with protein or glycoprotein stains. The protein content must have been too low, since plasma produced precipitation lines.

4.10 Biological Tests

These tests were carried out in collaboration with Dr. Heap (ARC Institute of Animal Physiology, Babraham).

Samples of peak I and II were collected and lyophilised. Peak I contained four times the protein of peak II. One milligram of each lyophilised peak was added to 3 ml medium (without leucine) then supplemented with ^{3}H-leucine and unlabelled leucine. To each sample of medium was added a section of pig blastocyst ready to implant. After incubation the ^{3}H-leucine incorporation into blastocyst protein and into proteins pre-

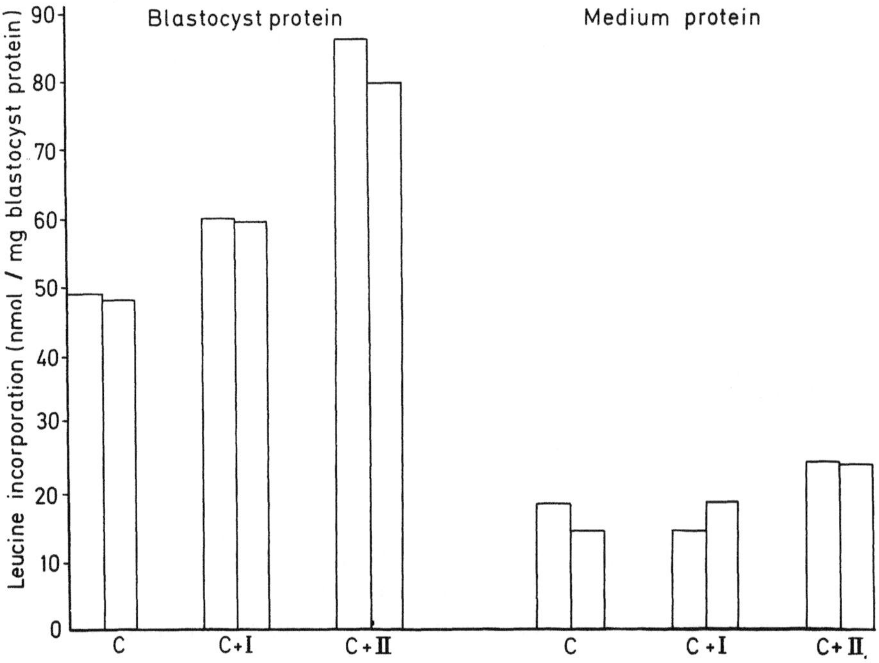

Fig. 8. ^{3}H-Leucine incorporation (nmol/mg blastocyst protein) into pig blastocyst protein and into proteins found in the culture medium in the presence of peaks I and II. *C*, plain culture medium; *I*, addition of peak I; *II*, addition of peak II

sent in the medium was counted. Figure 8 shows how peak I and II had little effect on proteins in the medium but peak I slightly increased amino acid incorporation by pig blastocyst into its proteins and peak II (although containing only one quarter of the protein) had a dramatic effect in increasing ^{3}H-leucine incorporation into pig blastocyst proteins. Also in the presence of peak II the number of protein bands detected at Babraham, by electrophoresis, was greatly increased. Thus the human material collected appears to have biological activity.

4.11 Spectral Analysis and Density Gradient Centrifugation

Samples of peak I and II were collected off the affinity gel and protein determinations carried out (Lowry et al. 1951). Peak I contained 28 μg/ml and peak II 52 μg/ml. Each sample was scanned on the spectrophotometer between 220 nm and 345 nm wavelength using distilled water as a reference. It was noted that there was no absorption peak at 280 nm but a peak at 265 nm. This does not disprove the absence of protein since a large peak at 265 nm could mask the 280 nm peak. The loss of the usual 280 nm peak could be due to lengthy preparation or to bound progesterone. Employing continuous sucrose gradient centrifugation (5%-10% or 5%-20%) to estimate the molecular weight of the peaks, similar results to those for rabbit uteroglobin were obtained. The estima-

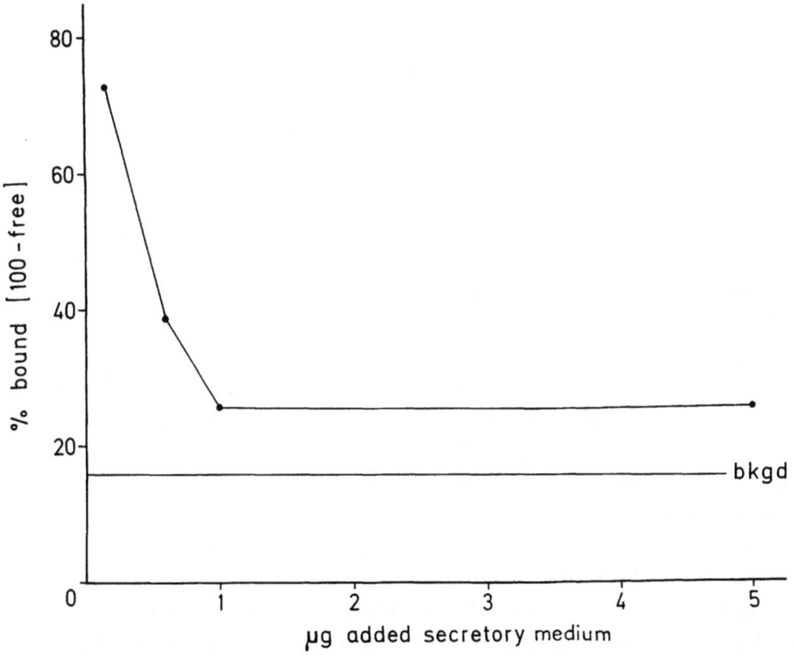

Fig. 9. Addition of secretory tissue medium in varying concentrations to a rabbit antigen-antibody system. Antiserum is 1/50, and bound and free are separated by dextran-coated charcoal. 0.015-5 μg of lyophilized secretory medium has been added to an overnight incubation at room temperature (*bkgd* binding to dextran-coated charcoal by labelled antigen)

ted molecular weights were: peak I 15 575 ± 2 974 and peak II 16 670 ± 3 242. Individual estimations on the two components of peak II have not been carried out due to lack of material.

4.12 Radioimmunoassay

Iodination of both peaks has been unsuccessful to date although various techniques have been tried, e.g. chloramine T, lactoperoxidase and the Bolton and Hunter reagent. Therefore, an alternative method has been used. Rabbit uterine washes have been collected 5 days post coitum, the washes lyophilised and eluted off G200. A peak was observed around the 15 000 molecular weight region, similarly an aliquot was applied to the affinity gel where both peaks I and II were observed. Thus the rabbit uterine wash and the culture medium from human secretory endometrium appear similar. The rabbit antigen also proved difficult to iodinate, but on the occasion it did it iodinate poorly, and had some binding to impure donated antiserum (Bullock). Using this poor system it was demonstrated that when human secretory medium was added to the antigen-antibody system the percentage binding fell rapidly, indicating a possible cross-reaction (Fig. 9).

5 Discussion

Thus it has been demonstrated that using gel permeation chromatography there is a peak specific to medium from the secretory phase tissue. It has a molecular weight of approximately 15 000 and could also be present in medium from mid-cycle tissue but at a much lower concentration. The material binds to Con-A-Sepharose, an affinity gel for glycoproteins. When eluted, the peak at the start of the sugar gradient is enlarged in the presence of secretory tissue and small peaks are seen which have no complement in other samples of medium. Incorporation of protein precursors proved these peaks to be glycoprotein, not pure polysaccharide. Also this material, specific to secretory endometrium, appears to contain the same glycoproteins as those affected quantitatively and possibly qualitatively by synthetic progestagens.

There is no reason why the endometrial protein uteroglobin should be drastically different in rabbits and humans. Differences may occur in a few amino acids or the peptide matching across the chains. Therefore, the criteria known for animal uteroglobin could be used to isolate and identify the human equivalent.

After removal of high-molecular-weight material, especially albumin, the peaks have been cleaned up but two still remain. The first (I), eluted by a weak sugar solution, has an equal affinity for progesterone and oestradiol ($\simeq$20%), and the second (II), eluted by a strong sugar solution, has a greater affinity for progesterone (70%) than oestradiol (50%). Although two distinct peaks are observed by affinity chromatography, there appear to be three components after further ion-exchange chromatography. Peak I consists of one component and peak II of two very similar components, although all three components are eluted by 0.26-0.34 M saline. These components are present in minute quantities since they are undectable by isoelectric focusing and just detectably by polyacrylamide gel electrophoresis, the medium from secretory tissue and the two peaks (I and II) moving similarly.

Both peaks increase the amino acid uptake by pig blastocyst tissue in vitro. The two -component peak (II) greatly enhances uptake; this increase is even more startling when one considers that peak II contains only one quarter of the protein content of peak I, and produces more bands, by electrophoresis, in the blastocyst protein. Thus these glycoproteins found around the time of implantation appear beneficial to blastocyst development.

Spectral analysis of both peaks I and II provided some puzzling facts. Even though small amounts of protein were present (detected by Folin's reaction) an absorption peak at 280 nm was not observed, but a peak at 265 nm. This does not prove the absence of protein; there could be an unusual sugar present, some nucleotide associated with these peaks, or it could be due to progesterone bound to the molecules. Since rabbit uteroglobin is low in tyrosine and tryptophan it is probable that human material will be similar and thus have a reduced 280 nm peak.

When protein content of identical aliquots of the peaks was estimated by Folin's reaction, optical density at 280 nm and biuret reagent, it was observed that the estimations varied widely. The Folin estimation (depending on the tyrosine content) gave similar results for peak I to the optical density method (which was dependent on the aromatic amino acid content), but an eight-fold lower result for peak II. The biuret test (depending on peptide linkages) gave much higher results for both peaks: 30 times higher for peak I and 200 times higher than Folin's for peak II. These peaks appear to vary in amino acid content, peak II being lower in tyrosine than peak I and both peaks low in aromatic amino acids.

Density gradient centrifugation provided evidence that these two peaks have molecular weights around 15 000, and other evidence has been gathered to indicate that these peaks cross-react to some degree in a rabbit antigen-antibody system.

6 Conclusions

Human uteroglobin appears to exist and has similar basic properties to those of animal uteroglobin. It is possible, that the three glycoproteins found in peaks I and II are the complete molecule plus its two subunits, acting as steroid carriers between the endometrium and the blastocyst. The two subunits, peak II, would then have a greater affinity for progesterone and Con-A-Sepharose than the whole molecule, the increased affinity being due to the opening of the disulphide bonds by lyophilisation, repeated chromatography or progesterone already bound.

Another interesting suggestion put forward by Dr. Heap and co-workers (personal communication and this volume) is that specific uterine glycoproteins not only carry steroids to the blastocyst but also transfer simuli from the blastocyst back to the maternal tissue. The latter stimuli act as signals for the recognition of pregnancy. In the pig the blastocyst metabolises basic steroids to oestrogens, particularly oestrone, which could be the recognition signal. Oestradiol is also converted to oestrone sulphate in pig endometrium and since oestrone sulphate is less metabolically active than oestradiol, this conversion could protect the pregnancy from the maternal oestradiol action. This conversion of oestradiol to oestrone also occurs in the human (Gurpide and Welch 1969).

The collection of further samples of culture medium and their consequent lack of exposure to multiple freezing and lyophilisation by using ultrafiltration, may provide samples with a higher protein content, thus enabling peaks to be collected for amino acid analysis, steroid binding, electrophoresis and isoelectric focusing. Even more important is the isolation of the three glycoproteins in large enough quantities to be injected into rabbits to raise antibody. Likewise the material needs to be iodinated by some method or even the antibody itself, so that these specific glycoproteins can be easily measured at all times throughout the cycle and steroid inhibition can also be monitored.

It seems unlikely that the recently described human uterine specific proteins with molecular weights around 50 000 (Sutcliffe et al. 1980) and 51 000 (Shapiro and Forbes 1978) are identical to the glycoprotein described here. The possibility however, that those proteins could be separated into subunits of molecular weight ranges of 15 000 to 20 000 is not ruled out. Immunological studies may help in further identification of the specific uterine protein components, consequently direct antibodies are needed.

If as in the rabbit, antiserum to uteroglobin can prevent pregnancy and if its effect is reversible, then similar antibodies to human uteroglobin may provide a new long-term contraceptive. In any event further studies on the secretory products of the human endometrium will help elucidate the events of implantation, a process which needs to be more fully understood.

References

Aitken RJ (1979) Uterine proteins. In: Finn AC (ed) Oxford reviews of reproductive biology vol. 1; Oxford University Press, Oxford, pp 351-382

Arthur AT, Chang MC (1974) Induction of blastokinin by oral contraceptive steroids: Implications fór fertility control. Fertil Steril 25:214-221

Arthur AT, Cowan BD, Daniel JC (1972) Steroid binding to blastokinin. Fertil Steril 23:85-92

Beato M, Saavedra A, Puigdomenech P, Tancredi T, and Temussi PA (1980) Progesterone binding to uteroglobin. In: Beato M (ed) Steroid induced uterine proteins. Elsevier/North-Holland Biomedical Press, Amsterdam New York Oxford, pp 105-119

Beier HM (1967) Veränderungen am Proteinmuster des Uterus bei dessen Ernährungsfunktion für die Blastocyste des Kaninchens. Verh Dtsch Zool Ges 31:139-148

Beier HM (1968) Uteroglobin: a horm ͏ ͏ nsitive endometrial protein involved in blastocyst development. Biochim Biophys Acta 160:289-291

Beier HM (1974) Oviducal and uterine Fluids. J Reprod Fertil 37:221-237

Beier HM, Beier-Hellwig K (1973) Specific secretory protein of the female genital tract. Acta Endocrinol [Suppl] (Kbh) 180:404-425

Beier HM, Bohn H, Müller W (1975) Uteroglobin-like antigen in the male genital tract secretions. Cell Tissue Res 165:1-11

Beier HM, Kirchner C, Mootz U (1978) Uteroglobin-like antigen in the pulmonary epithelium and secretion of the lung. Cell Tissue Res 190:15-25

Beier HM, Petry G, Kühnel W (1970) Endometrial secretion and early mammalian development. In: Gibian H, Plotz EJ (eds) Mammalian reproduction. Springer, Berlin Heidelberg New York, pp 264-285

Cooke ID, Lambadarios C (1974) The endometrium. Clin Obstet Gynaecol 1:369-393

Daniel JC (1973) A blastokinin-like component from the human uterus. Fertil Steril 24:326-328

El-Banna A, Daniel JC (1972) The effect of protein fractions from rabbit uterine fluids on embryo growth and uptake of nucleic acid and protein precursors. Fertil Steril 23:105-114

Feigelson M, Noske IG, Goswami AK, Kay E (1977) Reproductive tract fluid proteins and their hormonal control. Ann NY Acad Sci 286:273-286

Ferenczy A (1980) Morphological effects of exogenous gestagens on abnormal human endometrium. In: Dallenbach-Hellweg G (ed) Functional morphologic changes in female sex organs induced by exogenous hormones. Springer, Berlin Heidelberg New York, pp 101-110

Ferenczy A, Richart RM (1973) Scanning and transmission electron microscopy of the human endometrial surface epithelium. J Clin Endocrinol Metab 36:999-1008

Ferenczy A, Richart RM (1974) Female reproductive system: Dynamics of scan and transmission electron microscopy. Wiley, New York

Finn C ,Porter DG (1975) The uterus. In: Finn C (ed) Handbooks in reproductive biology, vol. 1. Elek Science, London, pp 57-73

Fridlansky F, Milgrom E (1976) Interaction of uteroglobin with progesterone, 5 α-pregnane-3,20-dione and estrogens. Endocrinology 99:1244-1251

Garcea N, Campo S, Caruso A, Porcelli F (1977) Effect of antiblastokinin on rabbits pregnancy. Experientia 33:1173-1175

Gurpide E, Welch M (1969) Dynamics of uptake of oestrogens and androgens by human endometrium. J Biol Chem 244:5159-5169

Johnson MH (1972) The distribution of a blastokinin-like uterine protein studied by immune fluoresence. Fertil Steril 23:929-939

Kirchner C (1972) Immuno histologic studies on the synthesis of a uterine specific protein in the rabbit and its passage through the blastocyst coverings. Fertil Steril 23:131-136

Kirchner C (1976) Uteroglobin in the rabbit I. Itracellular localization in the oviduct, uterus and preimplantation blastocyst. Cell Tissue Res 170:415-424

Lambadarios C, Hastings C, Abo-Darub J, Cooke ID (1976) Steroid effects on human endometrial glycoprotein biosynthesis. J Reprod Fertil 46:383-394

Lowry OH, Rosebrough NJ, Farr AL, Randall RJ (1951) Protein measurements with the folin-phenol reagent. J Biol Chem 193:265-276

McGaughey RW, Murray FA (1972) Properties of blastokinin: amino acid composition, evidence for subunits and estimation of isoelectric point. Fertil Steril 23:399-404
Murray RA, McGaughey RW, Tarus WJ (1972) Blastokinin: its size and shape, and an indication of the existence of subunits. Fertil Steril 23:69-77
Nieto A, Ponstingl H, Beato M (1977) Purification and quaternary structure of the hormonally induced protein uteroglobin. Arch Biochem Biophys 180:82-93
Noske IG, Feigelson M (1976) Immunological evidence of uteroglobin (blastokinin) in the male reproductive tract and in non-reproductive ductal tissues and their secretions. Biol Reprod 15: 704-713
Shapiro SS, Forbes SH (1978) Alterations in human endometrial protein synthesis during the menstrual cycle and in progesterone-stimulated organ culture. Fertil Steril 30:175-180
Sutcliffe RG, Bolton AE, Sharp F, Nicholson LVB, and MacKinnon R (1980) Purification of human alpha uterine protein. J Reprod Fertil 58:395-442
Trowell O (1959) The culture of mature organs in a synthetic medium. Exp Cell Res 16:118-147
Wolf DP, Mastroianni L (1975) Protein composition of human uterine fluid. Fertil Steril 26:246-247
Wynn RM (1979) Morphology of the endometrium. In: Beller FK, Schumacher GFB (eds) The biology of the fluids of the female genital tract. Elsevier/North-Holland, New York Amsterdam Oxford, pp 37-58

Studies on Uteroglobin Messenger RNA

D.W. Bullock and B.M. Bhatt

Department of Cell Biology, Baylor College of Medicine, Houston, Texas, USA

In our studies on the messenger RNA (mRNA) for uteroglobin, a progesterone-induced protein secreted by the rabbit uterus during early pregnancy, we have addressed some of the questions raised by Professor Karlson in his introduction to this book. Although an important question, the biological activity of uteroglobin remains an enigma and we have concentrated our efforts on the hormonal regulation of the protein. Uteroglobin provides a unique model for studies on the mechanism of action of progesterone on the uterus, for it is a discrete biochemical endpoint of the action of progesterone in this tissue.

Our work is directed toward distinguishing an action of progesterone on uteroglobin synthesis at the level of transcription, posttranscription or translation. To make this distinction, it is necessary to quantitate the levels of the specific mRNA for uteroglobin ($mRNA_{UG}$) under different hormonal conditions. This, in turn, requires methods for the extraction of $mRNA_{UG}$ and its translation in vitro into the corresponding specific protein by a heterologous, cell-free system. In this paper, we report the translation of $mRNA_{UG}$, in a system derived from wheat germ, and its quantitation during early pregnancy. We include also some observations on the uteroglobin-like protein in rabbit lung and seminal plasma and on the translation of its mRNA from the lung.

1 Translation of Uteroglobin mRNA

A total RNA extract was prepared from frozen endometrium, collected on day 4 or 6 of pregnancy, using methods described previously (Bullock et al. 1976). A poly-A-containing fraction was separated by chromatography on a column of oligo-dT-cellulose (Aviv and Leder 1972) and the mRNA activity in each of the total extract (TE), oligo-dT-bound (dT) and unbound (FT, flow-through) fractions was assessed by translation in the wheat germ cell-free system (Roberts and Paterson 1973). The conditions for the translation assay have been described (Bullock et al. 1976). Total mRNA activity was measured by incorporation of ^{35}S-methionine into peptides precipitated with trichloroacetic acid (TCA). Specific $mRNA_{UG}$ activity was quantitated by double-antibody precipitation of the total translation products, using uteroglobin antibodies purified by affinity chromatography (Bullock 1977).

Table 1 shows the recovery of TCA-precipitable activity in the different RNA fractions. There was about a ten-fold enrichment of mRNA specific activity (cpm/μg) in the dT fraction compared to the TE. This poly-A-rich fraction, however, contained only about 10% of the mRNA activity of the TE and 75% was recovered in the non-dT-bound RNA, suggesting that the majority of endometrial mRNA species lack poly-A sequences of

Table 1. Total mRNA activity of rabbit endometrium after separation of total RNA extract (TE) on oligo-dT-cellulose into bound (dT) and flow-through (FT) fractions

RNA fraction	mg	Specific activity cpm/μg	Activity %
TE	52.2	14011	100
dT	0.5	151372	10.3
FT	34.2	16011	74.9

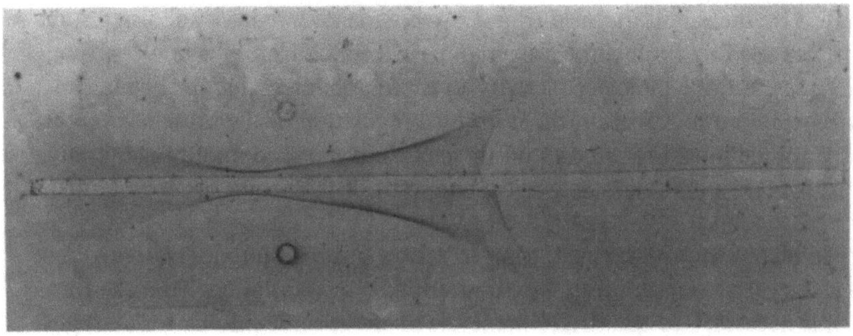

Fig. 1. Immunoelectrophoresis of uteroglobin (*UG*) and pregnant rabbit uterine flushings (*RUF*) against uteroglobin antibodies purified by affinity chromatography

greater than about 20 residues. No poly-A could be detected in the FT fraction by ^{3}H-poly-U hybridization (Bullock et al. 1976).

The specificity of the purified uteroglobin antibodies is demonstrated by the single, identical precipitin line seen on immunoelectrophoresis against authentic uteroglobin and whole uterine-flushing proteins (Fig. 1). When double-antibody precipitation was performed on the translation products of 4-day-pregnant endometrial RNA, under conditions which gave maximum precipitation, about 17.5% of the total mRNA activity was precipitated by the uteroglobin antibodies (Fig. 2). The specificity of immunoprecipitation is shown further by the low background radioactivity when the specific antibodies were replaced by normal goat serum or when immunoprecipitation was performed in the presence of excess nonradioactive uteroglobin (Fig. 2). A similar low trapping of nonspecific radioactivity was seen when products of translation of ovalbumin mRNA were precipitated with uteroglobin antibodies. Analysis of the immunoprecipitable cell-free product on 15% polyacrylamide gels containing 0.1% sodium dodecyl sulfate (SDS), by the method of Laemmli (1970), revealed a single peak of radioactivity (Fig. 3, D), which was not seen in the absence of added mRNA (Fig. 3, A) and was absent when uteroglobin antibodies were omitted (Fig. 3, B) or when excess nonradioactive uteroglobin was included (Fig. 3, C).

The translation assay is thus specific and can be used to quantitate the level of $mRNA_{UG}$ in comparison to the total mRNA population. Figure 4 shows the changes in endometrial $mRNA_{UG}$ during the first 8 days of pregnancy. There was an 11-fold

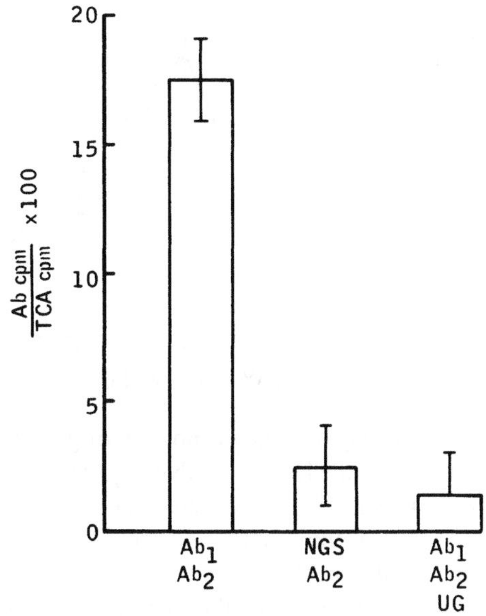

Fig. 2. Double-antibody immunoprecipitation of wheat germ translation products of 4-day pregnant endometrial poly-A-rich RNA. Ab_1 purified uteroglobin antibodies; Ab_2 rabbit anti-goat IgG antiserum; *NGS* normal goat serum; *UG* uteroglobin. Vertical lines show ± s.e.m. (d.f.= 3)

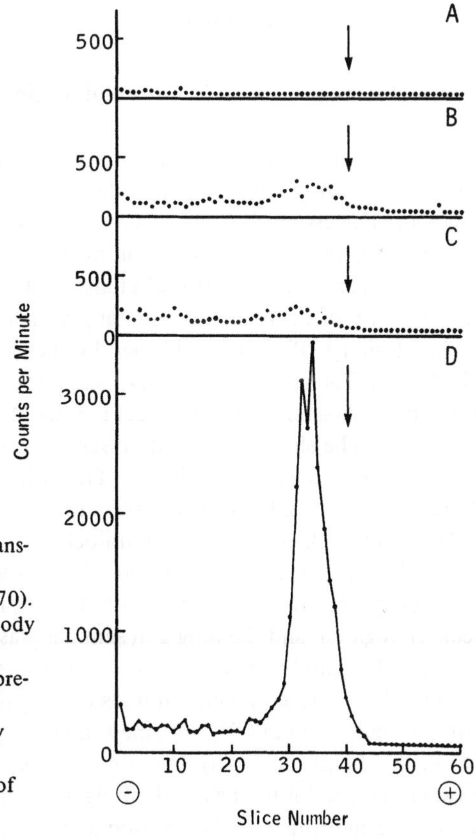

Fig. 3. Analysis of immunoprecipitate from translation products of endometrial mRNA on 15% polyacrylamide – 0.1% SDS gels (Laemmli 1970). *A* background, no added RNA; *B* double-antibody precipitation using normal goat serum in place of uteroglobin antibodies; *C* double-antibody precipitation in the presence of 20 μg authentic, nonradioactive uteroglobin; *D* double-antibody precipitation with uteroglobin antibodies and anti-goat IgG. The arrows mark the migration of authentic uteroglobin

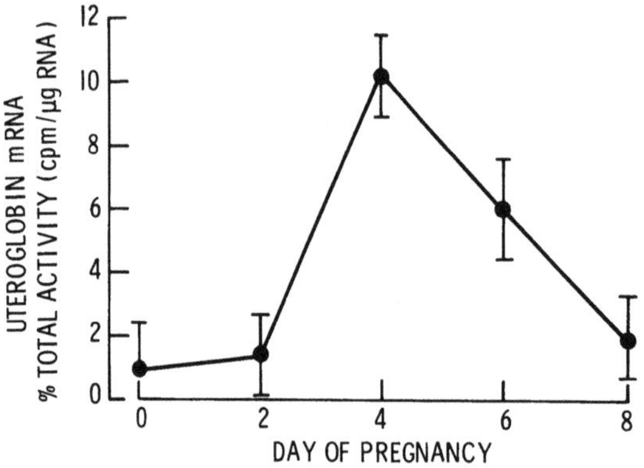

Fig. 4. Changes in proportion of poly-A-rich endometrial RNA represented by uteroglobin mRNA on different days of pregnancy. *Vertical lines* show ±s.e.m. (d.f.= 7)

increase on day 4 in the proportion of poly-A-rich mRNA accounted for by $mRNA_{UG}$ and by day 8 the level of $mRNA_{UG}$ had fallen to that of the nonpregnant state. We have not been able to calculate total tissue levels of $mRNA_{UG}$, due to variability of yield between animals, but it is clear that there are specific changes in this mRNA compared to other mRNA species. These changes reflect the changing hormonal status of the animal and resemble the pattern of uteroglobin secretion, but shorter sampling intervals will be required to establish the precise relationship between the synthesis of $mRNA_{UG}$ and of uteroglobin.

2 Uteroglobin-like Protein in Lung and Seminal Plasma

Hitherto regarded as a uterine-specific protein, recent work has indicated the presence of uteroglobin in seminal plasma, lung and other tissues of the rabbit (Beier et al. 1975; Noske and Feigelson 1976; Feigelson et al. 1977). The importance of this phenomenon for the biological role of uteroglobin prompted us to re-examine these observations, which were based on similarity of charge and immunological identity.

Size identity of the lung and seminal plasma proteins with uteroglobin in pregnant uterine flushings (PUF) is established by their comigration on 15% polyacrylamide-SDS gels (Fig. 5). The appropriate area of similar, unstained gels was cut out and the proteins eluted by homogenization of the gel in 0.01 M Tris-HCl buffer, pH 7.6, containing 0.05% SDS. The eluate was freed of SDS by precipitation with cold 0.25 M KCl, dialyzed, lyophilized and redissolved in Tris buffer.

Figure 6 shows results of immunoelectrophoresis against uteroglobin antibodies purified by affinity chromatography (Bullock 1977). Similar single precipitin lines were obtained with authentic uteroglobin, PUF, native lung and seminal plasma proteins and their SDS-gel eluates. These results confirm charge and immunological similarity between uteroglobin and the lung and seminal plasma proteins. Antigenic identity was examined by double diffusion in 1% agar against purified uteroglobin antibodies. As shown in Fig. 7, fused precipitin lines of the lung and seminal plasma SDS-gel eluates with uteroglobin and PUF confirmed the immunological identity of these proteins (Noske and Feigelson 1976). In earlier studies, we had observed the formation of spurs between the precipitin lines of the lung and semen proteins, but this phenomenon appears to be an artifact of the fixation and drying procedure, as it is not reproducible.

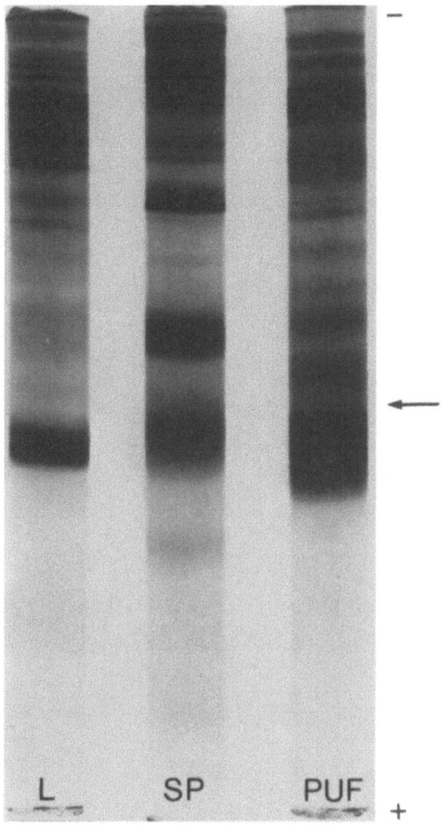

Fig. 5. Protein profiles of lung tissue extract (*L*), seminal plasma (*SP*) and pregnant uterine flushings (*PUF*) on 15% polyacrylamide-SDS gels. The *arrow* points to uteroglobin

To examine further the similarity of these proteins, we have subjected the lung and seminal plasma preparations to the purification procedure, involving gel filtration and DEAE-cellulose chromatography, that we use for uteroglobin (Bullock et al. 1976). The eluate of the columns was monitored for reaction with uteroglobin antibodies by the interfacial ring test. While cross-reaction with the antibodies followed the normal elution pattern for uteroglobin, the procedure did not effect purification of uteroglobin from the lung homogenate or from seminal plasma as it does from PUF. Fig. 8 shows the results of SDS-gel electrophoresis, coupled with immunodiffusion, of PUF and lung homogenate after the purification procedure for uteroglobin. While PUF has been reduced to a single uteroglobin band, several proteins in addition to uteroglobin are present in the lung fraction.

The amount of uteroglobin in the lung and semen preparations was estimated by rocket immunoelectrophoresis (Laurell 1966) in 1% agarose containing 0.5% uteroglobin antiserum. The slides were run at 20 v/cm in Tris-barbital buffer, pH 8.8, at 4°C for 3.5 h. After washing in 1% saline and staining in Coomassie blue, the heights of the rockets formed by immunoprecipitation (Fig. 9, b) were measured and compared to a standard curve (Fig. 9, a) of 10 to 100 ng of authentic uteroglobin. The amount of

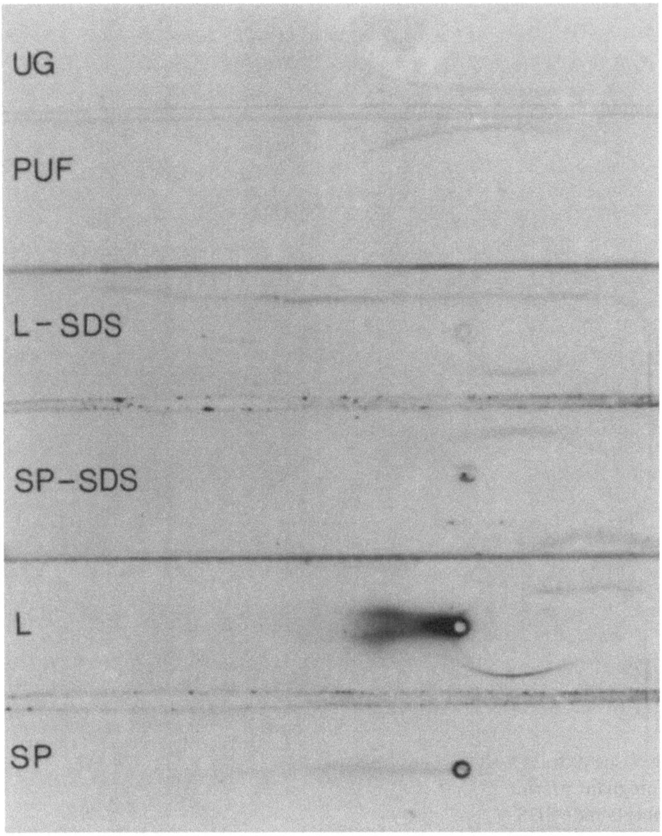

Fig. 6. Immunoelectrophoresis of uteroglobin (*UG*), pregnant uterine flushings (*PUF*), lung SDS-gel eluate (*L-SDS*), seminal plasma SDS-gel eluate (*SP-SDS*), lung homogenate (*L*), and seminal plasma (*SP*) against purified uteroglobin antibodies. The anode is to the left

uteroglobin was expressed as a percentage of the total protein present, determined by the method of Lowry et al. (1951).

The results (Fig. 9, a, b) showed that immunoprecipitable uteroglobin accounted for 0.3% of the protein in lung homogenate and 0.14% of the protein in seminal plasma. While these amounts are much lower than the amount of immunoprecipitable uteroglobin in the pregnant uterus (22% of secreted protein; Bullock and Bhatt, unpublished), it appears that uteroglobin may be one member of a family of very similar proteins common to secretory epithelium in the rabbit. Whether these proteins have similar or different functions in different tissues is a matter of speculation at present.

3 Comparison of Translation Products of Lung and Uterine mRNA

It was of interest to establish whether the lung was a site of synthesis of the uteroglobin-like protein. We have demonstrated this directly by extraction of the specific mRNA

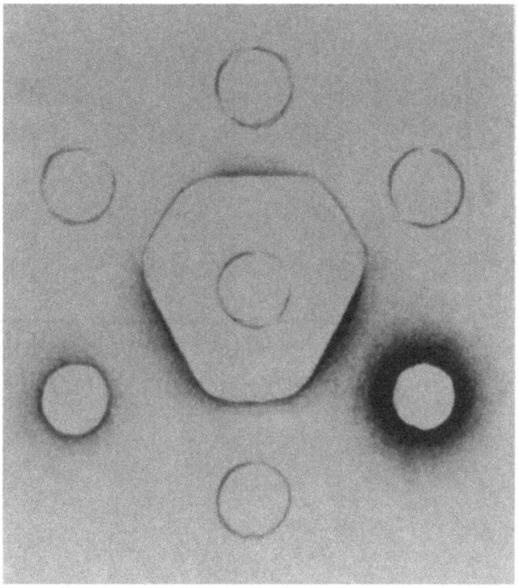

Fig. 7. Diagram of precipitin lines formed by immunodiffusion against purified uteroglobin antibodies (*center well*). Counting clockwise from the *top*, the *outer wells* contain uteroglobin, lung SDS-gel eluate, lung homogenate, pregnant uterine flushings, seminal plasma, and seminal plasma SDS-gel eluate

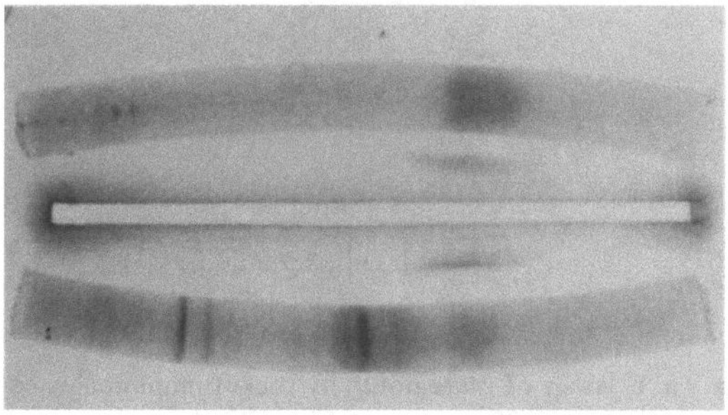

Fig. 8. Electrophoresis in 15% polyacrylamide – 0.1% SDS gels of pregnant uterine flushings (*upper gel*) and lung homogenate (*lower gel*) after fractionation by gel filtration and DEAE-cellulose chromatography (Bullock et al. 1976). After electrophoresis, the gels were placed on the surface of a layer of 1% agar and allowed to diffuse against purified uteroflobin antibodies in the central trough. The gels shown are companion gels stained with Coomassie blue to reveal the protein pattern. The anode is to the right

and its translation in vitro in the wheat germ system (Bullock 1977). Figure 10 shows the radioactivity, immunoprecipitable with uteroglobin antibodies, in the cell-free translation products of dT-bound RNA from rabbit lung on day 6 of pregnancy. The product is specifically precipitable with the uteroglobin antibodies, as shown by the displacement of radioactivity with authentic uteroglobin and its loss in the absence of specific antibodies (Fig. 10).

Analysis of the lung and endometrial immunoprecipitable products was carried out by electrophoresis on 12.5% acrylamide-10% cross-linked gels, containing 8M urea and 0.1% SDS, at pH 6.8 (Swank and Munkres 1971). The samples were dissolved in 1% SDS, 8M urea, 10mM dithiothreitol (DTT), 0.01 M H_3PO_4, titrated to pH 6.8 with

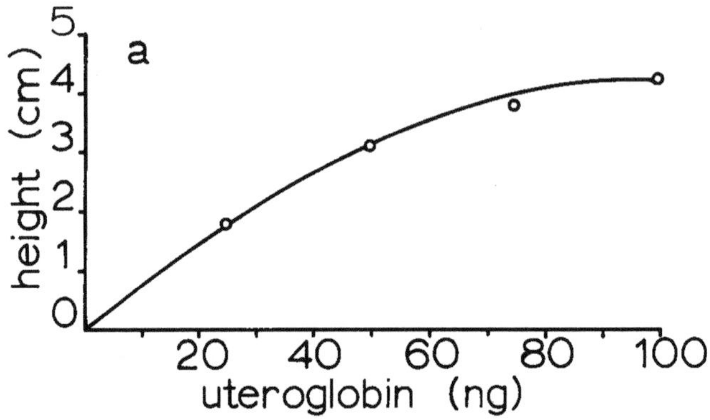

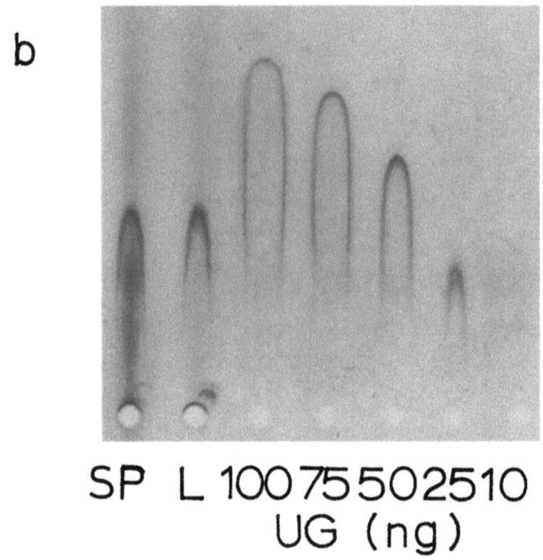

Fig. 9 a, b. Assay of uteroglobin by rocket immunoelectrophoresis. (**a**) standard curve, (**b**) rockets formed by 10 to 100 ng uteroglobin (*UG*), lung homogenate (*L*), and seminal plasma (*SP*)

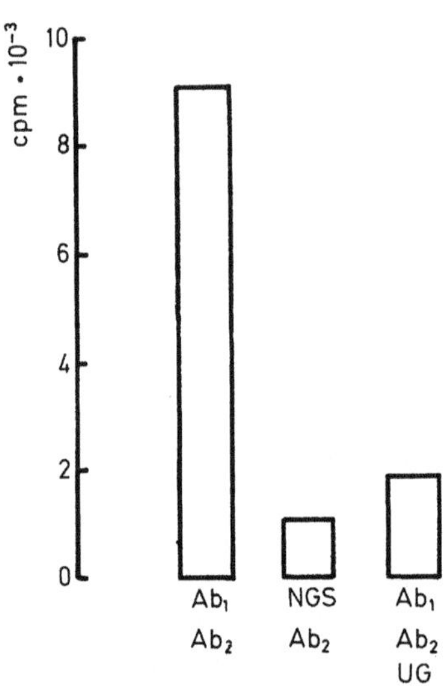

Fig. 10. Double-antibody precipitation of cell-free translation products of lung mRNA. Ab_1 purified uteroglobin antibodies; Ab_2 rabbit anti-goat IgG; *NGS* normal goat serum; *UG* excess, nonradioactive authentic uteroglobin. The TCA-precipitable radioactivity was 87 000 cpm

Tris base, and heated at 90°C for 5 min. Autoradiography was performed using Kodak XRP-1 X-ray film, exposed for 21 days at -20°C. Figure 11 shows that both immunoprecipitates gave a single radioactive band with the same electrophoretic migration. The cell-free products of both the lung and the uterine mRNA's were larger in size than authentic uteroglobin. The translation product of $mRNA_{UG}$ has been shown previously to be larger than the native protein (Beato and Nieto 1976; Bullock et al. 1976). The present results demonstrate that the specific lung mRNA codes for a precursor protein similar in size to the pre-uteroglobin of the uterus.

Estimation of molecular weight from a linear regression fitted to the relative migration of standard proteins (Fig. 11) gave values of 10 700 for the cell-free products and 9 500 for authentic uteroglobin. The precur[illegible] proteins both from lung and uterus are thus about 1 200 larger in molecular weight than the native proteins, in agreement with the finding of Beato and Nieto (1976) for uterine uteroglobin.

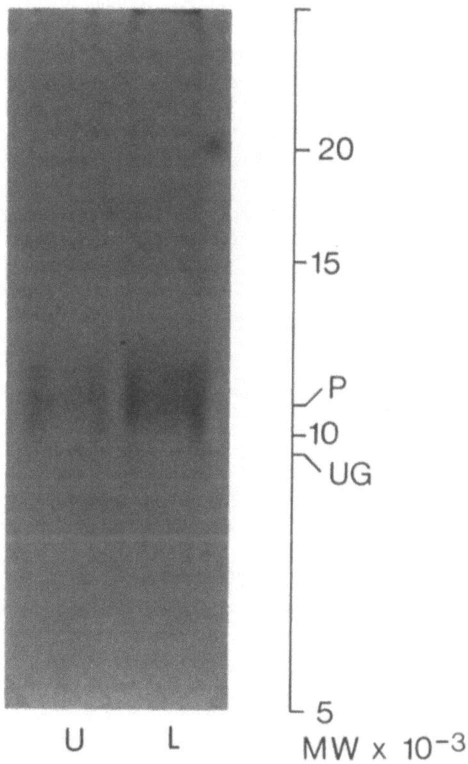

Fig. 11. Radioautography of immunoprecipitable cell-free translation products of lung (*L*) and uterine (*U*) mRNA after electrophoresis on 12.5% polyacrylamide gels containing 0.1% SDS and 8 M urea. The scale of molecular weight is drawn from a linear regression fitted to the relative migration of soya bean trypsin inhibitor (22 700), myoglobin (17 900), cytochrome C (13 370) and insulin (5 800). The migration of the cell-free products (*P*) and of authentic uteroglobin (*UG*) is indicated

Acknowledgments. We thank Dr. Savio Woo for his help and interest, and Ms. Rebecca Russell for technical assistance. The work was supported by NIH grant HD 09378 and Rockefeller Foundation grant 72062.

References

Aviv H, Leder P (1972) Purification of biologically active globin messenger RNA by chromatography on oligothymidilic acid cellulose. Proc Natl Acad Sci USA 69:1408-1412

Beato M, Nieto A (1976) Translation of the mRNA for uteroglobin in cell-free systems. Eur J Biochem 64:15-25

Beier HM, Bohn H, Müller W (1975) Uteroglobin-like antigen in the male genital tract secretions. Cell Tiss Res 165:1-11

Bullock DW, Woo SLC, O'Malley BW (1976) Uteroglobin messenger RNA: Translation in vitro. Biol Reprod 15:435-443

Feigelson M, Noske IG, Goswami AK, Kay E (1977) Reproductive tract fluid proteins and their hormonal control. Ann NY Acad Sci 286:273-286

Laemmli UK (1970) Cleavage of structural proteins during the assembly of the head of bacteriophage T4. Nature 227:680-688

Laurell CB (1966) Quantitative estimation of proteins by electrophoresis in agarose gel containing antibodies. Anal Biochem 15:45-52

Lowry OH, Rosebrough NT, Farr AL, Randall RT (1951) Protein measurement with the folinphenol reagent. J Biol Chem 193:265-275

Noske IG, Feigelson M (1976) Immunological evidence of uteroglobin (blastokinin) in the male reproductive tract and in non-reproductive ductal tissues and their secretions. Biol Reprod 15: 704-713

Roberts BW, Paterson BM (1973) Efficient translation of tobacco mosaic virus RNA and rabbit globin 9S RNA in a cellfree system from commercial wheat germ. Proc Natl Acad Sci USA 70: 2330-2334

Swank RT, Munkres KD (1971) Molecular weight analysis of oligopeptides by electrophoresis in polyacrylamide gel with sodium dodecyl sulfate. Anal Biochem 39:462-477

Structure and Binding Properties of Rabbit Uteroglobin. Search for a Similar Protein in Human Uterine Fluid

Miguel Beato

Institut für Physiologische Chemie der Philipps-Universität, 3550 Marburg, Germany

1 Introduction

The reports by Beier (1968) and Arthur and Daniel (1972) that uteroglobin can be induced by the administration of progesterone, in conjunction with the fact that this protein represents one of the major components of the uterine secretion, brought the hormonal control of uteroglobin biosynthesis to the attention of scientists interested in the mechanism of hormone action. An analysis of the induction process at the molecular level, however, requires the characterization of the messenger RNA for the induced protein, and the synthesis of a complementary DNA. Only then can the synthesis and degradation of the primary gene product be studied during the hormonal induction.

A prerequisite for the development of an assay for a particular mRNA is a good knowledge of the polypeptide which is encoded in the RNA. It is, therefore, necessary to unravel the structure of the induced protein, mainly in relation to the existence of subunits. In this contribution I will review our data on the structure of uteroglobin, and also describe an interesting property of this protein, namely its ability to bind the inducing steroid. I will then briefly review recent experiments directed to elucidate the presence in the human uterine fluid of proteins with structural or functional properties similar to those of uteroglobin.

2 Structure of Uteroglobin

2.1 Purification

Before starting a detailed structural analysis of uteroglobin it was necessary to obtain homogenous preparations of the protein. For this purpose we used as a first step column chromatography on Sephadex G-100 of the uterine fluid proteins of rabbits treated with estradiol and progesterone. If carefully performed, this procedure yields uteroglobin preparations up to 95% pure. An additional purification on columns of carboxymethyl cellulose was required to obtain homogeneous preparations of uteroglobin (Nieto et al. 1977). These preparations were pure not only according to gel electrophoretic criteria, under both denaturing and nondenaturing conditions, but also according to strict immunological criteria using antibodies directed either against total uterine fluid proteins or against purified uteroglobin (Nieto et al. 1977).

The isoelectric point of purified uteroglobin was shown to be 6.2 independently of the presence of 6 M urea during isoelectric focusing (Fig. 1). This finding provides further evidence for the purity of our uteroglobin preparations.

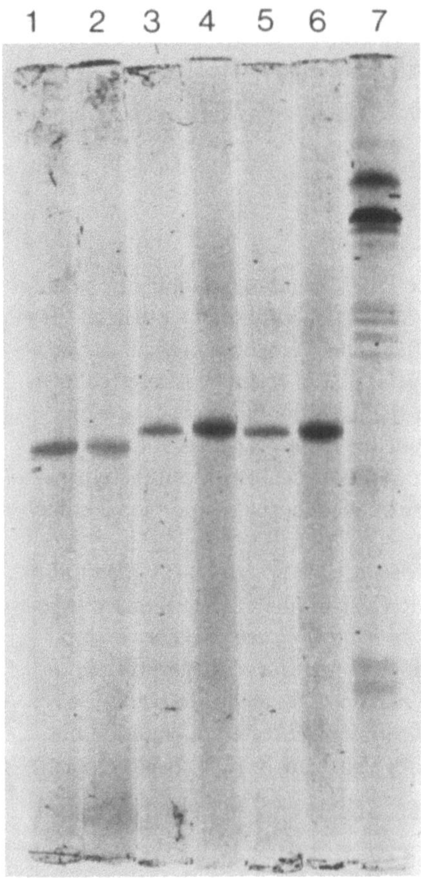

Fig. 1. Isoelectric focusing of native and reduced uteroglobin in polyacrylamide gels. Aliquots of either native uteroglobin or uteroglobin reduced with 5 mM dithioerythritol were electrofocused in polyacrylamide gels in the presence or absence of 6 M urea, according to the procedure of Utermann et al. (1975). After the run the gels were fixed in 10% trichloroacetic acid-10% sulfosalicylic acid (2h-4°C), and stained with Coomassie blue dissolved in fixative. *1,2* Native uteroglobin without urea (20 and 10 μg); *3,4* Native uteroglobin in 6 M urea (10 and 20 μg); *5,6* Reduced uteroglobin in 6 M urea (10 and 20 μg); *7* Human apo-V-LDL marker. The slight difference in the position of the band in gels 1 and 2 does not reflect a diffrence in isoelectric point but rather a shift of the pH gradient in the absence of urea

2.2 Molecular Weight

The molecular weight of native uteroglobin was determined by analytical ultracentrifugation using the meniscus depletion procedure (Yphantis 1964). A value of 14 000 was calculated assuming a specific volume of 0.738, as determined from the amino acid composition of uteroglobin. In velocity sedimentation experiments, the sedimentation coefficient of the native uteroglobin was calculated to be 1.56 $S_{20.w}$ at physiological ionic strength

Both the equilibrium and velocity sedimentation runs yielded linear plots, further supporting the purity of the uteroglobin preparation used.

In gel filtration experiments with Sephadex G-50 and Sephadex G-100, the Stoke's radius of native uteroglobin was determined to be 18.4 Å. This value in conjunction with the sedimentation coefficient, mentioned above, allows an independent calculation of the molecular weight using the Einstein modification of the Svedberg equation (Beato, 1977). The value obtained is 16 000, and therefore about twice as high as the minimum estimate of the molecular weight calculated from the amino acid analysis, suggesting that native uteroglobin may be composed of two similar subunits of molecular weight around 8000.

This suggestion was supported by the results of gel filtration experiments on Sepharose 6 B in the presence of 6 M guanidinium hydrochloride. Native uteroglobin elutes in this column in a position corresponding to a molecular weight of 17 000, whereas reduced and S-carboxymethylated uteroglobin elutes as a single peak in a position corresponding to molecular weight of around 8500 (Nieto et al. 1977). This finding suggested that uteroglobin is indeed composed of two similar subunits and that these subunits are held together by disulfide bonds.

These findings were confirmed by electrophoresis in polyacrylamide gels containing sodium dodecyl sulfate (SDS) and urea (Swank and Munkres 1971). Whereas the native protein migrates in a position corresponding to a molecular weight of around 12 000, after reduction and carboxymethylation with iodoacetate, a single band is observed migrating in a position corresponding to a molecular weight of 6000 (Nieto et al. 1977). Although the values obtained in gels containing SDS and urea are lower than those determined by gel filtration or by the amino acid analysis, the fundamental finding, namely that reduced and S-carboxymethylated protein exhibits half the apparent molecular weight of the native protein is obvious.

The similarity of the uteroglobin subunits was confirmed by the isoelectric focusing experiments (Fig. 1). After reduction with dithioerythritol and in the presence of 6 M urea, uteroglobin migrates as a single band with an isoelectric point 6.2, identical to that of the unreduced protein.

Further support for the hypothesis that uteroglobin is composed of two similar if not identical subunits of around 70 amino acid residues each was obtained by analyzing the peptides produced by a digestion of the protein either with trypsin or with cyanogen bromide. In both cases the number of peptides observed was compatible with a polypeptide chain of molecular weight 8000. Eleven tryptic peptides and six cyanogen bromide peptides were obtained as expected from the amino acid composition, seven lysines, two arginines and five methionines (Nieto et al. 1977).

2.3 Chemical Analysis

The purified preparations of uteroglobin were free of carbohydrate, as demonstrated by the anthrone-sulfuric acid reaction, and by the negativity of the PAS-reaction after polyacrylamide gel electrophoresis (Nieto et al. 1977).

An amino acid analysis obtained after reduction with 2-mercaptoethanol and S-carboxymethylation with iodoacetate is compatible with a polypeptide chain of a minimal length of some 70 amino acids. No tryptophan could be detected in purified uteroglobin either colorimetrically nor by the spectrophotometric technique (Nieto et al. 1977; Beato 1977). Besides, the lack of tryptophan, other interesting aspects of the amino acid analysis were the richness in sulphur-containing amino acids, cystein (2.87%) and methionine (5.66%). In addition, one tyrosine, one histidine, two arginins, and seven lysines are found for each polypeptide chain of the minimal length.

As mentioned above there are two cystein residues per polypeptide chain and, therefore, it was interesting to know whether native uteroglobin has any free sulfhydryl groups. Using 5,5-dithio-bis(2)-nitrobencoic acid (Ellmann 1959), no free sulfhydryl group could be detected on native uteroglobin under conditions where proteins having one tenth the cystein content of uteroglobin gave a positive result. This suggests that the two cystein residues of each uteroglobin subunit are in the oxidized form.

Determination of the N-terminal amino acid residue by the dansyl procedure (Gray 1972) yielded a single residue, namely glycine suggesting that uteroglobin is composed of homogenous polypeptide chains. Determination of the C-terminal residue with carboxypeptidase was less conclusive, but it appears that methionine is located at this position. The complete amino acid sequence of uteroglobin has been recently elucidated (Ponstingl et al. 1978), and is depicted in Table 1.

Table 1. Amino acid sequence of uteroglobin

10
Gly–Ile–Cys–Pro–Arg–Phe–Ala–His–Val–Ile–
20
Glu–Asn–Leu–Leu–Leu–Gly–Thr–Pro–Ser–Ser–
30
Tyr–Glu–Thr–Ser–Leu–Lys–Glu–Phe–Glu–Pro–
40
Asp–Asp–Thr–Met–Lys–Asp–Ala–Gly–Met–Gln–
50
Met–Lys–Lys–Val–Leu–Asp–Ser–Leu–Pro–Gln–
60
Thr–Thr–Arg–Glu–Asn–Ile–Met–Lys–Leu–Thr–
70
Gln–Lys–Ile–Val–Lys–Ser–Pro–Leu–Cys–Met–

No marked similarieties have been found between the sequence of uteroglobin and that of other known proteins. A minor homology is found between residues 22-29 and a sequence (Asp-Thr-Tyr/Phe-Lys-Glu-Phe-Glu) present in various somatotropins. Prediction of secondary structure yields four helical regions (residues 6-15, 25-29, 34-45 and 56-65) and four potential β-turns (3-6; 17-20; 29-32 and 66-69). The C-terminal helical region exhibits three lysines in one side of the helix, a structure which could interact with the phosphate groups of nucleic acids.

2.4 Quaternary Structures

All these findings indicate that uteroglobin is composed of two identical subunits held together by two disulfide bridges and by other noncovalent forces. This statement is based on the observation that reduction and S-carboxymethylation of uteroglobin does not lead to its dissociation into subunits. After this treatment uteroglobin still elutes on Sephadex columns in a position corresponding to a molecular weight of around

15 000-16 000, as does the native protein. It is necessary to add a denaturant (SDS or 6 M guanidinium hydrochloride) in order to separate the subunits (see above).

Reduction of the disulfide bonds, however, leads to conformational changes in the structure of the protein which can be demonstrated by differential ultraviolet spectroscopy (Beato 1977). A deflection in the region of 230 nm is observed after reduction, which has been interpreted as reflecting conformational changes of the polypeptide backbone. Nuclear magnetic resonance studies also show that reduction of the disulfide bonds produces conformational changes in the protein, which still keeps in part its tertiary and secondary structure (Puigdomenech and Beato 1977; Beato et al. 1980).

Additional evidence for conformational changes following reduction is based on differential sensitivity to digestion with a specific protease derived from *Staphilococcus aureus* strain V8 (Drapeau et al. 1972; Houmard and Drapeau 1972). This enzyme, which cleaves proteins specifically at the glutamoyl bonds, does not attack the native uteroglobin, but yields five clear peptides when incubated with reduced and S-carboxymethylated uteroglobin (Fig. 2). This finding suggests that large conformational changes take place after reduction making four glutamoyl bonds accessible to the protease.

The alignment of the subunits was studied with the diagonal peptide mapping technique (Bennett 1967). After cyanogen bromide digestion two cysteine-containing peptides were detected, each of which gave a single peptide following oxidation with performic acid (Beato 1977). This indicates that the two subunits are aligned in parallel, with the N-terminal oriented in the same direction.

The complete elucidation of the structure of uteroglobin will only be possible by crystallographic techniques. Recently we have obtained suitable crystals of the native protein with a resolution limit of 2.2 Å. A preliminary analysis of these crystals confirms the similarity and symmetry of the uteroglobin subunits (Buehner and Beato 1978; Atger et al. 1980).

3 Steroid Binding to Uteroglobin

3.1 Methodology and Identification of Uteroglobin as the Progesterone Binding Factor

The binding of steroids to proteins can be studied with a variety of methods, each of which has certain advantages and disadvantages. The most widely used methods involve radioactively labeled steroids, which allow measurements at very low concentrations of ligand. With the exception of the equilibrium dialysis technique, all the other routine procedures do not measure binding under equilibrium conditions. Most of them are based on the separation of the protein-bound and the free steroid, followed by independent measurements of the radioactivity in each fraction. A widely used procedure of this type is the adsorption of the free steroid to dextran-coated charcoal followed by centrifugation of the adsorbent and determination of the radioactivity in the supernatant, which corresponds to the protein-bound steroid. When certain controls are carried out, including incubations performed with an excess of nonlabeled steroid, this procedure allows a rapid determination of the binding parameters and is very convenient for studying the relative affinity of various steroids for a certain binding protein by means of competition studies (Beato 1976). Equilibrium dialysis techniques are more suitable

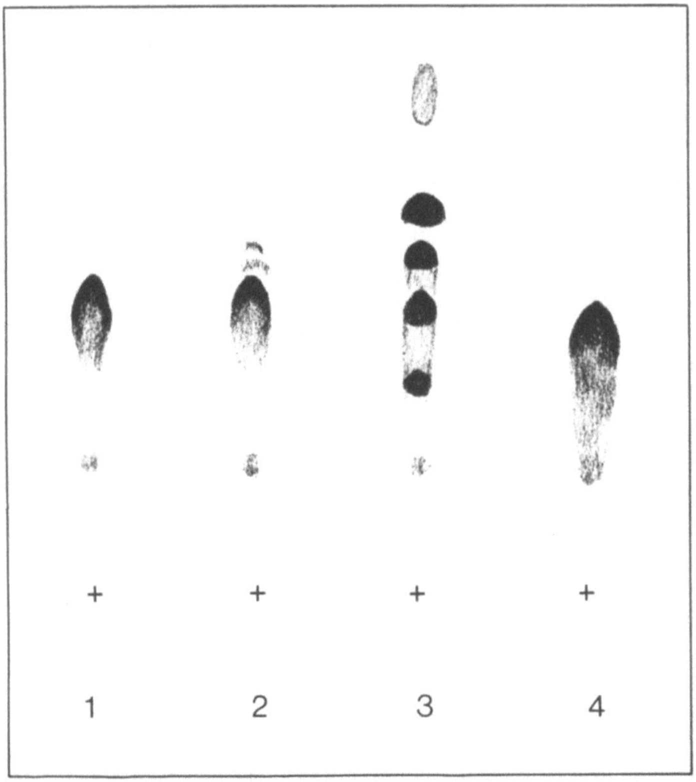

Fig. 2. Peptide pattern obtained after digestion of native and reduced uteroglobin with V8-protease. Native as well as reduced and *S*-carboxymethylated uteroglobin (500 μg) dissolved in 100 μl 50 mM ammonium acetate, pH 4.0, was incubated at 37°C for 48 h with V8-protease (17 μg). As a control, aliquots were also incubated under the same conditions but without added protease. The digested samples and the controls were lyophilized, dissolved in 50 μl 1 M acetic acid, and 40 μg aliquots were applied to 10 x 10 cm precoated thin-layer cellulose plates (E. Merck, Darmstadt). The plates were chromatographed for 2 h in butanol:acetic acid:pyridine:water (90:18:60:72), dried and sprayed with Ninhydrin. The figure is schematic drawing of the Ninhydrin-positive spots. *1* Native uteroglobin control; *2* Native uteroglobin incubated with V8-protease; *3* *S*-Carboxymethylated uteroglobin treated with V8-protease; *4* *S*-Carboxymethylated uteroglobin control

for a precise determination of the apparent equilibrium association constant and the number of steroid binding sites, but cannot be used with a mixture of proteins as they not distinguish between specific and unspecific binding. Another widely used procedure is the chromatographic separation of the free and protein-bound steroid on small Sephadex columns.

Using these conventional procedures it has been conclusively shown that uteroglobin is the main progesterone binding protein in the uterine fluid of progesterone-treated rabbits (Urzua et al. 1970; Arthur et al. 1972; Beato and Baier 1975; Fridlansky and Milgrom 1976). The affinity of uteroglobin for progesterone at low temperatures exhibits an equilibrium dissociation constant in the micromolar range and the number of binding sites per uteroglobin molecule (MW 16 000) varies around 0.5 (Beato and Baier 1975; Fridlansky and Milgrom 1976; Beato et al. 1980).

Given the high extinction coefficient of the progesterone chromophore (16 700 at 249 nm) and the known involvement of the steroid A-ring in the binding reaction, it could be expected that the interaction with uteroglobin will result in a measurable change of the absorption of the Δ^4-3-keto function, as has been described for the interaction of various 3-ketosteroids with serum proteins (Westphal 1971). In fact, the binding of uteroglobin to progesterone causes a very significant quenching of the absorbance of the steroid chromophore with the maximal deflection located at around 260 nm (Beato et al. 1977). If progesterone is completely bound to uteroglobin, its absorbance at this wavelength is only 22% of the absorbance of the free steroid (Beato et al. 1977). We have used this depression in the steroid absorbance to calculate the affinity of progesterone for uteroglobin, under equilibrium conditions (Beato et al. 1977). As shown in Fig. 3, the affinity of native uteroglobin for progesterone is relatively low, and increases markedly after reduction with either dithioerythritol or dithiothritol.

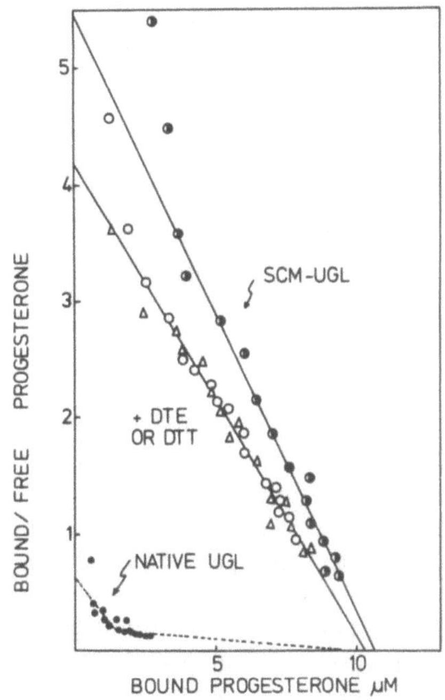

Fig. 3. Binding of [^{3}H]progesterone to native, reduced and *S*-carboxymethylated uteroglobin. Native uteroglobin (●—●), uteroglobin treated with 5 mM DTE (o—o) or 20 mM DTT (△—△), and *S*-carboxymethylated uteroglobin (◐—◐) were tested for progesterone binding using the spectrophotometric procedure (Beato et al. 1977). Incubations were performed at 10°C in Tris-saline buffer, pH 7.5. The binding data are plotted according to Scatchard, and the slope reflects the equilibrium association constant

This finding, already observed using radioactive progesterone (Beato and Baier 1975), has been confirmed by Fridlansky and Milgrom (1976), and can explain discrepancies in the results of previous binding studies. In addition, the results of Fig. 3 show that, if uteroglobin is carboxymethylated with iodoacetate after reduction, the affinity for progesterone increases slightly. This indicates that, although reduction of the disulfide bonds is necessary for binding, the sulfhydryl groups are not directly involved in the interaction with the steroid.

3.2 Properties of the Binding Reaction

Reduced and S-carboxymethylated uteroglobin shows an optimal pH for the binding of progesterone around 7.5 (Fig. 4), whereas the protein reduced with dithioerythritol exhibits optimal binding at pH 6.0 (Beato 1977). This difference is probably due to the fact that pH changes in the presence of dithioerythritol not only influence the binding reaction but also the extent of reduction and the conformational state of the protein.

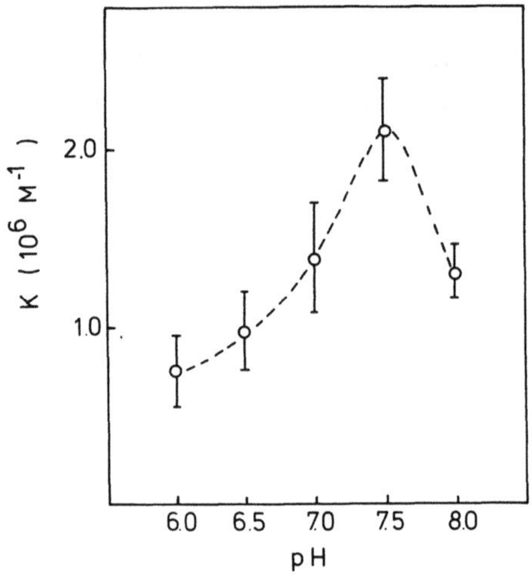

Fig. 4. Effect of pH on the binding of progesterone to *S*-carboxymethylated uteroglobin. Reduced and *S*-carboxymethylated uteroglobin was resuspended in 10 mM sodium phosphate buffer of the indicated pH, containing 15 mM NaCl, and used for spectrophotometric measurements of progesterone binding. The equilibrium association constant, *K*, calculated from the Scatchard analysis, is ploted versus the pH of the buffer. Measurements were carried out at 10°C, and the values represent the average of two determinations

Using the spectrophotometric procedure we have studied the influence of temperature on the binding reaction (Beato et al. 1977). The number of steroid binding sites per uteroglobin molecule (MW 16 000) is around 0.5, independently of the temperature of incubation, but the equilibrium association constant decreases with increasing temperature. The van't Hoff plot of these data allows the quantitation of the thermodynamic parameters of the binding reaction. There is a marked negative change in enthalpy (-3.46 kcal.mol^{-1}) and, interestingly, a small positive change in entropy (14.4 cal.K^{-1}.mol^{-1}). This can be interpreted either as consequence of the displacement of the hydration water from the steroid or the protein molecule, or alternatively as the result of a conformational change of the protein molecule leading to a higher degree of freedom. Since the positive change of entropy is observed with reduced and S-carboxymethylated uteroglobin, it cannot be attributed to the reduction itself and probably is a direct consequence of the steroid binding.

The spectrophotometric method also allows a determination of the kinetic parameters of the binding reaction. Initial measurements carried out with a routine spectrophotometer are depicted in Fig. 5, and clearly show that binding is a very rapid reaction with a half-life of around 1 s at 12°C. Precise measurements will require the use of "stopped flow" and temperature jump techniques. Approximate calculations from these experiments yield a rate constant of 1-2 x 10^4.M^{-1}.s^{-1} for the association reaction.

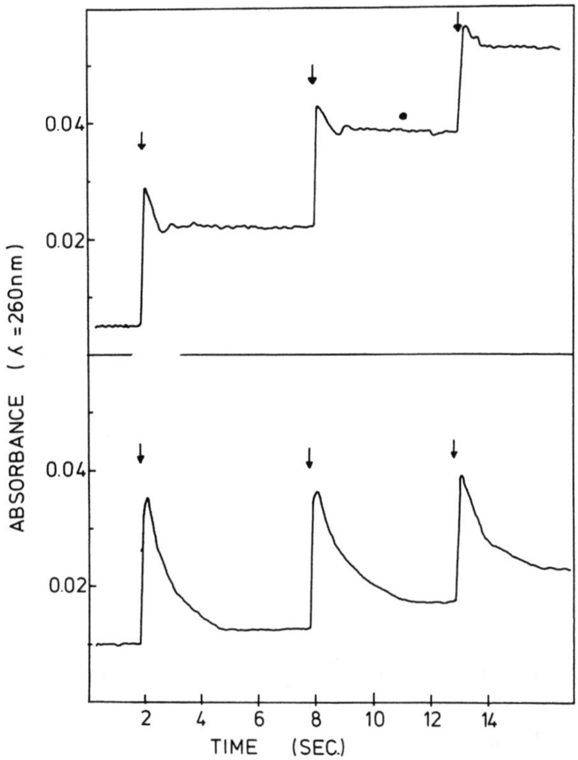

Fig. 5. Kinetics of progesterone binding to *S*-carboxymethylated uteroglobin. The measurements were carried out in a Beckman Acta II spectrophotometer equiped with magnetic stirrer. Three milliliter cuvettes were filled with 2.5 ml of either buffer (10 mM sodium phosphate pH 7.5, containing 15 mM NaCl) or a 20 μM solution of reduced and *S*-carboxymethylated uteroglobin in buffer. At the times indicated by the arrows, 50 μl progesterone (0.1 mM in buffer) was rapidly injected into the cuvette and the change in absorbance at 260 nm was registered. *Upper panel*, progesterone added to buffer. *Lower panel*, progesterone added to *S*-carboxymethylated uteroglobin

A comparison of the spectra of free versus progesterone-bound uteroglobin yields additional information concerning the involvement of certain amino acid residues in the binding reaction. As shown in Fig. 6, there are two major signals in the difference spectrum located at 287.5 and 281.5 nm indicating a change in the absorption properties of the tyrosine chromophore. These changes are evident both with the reduced protein and with the reduced and carboxymethylated uteroglobin. Such perturbations of the tyrosine spectrum are not detected when uteroglobin is incubated with cortisol, a steroid which does not bind to uteroglobin (Beato 1976). In addition, similar perturbations of a tyrosine spectrum are also observed with norethynodrel, a synthetic steroid with high affinity for uteroglobin, but without the Δ^4-3-keto group (Beato 1976). This indicates that the alteration of the tyrosine chromophore is not the consequence of changes in the absorption of the steroid molecule. Sequence data have shown that the only tyrosine residue of the uteroglobin subunit is located in position 21, and therefore close to the N-terminal (Fig. 1). We also know that one of the two cysteines of

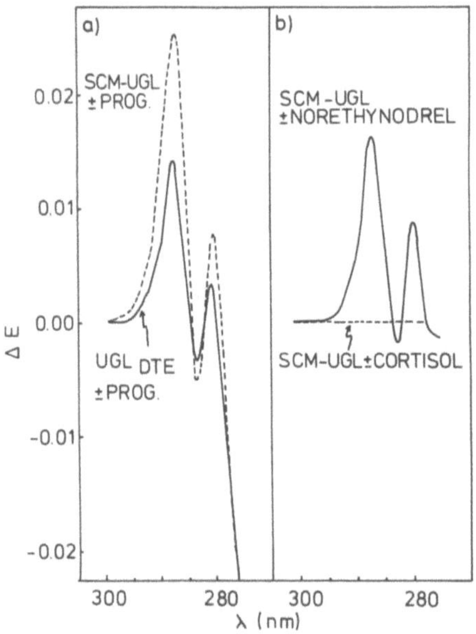

Fig. 6 a,b. Difference ultraviolet spectra of uteroglobin complexed with various steroids against the free protein. Difference ultraviolet spectra were obtained using a 0.2 mM solution of uteroglobin in Tris-saline and a 20 μM solution of progesterone, cortisol or norethynodrel. Incubations were at 10°C for 20 min, and the bandwith was 1 nm at 300 nm. **a** Reduced uteroglobin incubated with progesterone (—); *S*-carboxymethylated uteroglobin incubated with progesterone (- - -). **b** *S*-Carboxymethylated uteroglobin incubated with norethynodrel (—) and with cortisol (- - -)

the uteroglobin subunit is located at position 3, and therefore we can interpret the binding data by postulating that reduction of the disulfide bond at the N-terminus makes the binding site available for the steroid. Since the binding site is probably close to the tyrosine residue, its spectral properties change upon binding.

The tyrosine side chain, however, is not directly involved in binding as it can be modified chemically without reduction of the affinity for progesterone (Beato et al. 1980). The changes in the absorption spectrum induced by progesterone must, therefore, be the consequence of alterations in the microenvironment of the tyrosine chromophore following steroid binding. In fact, proton magnetic resonance spectra have shown that the steroid induces marked changes in the conformation of uteroglobin affecting mainly the aromatic aminoacids (Temussi et al. 1980). Evidence from these NMR studies and from chemical modification experiments have shown that the histidine residue (position 8) is indirectly involved in steroid binding (Beato et al. 1980; Saavedra and Beato 1980).

4 Uteroglobinlike Proteins in Human Uterine Fluid

Although the function of uteroglobin in the rabbit has not been established yet, the fact that it is induced by progesterone and present in the uterine lumen during the preimplantation phase of pregnancy, in conjunction with its affinity for progesterone and other steroids, has raised speculations concerning its involvement in the implantation process. Therefore it was interesting to know whether a protein with the structural or functional properties of uteroglobin was present in fertile women. The literature contains contradictory reports in this respect, based mainly on immunological techniques (Beier et al. 1970; Daniel 1971, 1973; Shirai et al. 1972; Wolf and Mastroianni 1975; Beier 1979; Aitken 1979; Lambadarios pp. 99).

We have examined the uterine fluid proteins of 50 fertile women using polyacrylamide gel electrophoresis, immunological procedures and various progesterone-binding assays (Voss and Beato 1977).

4.1 Polyacrylamide Gel Electrophoresis

A detailed analysis of the polyacrylamide gel pattern of uterine fluid proteins and the corresponding serum proteins did not indicate the presence in the human uterine fluid of a protein with the electrophoretic properties of rabbit uteroglobin. Although quantitative changes were observed in the pattern of proteins in the uterine fluid when compared with the corresponding serum proteins, there were no qualitative differences. In addition no significant differences were found between the uterine fluid proteins collected during the preovulatory phase of the cycle and those collected during the postovulatory phase.

In polyacrylamide gels containing urea and SDS (Swank and Munkres 1971) no indication for a protein of the molecular weight of rabbit uteroglobin was obtained.

The conclusion of these gel electrophoretic studies is that uterine fluid of humans contains mainly serum proteins, but the presence of small amounts of uterus-specific proteins cannot be excluded with this technique.

4.2 Immunological Studies

A monospecific antiserum directed against rabbit uteroglobin was used to detect the presence of cross-reacting proteins in the human uterine fluid. In a study involving 37 fertile women in preovulatory and postovulatory phases of the menstrual cycle, no immunoprecipitation line was detected in Ouchterlony double diffusion plates. The detection limit of this assay is such that a positive reaction would have been observed if an uteroglobin-like protein represented more than 1%-2% of the total protein mass.

4.3 Progesterone Binding

The charcoal absorption technique was used to study the binding of tritiated progesterone to proteins of the uterine fluid and serum. There is higher progesterone binding activity in the proteins of the uterine fluids than in serum (Table 2), but the qualitative aspects of binding were very similar. Whereas 5α-dehydehydroprogesterone and norethynodrel bind to rabbit uteroglobin with higher affinity than progesterone (Beato 1976) the binding of [^{3}H]progesterone to uterine and serum proteins was similarly competed by nonradioactive steroids including progesterone 5α-dehydroprogesterone and norethynodrel (Fig. 7). This indicates that the steroid specificity of the progesterone binding protein of human uterine fluid is identical to that of serum proteins, and very different from the steroid specificity of uteroglobin (Beato 1976).

The main progesterone binding activity in uterine fluid proteins elutes from a column of Sephadex G-100 in the same position as the main progesterone binding activity of serum, slightly ahead of albumin in the position corresponding to the corticosteroid

Table 2. Binding of progesterone to the proteins of human uterine fluid and serum

Protein fraction	n	[^{3}H]Progesterone [a] binding cpm/mg protein ($\bar{X} \pm S$)
Serum	24	2100 ± 1259
Uterine fluid	24	4216 ± 1841
Serum, preovul.	12	1947 ± 645
Uterine fluid, preovul.	12	4340 ± 1103
Serum, postovul.	12	3055 ± 1435
Uterine fluid, postovul.	12	5505 ± 1456

[a] (^{3}H)Progesterone, specific radioactivity 5 Ci/mmol

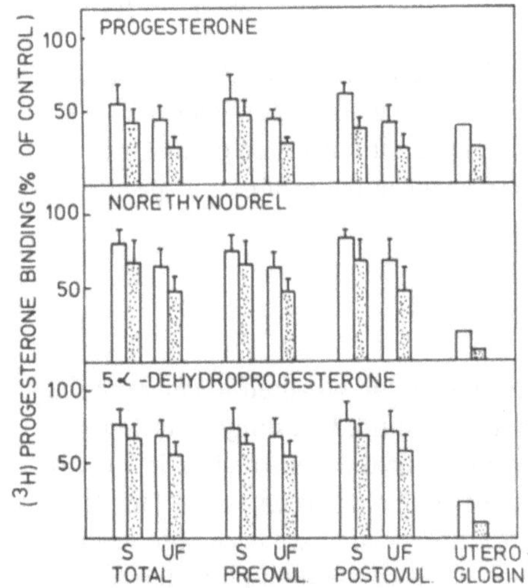

Fig. 7. Effect of various competing steroids on the binding of [^{3}H]progesterone to the proteins of human serum and uterine fluid, as compared with rabbit uteroglobin. Progesterone binding assays were performed with serum proteins and with proteins of the uterine fluid of humans in the absence of competing steroid, and in the presence of the indicated nonradioactive steroids. The competing steroids were added at two different concentrations: 4 μM (*open bars*) and 10 μM (*dotted bars*). The values are expressed as percentage of the binding observed in the absence of competing steroid, and the standard errors are indicated. *Total* refers to the 24 cases studied, of which half were in the preovulatory phase and half in the postovulatory phase of the cycle. For comparison, the effect of the same competing steroids on [^{3}H]progesterone binding to rabbit uteroglobin is also shown

binding globulin (Koblinsky et al. 1972). Neither progesterone binding activity, nor any protein peak could be detected in the position where rabbit uteroglobin would elute from the Sephadex column (Fig. 8).

These findings were confirmed using antibodies against purified uteroglobin. When either serum proteins or human uterine fluid proteins were incubated with these antibodies there was no change in their progesterone binding activity. However, incubation of the uterine fluid proteins of the rabbit with the same antibody results in a complete inhibition of their progesterone binding activity (Beato and Baier 1975). In conclusion, although there is a relative accumulation of the corticosteroid binding globulin, transcortin, in the uterine fluid, a protein with the steroid binding properties of rabbit uteroglobin was not detected in the human uterine cavity.

In this respect it should be mentioned that the exclusive uterine localization of uteroglobin in the rabbit has been recently disproved. A protein structurally identical to uterine uteroglobin has been purified from the lung of both male and female rabbits. Interestingly, this uteroglobin-like protein from the lung shows similar progesterone binding properties as uterine uteroglobin (Beato and Beier 1978).

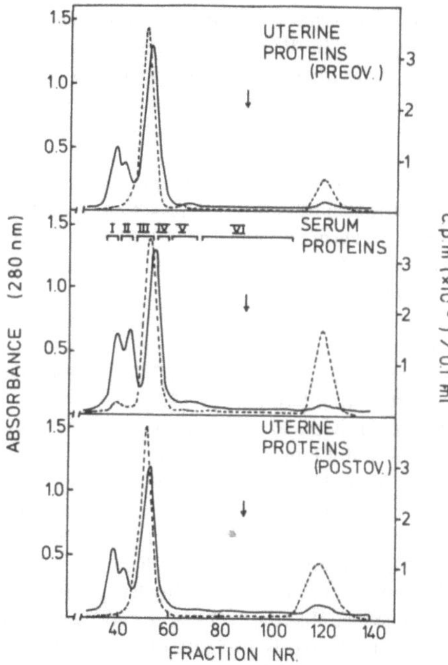

Fig. 8. Chromatography of the proteins from human serum (*middle panel*) or from uterine fluid during the preovulatory phase (*upper panel*) or postovulatory phase (*lower panel*) of the menstrual cycle were submitted to chormatography on similar columns of Sephadex G-100. The samples were incubated with [^{3}H]progesterone before application to the column. The absorbance at 280 nm (——) and the radioactivity in 0.1-ml aliquots (- - -) are depicted. The arrows indicate the position where rabbit uteroglobin would elute from these columns

Acknowledgments. Some of the experiments reported in this chapter were carried out in collaboration with J. Arnemann, H.M. Beier, M. Buehner, A. Nieto, N. Ponstingl, and H.-J. Voss. I thank the Behringwerke for providing the animals used in these studies, and the Schering AG, Berlin, for the gift of the nonradioactive steroids. The Deutsche Forschungsgemeinschaft provided financial support (SFB 103, B_1).

References

Arthur AL, Cowan BD, Daniel JC Jr (1972) Steroid binding to blastokinin. Fertil Steril 23:85-92

Arthur AL, Daniel JC Jr (1972) Progesterone regulation of blastokinin production and maintenance of rabbit blastocysts transferred into uteri of castrated recipients. Fertil Steril 23:115-122

Aitken RJ (1979) Uterine proteins. In: Finn CA (ed) Oxford Reviews of reproductive biology. Oxford University Press, Oxford, pp 351-382

Atger M, Mornon JP, Savouret JF, Loosfelt H, Fridlansky F, Milgrom E (1980) Uteroglobin: a model for the study of the mechanism of action of steroid hormones. In: Beato M (ed) Steroid induced uterine proteins. Elsevier/North-Holland Biomedical Press, Amsterdam New York Oxford, pp 341-350

Beato M (1976) Binding of steroids to uteroglobin. J Steroid Biochem 7:327-334

Beato M (1977) Physico-chemical characterization of uteroglobin and its interaction with progesterone. In: Johnson MH (ed) Development in mammals, vol 2. Elsevier/North-Holland Biomedical Press, Amsterdam Oxford New York, pp 173-198

Beato M, Baier R (1975) Binding of progesterone to the proteins of the uterine luminal fluid. Identification of uteroglobin as the binding protein. Biochim Biophys Acta 392:346-356

Beato M, Beier HM (1978) Characteristics of the purified uteroglobin-like protein from rabbit lung. J Reprod Fertil 53:305-314

Beato M, Arnemann J, Voss HJ (1977) Spectrophotometric study of progesterone binding to uteroglobin. J Steroid Biochem 8:725-730

Beato M, Saavedra A, Puigdomenech P, Tancredi T, Temussi PA (1980) Progesterone binding to uteroglobin. In: Beato M (ed) Steroid induced uterine proteins. Elsevier/North-Holland Biomedical Press, Amsterdam New York Oxford, pp 105-119
Beier HM (1968) Uteroglobin: a hormone-sensitive endometrial protein involved in blastocyst development. Biochim Biophys Acta 160:289-291
Beier HM (1979) Endometrial secretion proteins – Biochemistry and biological significance. In: Beller FK, Schumacher GFB (eds) The biology of the fluids of the female genital tract. Elsevier/North-Holland Biomedical Press, New York Amsterdam Oxford, pp 89-114
Beier HM, Petry G, Kühnel W (1970) Endometrial secretion and early mammalian development. In: Gibian H, Plotz EJ (eds) Mammalian reproduction. Springer, Berlin Heidelberg New York, pp 264-285
Bennett C (1967) Paper chromatography and electrophoresis; special procedure for peptide map. Methods Enzymol 11:330-339
Bernstein GS, Aladjem F, Chen S (1971) Proteins in human endometrial washings – a preliminary report. Fertil Steril 22:722-730
Buehner M, Beato M (1978) Crystallization and preliminary crystallographic data of rabbit uteroglobin. J Mol Biol 120:337-341
Daniel JC Jr (1971) Uterine proteins and embryonic development. Adv Biosci 6:191-203
Daniel JC Jr (1973) A blastokinin-like component from the human uterus. Fertil Steril 24:326-328
Drapeau GR, Boily Y, Houmard J (1972) Purification and properties of an extracellular protease of *Staphylococcus aureus*. J Biol Chem 247:6720-6726
Ellmann GL (1959) Tissue sulfhydryl groups. Arch Biochem Biophys 82:70-77
Fridlansky F, Milgrom E (1976) Interaction of uteroglobin with progesterone, 5α pregnane-3, 20-dione and estrogens. Endocrinology 99:1244-1251
Gray WK (1972) Endgroup analysis using dansyl chloride. Methods Enzymol 25:121-138
Houmard J, Drapeau GR (1972) Staphylococcal protease: A proteolytic enzyme specific for glutamoyl bonds. Proc Natl Acad Sci USA 69:3506-3509
Koblinsky M, Beato M, Kalimi M, Feigelson P (1972) Glucocorticoid-binding proteins of rat liver. II. Physical characterization and properties of the binding proteins. J Biol Chem 247:7897-7904
Nieto A, Ponstingl H, Beato M (1977) Purification and quaternary structure of the hormonally induced protein uteroglobin. Arch Biochem Biophys 180:82-92
Ponstingl H, Nieto A, Beato M (1978) Amino acid sequence of progesterone induced rabbit uteroglobin. Biochemistry 17:3908-3912
Puigdomenech P, Beato M (1977) Nuclear magnetic resonance studies on rabbit uteroglobin. FEBS Lett 83:217-221
Saavedra A, Beato M (1980) Influence of chemical modifications of amino acid side chains on the binding of progesterone to uteroglobin. J Steroid Biochem 13:1347-1353
Shirai E, Iizuka R, Notake Y (1972) Analysis of human uterine fluid proteins. Fertil Steril 23:522-528
Swank RT, Munkres KD (1971) Molecular weight analysis of oligopeptides by electrophoresis in polyacrylamide gel with sodium dodecyl sulphate. Anal Biochem 39:462-477
Temussi PA, Tancredi T, Puigdomènech P, Saavedra A, Beato M (1980) Interaction of S-carboxymethylated uteroglobin with progesterone. Biochemistry 19:3287-3293
Urzua MA, Stambaugh R, Flickinger G, Mastroianni L Jr (1970) Uterine and oviduct fluid protein patterns in the rabbit before and after ovulation. Fertil Steril 21:860-865
Utermann G, Menzel HJ, Schoenborn W (1975) Plasma lipoprotein abnormalities in a case of primary high-density-lipoprotein (HDL) deficiency. Clin Genet 8:258-268
Voss HJ, Beato M (1977) Human uterine fluid proteins: gel electrophoretic pattern and progesterone-binding properties. Fertil Steril 28:972-980
Westphal U (1971) Steroid-protein interaction. Springer, Berlin Heidelberg New York
Wolf DP, Mastroianni L Jr (1975) Protein composition of human uterine fluid. Fertil Steril 26:240-247
Yphantis DA (1964) Equilibrium ultracentrifugation of dilute solutions. Biochemistry 3:297-317

Steroid Binding to Uteroglobin

F. Fridlansky and E. Milgrom

Groupe de Recherches sur la Biochimie Endocrinienne et la Reproduction (INSERM U 135), Faculté de Médicine Paris-Sud, Bicêtre, France

Conflicting reports have appeared in the literature on the binding of steroids by uteroglobin (Beier 1966, 1968) or blastokinin (Krishnan and Daniel 1967). Either binding of both progesterone and estradiol was described (Urzua et al. 1970; Arthur et al. 1972; Garcea et al. 1974; Beato and Baier 1975; Beato 1976) or no actual binding of steroids was observed (Goswami and Feigelson 1974; Rahman et al. 1975).

In this study we show that uteroglobin does bind progesterone and 5α-pregnanedione, whereas the binding of estradiol has the characteristics of a nonspecific interaction.

1 Steroid Binding Is Dependent on the Presence of SH Reduced Groups in Uteroglobin

Uterine fluid was obtained from 5-day-pregnant rabbits; uteroglobin was purified to homogeneity (Mayol and Longenecker 1974; Fridlansky and Milgrom 1976) by chromatography on Sephadex G 75 followed by DEAE-cellulose chromatography. The purified protein binds progesterone (especially if no lyophilization has been performed during the purification procedure); however, the binding can be markedly increased by SH-reducing agents (Fig. 1). The most effective was Cleland's reagent (dithiothreitol =DTT). At optimal concentration (20 mM) it increased the binding 14-fold. Dithioerythritol was slightly less effective whereas β-mercaptoethanol was active at concentrations varying between 25 and 100 mM. SH-blocking reagents had an inhibitory effect (Fig. 2). Iodoacetate and iodoacetamide were more effective in preventing binding of progesterone to uteroglobin than ρ-hydroxymercuribenzoate and dithiobisnitrobenzoic acid.

The effect of SH-reducing agents was found to bear only on the number of steroid binding sites and not on the affinity constant for progesterone. In non-SH-reduced uteroglobin only about one in every 500 molecules bound progesterone, in SH-reduced uteroglobin (dithiothreitol 20 mM) one in every two molecules had the steroid binding properties.

2 Specificity of Steroid Binding to Uteroglobin

Uteroglobin was reduced by dithiothreitol (20 mM) and dialyzed with ^{3}H-progesterone in the presence of competing unlabeled steroids. As shown in Table 1, 5α-pregnanedione was a very effective inhibitor, about three times more potent than progesterone itself. Synthetic progestins (R5020, norethisterone, chlormadinone acetate), which were

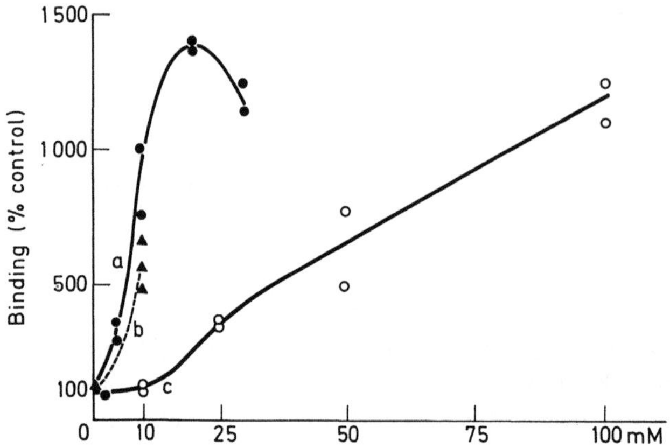

Fig. 1. Effect of SH-reducing agents on progesterone binding to uteroglobin. Dithiothreitol (**a**), dithioerythritol (**b**), and β-mercaptoethanol (**c**) were added to uteroglobin (50 μg/ml). Incubation was for 60 min at 25°C, then dialysis was performed with 10 nM ^{3}H-progesterone for 40 h at 0°C. (100= binding in the absence of SH-reducing agent

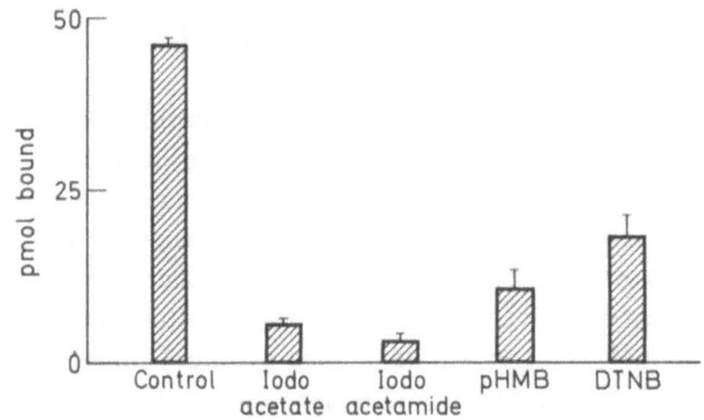

Fig. 2. Effect of SH-blocking agents on progesterone binding to uteroglobin. Iodoacetate (10 mM), iodoacetamide (10 mM), *p*-hydroxymercuribenzoate (*pHMB* 1 mM) and dithiobisnitrobenzoic acid (*DTNB*, 10 mM) were added to uteroglobin (100 μg/ml). Incubation and dialysis were performed as in Fig. 1. Results are the mean ± s.e. of three experiments

bound very strongly to the uterine progesterone receptor, had little affinity for uteroglobin. Natural estrogens (estradiol, estrone, estriol) were also bound very poorly, whereas diethylstilbestrol had some affinity. Cortisol and testosterone did not bind to uteroglobin. In order to measure precisely the equilibrium dissociation constants, radioactive progesterone, 5α-pregnanedione and estradiol were dialyzed with uteroglobin. As shown in Fig. 3 ^{3}H-5α-pregnanedione and ^{3}H-progesterone bound to the same site with $K_D = 1.3 \times 10^{-7}$M and 4.1×10^{-7}M, respectively. ^{3}H-Estradiol showed a very weak and nonsaturable interaction.

To compare in another manner the interaction of progesterone and estradiol with uteroglobin, their binding to this protein was studied before and after SH-group reduction. As shown in Fig. 4, whereas binding of ^{3}H-progesterone was increased about

Table 1. Comparison of steroid specificities of uteroglobin [a] and progesterone receptor from rabbit uterus [b]

	Relative binding affinity	
	Uteroglobin	Receptor
Progesterone	100	100
5α-pregnane-3,20-dione	317	108
R5020	10	158
Norethisterone	2.5	141
Chlormadinone acetate	1.8	127
17-hydroxyprogesterone	6.8	3
Testosterone	0	1.5
5α-dihydrotestosterone	2.1	–
Cortisol	0	–
Estradiol	0.8	2.5
Estriol	0	– –
Estrone	1	–
Diethylstilbestrol	15	0

[a] Uteroglobin was reduced by 20 mM dithiothreitol. Binding of ^{3}H-progesterone was studied by equilibrium dialysis in the presence of various concentrations of unlabeled steroids

[b] Uterine cytosol was obtained from immature rabbits primed with estradiol. Binding of ^{3}H-progesterone was studied by dextran-coated charcoal adsorption in presence of 1 μM cortisol and 30% glycerol.

Relative binding affinities were established from the comparison of the concentration of competing steroid necessary to inhibit 50% of ^{3}H-progesterone binding

100-fold by this treatment, no significant effect was observed on ^{3}H-estradiol binding. Moreover the binding of estradiol to the reduced protein was about 500-fold lower than that of progesterone.

3 Discussion

The main unresolved problem is that of the physiological role of uteroglobin and especially how it can be related to its hormone binding properties. It is very difficult even to evaluate how much progesterone is bound to uteroglobin in vivo (Atger et al. 1980b). The concentration of progesterone in the uterine fluid is unknown, since the methods of collection lead to a nonquantitated dilution. On the other hand, the proportion of uteroglobin which is reduced and thus able to bind steroids is also not known. Finally the relatively low affinity of uteroglobin for steroids (two orders of magnitude lower than that of receptor) could impede a quantitatively important binding of progesterone by this protein. However, the extremely high concentration of uteroglobin (40%-60% of total proteins in the uterine fluid at day 5 of pregnancy) probably compensates for this low affinity.

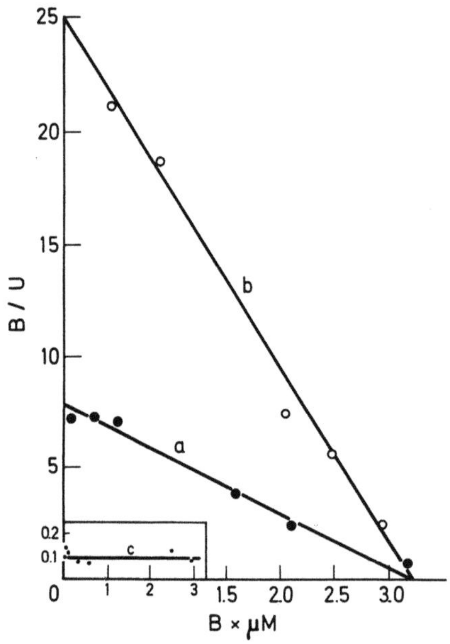

Fig. 3. Binding of ^{3}H-progesterone (*a*), ^{3}H-5α-pregnanedione (*b*) and ^{3}H-estradiol (*c*) uteroglobin. Uteroglobin (105 μg/ml) was incubated for 60 min at 25°C with 20 mM DTT. Dialysis was then performed with radioactive steroids as described in Fig. 1

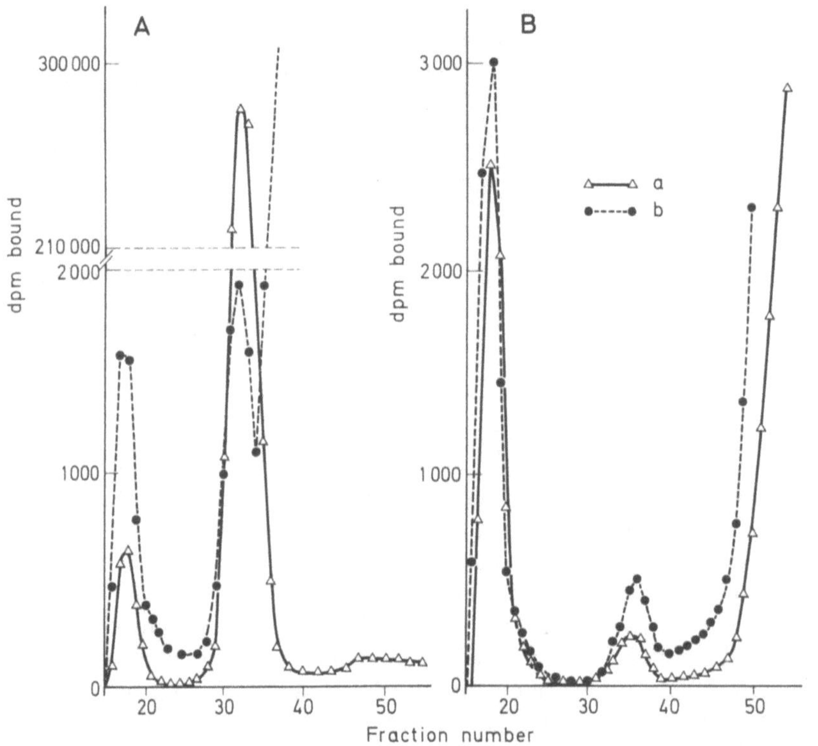

Fig. 4, A,B. Effect of SH-group reduction on the binding of progesterone (**A**) and estradiol (**B**) to uteroglobin. Uterine fluid was incubated with ^{3}H-progesterone (SA= 47.6 Ci/mmol) or ^{3}H-estradiol (SA= 43 Ci/mmol) 10 nM either in the presence (*b*) or in the absence (*a*) of 20 mM dithiothreitol. It was then chromatographed on Sephadex G75. Uteroglobin is eluted between fractions 30 and 40

If in vivo uteroglobin does bind progesterone, what could be the physiological significance of this binding? A K_D of 10^{-7}M, only slightly lower than that of many steroid-metabolizing enzymes, suggested to us the possibility that uteroglobin could have such a catalytic activity. The protein (SH-reduced or not) was thus incubated with progesterone in the presence of NADPH or NADH. In all cases only unchanged progesterone was recovered at the end of the incubation. It still remains possible that uteroglobin is an enzyme but metabolizes a steroid other than progesterone. The relatively high affinity of uteroglobin for 5α-pregnane-3,20-dione may suggest a possible physiological role for this steroid.

Alternative possibilities are that progesterone is carried to some cellular or subcellular structure by uteroglobin, or on the contrary, that the binding prevents progesterone from reaching this structure. For example, it is possible that progesterone is necessary for the maintenance of the uterus in a decidual state; however, the high levels of hormone which diffuse into the uterine lumen could have an adverse effect on the blastocyst. Binding by uteroglobin could then prevent these high levels of progesterone from reaching the blastocyst while still allowing the action of the hormone on the uterus (Atger et al. 1980a).

Acknowledgments. This work was supported by the Institut National de la Sante et de la Recherche Medicale, by the Centre National de la Recherche Scientifique and the Fondation pour la Recherche Medicale Francaise.

Abbreviations

Chlormadinone acetate: 6-chloro-17-acetoxy-pregna-4,6diene-3,20-dione
norethisterone: 17α-ethinyl-17-hydroxyestra-4-en-3-one
R5020: 17,21-dimethyl-19-nor-pregna-4,9-diene-3,20-dione
5α-pregnanedione: 5α-pregnane-3,20-dione
17-hydroxyprogesterone: 17α-hydroxy-pregn-4-ene-3,20-dione
dihydrotestosterone: 17β-hydroxy-5α-androstan-3-one
DTT: dithiothreitol.

References

Arthur AT, Cowan BO, Daniel JC (1972) Steroid binding to blastokinin. Fertil Steril 23:85-92
Atger M, Fridlansky F, Savouret JF, Loosfelt H, Milgrom E (1980a) L'uteroglobine. Reprod Nutr Develop 20:553-555
Atger M, Mornon JP, Savouret JF, Loosfelt H, Fridlansky F, Milgrom E (1980b) Uteroglobin: a model for the study of the mechanism of action of steroid hormones. In: Beato M (ed) Steroid induced uterine proteins. Elsevier/North-Holland Biomedical Press, Amsterdam New York Oxford, pp 341-350
Beato M (1976) binding of steroids to uteroglobin. J Steroid Biochem 7:327-334
Beato M, Baier R (1975) Binding of progesterone to the proteins of the uterine luminal fluid. Identification of uteroglobin as the binding protein. Biochim Biophys Acta 392:346-356
Beier HM (1966) Das Proteinmilieu in Serum, Uterus und Blastocysten des Kaninchens vor der Nidation. In: Beermann W (Chairman) Coll. Biochemie der Morphogenese. Deutsche Forschungsgemeinschaft, Konstanz

Beier HM (1968) Uteroglobin: A hormone-sensitive endometrial protein involved in blastocyst development. Biochim Biophys Acta 160:189-291

Fridlansky F, Milgrom E (1976) Interaction of uteroglobin with progesterone, 5α-pregnane-3, 20-dione and estrogens. Endocrinology 99:1244-1251

Garcea N, Campo S, Caruso A, Milano L, Bompiani A (1974) Binding of steroids to blastokinin. Int J Fertil 19:197-201

Goswami A, Feigelson M (1974) Differential regulation of a low-molecular weight protein in oviductal and uterine fluid. Endocrinology 95:669-675

Krishnan RS, Daniel JC (1967) "Blastokinin": Inducer and regulator of blastocyst development in the rabbit uterus. Science 158:490-492

Mayol RF, Longenecker ME (1974) Development of a radioimmunoassay for blastokinin. Endocrinology 95:1534-1542

Rahman SS, Velayo N, Domres P, Billiar RB (1975) Evaluation of progesterone binding to uteroglobin. Fertil Steril 26:991-995

Urzua MR, Stambaugh R, Flickinger G, Mastroianni L (1970) Uterine and oviduct fluid protein patterns in the rabbit before and after ovulation. Fertil Steril 21:860-865

Steroids Receptors in the Genital Tract and the Monitoring of Steroid Hormone Action

P.W. Jungblut

Max-Planck-Institut für experimentelle Endokrinologie, Hannover, Germany

Organs of the mammalian genital tract were the first to be recognized as targets for steroid hormones. Hormonal action is mediated through so-called receptors, which are in fact transcription-regulating proteins. The earlier notion, that receptors are not only hormone-specific but could also occur each in but one cell type was short-lived (Jungblut et al. 1971, Wagner et al. 1972). Receptors for both male and female sex steroids and for glucocorticoids do exist simultaneously within one and the same cell (Table 1). Since the presence of steroid receptors in target cells is a prerequisite for hormone-sensitivity, the quantitative assay of receptors has become a widely practized laboratory handicraft. The apparently simple techniques are however complicated by numerous intricacies, and moreover, the interpretation of results frequently dwells on misconceptions.

Table 1. Steroid Receptors found in the genital tract

	Estrogen R	Gestagen R	Androgen R	Glucocorticoid R
Ovary	Scott and Rennie 1971	Leymarie and Guerignian 1970		
Oviduct	Stone and Baggett 1967	O'Malley et al. 1971		
Uterus	Taft and Gorski 1966	Falk and Bardin 1970	Jungblut et al. 1971	Wagner et al. 1972
Vagina	Jungblut et al. Stone and Baggett 1967			
Testis	Brinkmann et al. 1972		Bruchovsky and Wilson 1968	
Prostate	Jungblut et al. 1971		Bruchovsky and Wilson 1968	
Seminal vesicle	Jungblut et al. 1967		Stern and Eisenfeld 1969	

1 Peculiarities of Receptor Assay

Steroid receptors are intracellular proteins, which bind their respective steroids non-covalently, but with high affinity and specificity. Their assay thus "only" requires

(1) a quantitative extraction, (2) the employment of a suitable, sufficiently labelled steroid for occupying all available binding sites, and (3) the clear separation between specific steroid-receptor complexes, unspecific complexes, and the excess free steroid.

1.1 Preparation of Tissue Extracts

The quantitative extraction of cell constituents from a tissue demands a quantitative cell rupture. This can be achieved by various procedures and checked by counting the ratio of nuclei to unbroken cells in the homogenate. Considering the accuracy of all subsequent steps, a ratio of 98:2 usually suffices. For the extraction of receptors, one other aspect ought to be considered. These proteins are heat-labile and local overheating during homogenization, therefore, must be avoided. Immersion of Potter-Elvejhem homogenizers in ice-water is no safeguard against this when smooth plastic pestles (Teflon, PVC, Nylon) are used, which are poor heat dissipators. The less gentle glass pestle, here, is the better choice. Furthermore, the homogenization procedure must be adapted to the particular purpose. Receptors are localized in the soluble phase of the cytoplasm, in cytoplasmic structures, and in the cell nucleus. Homogenization of frozen tissue, while giving the highest yields for cytosol receptor (Table 2) prohibits the analysis of cell structures which do not withstand freezing and thawing. For most investigators, neither the membrane-bound precursors nor the nuclear ("in-action") receptor are of prime interest, but rather the "free" receptor in the "cytosol". This fraction, the particle-free supernatant of homogenates prepared with hypotonic buffers or isotonic sucrose solutions, should not be mistaken for the soluble phase of the cytoplasm, which cannot be prepared in a pure state, unless from isolated cells. Even the thorough perfusion of an organ through a main artery or vein leaves some admixture of extracellular proteins in its "cytosol", due to incomplete drainage of lymph capillaries and intercellular spaces.

Table 2. Yields of estradiol receptor from calf uterus after different homogenization procedures

Method	E-2-R extracted per mg tissue protein	
	mol x 10^{-13}	%
Ultraturrax	1.1	78
All-glass Potter	1.2	86
Dismembrator	1.4	100

1.2 Assay of Cytosol Receptors

I am stressing this point, because the admixtures – besides unspecific binders – can contain specific steroid-binding plasma proteins, the presence of which complicate the assay of some receptors. Keeping in mind that receptors for more than one steroid hormone occur simultaneously within the same cell (Jungblut et al. 1971; Wagner et.

al. 1972), one is thus confronted with the necessity of discriminating. The best tools for this are steroids, which would bind exclusively to only one of the high-affinity proteins present, as exemplified by the first four compounds in Table 3. With the exception of estradiol, none of the other natural steroids fully qualifies for the assay of its own receptor. CBG (transcortin), is a good binder for progesterone and its affinity for cortisol resembles that of the glucocorticoid receptor. There are remarkable cross-affinities between dihydrotetosterone and progesterone vis-a-vis their respective receptors (Wagner et al. 1972), which might be of physiological importance but undoubtedly interfere with receptor assay.

Table 3. Steroids best suited for the quantitative assay of specific binders

Protein	Steroid
Estrogen receptor	Estradiol
Androgen receptor	R 1881 (Roussel/Uclaf) Dihydrotestosterone [a]
Gestagen receptor	R 5020 (Roussel/Uclaf), ORG 2058 (Organon) *d*-Norgestrel (Schering AG)
Glucocorticoid receptor	Dexamethasone
Corticoid binding globulin	Cortisol
Sex hormone binding globulin	Dihydrotestosterone

[a] In presence of NAD-nucleosidase with technique discriminating against SHBG in target tissue extracts of primates

It is beyond the scope of this lecture to reembark on a comparison of the numerous methods employed for the measurement of steroid-receptor complexes (Jungblut et al. 1972). All of them rest on one commonplace: the labeled compounds should be better than 95% pure at the time of use ! Radiolysis products can react covalently with proteins and counterfeit specific binding. Tetra-^{3}H-labeled compounds are more accident-prone than those containing only two tritium atoms.

In briefly mentioning our method of choice, the agar gel electrophoresis at low temperature (Wagner 1972), I pursue a dual purpose: First, I wish to demonstrate that this method, besides the obligatory detachment of labeled steroid from unspecific binders and the removal of free steroid, can differentiate in a single-step analysis between intracellular receptors and specific extracellular binders present in the same extract (Wagner and Jungblut 1976). CBG and SHBG (TeBG) contain terminal sialic acid residues which can be removed enzymatically. The resulting decrease in electrophoretic mobility allows, for example, the discrimination between CBG and the progestin receptor (Fig. 1). Second, I wish to emphasize a frequently overlooked phenomenon: in spite of equilibrium constants of dissociation in the range of 10^{-9}-10^{-10}M, steroid-protein complexes can still display dissociation rates at low temperature which complicate their quantitation. The examples given in Fig. 2 demonstrate that while estradiol is a suitable reagent for the assay of its own receptor, progesterone is not as well suited for progestin receptor assay. The synthetic progestagens R 5020 (Roussel/Uclaf), ORG 2058 (Orga-

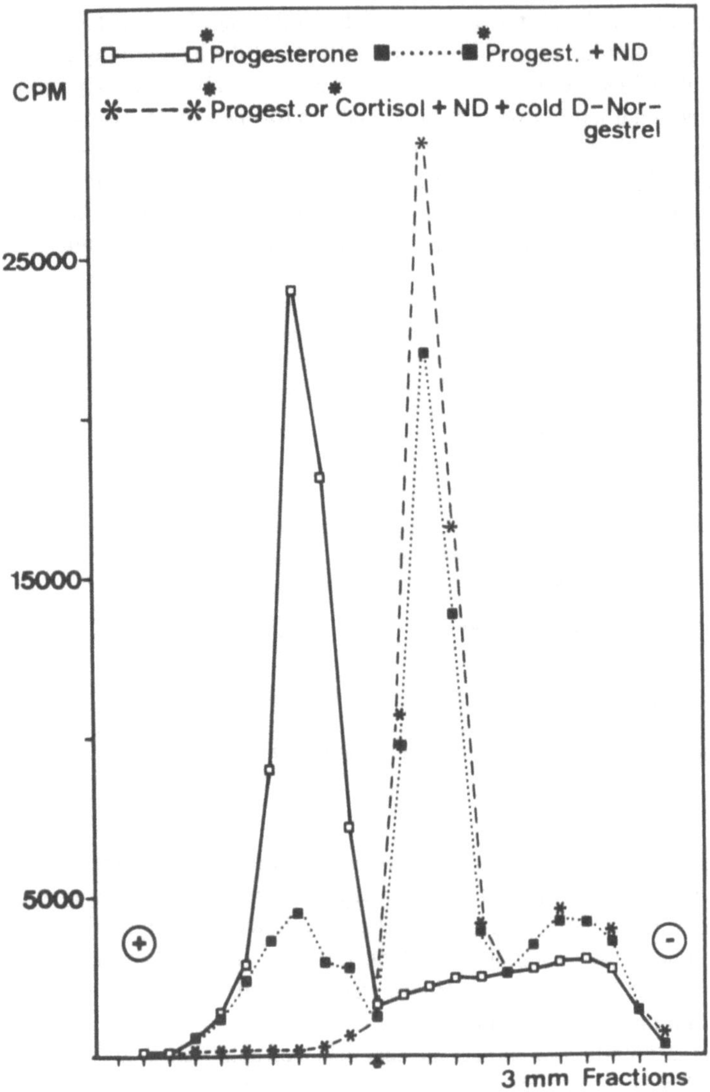

Fig. 1. Human uterus extract: discrimination between CBG and progesterone receptor after incubation with neuraminidase (ND). Uterus from secretory phase of menstrual cycle. Extract incubated with [^{3}H]cortisol (5×10^{-8}M) for 4 h at 0-2°C (excess steroid removed by charcoal treatment). □—□ control; ■---■ plus 80 μg/ml neuraminidase; *----* plus 80 μg/ml neuraminidase and 1.25×10^{-5} M cold *d*-norgestrel. Electrophoresis: 50 μl samples run in 0.9% agar gel in 0.05 M sodium diethylbarbiturate acetate buffer, pH 8.2 for 2 h at 2-3°C gel temperature

non, OK) and *d*-norgestrel (Schering AG, not shown), in contrast, form progestin receptor complexes of a stability similar to that of the estradiol-estrogen receptor complex. If synthetic steroids are not available for correcting the "deficiency" of the natural hormone, extrapolation to zero time is necessary in order to arrive at the correct value

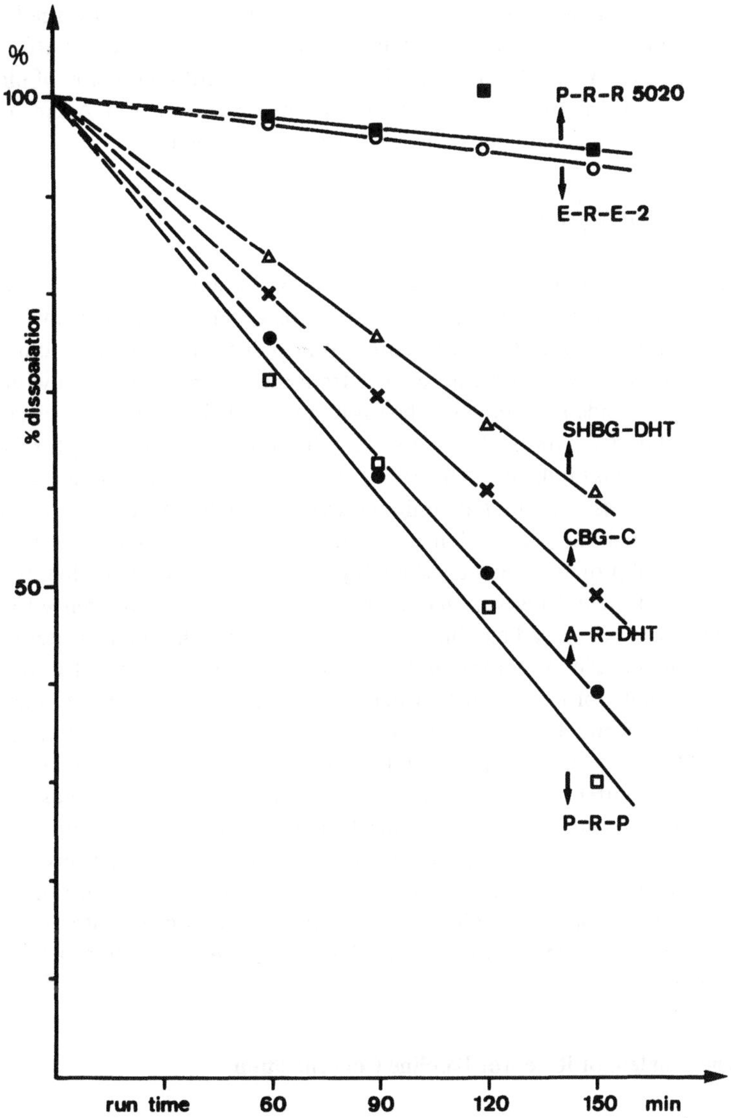

Fig. 2. Dissociation of various "high affinity" complexes between steroid hormones and steroid binding proteins during analysis in agar electrophoresis. *P-R,* progesterone receptor; *E-R,* estrogen receptor; *A-R,* androgen receptor; *SHBG,* sex hormone binding globulin; *CBG,* corticosteroid binding globulin; *R 5020,* 17,21-dimethyl-19-nor-4,9-pregnadiene-3,20-dione; *E-2,* estradiol; *DHT,* dihydrotestosterone; *C,* cortisol

of receptor concentration. It is not important, in this context, whether a true, reversible dissociation occurs or a dissociation followed by steroid metabolism and finally receptor denaturation (loss of binding ability).

The apparently high rate of dissociation of the dihydrotestosterone/androgenreceptor complex, even at low temperature, has recently been explained by a reduction of the steroid in tissue extracts to 3α-and 3β-ol derivatives of low receptor affinity. Cosubstrate depletion by NAD-nucleosidase prohibits the enzymatic conversion of dihydrotestosterone, which then serves as a very suitable ligand (Moeller 1981). A similar situation could prevail for the progesterone/gestagenreceptor complex.

1.3 Choice of Reference

The predominant unspecific binder in "cytosols" is albumin, the major protein constituent of blood plasma and lymph. Its interference with receptor assay is ruled out in high-affinity site saturation procedures by stripping off the weakly bound steroids. (In stored extracts, a problem can arise from the coating of albumin with saponification products of lipids.) Albumin can be easily quantitated by immunological methods. We have used its assay routinely for assessing the contamination of "cytosols" with extracellular proteins (Jungblut et al. 1967). The formula: Total protein –albumin x 1.67= tissue protein assumes that albumin represents roughly 60% of the contaminating proteins. Although this assumption is not quite correct, receptor concentrations based on "soluble cell protein" come considerably closer to the true value than those based on "cytosol" protein. However, in extreme situations, even the reference to "soluble cell protein" can be misleading. Such a situation arises for the assay of uterine receptors during the so-called water imbibition phase caused by estradiol, when the proportion of extracellular proteins in the particle-free supernatant dramatically increases. The data of Fig. 3 are taken from our experiments on estradiol receptor synthesis in pig uteri (Hughes et al. 1976a). The animals used in these experiments are ovariectomized and their left uterus horn is surgically detached from the small corpus. This horn serves as an internal control for the receptor depletion/replenishment sequence which is initiated in the open horn by local administration of estradiol via the cervix and which is monitored by quantitative receptor assay. By referring the concentration of receptor to total protein or to "soluble tissue protein", the onset of receptor synthesis at 90 min after hormone pulsing is not apparent, but becomes evident when DNA is used as the reference.

2 The Problem of Receptor Baseline Concentration

The experimental design just mentioned – animals deprived of endogenous hormone, pulsing of one half of a responsive organ while the other half remains unaffected by the administered dose – appears to be optimal for the investigation of hormone action. Its flaws are some "unreasonable" physiological complications. We observed that the shape of the receptor depletion/replenishment response curve depends on the starting level of receptor concentration. This level is subject to a seasonal fluctuation (Fig. 4) and also shows a circadian rhythm (Hughes et al. 1976b). Both might be due to peripherally aromatized steroids, eliciting a receptor turnover without giving rise to measurable amounts of estrogen in the blood. The presence of estradiol in uterus nuclei of ovariectomized pigs supports this view (Jungblut et al. 1976). That the same nuclei contained a several-fold higher number of estradiol binding sites came as a surprise. The

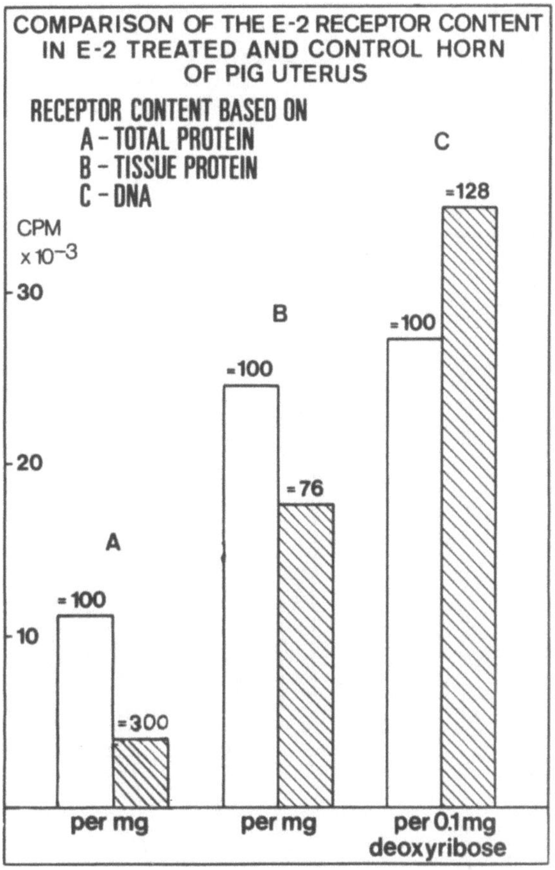

Fig. 3. Receptor content at 120 min after injection of 20 ml of a 2 x 10^{-6}M estradiol solution into the open horn *(shaded)*

nuclear uptake of steroids requires their prior binding to receptors in the cytoplasm. It was therefore reasonable to assume that receptors, in turn, enter the nucleus only in combination with steroids. Unless the release of the steroid from the nucleus proceeds at a faster rate than that of the receptor, we must now consider the possibility of a steroid-independent translocation of receptor from the cytoplasm into the nucleus and, moreover, the possibility of a steroid-independent action, since most of the excess receptor extracted from the nuclei was in its "activated" dimer form. The action, an enhancement of transcription, includes, in the case of the estradiol receptor, the synthesis of its own message. Whether, in a steroid-free system, receptor synthesis and degradation proceed at a steady rate must be doubted. The uterine estradiol receptor content of ovariectomized/hypophysectomized rats fluctuates with a periodicity of 9-12 days (Little et al. 1975). Receptor baseline levels, thus, do not exist.

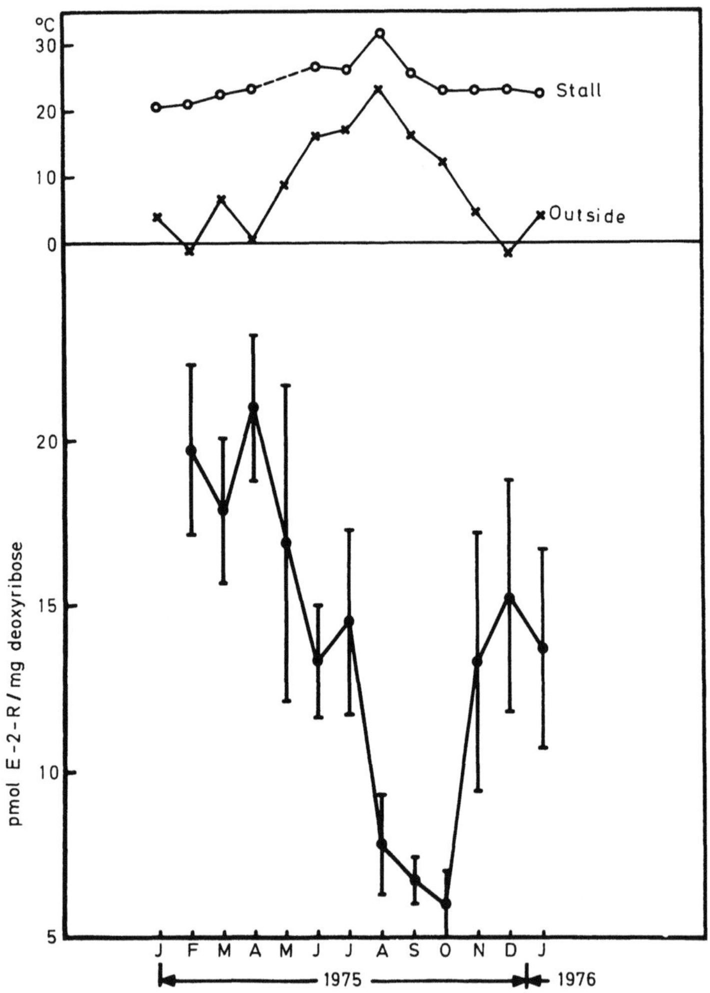

Fig. 4. Fluctuation in uterine estradiol receptor content of ovariectomized pigs during 12 consecutive months. Each point represents the average ±s.d. of ten pools of thirty blind control horns. Temperatures on the day of experiments are plotted

3 The Monitoring of Steroid Hormone Action

Regardless of these new aspects (Jungblut et al. 1976), the assay of receptors remains useful for endocrine research, if properly amended by other parameters. They should be chosen with the intention of covering the complete chain of events (Fig. 5), which is accelerated by the steroid via a "physical catalysis" [1] exerted on receptor conformation. Since the "catalyst" travels with the transcription-regulating protein to its site

1 A term coined in 1959 by Elwood Jensen and Herbert Jacobson in a seminar given at Northwestern University, Chicago, to explain the action of estradiol on the uterus of immature Sprague-Dawley rats, in which organ the hormone is not metabolized

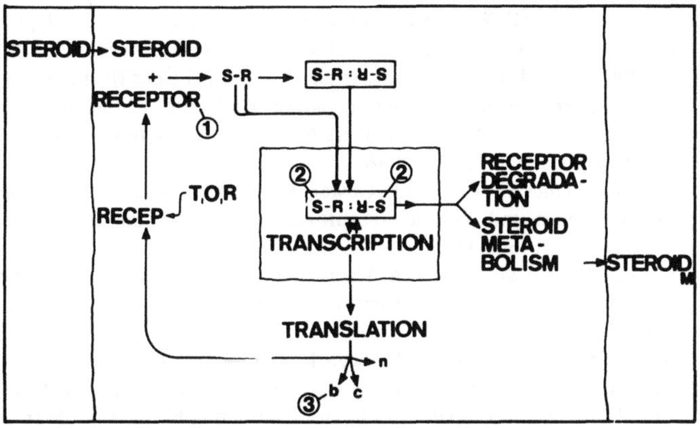

Fig. 5. Basic mechanism of action of steroid hormones. *1* cytosol estradiol receptor; *2* estradiol; *3* cytosol progesterone receptor

of action, this step can be checked by the assay of nuclear-bound steroid, the active moiety of the hormone. The further assay of a specific end product, like the progestin receptor in the estradiol receptor system, then allows for a safe statement on the hormone sensitivity of the tissue investigated.

Abbreviations and Trivial Names

Estradiol	1,3,5 (10)-estratrien-3,17β-diol
Dihydrotestosterone	17β-hydroxy-5α-androstan-3-one
Dexamethasone	9α-fluoro-11β,17,21-trihydroxy-16α-methyl-pregna-1,4-diene-3,20-dione
Cortisol	11β,17α21-trihydroxy-4-pregnene-3,20-dione
d-norgestrel	*d*-17-hydroxy-18-methyl-14-nor-17α-pregn-4-en-20-in-3-one
ORG 2058	16α-ethyl-21-hydroxy-19-nor-pregn-4-ene-3,20-dione
R 1881	17β-hydroxy-17α-methyl-estra-4,9,11-trien-3-one
R 5020	17,21-dimethyl-19-nor-4,9-pregnadiene-3,20-dione
CBG	corticoid binding globulin, transcortin
SHBG	sex hormone binding globulin, testosterone binding globulin

References

Brinkmann AO, Mulder E, Lamers-Stahlhofen GJM, Mechielsen MJ, Van der Molen HJ (1972) An oestradiol receptor in rat testis interstitial tissue. FEBS Lett 26:301-305

Bruchovsky W, Wilson SD (1968) The intranuclear binding of testosterone and 5α-Androstan-17β-01-3-one by rat prostate. J Biol Chem 243:5953-5960

Falk RJ, Bardin CW (1970) Uptake of tritiated progesterone by the uterus of the ovariectomized guinea pig. Endocrinology 56:1059-1063

Hughes A, Szendro P, Teran C, Kielhorn J, Sierralta W, Stone G, Little M, Jungblut PW (1976) Biosynthesis of steroid-hormone receptors. In: Vermeulen A, Breuer H, Jungblut PW (eds) Research on steroids vol VII. North Holland, Amsterdam

Hughes A, Jacobson HI, Wagner RK, Jungblut PW (1976b) Ovarian-independent fluctuations of estradiol receptor levels in mammalian tissues. Mol Cell Endocrinol 5:379-388

Jungblut PW, Hätzel I, Desombre ER, Jensen EV (1967) Über Hormon;"Receptoren". Die oestrogenbindenden Prinzipien der Erfolgsorgane. In: Karlson P (ed) Wirkungsmechanismen der Hormone. Springer Verlag, Berlin Heidelberg New York (Colloquium der Gesellschaft für Biologische Chemie in Mosbach vol 18, pp 58-86)

Jungblut PW, Hughes SF, Görlich L, Gowers U, Wagner RK (1971) Simultaneous occurrence of individual estrogen- and androgen receptors in female and male target organs. Hoppe Seylers Z Physiol Chem 352:1603-1610

Jungblut PW, Hughes SF, Hughes A, Wagner RK (1972) Evaluation of various methods for the assay of cytoplasmic oestrogen receptors in extracts of calf uteri abd human breast cancers. Acta Endocrinol (Kbh) 70:185-195

Jungblut PW, Gaues J, Hughes A, Kallweit E, Sierralta W, Szendro P, Wagner RK (1976) Activation of transcription-regulating proteins by steroids. J Steroid Biochem 7:1109-1116

Leymarie P, Gueriguian JL (1970) Progesterone-binding by the soluble fraction of corpus luteum from the pregnant cow. In: Finkelstein M, Klopper A, Conti C, Cassano C (eds) Research on steroids vol IV, Pergamon Press, Oxford New York, pp 239-256

Little M, Szendro P, Teran C, Hughes A, Jungblut PW (1975) Biosynthesis and transformation of microsomal and cytosol estradiol receptors, J Steroid Biochem 6:493-500

O'Malley BW, Sherman MR, Toft DO, Spelsberg TC, Schrader WJ, Steggles AW (1971) A specific oviduct target-tissue receptor for progesterone identification, characterization, partial purification, inter-compartmental transfer kinetics and specific interaction with the genome. Adv Biosciences 7:213-230

Moeller H (1981) Bestimmung von Androgenrezeptoren, Habilitationsschrift Tübingen

Scott RS, Rennie PIC (1971) An estrogen receptor in the corpora lutea of pseudopregnant rabbits. Endocrinology 89:297-301

Stern J, Eisenfeld A (1969) Androgen accumulation and binding to macromolecules in seminal vesicles: Inhibition by cyproterone. Science 166:233-235

Stone GM, Baggett B (1967) Abstract of lecture presented at the 10th meeting of the Austr Soc for Endocrin, Sydney, Australia (Edited by Austr Soc Endocr)

Toft D, Gorski J (1966) A receptor molecule for estrogens: Isolation from the rat uterus and preliminary characterization, Proc Natl Acad Sci USA 55:1574-1581

Wagner RK (1972) Characterization and assay of steroid hormone receptors and steroid-binding serum proteins by agargel electrophoresis at low temperature. Hoppe Seylers Z Physiol Chem 353:1235-1245

Wagner RK, Jungblut PW (1976) Oestradiol- and dihydrotestosterone receptors in normal and neoplastic human mammary tissue. Acta Endocrinol (Kbh) 82:105-120

Wagner RK, Görlich L, Junglbut PW (1972) Multiple steroid hormone receptors in calf uterus. Binding specifities and distribution. Hoppe Seylers Z Physiol Chem 353:1654-1656

The Synthesis of Steroids and Proteins in the Pig Blastocyst

R.B. Heap, J.E. Gadsby, Catherine Rice, and J.S. Perry

A.R.C. Institute of Animal Physiology, Babraham, Cambridge, U.K.

1 Introduction

The notion has grown in recent years that tne olastocyst may synthesize substances that participate actively in the process of implantation and in the maternal recognition of pregnancy. Deanesly (1967) showed in guinea pigs that fertilized eggs have "... a specific capacity to induce changes in the endometrium..." and that this property is appreciably greater than that of artificial traumatization. The difference in response was not affected by exogenous progesterone treatment. The capacity of the blastocyst to induce changes in the endometrium is also apparent from the localized increase in capillary permeability that occurs in close proximity to blastocysts in rats, hamsters, mice, and sheep (Psychoyos 1960; Orsini 1963; Finn and McLaren 1967; Boshier 1970). This increase in permeability occurs on the afternoon of day 5 in rats, on day 4 in hamsters and mice, and on day 15 in sheep, that is, about 20-24 h before morphological evidence of implantation. The nature of the stimulus given to the uterus by the blastocyst has been intensively studied in laboratory rodents, and experimentation has led to theories about the release of histamine from endometrial mast cells activated by the presence of the blastocyst (Shelesnyak 1960) and the local production of CO_2 associated with the respiratory activity of the blastocyst (Menke and McLaren 1970a, b).These hypotheses, however, cannot explain the deciduagenic action of oil injected into the uterine lumen which Finn and Porter (1976) consider may be due to a complex molecular change at the epithelial cell surface.

Endometrial sensitivity in early pregnancy seems to differ among species. In the rat, a species in which a decidual response is readily elicited by viable blastocysts and by a range of non-specific stimuli, the response does not occur with unfertilized ova, two-cell eggs, or blastocysts with no viable blastomeres. In the mouse, a species with a relatively high threshold of sensitivity as indicated by the inability of inert objects such as plastic beads to cause decidualization, a response can be produced by blastocysts from various species (Kirby 1970). Finn and Porter (1976) suggest that since a decidual response can be evoked in several species, and in some instances by non-specific stimuli such as saline, air, and arachis oil, the physicochemical property of the stimulus given by the blastocyst is probably similar, although the nature of the endometrial response may vary considerably.

The process of implantation is regulated by endocrine mechanisms. In the rat, implantation depends on a hormonal sequence during which the uterus is exposed to a 48-h period of progesterone followed by a trace amount of oestradiol-17β (Psychoyos 1976). When ovariectomy is performed at any time before noon of day 4, injections of progesterone alone will maintain the blastocyst but will not permit implantation until the addition of 20-50 ng oestradiol-17β (Psychoyos 1973). These experimental observations have been verified by the determination of oestradiol-17β in peripheral plasma, though it should be noted that contradictory reports have been published about the existence of a prenidatory oestrogen surge (see Heald 1976). According to the findings of Watson et al. (1975) the levels of circulating oestradiol-17β reach a peak between 2000 hours on day 3 and 0400 hours on day 4 and do not show a second rise by 1600 hours on day 4. A physiological role for ovarian oestrogen in the process of implantation has also been demonstrated in the mouse, but there is no evidence that this is a universal phenomenon. Implantation and normal gestation will occur in ovariectomized pregnant animals treated with progesterone alone in the rabbit and guinea pig (Deanesly 1960; Chambon 1949; Hafez and Pincus 1956), hamster (Orsini and Meyer 1959; Orsini and Psychoyos 1965), monkey (Meyer 1972), sheep (Foote et al. 1957; Bindon 1971), and pig (Gentry et al. 1973). Whereas a prenidatory release of ovarian oestrogen may be indispensable for implantation in murine rodents, it may be that endometrial sensitivity in some species is determined by the release of oestrogen at pro-oestrus or by the onset of progesterone secretion (see Finn and Porter 1976).

In this paper we shall be concerned with evidence of an alternative source of oestrogen in early pregnancy, namely, the capacity of blastocyst tissue to produce oestrogens. In the pig this has been found to occur at the time of attachment of fetal and maternal tissues, and coincident with the time of the maternal recognition of pregnancy. The pig is a species in which the zona pellucida is lost about day 6-8, by day 10 the blastocyst is about 2 mm in diameter, by day 11 it is a flaccid sac about 5 mm in diameter, and by day 14 it has elongated to form a tubular, bilaminar structure of the order of 1000 mm in length. By this time the blastocysts, typically 10 to 12 in number, are distributed along the length of the uterine horns and the trophoblast is in contact with the mesometrial surface of the endometrium. Failure to establish this contact in both uterine horns leads to pregnancy loss. Points of loose attachment occur on day 14, but intimate contact between fetal and maternal epithelia by interlocking villi is not established until day 18 (see Perry et al. 1976). Attachment in the pig, unlike implantation in murine rodents, occurs along the mesometrial side of the uterine lumen and placentation is non-invasive. This non-invasive property permits the study of endometrial tissue immediately adjacent to the blastocyst tissue yet uncontaminated by invading fetal cells. We shall also report work on the interaction between blastocyst and endometrial tissues as reflected by patterns of protein synthesis.

2 Blastocyst Steroid Synthesis

2.1 Pathways of Steroid Synthesis

In 1973 we reported evidence for the production of oestrogens by pig blastocyst tissue incubated in vitro with labelled neutral steroid precursors (Perry et al. 1973). Blastocyst tissue was recovered from the reproductive tract which had been removed from animals killed by stunning and exsanguination. The procedures used for incubation studies, the extraction and purification of labelled products, and the presentation of results are given in previous papers (Perry et al. 1973, 1976; Heap et al. 1975).

Table 1. Oestrogen synthesis from [7α-^{3}H]dehydroepiandrosterone in vitro by 14 to 16 day pig blastocyst tissue. Figures given as percentage incorporation per 300 mg tissue during a 3 h incubation of elongated, unattached blastocysts (mean ± s.e.m.)

Fraction	No. of animals	Percentage incorporation of radioactivity per 300 mg tissue
Phenolic	6	39.1 ± 8.8
Oestrone	6	16.3 ± 2.4
Oestradiol-17β	6	2.5 ± 0.8
Aqueous	5	17.1 ± 4.1

The percentage incorporation of [7α-^{3}H]dehydroepiandrosterone by 14-15 day blastocyst tissue into various components is shown in Table 1. Approximately 40% of the radioactivity was found in a phenolic fraction, 16.3% and 2.5% was associated with oestrone and oestradiol-17β respectively, and 17.1% was recovered in an aqueous fraction indicating the occurrence of highly polar or conjugated metabolites. In control incubations with endometrial or myometrial tissue, or in the absence of tissue, the amount of radioactivity recovered as oestrone or oestradiol-17β was negligible. The results obtained with [7α-^{3}H]androstenedione were similar (Gadsby et al. 1980).

The analytical procedure used in experiments shown in Table 1 was that described by Perry et al. (1973) and involved solvent partition, repeated thin layer chromatography, and chromatography of acetylated derivatives of the isolated oestrogens. Table 2 shows the constancy of the ^{3}H:^{14}C ratio after an initial chromatographic separation suggesting that the ^{3}H-radioactivity corresponded with authentic oestrone and oestradiol-17β. In further experiments definitive evidence was obtained by recrystallization of the products (Table 3), confirming the capacity of the pig blastocyst to synthesize oestrogens from neutral steroid precursors. Radioimmunoassay of pig blastocyst tissue has revealed the presence of immunoreactive progestogens as well as oestrogens (Perry et al. 1973, Gadsby et al. 1976). The possibility that C_{21} steroids may provide a precursor source for blastocyst oestrogen synthesis was explored. Oestrone and oestradiol-17β could not be positively identified in a labelled form from incubations of 16-day pig blastocyst tissue with [^{3}H]progesterone unless the medium was supplemented with co-factors (Table 4). Although the percentage incorporation was small, definitive evidence was obtained by recrystallization for the formation of oestradiol-17β from progesterone (and pregnenolone) and of oestrone from progesterone. When the co-factors were added the amount of labelled precursor incorporated into oestradiol-17β was substantially greater than that recovered as oestrone (Table 4), contrasting with the results found with neutral steroid precursors (Table 1). This finding is probably related to the reducing conditions provided by the inclusion of co-factors.

Histochemical localization of the enzyme 3β-hydroxysteroid dehydrogenase in pig blastocyst tissue has been reported by Flood (1974). The substrates used in these experiments were dehydroepiandrosterone (DHA), 3β-hydroxy-5β-androstan-17-one, testosterone, and oestradiol-17β, and NAD^+ and $NADP^+$ were included as co-factors. Metabolism of the 3β-hydroxy substrates occurred at day 12 post coitum and the intensity

Table 2. Investigation of radiochemical purity of oestrone and oestradiol-17β isolated from incubations of blastocyst tissue

Substrate	Product	Ratio of $^3H:^{14}C$ after thin-layer chromatography		
		1st chromatogram	2nd chromatogram	3rd chromatogram
[^{3}H]Androstenedione	Oestrone	119	111	117
		153	165	171
	Oestradiol-17β	20	23	21
		38	36	32
[^{3}H]Dehydroepiandrosterone	Oestrone	319	340	316
		205	210	215
	Oestradiol-17β	124	129	119
		30	39	32

$^3H:^{14}C$ ratios of oestrogens after extraction, solvent partition, repeated thin layer chromatography (TLC), and derivative formation. After the addition of [^{14}C]oestrone and [^{14}C]oestradiol-17β, blastocyst tissue was extracted with diethyl ether, a phenolic fraction was prepared by solvent partition, and oestrone and oestradiol-17β were separated by TLC (ethyl acetate:cyclohexane, 1:1). After acetylation, the oestrogen-3-monoacetates were chromatographed again (ethyl acetate:cyclohexane, 3:7), then hydrolysed, and the free oestrogens chromatographed a third time (ethyl acetate: cyclohexane, 1:1) (from Perry et al. 1973)

Table 3. Formation of oestrone and oestradiol-17β from [7α-^{3}H]androstenedione and [7α-^{3}H]dehydroepiandrosterone by 16-day pig blastocyst tissue. Recrystallization of oestrone and oestradiol-17β from aqueous methanol after isolation by extraction, solvent partition and a thin-layer chromatographic separation

Substrate	Crystallization No.	Oestrone C	Oestrone ML	Oestradiol-17β C	Oestradiol-17β ML
Androstenedione	1	646	52750	1176	2581
	2	514	952	1081	1263
	3	587	659	1233	1265
	4	496	400	1356	1365
Dehydroepiandrosterone	1	881	27889	995	2837
	2	601	931	924	1081
	3	725	908	857	1407
	4	646	647	894	1404

Table 4. Oestrogen synthesis from [^{3}H]progesterone in vitro by 14 to 17 day pig blastocyst tissue. Figures given as percentage incorporation per 300 mg tissue during a 3 h incubation of elongated, unattached blastocysts (mean ± s.e.m., number of observations in parentheses)

Fraction	Co-factors [a]	
	Added	Not added
Phenolic	12.8 (1)	8.8 ± 1.3 (4)
Oestrone	0.5 (1) [b]	n.d. (4)
Oestradiol-17β	3.1 (1) [b]	n.d. (4)
Aqueous	8.6 (1)	17.4 ± 6.7(4)

[a] 10 μmol NAD, 10 μmol NADP$^+$, 60 mol glucose-6-phosphate, and 1.5 units (Kornberg) glucose-6-phosphate dehydrogenase

[b] Values and identity of oestrone and oestradiol-17β were established by recrystallization

n.d., not detectable

of the histochemical reaction increased by days 15 and 19 post coitum. In our studies we have failed to obtain definitive evidence for the conversion of pregnenolone to progesterone by minced blastocyst tissue incubated in the absence of added co-factors. This finding is unexpected in view of the conversion of dehydroepiandrosterone to oestrogens, presumably after its conversion to androstenedione (Fig. 1), and that of pregnenolone to oestradiol-17β. An explanation may be that under the conditions of our incubation, any progesterone formed from pregnenolone is rapidly metabolized by blastocyst tissue, as reported elsewhere in this volume by Gadsby (p. 173).

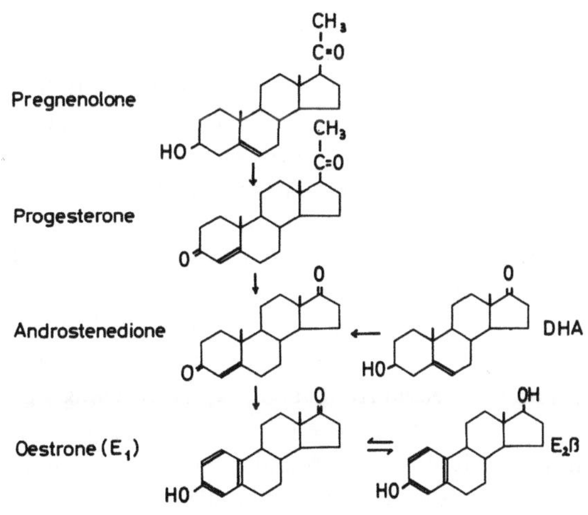

Fig. 1. Pathways of steroid metabolism and oestrogen synthesis in the pig blastocyst. Definitive evidence has been obtained for the synthesis of oestrone and oestradiol-17β from pregnenolone, progesterone, androstenedione and dehydroepiandrosterone (*DHA*), and for the conversion of oestrone to oestradiol-17β

The occurrence of significant quantities of endogenous progesterone in blastocyst tissue implies that the steroid may be protected from metabolism by a binding component, or that under the conditions of our incubation technique the rates of metabolism normally found in vivo had been greatly accelerated.

Pathways of steroid synthesis in pig blastocyst tissue in vitro are summarized in Fig. 1. In addition, evidence has been obtained for the formation of a small quantity of oestradiol-17α, and of an unknown phenolic compound identical after repeated chromatography and after acetylation to oestrone and oestrone-3-monoacetate respectively, but distinguishable from oestrone by recrystallization. Thus, whereas estimates for the incorporation of labelled precursors into oestradiol-17β are comparable whether determined after a chromatographic or a recrystallization procedure, those obtained for oestrone probably overestimate the percentage incorporation of labelled precursor on account of the formation of an additional phenolic compound.

2.2 Onset of Steroid Synthesis

The majority of experiments reported here involve animals between days 14 and 18 of pregnancy when adequate amounts of tissue are available for studies of steroid synthesis. By day 14 the elongation of the blastocyst is complete and the extra-embryonic tissue consists almost entirely of trophectoderm lined by a tenuous layer of endoderm. By day 16 there is a considerable extension of mesoderm and development of the yolk sac. Figure 2 shows that throughout this period there was active incorporation of [^{3}H] dehydroepiandrosterone into oestrone and oestradiol-17β, and this was also true when [^{3}H]androstenedione was used as substrate.

In incubations of spherical and enlarging blastocysts on days 10 and 12 neither of the C_{19} neutral steroid precursors was converted to significant amounts of oestrogens by blastocysts recovered on day 10. At day 12 the incorporation was appreciable, though variable, suggesting that enzymes involved in steroid synthesis appear about this time. However, the precise time of onset of this activity is difficult to define as detection at day 10 may have been precluded by the small amount of tissue available (see Flint, Burton, Gadsby, Saunders and Heap 1979). The percentage incorporation of neutral steroid precursors into oestrogens per unit weight of trophectoderm tissue appeared not to increase after about day 18, but declined during the next week of gestation. This decline may be associated with fusion of the allantois and chorion which begins after the definitive microvillous attachment at day 18. Nevertheless the enormous increase in the amount of tissue would be expected to result in a marked rise in the total production of oestrogens, and this is in accord with the results of Robertson and King (1974). These workers showed that the concentration of oestrone sulphate in maternal circulation increased from about day 16 reaching a peak between day 23 and day 30, and declined rapidly during the next 10 days (Fig. 3). A similar pattern has been found in urine and it was concluded that the steroid was uterine in origin as indicated by high levels in uterine venous blood and allantoic fluid (Lunaas et al. 1973; Robertson and King 1974). This pattern of urinary oestrone excretion was not disturbed by maternal ovariectomy or hypophysectomy in animals in which pregnancy was maintained by exogenous progesterone. The pattern of excretion in operated and control animals

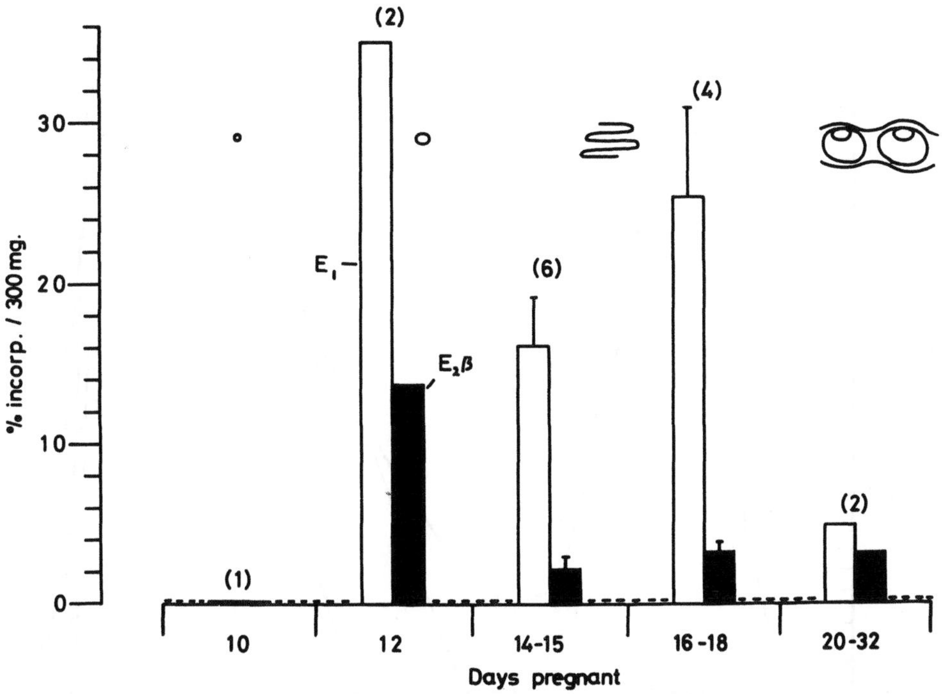

Fig. 2. Oestrogen synthesis from [7α-^{3}H]dehydroepiandrosterone (*DHA*) in vitro by early embryonic tissue in the pig, between day 10 and days 20 to 32 of pregnancy. The percentage incorporation of DHA per 300 mg fresh tissue (except at days 10 and 12 when 7 and 3 blastocysts per flask were used, respectively) into oestrone (*E_1 open columns*) and oestradiol-17β (*$E_2\beta$ black columns*) was measured. The horizontal *dotted line* indicates the level of incorporation found in control incubations in the absence of trophoblast tissue, or in the presence of endometrial or myometrial tissue. Symbols above the columns indicate the stages of embryo development from a blastocyst of 2 mm diameter at day 10 post coitum to an elongated bilaminar blastocyst of 1 m length at days 14 and 15 post coitum, and to a distended chorionic sac at days 20 to 32 post coitum. Numbers of animals are given in parentheses

consisted of an initial peak recorded towards the end of the first month of pregnancy, and a second, much larger, peak after day 80 of pregnancy (Fig. 3; Lunaas 1962; Raeside 1963; Fevre et al. 1968).

2.3 Steroid Metabolism in Uterine Tissue

Further consideration is required of the apparent contradiction between our findings wnich show that the pig blastocyst has the enzymatic capacity to synthesize unconjugated oestrogens, and those of Robertson and King (1974) already referred to, showing that the oestrogen present in circulation was predominantly in a conjugated form. An appreciable degree of arylsulphatase activity has been detected in pig blastocyst tissue (Perry et al. 1973). Sulphotransferase activity, however, seems to be located predominantly in uterine tissue. Figure 4 shows a comparison of the metabolism of oestrone,

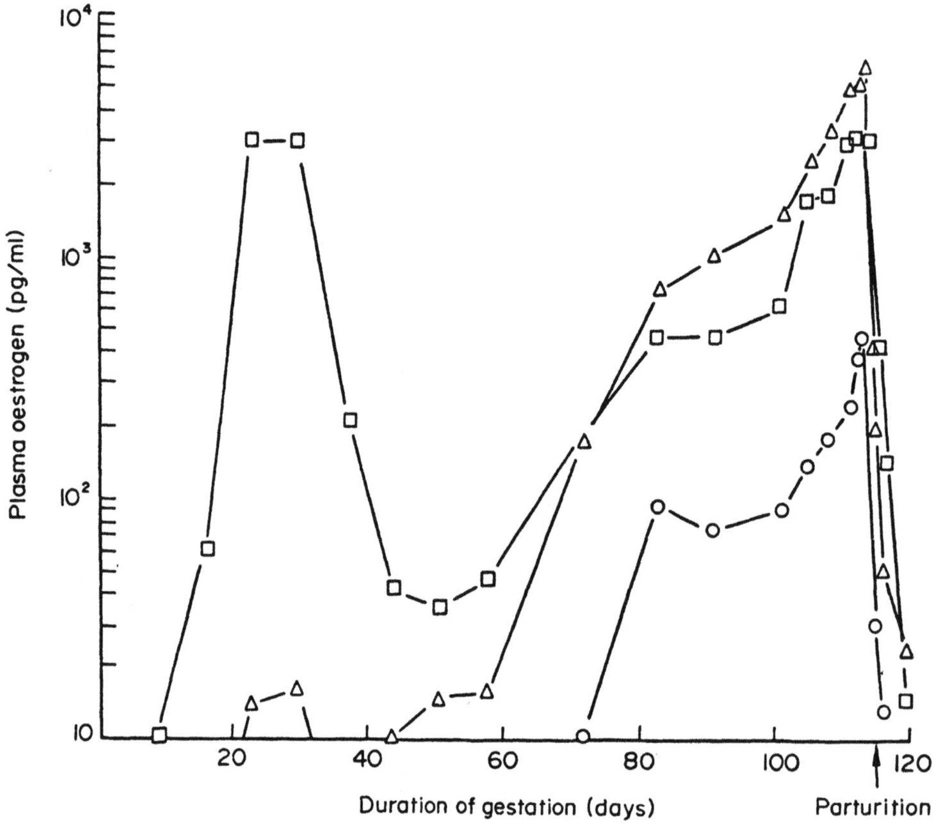

Fig. 3. The concentration of unconjugated oestrone (△), oestradiol-17β (0), and oestrone sulphate (□) in the plasma of a gilt at implantation, throughout gestation, and at parturition (Robertson and King 1974)

oestradiol-17β, and oestrone sulphate by minces of endometrial tissue prepared from pregnant animals 12-17 days post coitum. Oestrone and oestradiol-17β were converted into aqueous soluble compounds. After solvolysis and solvent partition, over half the radioactivity was recovered in a phenolic fraction, indicating the formation of oestrogen sulphoconjugates. There was a relatively low conversion of oestrone sulphate into oestrone. Similar results were obtained with minces of myometrial tissue, though the percentage incorporation of unconjugated oestrogens into aqueous soluble compounds was lower than that obtained with endometrial tissue.

The results suggest that oestrogens synthesized in the trophectoderm may be conjugated in the endometrium and myometrium, and secreted into maternal circulation as oestrogen sulphoconjugates. Pack and Brooks (1974) have described a cyclic variation in oestrogen sulphotransferase and 17β-oxidoreductase activities in the pig uterus (Fig. 5). The conversion of oestradiol-17β to aqueous soluble compounds was 56% and 59% respectively at 12 to 17 days post coitum. It is notable that uterine sulphotransferase activity seems to be high, and uterine arylsulphatase is low at a time when the concentration of oestrone sulphate in plasma is rising sharply.

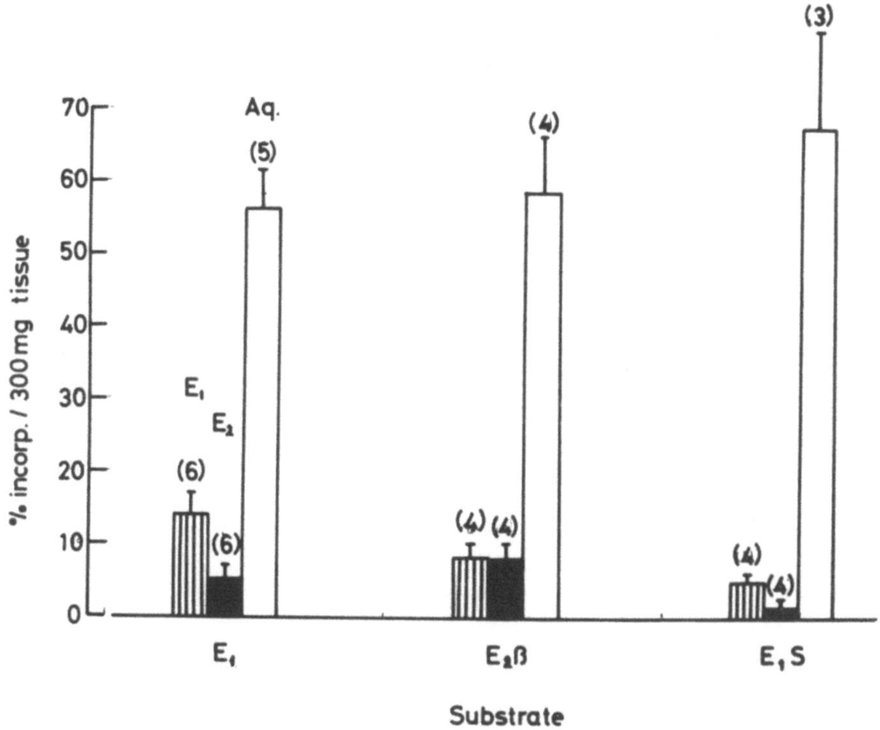

Fig. 4. Metabolism of oestrone (E_1), oestradiol-17β ($E_2\beta$) and oestrone sulphate (E_1S) by endometrial minces prepared from pregnant sows, 12 to 17 days post coitum (mean $\pm$ s.e.m., number of sows in parentheses)

2.4 Role of Blastocyst Oestrogen Secretion

The demonstration of blastocyst oestrogen synthesis by day 12 in the pig suggests that it may act as an embryonic signal of luteal transformation coinciding with the time of maternal recognition of pregnancy. In previous papers (Perry et al. 1976; Flint et al. 1979; Heap, Flint and Gadsby 1979) we discussed the manner in which blastocyst oestrogen synthesis may function as an embryonic signal. Up to about day 11 the corpus luteum of the pig is relatively insensitive to prostaglandin $F_{2\alpha}$($PGF_{2\alpha}$) as indicated by experiments with prostaglandin analogues, ICI 79,939 and ICI 80,996 (Guthrie and Polge 1976) but after day 14 it requires luteotrophic support even in the absence of a uterine luteolysin (du Mesnil du Buisson and Denamur 1968). Oestrone sulphate has been described as a prohormone, a substance with a low intrinsic biological activity but one that is readily hydrolysed to more potent oestrogens by sulphatases. Tissues in which sulphatase activity has been found in early pregnancy in the pig include the hypothalamus, pituitary, and corpus luteum (Heap et al. 1977), so that if oestrogens produced by embryonic tissue are involved in the maternal recognition of pregnancy their regulatory activity may be exerted at more than one site. Exogenous oestrogens are known to be luteotrophic at this time both in vivo (Gardner et al. 1963) and in vitro (Golden-

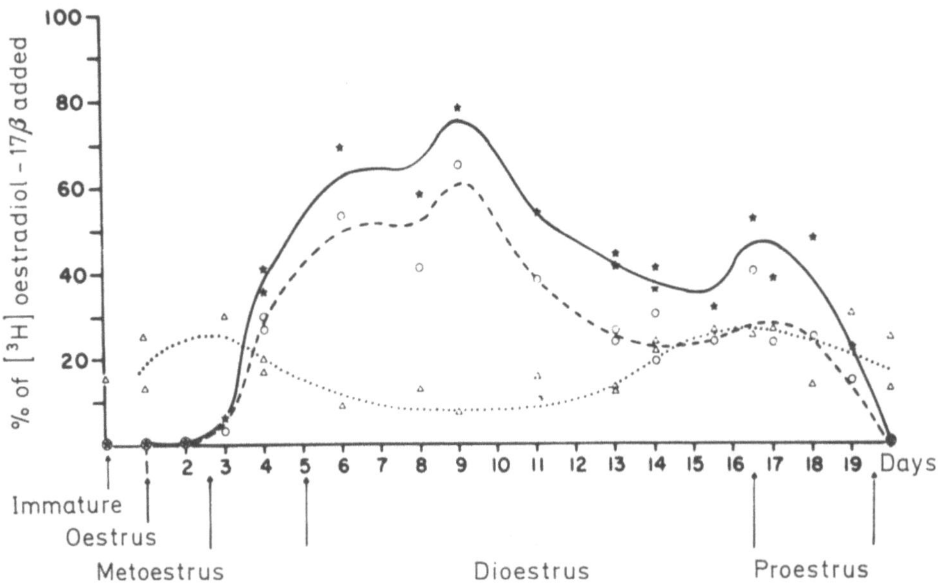

Fig. 5. Pattern of sulphation and oxidation of oestrogens by uterine minces at various times during the oestrous cycle of the pig. Tissue (400 mg) was incubated for 2 h with [6,7-^{3}H]-oestradiol-17β and Na_2 [$^{35}SO_4$]. Diagram shows the percentage sulphation of the total ^{3}H-labelled oestrogens (*), the percentage represented by oestrone sulphate (0) and oestrone (△) (Pack and Brooks 1974)

berg et al. 1972). In the latter work oestradiol-17β acted synergistically with human chorionic gonadotrophin (HCG) in increasing progesterone synthesis by pig granulosa cells in tissue culture, and a maximal stimulation was observed when HCG treatment was followed by oestradiol. It is notable that plasma luteinizing hormone concentrations were significantly greater in early pregnancy 3 to 24 days post coitum compared with those found during the oestrous cycle (excluding the pre-ovulatory period) (Guthrie et al. 1972). Further evidence that the luteotrophic action of oestrogen may be a direct one on the corpus luteum comes from the work of Norris and Kohler (1974) who found that pig granulosa cells specifically concentrate labelled oestrogens, and in particular they accumulate oestrone. More recently, however, the studies of Bazer and his colleagues indicate that oestrogens may act locally on the endometrium thereby protecting the corpus luteum by redirecting $PGF_{2\alpha}$ secretion into the uterine lumen rather than into the uterine vein (Bazer and Thatcher 1977).

The luteotrophic action of oestrogen in the sow contrasts with its luteolytic effect in species such as the sheep in which oestradiol-17β has been shown to stimulate the release of uterine $PGF_{2\alpha}$ (Caldwell et al. 1972; Harrison and Heap 1974; Barcikowski et al. 1974). If oestrogens of embryonic origin have a similar effect in the sow, it is conceivable that their conjugation in the uterine wall reduces their luteolytic potential. Alternatively, oestradiol of embryonic origin may exert a luteotrophic effect which overrides or neutralizes the effect of uterine $PGF_{2\alpha}$ secretion, as we suggested in an earlier paper (Perry et al. 1976). This interpretation is supported by the results of Kraeling et al. (1975) who showed in hysterectomized sows that oestradiol benzoate protects the corpus luteum from the luteolytic action of administered $PGF_{2\alpha}$. Du Mesnil du Buisson (1961) showed that the presence of a sterile horn caused the regression of the corpora lutea after about day 16 of gestation. Furthermore, when only 15-20 cm of sterile horn remained, pregnancy continued in six sows out of 11 and by 55-70 days post coitum the corpora lutea persisted on the side of the gravid horn in four cases out of six, indicating that a relatively short section of the horn exerted a local luteolytic effect. The deleterious action of the sterile horn upon the corpora lutea and pregnancy seemed to be counteracted by the injection of oestrogen. Thus the secretion of oestrogens by the gravid uterus before day 16 of pregnancy may be essential in the maintenance of the corpus luteum and the establishment of pregnancy. Endogenous oestrogen secretion at this time appears inadequate to maintain the corpora lutea adjacent to a non-gravid horn against the local luteolytic action of uterine $PGF_{2\alpha}$.

3 Blastocyst-Endometrial Interactions and Protein Synthesis

In the preceding section we were concerned with the physiological role of blastocyst oestrogen synthesis and its involvement in the transformation of the corpus luteum from that of the normal oestrous cycle to that of pregnancy. Although gestation procedes normally in ovariectomized pregnant pigs treated with progesterone alone as described previously, the possibility cannot as yet be ruled out that blastocyst oestrogen synthesis is involved in the process of attachment. An early effect of oestrogen on the uterine endometrium is the stimulation of new protein synthesis, subsequently followed by tissue growth, and studies on the pig blastocyst in culture have revealed an interaction between protein synthesis in blastocyst and endometrial tissue in co-culture (Wyatt 1976; Wyatt et al. 1977).

An organ culture method was developed for the co-culture of blastocyst explants with various maternal tissues (Wyatt 1976). Explants of blastocyst tissue were prepared as 5-mm lengths of trophectoderm, and explants of maternal tissue as 2-mm^3 pieces. A grid technique was used to culture explants of blastocyst grown alone or in co-culture with maternal tissues for 48 h at 37°C. Blastocyst growth was monitored by the incorporation of labelled leucine into proteins recovered from blastocyst tissue and from the culture medium. Medium 199 without leucine was supplemented with L-[4,5-^{3}H]-leucine 0.33 μCi/ml (a concentration that minimized the influence of variations in the endogenous pool of leucine), insulin 5 μg/ml, penicillin 200 μg/ml, and streptomycin 200 μg/ml.

Figure 6 shows the incorporation of labelled leucine into blastocyst tissue proteins, and into proteins recovered from the culture medium. In four experiments on 16-day pig blastocysts, the mean leucine incorporation into blastocyst tissue proteins was 29.2 ± 5.2 nmol/mg (blastocyst alone). When blastocyst tissue was co-cultured with peritoneum, striated muscle, kidney, or liver the incorporation varied from 40.6 to 53.7 nmol/mg, and when co-cultured with endometrium 137 ± 18.4 nmol/mg ($P \langle 0.002$). The pattern of response was similar when protein synthesis was determined in the culture medium.

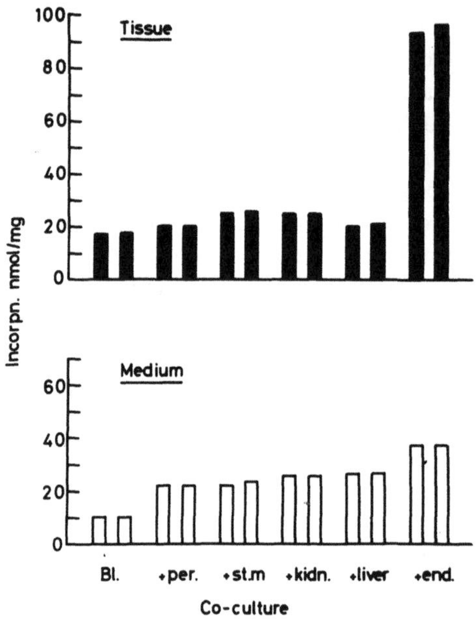

Fig. 6. Protein synthesis in 16-day pig blastocyst tissue determined by the incorporation of [^{3}H]leucine into blastocyst tissue proteins and into the culture medium. In this experiment blastocyst tissue was cultured alone or in co-culture with peritoneum (*per.*), striated muscle (*st.m.*), kidney (*kidn.*), liver or endometrium (*end.*). Diagram shows results of duplicate determinations (Wyatt 1976)

The nature of protein synthesis in blastocysts stimulated in co-culture was examined by disc gel electrophoresis and gel slicing techniques. Blastocyst tissue co-cultured with endometrial explants contained a labelled pre-albumin component(s) (Rf 0.85) which was not present in blastocyst tissue cultured alone or with any of the other control tissues shown in Fig. 6. The incorporation of [^{3}H]leucine by endometrial explants co-cultured with blastocyst tissue was appreciably greater than that of endometrial tissue cultured alone, indicating that the blastocyst stimulated the protein-synthetic activity of the endometrium. Whether this activity is oestrogen-dependent is not known, but Table 5 shows that explants of blastocyst tissue grown either alone, or in co-culture with endometrial tissue will synthesize labelled oestrone and oestradiol-17β from [^{3}H] androstenedione. The apparent conversion was lower in co-cultures because aqueous soluble compounds were formed by sulphoconjugation in endometrial tissue. There was a significant conversion of oestrone to oestradiol-17β in blastocyst explants.

The site of synthesis of a labelled pre-albumin component recovered from blastocyst-endometrial co-cultures has not yet been resolved. The component was found in blastocyst tissue and in the spent medium of co-cultures (Rf 0.845 ± 0.013, $n = 4$). It was first detectable after 8 h culture, and increased in quantity up to 48 h. It was thermolabile at 60° for 20 min.

4 Summary

We have presented evidence for oestrogen synthesis by pig blastocyst tissue in vitro from about day 12 of pregnancy, a stage when there is the rapid elongation of the blastocyst prior to definitive attachment on day 18. The evidence for oestrogen synthesis

Table 5. Evidence of oestrogen synthesis in 16-day blastocyst tissue cultured alone, or in co-culture with endometrial tissue for 48 h at 37°C. Percentage incorporation of [7α-^{3}H]androstenedione into oestrone and oestradiol-17β

Culture	% incorporation/mg tissue	
	Oestrone	Oestradiol-17β
Blastocyst alone	7.5 [a]	5.0 [a]
Blastocyst + endometrium	5.6 [a]	1.6 [a]

[a] Identity confirmed by recrystallization

consists in the isolation, purification, and recrystallization of labelled oestrone and oestradiol-17β formed from C_{19} neutral steroid precursors, androstenedione and DHA, and from C_{21} steroids, pregnenolone and progesterone, in the presence of added co-factors. Immunoreactive oestrogens and progestogens have been detected in blastocyst tissue. It is not yet known whether progesterone present in blastocyst tissue is of ovarian origin and constitutes a major precursor of blastocyst oestrogen synthesis, or whether steroidogenesis de novo can occur in this early embryonic tissue. The onset and pattern of blastocyst oestrogen synthesis corresponds closely with the occurrence of oestrone sulphate in peripheral circulation, and it seems probable that the presence of this conjugate in the circulation arises from blastocyst oestrogens that are sulphoconjugated in the uterine wall prior to their entry into the circulation. A physiological role for oestrogen sulphoconjugates has been discussed in terms of the maternal recognition of pregnancy, an event that precedes blastocyst attachment in this species by about 6 days. The interaction between blastocyst and endometrial tissue has been studied in terms of protein synthesis in vitro, and it is suggested that the endocrine properties of the blastocyst, at least in the pig, may be associated with the process of attachment.

Acknowledgments. We thank Mr R.W. Ash and Mr J. Smith for providing animals at known stages of pregnancy. J.E.G. was supported by an A.R.C. Research Scholarship. This investigation received financial support form the World Health Organization.

References

Barcikowski B, Carlson JC, Wilson L, McCracken JA (1974) The effect of endogenous and exogenous estradiol-17β on the release of prostaglandin $F_{2\alpha}$ from the ovine uterus. Endocrinology 95:1340-1349

Bazer FW, Thatcher WW (1977) Theory of maternal recognition of pregnancy in swine based on estrogen controlled endocrine versus exocrine secretion of prostaglandin $F_{2\alpha}$ by the uterine endometrium. Prostaglandins 14:397-401

Bindon BM (1971) Role of progesterone in implantation in the sheep. Aust J Biol Sci 24:149-158

Boshier DP (1970) The Pontamine blue reaction in pregnant sheep uteri. J Reprod Fertil 22:595-596

Caldwell BV, Tillson SA, Brock WA, Speroff L (1972) The effects of exogenous progesterone and estradiol on prostaglandin F levels in ovariectomized ewes. Prostaglandins 1:217-228

Chambon Y (1949) Besoins endocriniens qualitatifs et quantitatifs de l'ovoimplantation chez la lapine. C R Soc Biol (Paris) 143:1172-1175

Deanesly R (1960) Implantation and early pregnancy in ovariectomized guinea-pigs. J Reprod Fertil 1:242-248
Deanesly R (1967) The role of the fertilized egg: reactions in the guinea-pig uterus at ovo-implantation and after thread traumatization. J Reprod Fertil 14:243-248
du Mesnil du Buisson F (1961) Possibilite d'un fonctionnement dissemblable des ovaires pendant la gestation chez la truie. CR Acad Sci (Paris) 253:727-729
du Mesnil du Buisson F, Denamur R (1968) Mecanismes du controle de la fonction luteale chez la truie, la brebis et la vache. Excerpta Med Int Congr Ser 184:927-934
Fevre J, Leglise PC, Rombauts P (1968) Du role de l'hypophyse et des ovaires dans la biosynthese des oestrogenes au cours de la gestation chez la truie. Ann Biol Anim Biochim Biophys 8:225-233
Finn CA, McLaren A (1967) A study of the early stages of implantation in mice. J Reprod Fertil 13:259-267
Finn CA, Porter DG (1976) The uterus Flek Science, London (Handbooks in reproductive biology, vol 1)
Flint APF, Burton RD, Gadsby JE, Saunders PTK, Heap RB (1979) Blastocyst oestrogen synthesis and the maternal recognition of pregnancy. In: Whelan J (ed) Maternal recognition of pregnancy (Ciba Fdn Symp NS 64) Excerpta Medica, Amsterdam, pp 209-228
Flood PF (1974) Steroid-metabolizing enzymes in the early pig conceptus and in the related endometrium. J Endocrinol 63:413-414
Foote WD, Gooch LD, Pope AL, Casida LE (1957) The maintenance of early pregnancy in the ovariectomized ewe by injections of ovarian hormones. J Anim Sci 16:986-989
Gadsby JE, Burton RD, Heap RB, Perry JS (1976) Steroid metabolism and synthesis in early embryonic tissue of pig, sheep and cow. J Endocrinol 71:45P-46P
Gadsby JE, Burton RD, Heap RB, Perry JS (1980) Oestrogen production by blastocyst and early embryonic tissue of various species. J Reprod Fert 60:409-417
Gardner ML, First NL, Casida LE (1963) Effect of exogenous estrogen on corpus luteum maintenance in gilts. J Anim Sci 22:132-134
Gentry BE, Anderson LL, Melampy RM (1973) Exogenous progesterone and estradiol benzoate on early embryonic survival in the pig. J Anim Sci 37:722-727
Goldenberg RL, Bridson WE, Kohler PO (1972) Estrogen stimulation of progesterone synthesis by porcine granulosa cells in culture. Biochem Biophys Res Commun 48:101-107
Guthrie HD, Polge C (1976) Luteal function and oestrus in gilts treated with a synthetic analogue of prostaglandin-$F_{2\alpha}$ (ICI 79,939) at various times during the oestrous cycle. J Reprod Fertil 48: 423-425
Guthrie HD, Henricks DM, Handlin DL (1972) Plasma estrogen, progesterone and luteinizing hormone prior to estrus and during early pregnancy in pigs. Endocrinology 91:675-679
Hafez ESE, Pincus G (1956) Hormonal requirements of implantation in the rabbit. Proc Soc Exp Biol Med 91:531-534
Harrison FA, Heap RB (1974) Prostaglandin secretion by the autotransplanted uterus in sheep. J Endocrinol 64:13P-14P
Heald PJ (1976) Biochemical aspects of implantation. J Reprod Fertil [Suppl] 25:29-52
Heap RB, Flint AP, Gadsby JE (1979) Role of embryonic signals in the establishment of pregnancy. Brit Med Bull 35:129-135
Heap RB, Perry JS, Gadsby JE, Burton RD (1975) Endocrine activities of the blastocyst and early embryonic tissues in the pig. Biochem Trans 3:1183-1188
Heap RB, Perry JS, Burton RD, Gadsby JE, Wyatt C, Jenkin G (1977) Blastocyst steroidogenesis and embryo-maternal interactions in the establishment of pregnancy. In: Calaby JH, Tyndale-Biscoe CH (eds) Reproduction and Evolution. Aust Acad Sci, pp 341-347
Kirby DRS (1970) Immunological aspects of implantation. In: Hubinont PO (ed) Ovoimplantation, human gonadotrophins and prolactin. Karger, Basel pp 86-100
Kraeling RR, Rampacek GB, Ball GD (1975) Estradiol inhibition of $PGF_{2\alpha}$ luteolysis in the pig. J Anim Sci 41:466A
Lunaas T (1962) Urinary oestrogen levels in the sow during oestrous cycle and early pregnancy. J Reprod Fertil 4:13-20
Lunaas T, Refsdal AO, Schultz RH (1973) Conjugated oestrone in uterine vein blood during early pregnancy in the pig. Acta Endòcrinol [Suppl] (Kbh) 177:420

Menke TM, McLaren A (1970a) Carbon dioxide production by mouse blastocysts during lactational delay of implantation or after ovariectomy. J Endocrinol 47:287-294

Menke TM, McLaren A (1970b) Mouse blastocysts grown *in vivo* and *in vitro*: carbon dioxide production and trophoblast outgrowth. J Reprod Fertil 23:117-127

Meyer RK (1972) Chorionic gonadotrophin, corpus luteum function and embryo implantation in the rhesus. In: Diczfalusy E, Standley CC (eds) The use of non-human primates in research on human reproduction. WHO Research and Training Centre, Karolinska Institute, Stockholm. Acta Endocrinologica Suppl No. 166, pp 214-217

Norris JS, Kohler PO (1974) Estrogen accumulation by porcine granulosa cells in culture. Endocr Res Commun 1:133-144

Orsini MW (1963) Induction of deciduomata in hamster and rat by injected air. J Endocrinol 28: 119-121

Orsini MW, Meyer RK (1959) Implantation of the castrate hamster in the absence of exogenous estrogen. Anat Rec 134:619-620

Orsini MW, Psychoyos A (1965) Implantation of blastocysts transferred into progesterone treated virgin hamsters previously ovariectomized. J Reprod Fertil 10:300-301

Pack BA, Brooks SC (1974) Cyclic activity of estrogen sulfotransferase in the gilt uterus. Endocrinology 95:1680-1690

Perry JS, Heap RB, Amoroso EC (1973) Steroid hormone production by pig blastocysts. Nature 245:45-47

Perry JS, Heap RB, Burton RD, Gadsby JE (1976) Endocrinology of the blastocyst and its role in the establishment of pregnancy. J Reprod Fertil [Suppl] 25:85-104

Psychoyos A (1960) La reaction deciduale est precedee de modifications precoces de la permeabite capillaire de l'uterus. CR Soc Biol (Paris) 154:1384-1387

Psychoyos A (1973) Endocrine control of egg implantation. In: Greep RO (ed) Female reproductive system, part 2. American Physiological Society (Washington), (Handbook of physiology, sec. 7, Endocrinology, vol 2, pp 187-215)

Psychoyos A (1976) Hormonal control of uterine receptivity for nidation. J Reprod Fertil [Suppl] 25:17-28

Raeside JI (1963) Urinary oestrogen excretion in the pig during pregnancy and parturition. J Reprod Fertil 6:427-431

Robertson HA, King GJ (1974) Plasma concentration of progesterone, oestrone, oestradiol-17β and of oestrone sulphate in the pig at implantation, during pregnancy and at parturition. J Reprod Fertil 40:133-141

Shelesnyak MC (1960) Nidation of the fertilized ovum. Endeavour 19:81-86

Watson J, Anderson FB, Alam M, O'Grady JE, Heald PJ (1975) Plasma hormones and pituitary luteinizing hormone in the rat during the early stages of pregnancy and after post-coital treatment with Tamoxifen (ICI 46,474). J Endocrinol 65:7-17

Wyatt C (1976) Endometrial components involved in protein synthesis in 16-day pig blastocyst tissue in culture. J Physiol (Lond) 260:73P-74P

Wyatt C, Heap RB, Perry JS (1977) Protein synthesis in co-cultures of blastocyst and endometrium explants during blastocyst steroidogenesis in the pig. Proc V Int Congr Endocr, Hamburg, 266A

Steroid Hormones in Blastocyst Tissue, Uterine Flushings, and Endometrium of Pig, Sheep, and Cow

J.E. Gadsby *

A.R.C. Institute of Animal Physiology, Babraham, Cambridge, U.K.

1 Introduction

Pig blastocyst tissue as early as the 12th day of gestation displays a marked capacity to synthesize oestrogens in vitro (Heap et al., this volume, p. 157). Definitive evidence for oestrogen synthesis by recrystallization has recently been obtained from steroid precursors pregnenolone, progesterone, dehydroepiandrosterone (DHA), and androstenedione (Gadsby et al. 1976)suggesting the presence of these pathways of oestrogen synthesis shown in Fig. 1 (Heap et al., p. 161).

With regard to the possible physiological significance of this oestrogen synthesis, it has been proposed as a luteotrophic agent in the maternal recognition of pregnancy, which occurs on about day 12 (Heap et al. 1975; Perry et al. 1976). However, this does not rule out a possible influence on the definitive attachment of the trophoblast to the uterine epithelium which occurs on day 18. In the cow these two events are temporally more distinct, the maternal recognition of pregnancy occurring at about day 15 whereas definitive attachment of the allantochorion to endometrial caruncles does not begin until about day 40-45. This species thus provides an opportunity to examine whether embryonic oestrogen synthesis is related to the maternal recognition of pregnancy or to attachment. Experiments were also carried out in sheep, in which the timing of the events of early pregnancy are similar to those in the pig (Gadsby and Heap 1978).

2 Steroid Metabolism

Blastocyst and embryonic tissue from pigs, sheep, and cows were incubated at various stages of gestation, to investigate the transformation of tritiated progesterone, androstenedione, and DHA into oestrogens in vitro. Details of the methods used have been described elsewhere (Heap et al. 1975).

2.1 Pig

Pig blastocyst tissue taken between day 14 and day 18 of gestation converted about 30% of added [^{3}H]androstenedione or [^{3}H]DHA into phenolic compounds (the fraction of the ether-soluble radioactivity which contained oestrogens) (Table 1). However,

---

* Present address: Reproductive Endocrinology Program, Department of Pathology, University of Michigan, Ann Arbor, Michigan 48109, USA

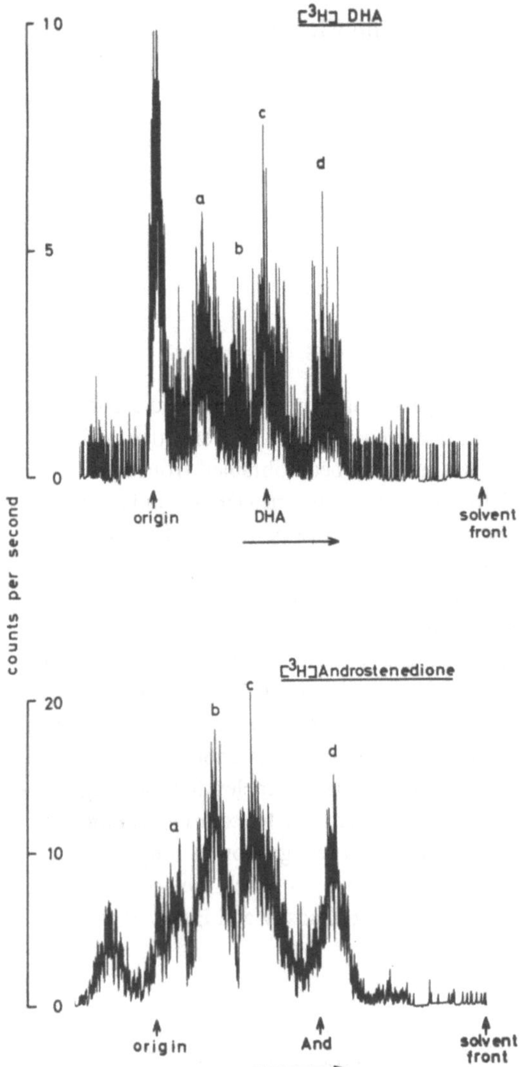

Fig. 1. Metabolism of [^{3}H]dehydroepiandrosterone (*DHA*) and [^{3}H]androstenedione by 16-day pig blastocyst tissue. The neutral steroid fraction was examined by thin-layer chromatography (*TLC*) in benzene:methanol (9:1). Horizontal arrows show the direction of TLC run from origin to solvent front. The relative positions of authentic androstenedione and DHA, run on the same plate, are indicated by the small vertical arrows. Metabolites labelled *a-d* are unidentified

there was a considerable amount of radioactivity remaining in the neutral steroid fraction, (i.e. radioactivity in ether fraction −radioactivity in phenolic fraction), about 50% of the total amount added, and it was necessary to establish whether this could all be accounted for by unmetabolised precursor.

The constituents of the neutral steroid fraction were investigated by thin-layer chromatography. A radiochromatogram scan (Fig. 1) showed that substantial metabolism of

Table 1. A comparison of the ability of early embryonic tissues from pig, sheep, and cow to produce phenolic and aqueous soluble steroids from C_{19} steroid precursors, [^{3}H]androstenedione (And.) and [^{3}H]dehydroepiandrosterone (DHA). Figures given as percentage incorporation per 300 mg tissue during a 3-h incubation (mean ± s.e.m.; number of observations in parentheses)

	% Radioactivity recovered in various fractions		
	Ether	Phenolic	Aqueous
Pig (14-18 days post coitum)			
And.	79.2 ± 7.9 (6)	26.7 ± 3.0 (6)	21.0 ± 7.8 (6)
DHA	84.8 ± 2.9 (8)	37.3 ± [illegible] (8)	15.2 ± 5.4 (8)
Total	82.5 ± 3.7 (14)	32.7 ± 4.2 (14)	17.7 ± 3.6 (14)
Sheep (14-16 days post coitum) Cow (21-65 days post coitum)			
And.	95.6 ± 0.5 (14)	1.0 ± 0.3 (17)	4.3 ± 0.5 (14)
DHA	98.5 (1)	1.3 (1)	1.5 (1)
Total	95.8 ± 0.5 (15)	1.0 ± 0.3 (18)	4.2 ± 0.5 (15)

DHA and almost complete metabolism of androstenedione had occurred. Most of the metabolites formed were more polar than both precursors. In another experiment, pig blastocyst tissue was shown to be capable of metabolising [^{3}H]pregnenolone and [^{3}H] progesterone to more polar compounds, and as a result the original precursors had almost completely disappeared. The identity of these natural steroid metabolites has not been positively established so far, but their chromatographic mobilities suggest that they are the products of enzymic reduction and hydroxylation. Thus pig blastocyst tissue seems to be capable of extensive steroid metabolism, in addition to oestrogen synthesis from pregnenolone, progesterone, DHA, and androstenedione.

2.2 Cow and Sheep

Under incubation conditions identical to those used for the pig, embryonic tissue of cows from day 16 to day 65 post coitum and sheep blastocyst tissue from day 14 to day 16 post coitum did not display a very marked ability to convert neutral steroids into oestrogens. The transformation into phenolic compounds was only 1%, showing that in this respect embryonic tissue from these two species was considerably less active than in the pig (Table 1). Definitive evidence for oestrogen synthesis has not yet been obtained in these two species and with about 95% of the radioactivity remaining in the neutral steroid fraction it became essential to determine whether or not these tissues are completely inert in their ability to metabolise or synthesise steroids.

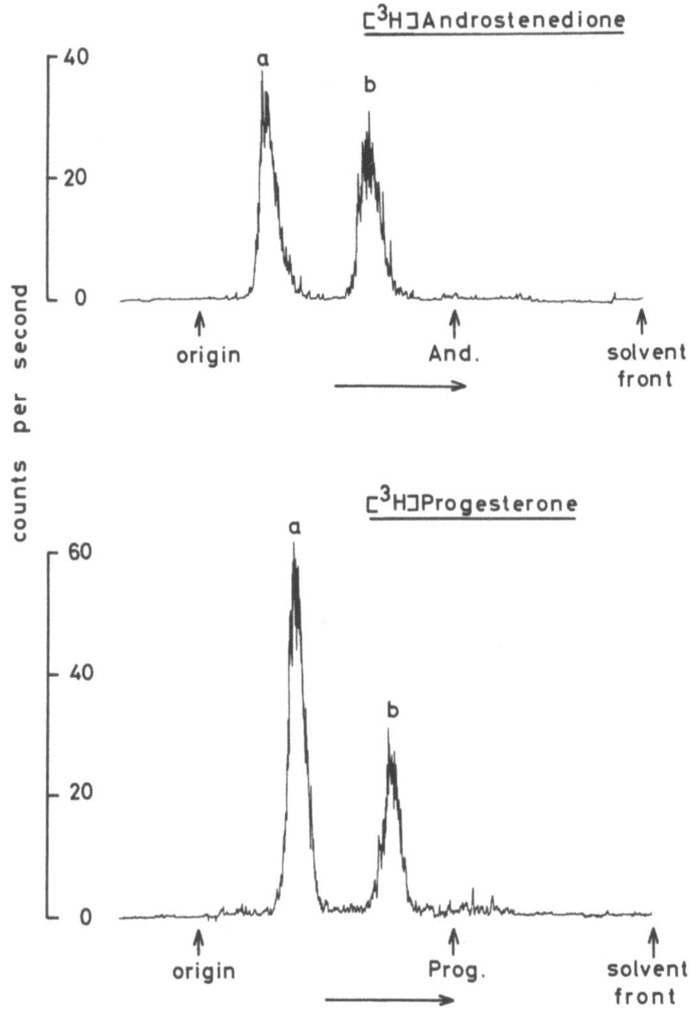

Fig. 2. Metabolism of [^{3}H]androstenedione and [^{3}H]progesterone by 29-day cow, pre-attachment blastocyst tissue. The neutral steroid fraction was chromatographed in benzene:methanol (9:1). Metabolites labelled *a* and *b* are unidentified

Radiochromatogram scans (Fig. 2) showed complete metabolism of added androstenedione and progesterone by the pre-attachment cow blastocyst at day 29 post coitum yielding two discrete and more polar metabolites in each case. In addition evidence has been obtained to suggest that embryonic tissues from both attachment and post-attachment stages of pregnancy possess similar neutral steroid metabolising activities. An initial study of neutral steroid metabolism by sheep blastocyst tissue was also carried out by thin-layer chromatography. The radiochromatogram scans obtained (Fig. 3) again demonstrate the extensive metabolism of both androstenedione and progesterone leading to their near-complete disappearance.

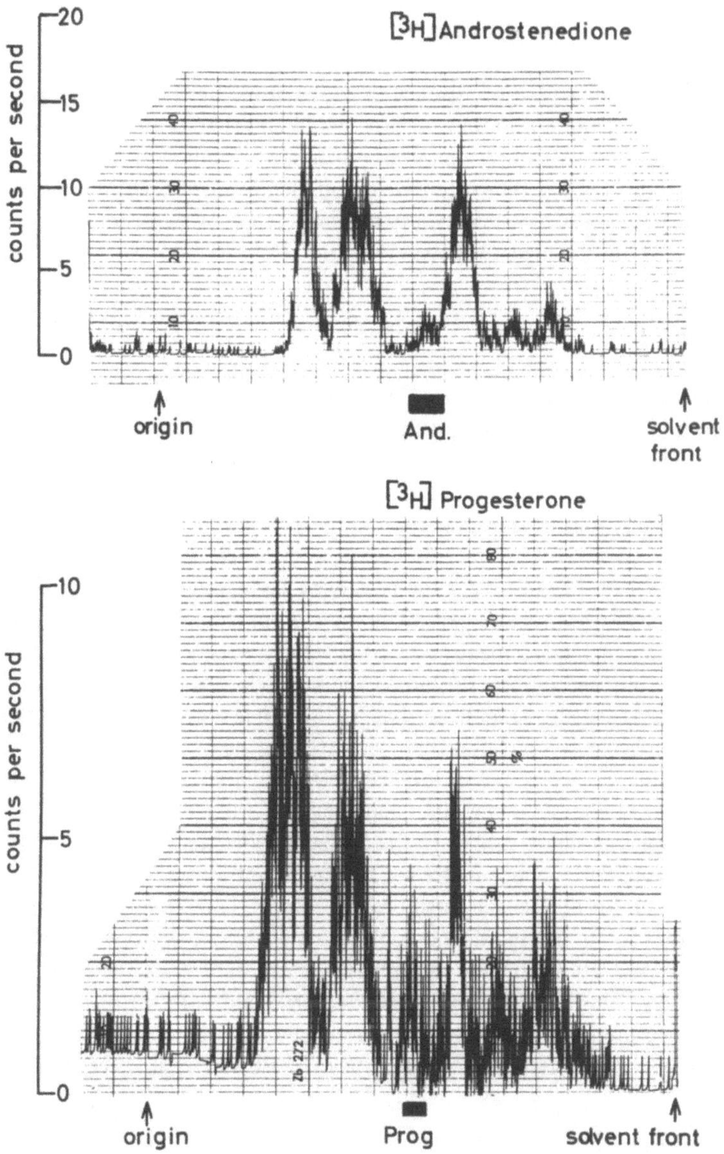

Fig. 3. Metabolism of $[^3H]$androstenedione and $[^3H]$progesterone by 15-day sheep blastocyst tissue. The neutral steroid fraction was chromatographed in benzene:methanol (9:1). Positions of authentic steroid markers are shown by the black bars

In summary, these findings indicate that in spite of a low conversion of neutral steroid precursors into oestrogens, early embryonic tissue from both cows and sheep shows a marked capacity to metabolise androstenedione and progesterone. As in the case of the pig the metabolites appear to be mainly more polar than the added precursors.

3 Steroids in the Gravid Uterus

An investigation of the occurrence of endogenous steroids in the blastocyst, uterine flushings, and endometrium is in progress in an attempt to identify oestrogens and locate the possible natural source of precursor of their synthesis. A preliminary study, reported here, was designed to detect and measure, by radioimmunoassay, the major cross-reacting oestrogens and progestogens in each of these compartments in the gravid uterus of both the pig and sheep.

3.1 Pig

Extracts of total blastocyst and uterine flushings, and of a small sample of endometrium, from a 17-day-pregnant sow, were subjected to thin-layer chromatography, and radioimmunoassay of oestrogens and progesterone was performed.

The antiserum used for the oestrogen assay (kindly supplied by Dr. B.J.A. Furr) showed an equally high cross-reactivity with both oestrone and oestradiol-17β, these being the major cross-reacting steroids. The samples were assayed against known amounts of oestradiol-17β. The results of the oestrogen assay, expressed as the concentration of oestradiol-17β equivalents per milligram protein (Fig. 4), show that pig blastocyst tissue contained a high concentration of both oestrone and oestradiol-17β compared with the endometrium and uterine flushings. The same eluates were also assayed for progesterone and the results (Fig. 5) indicate that both the blastocyst and the endometrium contained progesterone but that there was little in the uterine flushings. However, when these results are expressed as total progesterone content (Fig. 6) the uterine flushings do appear to contain nanogram quantities of a cross-reacting steroid which has a chromatographic mobility similar to the 5-pregnanediones. Figure 6 is obtained when the results from the previous figure are presented without correction for the protein content, which for the uterine flushings was 63 mg. This finding of a low progesterone concentration in uterine flushings contrasts with that of Fowler et al. (1976) in which a high progesterone concentration was found in rabbit uterine flushings during pseudopregnancy. This high concentration was correlated with the protein content of the flushings, and presumably with the presence of uteroglobin which has been demonstrated to have progesterone binding properties (Fowler et al. 1976; Beato and Baier 1975).

3.2 Sheep and Cow

A study of the endogenous steroids in the uterus of a 15-day-pregnant ewe was carried out as described above and the concentrations of oestradiol-17β and progesterone in the blastocyst, uterine flushings, and endometrium were determined. The blastocyst and uterine flushings contained a low concentration of immunoreactive oestrogens decreasing in concentration from blastocyst across the lumen to the endometrium. Conversely, progesterone was found predominantly in the endometrium with decreasing concentrations in the uterine flushings and blastocyst tissue.

It has been reported recently that in the cow, both conjugated and unconjugated oestrogens are present in amniotic and allantoic fluids, increasing in concentration from

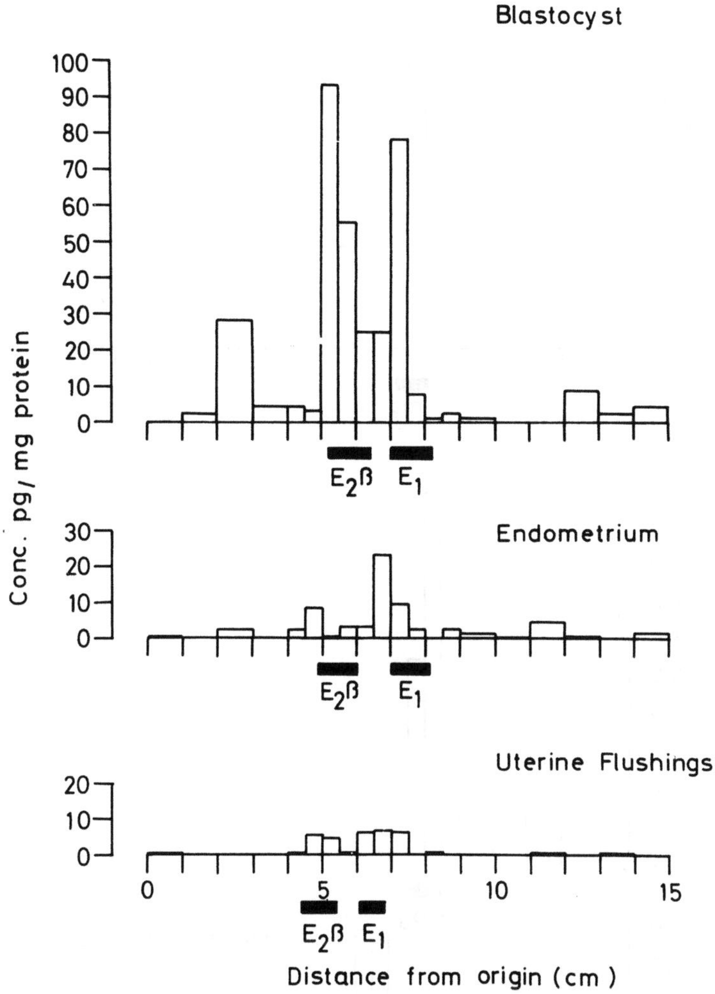

Fig. 4. Oestrogen concentration in the blastocyst, endometrium, and uterine flushings of a 17-day-pregnant sow. Ether extracts of the total blastocyst and uterine flushings and of a sample of endometrium were subjected to thin-layer chromatography (*TLC*) in cyclohexane:ethylacetate (1:1) on separate plates. The sample lane and an adjacent blank lane were eluted in 1 cm band; 0.5 cm bands were taken in the region corresponding to the positions of oestrone (E_1) and oestradiol-17β($E_2\beta$). Aliquots of the eluates were assayed for oestrogen by radioimmunoassay using the antiserum BF 461/6 described in the text. The results shown above are expressed as concentration of oestradiol-17β equivalents in each band eluted from the TLC plate, corrected for the protein of the sample. The black bars represent the relative positions of authentic E_1 and $E_2\beta$ markers run on the same plate

day 41 up to day 68, at about the time of caruncular attachment, which was visible by day 52 (Robertson and King 1975).

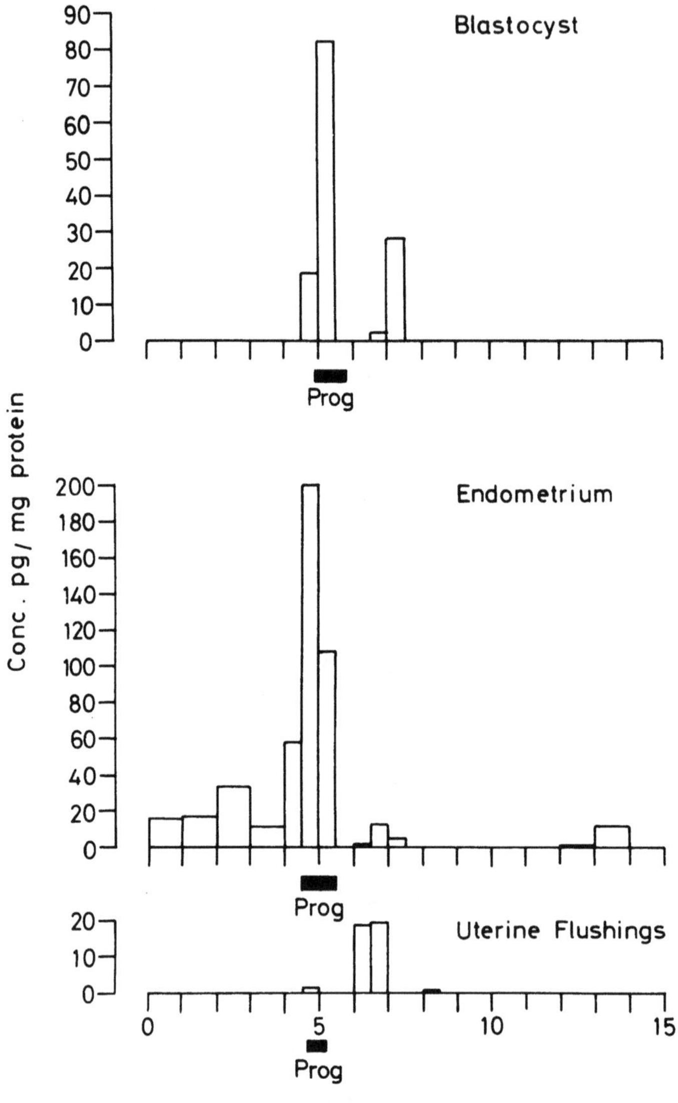

Fig. 5. Progesterone concentration in the blastocyst, endometrium, and uterine flushings of a 17-day-pregnant sow. The eluates described in Fig. 4 were assayed for progesterone by radioimmunoassay. The results were calculated as described in Fig. 4 and are expressed as concentration of progesterone for each of the TLC plate eluates, corrected for the protein content of the sample. Authentic progesterone marker position indicated by the black bars

4 Conclusion

The evidence presented here, taken together with the results of Robertson and King (1975), suggests that embryonic tissue from pigs, sheep, and cows both contains and

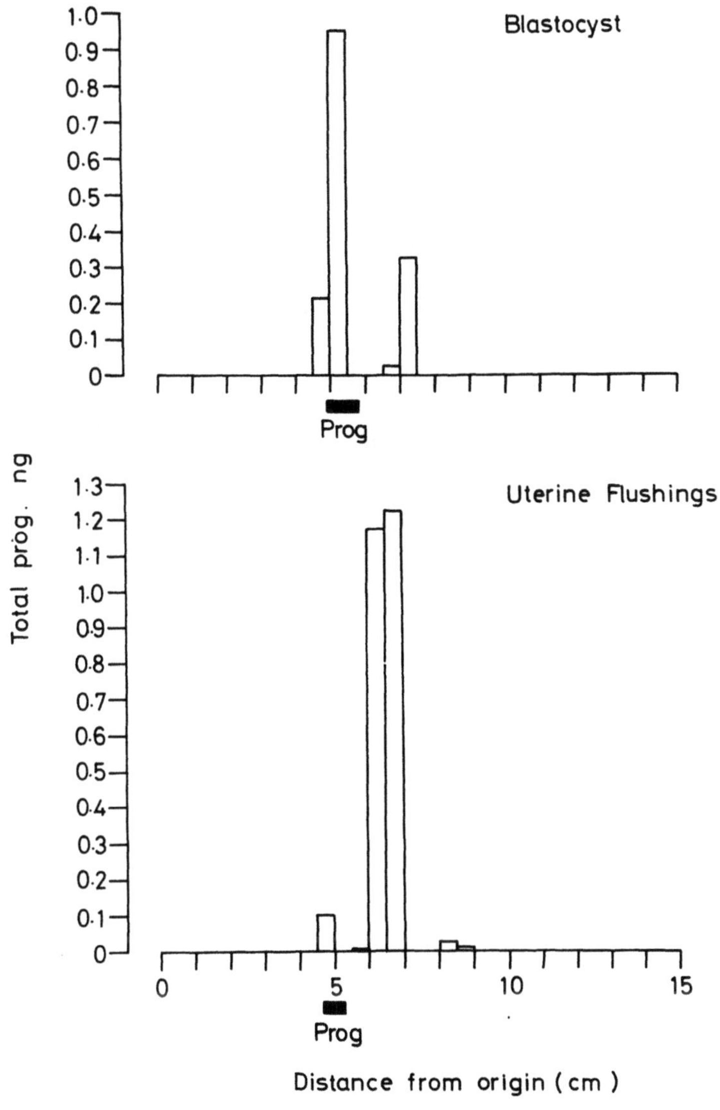

Fig. 6. Blastocyst and uterine flushings total progesterone content. The results shown above are those presented in Fig. 5, *without* correction for the protein contents of the blastocyst (11.5 mg) and the uterine flushings (63.0 mg)

synthesises oestrogens. We have only been able so far to provide definitive evidence of oestrogen synthesis in the pig. However, both sheep and cow embryonic tissue displays a striking capacity for steroid metabolism and this may reflect a much slower rate of oestrogen synthesis. This is illustrated in the cow, in which the accumulation of oestrogens in allantoic and amniotic fluids over a period of about 20 days coincides with onset of caruncular attachment of the allantochorion.

Acknowledgments. I wish to acknowledge the financial support of the Agricultural Research Council.

References

Beato M, Baier R (1975) Binding of progesterone to the proteins of the uterine luminal fluid. Identification of uteroglobin as the binding protein. Biochim Biophys Acta 392:346-356

Fowler RE, Johnson MH, Walters DE, Pratt HPM (1976) The progesterone and protein composition of rabbit uterine flushings. J Reprod Fertil 46:427-430

Gadsby JE, Burton RD, Heap RB, Perry JS (1976) Steroid metabolism and synthesis in early embryonic tissue of pig, sheep and cow. J Endocrinol 71:45P-46P

Gadsby JE, Heap RB (1978) Steroid hormones and their synthesis in the early embryo. In: Spilman CH, Wilks JW (eds) Novel aspects of reproductive physiology, Spectrum Publ, New York London, pp 263-285

Heap RB, Perry JS, Gadsby JE, Burton RD (1975) Endocrine activities of the blastocyst and early embryonic tissue of the pig. Biochem Soc Trans 3:1183-1188

Perry JS, Heap RB, Burton RD, Gadsby JE (1976) Endocrinology of the blastocyst and its role in the establishment of pregnancy. J Reprod Fertil [Suppl] 25:85-104

Robertson HA, King GJ (1975) Oestrogens and placental attachment in the cow. J Anim Sci 41:377

Proteases of the Blastocyst and of the Uterus

H.-W. Denker

Abteilung Anatomie der Medizinischen Fakultät, Rheinisch-Westfälische Technische Hochschule, Aachen, Germany

1 Introduction

The physiological role which certain proteinases seem to play in embryo implantation in the uterus is recently receiving increasing interest. Analytical studies performed in the rabbit, the mouse and the rat have provided evidence that proteinase systems with peculiar properties are not only present at these sites but certain enzymes also show remarkable changes in activity around implantation (Andary 1974; Dabich and Andary 1976; Denker 1969a, 1972, 1974b, 1975, 1976a; Denker and Fritz 1979; Denker and Petzold 1977; Kirchner 1972; Kirchner et al. 1971; Pinsker et al. 1974; Rosenfeld and Joshi 1977; Strickland et al. 1976). From these studies the hypothesis was derived that the action of proteinases may form an essential part of the mechanisms involved in implantation, and has received support from studies in which proteinase inhibitors were administered in vivo (Dabich and Andary 1974; Denker 1977, 1978a,b). Furthermore, evidence was presented suggesting that the physiological regulation of implantation initiation as governed by maternal hormones may in part be mediated by changes in proteinase and/or proteinase inhibitor activity of uterine tissues or the trophoblast.

Recent cell biological and oncological studies performed in other systems have provided data which indicate that *limited proteolysis* by certain proteinases may form an important element of such phenomena as cell adhesion and invasion (cf. Ribbons and Brew 1976). If the hypothesis is correct, namely that proteinases have the suggested role in embryo implantation, their cell physiological mechanism of action must be looked for in the framework of concepts on the physiological elements of the process of implantation, which appears to be a complex interaction between embryonic and maternal tissues (Denker 1977; Leroy et al. 1980). The essentials of these concepts may be summarized as follows:

1. Implantation is the formation of a specialized intimate cellular contact between trophoblast and uterine tissues.
2. Initiation of implantation requires a special cell physiological state of maturation of the *trophoblast*. The maturated trophoblast is *potentially* invasive probably in all eutherian species.
3. The physiologically predominant role of the *uterine epithelium* (and uterine secretion?) with regard to implantation is an *inhibitory* one rather than a stimulatory one. The maturated trophoblast can implant as soon as the inhibiting principles are sufficiently reduced. "Induction" of implantation from the maternal side is basically a de-inhibition.
4. The cell physiological processes which take place in the uterine tissues are largely

preprogrammed: The luminal epithelium is preprogammed for undergoing degeneration (being sloughed off: mouse, rat; or cell junctions becoming penetrable: ferret) or for fusing (uterine epithelial cells with each other and with the trophoblast: rabbit). This program is relatively rigid. It is part of a general program imprinted on the cells under appropriate hormonal control, and needs then only minor (even relatively nonspecific, like mechanical) stimuli in order to become completed.

5. *Decidualization* is likewise a preprogrammed process. Many of the cell biological phenomena connected with it are closely related to those shown by connective tissue elements in certain forms of *inflammation.* As in inflammation, reactions of the blood vessels form an integral part.

2 Aminopeptidases (Amino Acid Arylamidases)

Remarkable activities of various aminopeptidase-type enzymes were found to be present in rabbit uterine tissues, uterine secretion and blastocyst tissues (Denker 1969; Denker and Stangl 1976). Since this group of enzymes was found to split particularly well certain synthetic arylamide substrates (amino acid βnaphthylamides or p-nitroanilides) we refer to them as amino acid arylamidases, as has become customary in recent biochemical literature (Denker and Van Hoorn 1974; Van Hoorn and Denker 1975; Denker 1977).

Of particular interest is an enzyme (arylamidase I) which appears in remarkably high quantity in preimplantation-stage uterine secretion of the rabbit (Denker 1969; Beier et al. 1972; Denker and Van Hoorn 1974; Van Hoorn and Denker 1975; Denker 1977). Since the substrate leucine-β-naphthylamide had been used in the original investigations, this enzyme was initially referred to as "leucine aminopeptidase" (LAP) (Denker 1969; Beier et al. 1972). More detailed biochemical investigations revealed, however, that there is only very little activity of classical leucine aminopeptidase (α-aminoacylpeptide hydrolase, cytosol, EC 3,4,11,1) in rabbit uteri, and that the described arylamidase differs from LAP in several biochemical properties: it shows only little activity toward leucine hydrazide, is activated by Co^{2+}, is inhibited by puromycin, and shows a pH optimum between pH 7.0 and 8.0 (substrate:leucine-β-naphthylamide) (Van Hoorn and Denker, unpublished). In addition to the histochemical studies mentioned above (Denker and Stangl 1976), these biochemical investigations also provided strong evidence that rabbit endometrial tissues contain more than one arylamidase capable of hydrolyzing the substrates used (leucine-β-naphthylamide.,leucine-p-nitroanilide, alanine-β-naphthylamide).

Most fascinating about rabbit uterine arylamidase I are the immense changes in activity which take place in the uterine epithelium and secretion during the preimplantation phase. In the *uterine epithelium,* the activity is nearly nondetectable, by histochemical tests, before coitus and at 2 days post coitum, but increases at 3 days post coitum, i.e., at the time when embryos enter the uterus in regular pregnancy (Denker 1969). This is, however, independent of the presence of embryos, because the same happens in pseudopregnancy (Van Hoorn and Denker 1975; Denker 1977, 1978a). Peak levels of activity are found in the uterine epithelium at 5-6 days post coitum, and the activity declines thereafter. The activity disappears quickly when copper IUDs are implanted

(Denker 1976b). The histochemical findings have been confirmed by biochemical investigations (Van Hoorn and Denker 1975 and unpublished data; Denker 1977, 1978a). Differences in histochemical reactivity between nonpregnant and 6-day-pregnant endometria have also been described by Petry et al. (1970).

Even more prominent and probably more significant for the physiology of preimplantation development are the changes in arylamidase I activity in the *uterine secretion* (Fig. 1; Van Hoorn and Denker, unpublished). The activity is barely detectable in nonpregnant uteri and until 2 days post coitum but increases at 3 days post coitum (again the time when embryos enter the uterus in normal pregnancy) and reaches a *sharp and high peak at 5 days post coitum.* Interestingly, *peak activity lasts for only approximately 1 day.* From 6 days post coitum on, the decline is slowed down considerably so that a sort of a plateau is formed (lasting only until 8 1/2 days post coitum).

This pattern of changes of uterine secretion arylamidase I activity (increase after the 2nd day, peak around the 5th day, decrease until the 8th day) can be mimicked by treating nonpregnant females with *progesterone* (4 mg in 1 ml sesame oil, one daily dose subcutaneously) (Fig. 1; Van Hoorn and Denker, unpublished). In endometrial homogenates, an increase in activity can also be shown until the 8th day of progesterone treatment. *Estradiol-17β-benzoate* (50μg in 1 ml sesame oil, one daily dose subcutaneously) induces an increase in endometrial activity but does not cause any significant changes in activity in the uterine secretion (Fig. 1). It should be noted that the animals used in these experiments had not been ovariectomized so that endogenous estrogen levels had not been eliminated. Therefore it appears reasonable to assume that estrogens play a role in induction of the enzyme even in the experiments with progesterone. However, it becomes very clear from these experiments that the observed pattern of increase in arylamidase I activity in the endometrial tissue and particularly the extrusion into the uterine lumen are progesterone-dependent. This fits well with observations made in histochemical tests and in zymograms of uterine flushings (Kühnel et al. 1971; Beier et al. 1972; Van Hoorn and Denker, unpublished).

Actinomycin, injected intravenously (three injections, i.e., at 2, 3 and 4 days post coitum, 0.1 mg/kg each), partially inhibits the increase in endometrial and uterine secretion arylamidase I activity as occurring normally until 5 days post coitum (Van Hoorn and Denker, unpublished). Although results from such experiments, in which inhibitors are administered systemically, should be evaluated with caution, keeping in mind the possibility of side effects and indirect actions via changes of, e.g., hormone synthesis, the results suggest that the progesterone action possibly involves changes in arylamidase I messenger synthesis. On the other hand, progesterone certainly regulates the process of extrusion into the uterine lumen.

The *short duration of the arylamidase wave* in the uterine secretion is most interesting. Uteroglobin, for example, which appears in the uterine secretion at approximately the same stage as arylamidase I, seems to keep a high concentration there for a few more days (Beier et al. 1972; Beier 1974). However, exact comparison is hampered by the fact that most data on uteroglobin levels are derived from semiquantitative determinations based on electropherograms, while the used photometric tests of arylamidase activity may monitor changes more sensibly. If the difference in the profiles of both proteins should prove to be real, it will have to be investigated further whether this reflects a higher turnover rate of the enzyme.

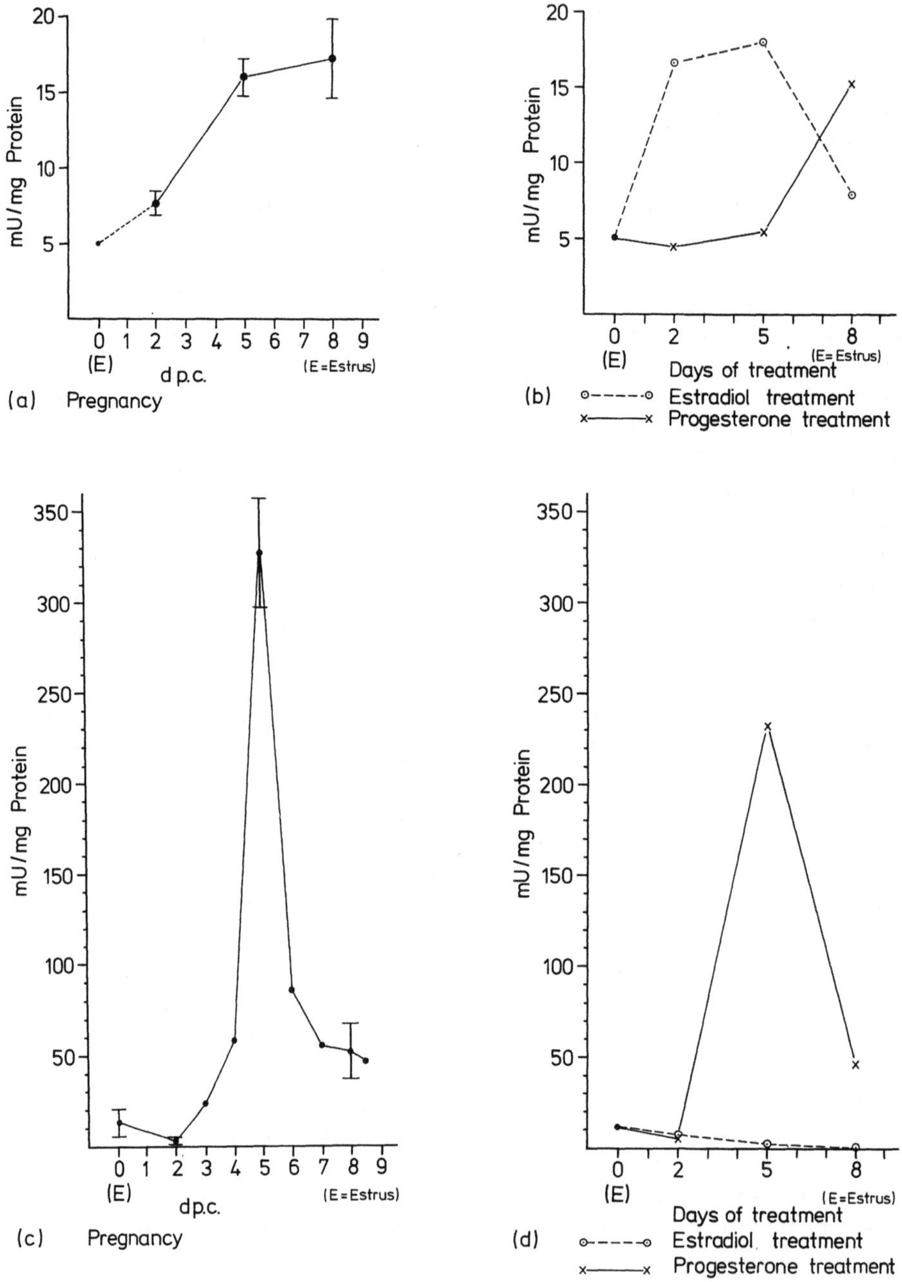

Fig. 1 a-d. Amino acid arylamidase activity in rabbit uteri in pregnancy and after estradiol and progesterone treatment. Substrate: 0.8 mM L-leucine-*p*-nitroanilide, 0.05 M phosphate buffer pH 7.2, 0.005 M Mg^{+2}, 25°C. For details of methods of preparation of endometria and uterine flushings see Van Hoorn and Denker (1975). **a** Endometrial homogenates, estrus and early pregnancy; **b** Endometrial homogenates, hormone treatments; **c** Uterine flushings, estrus, and early pregnancy; **d** Uterine flushings, hormone treatments. Note that progesterone treatment results in a sharp peak of arylamidase activity in the uterine lumen which is roughly comparable to the peak seen in pregnancy (**c**). Values (mU/mg protein, mean ± s.e. *n*= number of animals): **a** Estrus: 5.1 (*n*= 2); 2 d p.c.: 7.6 ± 0.7 (*n*= 5); 5 d p.c.: 16.0 ± 1.2 (*n*= 5); 8 d p.c.: 17.3 ± 2.6 (*n*= 5); **b** Estrus: see **a**. Estradiol treatment: 2 days: 16.7 (*n*= 2); 5 days: 17.9 (*n*= 2); 8 days: 8.0 (*n*= 2). Progesterone treatment: 2 days: 4.4 (*n*= 2); 5 days: 5.5 (*n*= 2); 8 days: 15.3 (*n*= 2); **c** Estrus: 13.1 ± 7.3 (*n*= 4); 2 d p.c.: 3.2 ± 1.6 (*n*= 5); 3 d p.c.: 23.2 (*n*= 2); 4 d p.c.: 58.8 (*n*= 1); 5 d p.c.: 329.9 ± 29.4 (*n*= 4); 6 d p.c.: 87.3 (*n*= 2); 7 d p.c.: 55.1 (*n*= 2); 8 d p.c.: 51.8 ± 14.7 (*n*= 5); 8 d 12 h p.c.: 46.3 (*n*= 2). **d** Estrus see **c**. Estradiol treatment: 2 days: 6.4 (*n*= 2); 5 days: 2.5 (*n*= 2); 8 days: 1.0 (*n*= 2). Progesterone treatment: 2 days: 6.0 (*n*= 2); 5 days: 232.6 (*n*= 2); 8 days: 45.8 (*n*= 2)

The *physiological role* of arylamidase I is unknown so far. Being an exopeptidase, the enzyme possibly participates in the conversion of uterine proteins and peptides into active or inactive forms (Denker 1970b, 1977). Since there is a physiological need for uptake of fixed nitrogen in the form of amino acids or protein by the blastocyst at this phase (Gwatkin 1966; Daniel and Krishnan 1967; Maurer et al. 1970; Jaszczak et al. 1970; Weitlauf 1971; Van Winkle et al. 1972; Spindle and Pedersen 1973), liberation of amino acids by endometrial and uterine secretion arylamidases may also be physiologically significant. On the other hand there is a considerable change in the composition of the extracellular blastocyst coverings around 5 days post coitum (Gerdes and Denker 1979; Denker and Gerdes 1979), i.e., at the time of the arylamidase I peak. At present it is impossible to say whether this coincidence has any physiological meaning. The enzyme is also found in the trophoblast and it would be interesting to see whether it is taken up from the uterine secretion.

In postimplantation stages, considerable leucine-β-naphthylamide splitting activity appears in the *entoderm* of the abembryonic pole (Fig. 2, Denker, unpublished). It becomes detectable, histochemically, around 8 1/2 days post coitum and reaches a very high level at this site at 9 1/2 days post coitum. The entoderm of the embryonic pole and the lateral parts remain negative in these tests. Low to moderate activity is found in the trophoblast cells, predominantly in attached trophoblastic knobs. The entodermal arylamidase activity appears always at the *abembryonic pole*, even in cases of malorientation of the blastocyst in the uterus, i.e., when the abembryonic pole faces the mesometrial part of the endometrium (endopeptidase activity in maloriented blastocysts has been described and discussed by Denker 1974b, 1977). Since at this phase most of the uterine arylamidase I activity has disappeared from the implantation sites (Denker and Van Hoorn 1974; Van Hoorn and Denker 1975; Denker 1976c), it seems improbable that the enzyme of the abembryonic entoderm cells is taken up from the uterine environment, but rather may be synthesized in the entoderm cells. There is also no correlation between the appearance of the enzyme in the entoderm and attachment of the trophoblast and formation of contact with the maternal circulation: the abembryonic entoderm develops its arylamidase acitvity even when implantation is blocked by application of proteinase inhibitors in vivo (see below).

The mechanism behind the rapid decline of uterine secretion arylamidase activity after 5 days post coitum is not clear. Interestingly there is evidence that the *preimplantation blastocyst* of the rabbit *stimulates the discharge from the uterine epithelium*, as suggested by comparison of acitvity in the endometrium and the uterine secretion of normally pregnant (blastocysts present) and pseudopregnant (blastocysts absent) animals

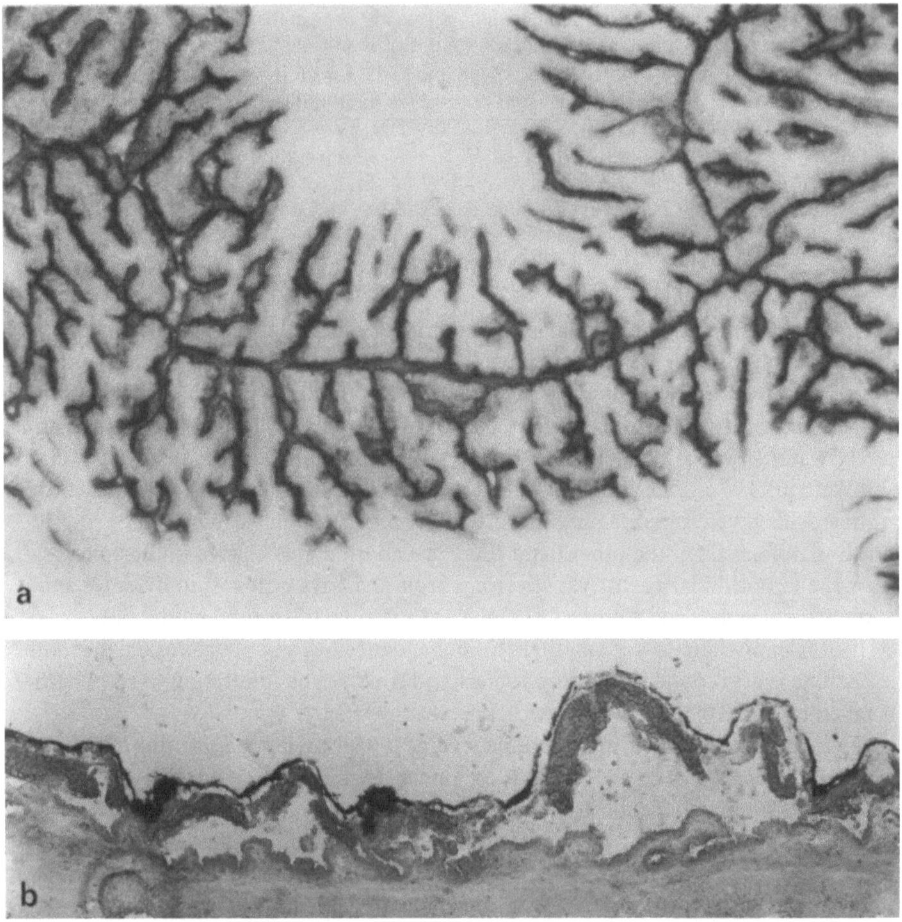

Fig. 2 a,b. Histochemical localization of amino acid arylamidase activity in rabbit uterus and conceptus. Substrate: L-leucine-4-methoxy-β-naphthylamide. **a** Endometrium, 5 d p.c., x 88. A membrane method (Denker and Stangl 1976) is used here which avoids loss of easily diffusible forms of enzyme as present in uterine secretion. High activity is found in the apical parts of the uterine epithelium. The lumina of crypts and urrows are filled with strongly positive uterine fluid. **b** Antimesometrial uterine wall and abembryonic part of the conceptus, 9 1/2 d p.c., x 60. High activity is found in the abembryonic entoderm (*dark line, above*). Below it, the common feto-maternal syncytium is seen which is formed by fusion of trophoblast and uterine epithelial syncytium and which shows only moderate activity. At this stage, it is already degenerating and being sloughed off. The cellular uterine epithelium (*below*) which has already largely replaced it by growing out from the depths of the crypts has only very weak arylamidase activity

(Denker and Van Hoorn 1974; Van Hoorn and Denker 1975; Denker 1977, 1978a,b). The blastocyst obviously signalizes its presence in the uterus via diffusible substances: it induces a decline in uterine epithelial arylamidase I activity in its vicinity already before the blastocyst coverings are dissolved. This is shown even more impressively after blockage of trophoblast attachment by administration of proteinase inhibitors (like Trasylol and antipain, see below) in vivo: even in this case the blastocyst exerts its influence on arylamidase activity of the surrounding endometrium (Denker, unpublished).

3 Endopeptidases

3.1 Trophoblast-Dependent Blastocyst Proteinase (Blastolemmase)

3.1.1 Occurrence and Physiological Function

A peculiar gelatin-dissolving proteinase which is found in the trophoblast and at the surface of implanting rabbit blastocyst has recently been receiving much attention. In fact, this enzyme seems to be of great importance for initiation of implantation, as was already assumed at the time of its first description (Denker 1969) and as a number of investigations confirmed in subsequent years (for review see Denker 1977). In particular it seems to be involved in dissolution of the extracellular blastocyst coverings and it is therefore named "*blastolemmase*".

When the highly sensitive gelatin substrate film test is applied which was designed and optimized for this enzyme (Denker 1971, 1974a), blastolemmase is nondetectable before 6 days post coitum (sometimes traces are found at 5-6 days post coitum) but rises abruptly at 6 2/3 days post coitum, i.e., few hours before the dissolution of the blastocyst coverings begins. At 7-7 1/2 days post coitum when the lysis of the coverings is under way and the trophoblast attaches to the uterine epithelium (Böving 1963; Denker 1970 a,b), the proteinase activity is restricted to the area in which these processes are going on, i.e., the abembryonic hemisphere of the blastocyst (Fig. 4). On contrast, the embryonic pole where the blastocyst coverings remain still intact and where no attachment takes place in this phase, does not show any activity worthy of mention. After completion of the abembryonic dissolution of the coverings and attachment of the trophoblast, the enzyme activity disappears abruptly so that only minor remnants of activity can be traced irregularly in some places between trophoblast and uterine epithelium from 8 days post coitum on (Denker 1969).

Some controversy has developed on the question whether this proteinase derives from the trophoblast, as suggested by the observation described above, or from the uterine secretion as proposed by Kirchner (1972); Kirchner et al. (1971); and Kirchner and Mootz (1974). There is much evidence, however, that this proteinase at least *depends* in some way on the *abembryonic trophoblast:* The maximum of enzyme activity and the beginning of the dissolution of the blastocyst coverings are always observed at the abembryonic pole of the blastocyst, even if it is abnormally oriented in the uterus, i.e., facing the mesometrial instead of the antimesometrial endometrium (maloriented blastocysts) (Denker 1974b, 1977, 1978a,b). Models of blastocyst coverings without trophoblast do not gain any comparable proteinase activity and are not dissolved in the 7-8 day uterus (Fig. 3; Denker 1975; Denker and Hafez 1975). It became very clear from these investigations that the trophoblast has to provide a factor (or factors) essential for the development of this proteinase activity. It has not been possible so far, however, to decide whether this factor is an activator for a uterus-derived proteinase or whether it is the gelatinolytic enzyme (or proenzyme) itself. The latter assumption is probably correct

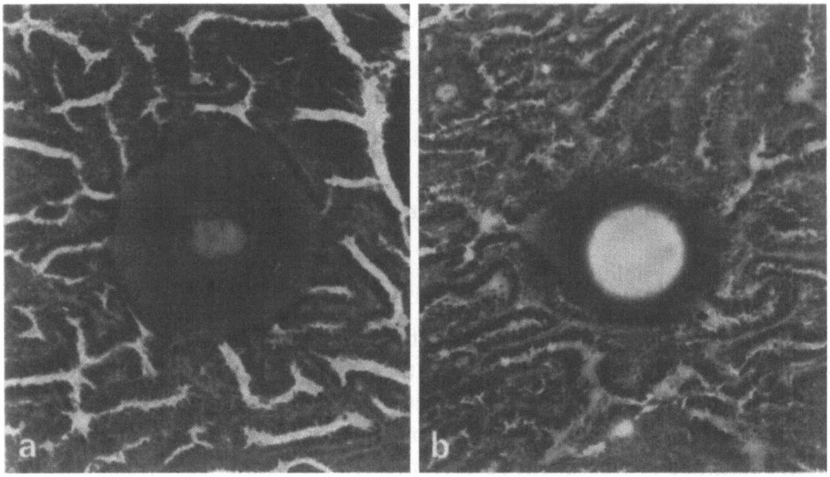

Fig. 3 a,b. The trophoblast-dependence of implantation-associated blastocyst proteinase in the rabbit is demonstrated by the lack of such proteinase activity in models which are devoid of trophoblast tissue. Morulae as shown in **a** or Sephadex beads covered with a mucoprotein layer (deposited in the tube) as shown in **b** are exposed to an implantation stage uterine milieu from 7 to 8 d p.c. (Denker and Hafez 1975). The gelatin substrate film test was performed with cryostat sections of these models and the surrounding endometrium. Proteinase acitvity as found in disintegrating blastocyst coverings of this stage (cf. Fig. 4) is noct detectable in this case. x 110

because it was possible to demonstrate very high gelatinolytic proteinase activity of the same electrophoretic mobility in both the trophoblast and the disintegrating blastocyst coverings of the rabbit, while uterine secretion proteinase appears to be a different entity (Denker and Petzoldt 1977).

The main physiological function of this trophoblast-dependent proteinase seems to be connected with the *dissolution of the blastocyst coverings.* One may be tempted, therefore, to compare it with hatching enzymes of lower animals (Denker 1977). This function is most impressively demonstrated by *in vivo inhibition experiments* (Denker 1977, 1978b). Specific proteinase inhibitors which had previously been shown to inhibit this enzyme strongly in vitro (Denker 1976a) were administered intrauterally to rabbits half a day before implantation (i.e., at 6 1/2 days post coitum). Detailed morphological and histochemical studies showed that this treatment inhibits the blastocyst proteinase activity and effectively blocks the dissolution of the blastocyst coverings (Figs. 5, 6). As a result, attachment of the trophoblast to the endometrium does not occur. Certain differences in the efficacy of the various inhibitors were observed and could in part be correlated with differences in elimination rates (Denker 1977). The most effective inhibitor in the series tested was aprotinin (Trasylol). ∈-Aminocaproic acid, an inhibitor of plasminogen activation which does not affect the gelatinolytic blastocyst proteinase in vitro, was administered as a control (Fig. 4). It was found to have no effect on the dissolution of the blastocyst coverings or on the attachment of the trophoblast even when administered in 100 times higher molar doses than those proven effective in the case of Trasylol, suggesting that *plasminogen activators* and *plasmin* probably do not play a major role in these processes.

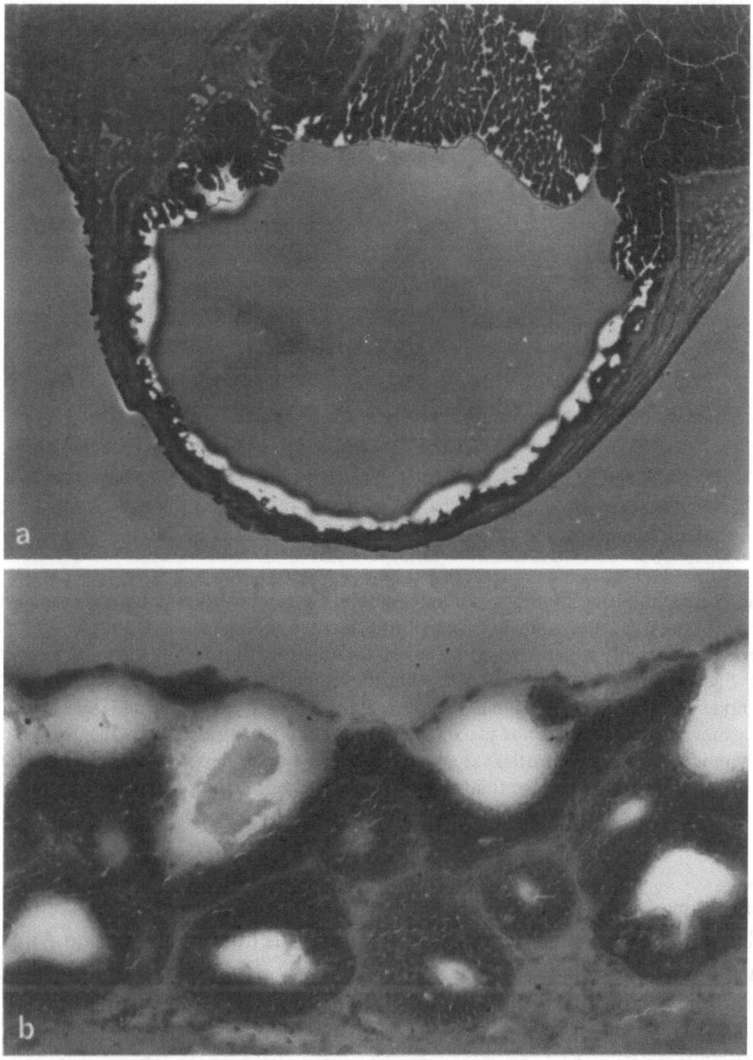

Fig. 4 a,b. Implantation-associated proteinase activity (blastolemmase) in the rabbit, 7 1/2 d p.c. Gelatin substrate film test, longitudinal cryostat sections through implantation site, **a** overview, x 11.5, mesometrial side above, **b** detail from the antimesometrial-abembryonic region, x 140. In this animal, ϵ-aminocaproic acid (in **a** 0.78 mg, in **b** 6 mg) had been injected into the uterine lumen at 6 1/2 d p.c., but this treatment does not interfere with regular enzyme acitvity, dissolution of blastocyst coverings and attachment of the trophoblast. The typical proteinase pattern is seen with high activity (*bright lysis zones*) in the abembryonic-antimesometrial region (**a**), predominantly at the site of disintegrating blastocyst coverings, i.e., between trophoblast and uterine epithelium (**b**)

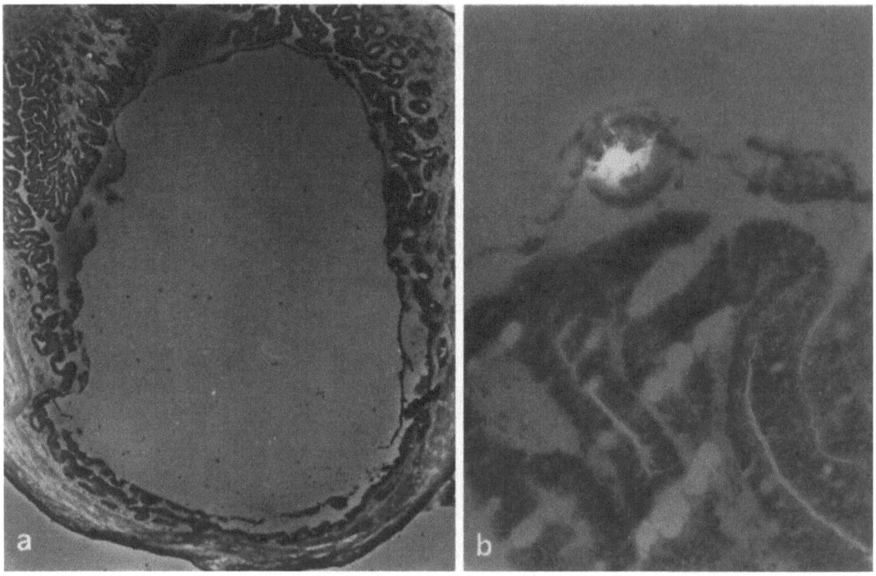

Fig. 5 a,b. Inhibition of implantation by administration of proteinase inhibitors in vivo in the rabbit. In this case, 6 mg antipain was injected into the uterine lumen at 6 1/2 d p.c. At 7 1/2 d p.c., no proteinase activity is detected with the gelatin substrate film test as shown in **a** (overview, x 15.5, cf. Fig. 4a), and the dissolution of the blastocyst coverings (*dark band*) is inhibited. Since the intrauterine antipain concentration decreases rapidly after a single injection (Denker 1977), lysis of the blastocyst coverings does not place take until 8 1/2 d p.c. as seen in **b** (abembryonic region, x 140). There is still not much proteinase activity between trophoblast and uterine epithelium; only a few trophoblastic knobs manage to attach to the uterine epithelium. Unattached knobs often show considerable intracellular proteinase activity

Proteinase inhibitors (TLCK, TPCK, aprotinin) have been administered intrauterally also in the *mouse*, and an antifertility effect was noted (Dabich and Andary 1974). Although no analysis of implantation and early postimplantation stages has been carried out in this case, the authors assumed that it was the implantation process which was disturbed also in this species.

The inhibition of implantation by administration of specific proteinase inhibitors is considered a new concept possibly applicable in fertility regulation. More extended studies involving various synthetic low-molecular-weight inhibitors are in progress in our laboratory. Possible side effects, e.g., increased incidence of dystopic implantation, and ethical aspects should be taken into consideration (Denker 1977). It should further be kept in mind that interference with implantation may be involved in all experiments in which proteinase inhibitors are administered to the genital tract, as in attempts to interfere with fertilization (Schumacher et al. 1971) or in treatment of hypermenorrhea in women using IUDs (Tauber et al. 1977).

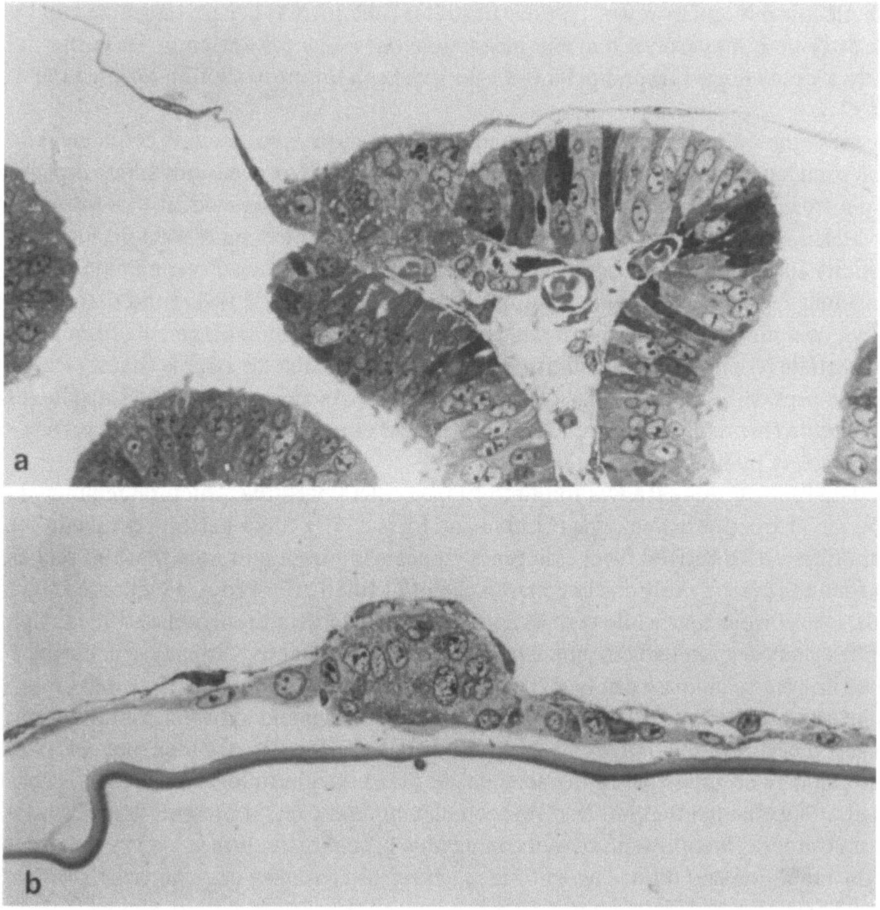

Fig. 6 a,b. Inhibition of implantation by administration of proteinase inhibitors in vivo in the rabbit. Morphology, 7 1/2 d p.c., semithin sections, toluidine blue, x 250. **a** Control. The blastocyst coverings are dissolved in this abembryonic region. A trophoblastic knob is shown which has invaded the endometrium and made contact to the subepithelial capillary. **b** Antipain treatment as in Fig. 5. Blastocyst coverings are still present, but arrosion seems to have just begun below this trophoblastic knob

3.1.2 Biochemical Properties

Until very recently, investigations of blastocyst proteinases were generally based on gelatin substrate film tests, particularly on one version which had been specifically designed and optimized for this purpose (Denker 1971, 1974a). Although the lack of a kinetic assay was a severe obstacle to detailed biochemical investigations, many interesting data could, nevertheless, be collected. Experiments on the inhibition of rabbit *trophoblast-dependent blastocyst proteinase (blastolemmase)* by specific proteinase inhibitors in vitro (Denker 1976a) revealed that the *active center* of this enzyme is closely related to that of trypsin. However, its *substrate specificity* is much more restricted:

synthetic low-molecular-weight trypsin substrates (like BANA, benzoyl-arginine ethyl ester, etc.) are not hydrolyzed in any measurable rates. The pH optimum lies in the slightly alkaline range (around pH 8.0-8.5 with the gelatin substrate film test, see Denker 1977).

Recently, these findings could be substantiated and more accurate data could be obtained using quantitative biochemical assays based on the hydrolysis of various tripeptide-p-nitroanilide substrates (Denker and Fritz 1979). It became evident that the enzyme which is extracted from rabbit trophoblast tissue (blastolemmase) exhibits high specificity for arginyl bonds but also recognizes, in addition, the adjacent amino acids in positions P_2 and P_3 (P_1 being arginine). Since, furthermore, it was found that this enzyme does not release any measurable amounts of TCA-soluble fragments from protein substrates (such as casein and others), it seems reasonable to assume that it cleaves only few peptide bonds in its physiological substrates. As mentioned in the Introduction, such a *limited proteolysis* may induce, in vivo, even more significant biological effects than complete hydrolysis of proteins does.

The Gly-Pro-Arg sequence was found to fit particularly well the structural requirements of the rabbit trophoblast enzyme (Denker and Fritz 1979). Since gelatin (as used in the gelatin film test) is derived from collagen, it appears interesting to note that this sequence is found in several positions in collagen chains (Table 1; Gly-Pro-Arg sequences underlined). It is obvious that while arginyl bonds (underlined with broken line), which might be split by any trypsin-like enzyme, are present more frequently, a highly specific proteinase like blastolemmase can be expected to cleave only a few of them, thereby predominantly producing large fragments of the protein chain and allowing conformational changes to occur. In the native state, not even all of the Gly-Pro-Arg sequences of such a collagen may be accessible to the enzyme. As far as the physiological situation is concerned, it must be pointed out that it is completely unknown, at present, whether blastocyst coverings glycoproteins have a comparable primary structure.

In the rabbit implantation stage *uterine secretion,* a trypsinlike enzyme was found which, in contrast to blastolemmase, does show BANA-splitting activity and which, in this respect, resembles trypsin more closely than the blastocyst-derived enzyme (Denker and Petzoldt 1977). Like the latter, it is inhibited by the specific high-molecular-weight trypsin inhibitors. Some first evidence was found that *uteroglobin* which had recently been shown to have trypsin-inhibiting activity in vitro (Beier 1977) possibly acts as an inhibitor of this enzyme in vivo (Denker and Petzold 1977; Denker, unpublished data). This would provide a way for the physiological regulation of uterine secretion trypsin-like proteinase activity.

Minor activities of *chymotrypsinlike enzymes* have also been detected in rabbit uterine secretion and blastocysts (Denker and Petzoldt 1977), but have not been studied in detail so far.

As far as other species are concerned, some comparable studies have only been performed in the *mouse.* In mouse uterine secretion, a zona pellucida dissolving activity was reported to increase before implantation ("implantation initiating factor", Mintz 1971) and was attributed to a casein-splitting endopeptidase showing parallel changes in activity (Pinsker et al. 1974). Several trypsinlike (BANA-splitting) and chymotrypsinlike (GPNA-splitting) enzymes as well as casein and hemoglobin hydrolyzing enzymes, have been detected biochemically in blastocysts and uterine flushings of this species (Andary

Table 1. Amino acid sequence in the helical region of collagen chain α1 (I) [rat, (bovine)] (after Bornstein and Traub 1979). –Gly–Pro–Arg–sequences, preferably recognized by blastolemmase in synthetic oligopeptide substrates, are underlined

1 5 10 15 20 25
Gly–Pro–Met–Gly–Pro–Ser–*Gly–Pro–Arg*–Gly–Leu–Hyp–Gly–Pro–Hyp–Gly–Ala–Hyp–Gly–Pro–Gln–Gly–Phe–Gln–Gly–Pro–Hyp–Gly–Glu–

30 35 40 45 50 55
Hyp–Gly–Glu–Hyp–Gly–Ala–Ser–Gly–Pro–Met–*Gly–Pro–Arg*–Gly–Pro–Hyp–Gly–Pro–Hyp–Gly–Lys–Asn–Gly–Asp–Asp–Gly–Glu–Ala–Gly–Lys–

60 65 70 75 80 85
Pro–Gly–*Arg*–Hyp–Gly–Gln–*Arg*–Gly–Pro–Hyp–Gly–Pro–Gln–Gly–Ala–*Arg*–Gly–Leu–Hyp–Gly–Thr–Ala–Gly–Leu–Hyp–Gly–Met–Hyl–Gly–His–

90 95 100 105 110 115
Arg–Gly–Phe–Ser–Gly–Leu–Asp–Gly–Ala–Lys–Gly–Asn–Thr–Gly–Pro–Ala–Gly–Pro–Lys–Gly–Glu–Hyp–Gly–Ser–Hyp–Gly–Glu–Asn–Gly–Ala–

120 125 130 135 140 145
Hyp–Gly–Gln–Met–*Gly–Pro–Arg*–Gly–Leu–Hyp–Gly–Glu–*Arg*–Gly–*Arg*–Hyp–Gly–Pro–Hyp–Gly–Ser–Ala–Gly–Ala–*Arg*–Gly–Asp–Asp–Gly–Ala–

150 155 160 165 170 175
Val–Gly–Ala–Ala–Gly–Pro–Hyp–Gly–Pro–Thr–Gly–Pro–Thr–Gly–Pro–Hyp–Gly–Phe–Hyp–Gly–Ala–Ala–Gly–Ala–Lys–Gly–Glu–Ala–Gly–Pro–

180 185 190 195 200 205
Gln–Gly–Ala–*Arg*–Gly–Ser–Glu–Gly–Pro–Gln–Gly–Val–*Arg*–Gly–Glu–Hyp–Gly–Pro–Hyp–Gly–Pro–Ala–Gly–Ala–Ala–Gly–Pro–Ala–Gly–Asn–

210 215 220 225 230 235
Hyp–Gly–Ala–Asp–Gly–Gln–Hyp–Gly–Ala–Lys–Gly–Ala–Asn–Gly–Ala–Hyp–Gly–Ile–Ala–Gly–Ala–Hyp–Gly–Phe–Hyp–Gly–Ala–*Arg*–Gly–Pro–

240 245 250 255 260 265
Ser–Gly–Pro–Gln–Gly–Pro–Ser–Gly–Ala–Hyp–Gly–Pro–Lys–Gly–Asn–Ser–Gly–Glu–Hyp–Gly–Ala–Hyp–Gly–Asn–Lys–Gly–Asp–Thr–Gly–Ala–

270 275 280 285 290 295
Lys–Gly–Glu–Hyp–Gly–Pro–Ala–Gly–Val–Gln–Gly–Pro–Hyp–Gly–Pro–Ala–Gly–Glu–Glu–Gly–Lys–*Arg*–Gly–Ala–*Arg*–Gly–Glu–Hyp–Gly–Pro–

300 305 310 315 320 325
Ser–Gly–Leu–Hyp–Gly–Pro–Hyp–Gly–Glu–*Arg*–Gly–Gly–Hyp–Gly–Ser–*Arg*–Gly–Phe–Hyp–Gly–Ala–Asp–Gly–Val–Ala–Gly–Pro–Lys–Gly–Pro–

330 335 340 345 350 355
Ala–Gly–Glu–*Arg*–Gly–Ser–Hyp–Gly–Pro–Ala–Gly–Pro–Lys–Gly–Ser–Hyp–Gly–Glu–Ala–Gly–*Arg*–Hyp–Gly–Glu–Ala–Gly–Leu–Hyp–Gly–Ala–

360 365 370 375 380 385
Lys–Gly–Leu–Thr–Gly–Ser–Hyp–Gly–Ser–Hyp–Gly–Pro–Asp–Gly–Lys–Thr–Gly–Pro–Hyp–Gly–Pro–Ala–Gly–Glx–Asx–Gly–*Arg*–Hyp–Gly–Pro–

Table 1 (continued)

390 395 400 405 410 415
Ala–Gly–Pro–Hyp–Gly–Ala–*Arg*–Gly–Gln–Ala–Gly–Val–Met–Gly–Phe–Hyp–Gly–Pro–Lys–Gly–Thr–Ala–Gly–Glu–Hyp–Gly–Lys–Ala–Gly–Glu–

420 425 430 435 440 445
Arg–Gly–Val–Hyp–Gly–Pro–Hyp–Gly–Ala–Val–Gly–Pro–Ala–Gly–Lys–Asp–Gly–Glu–Ala–Gly–Ala–Gln–Gly–Ala–Hyp–Gly–Pro–Ala–Gly–Pro–

450 455 460 465 470 475
Ala–Gly–Glu–*Arg*–Gly–Glu–Gln–Gly–Pro–Ala–Gly–Ser–Hyp–Gly–Phe–Gln–Gly–Leu–Hyp–Gly–Pro–Ala–Gly–Pro–Hyp–Gly–Glu–Ala–Gly–Lys–

480 485 490 495 500 505
Hyp–Gly–Glx–Glx–Gly–Val–Hyp–Gly–Asp–Leu–Gly–Ala–Hyp–Gly–Pro–Ser–Gly–Ala–*Arg*–Gly–Glu–*Arg*–Gly–Phe–Hyp–Gly–Glu–*Arg*–Gly–Val–

510 515 520 525 530 535
Gln–Gly–Pro–Hyp–Gly–Pro–Ala–*Gly–Pro–Arg*–Gly–Asn–Asn–Gly–Ala–Hyp–Gly–Asx–Asx–Gly–Ala–Lys–Gly–Asp–Thr–Gly–Ala–Hyp–Gly–Ala–

540 545 550 555 560 565
Hyp–Gly–Ser–Gln–Gly–Ala–Hyp–Gly–Leu–Glx–Gly–Met(Hyp–Gly–Glu–*Arg*–Gly–Ala–Ala–Gly–Leu–Hyp–Gly–Pro–Lys–Gly–Asp–*Arg*–Gly–Asp–

570 575 580 585 590 595
Ala–Gly–Pro–Lys–Gly–Ala–Asp–Gly–Ala–Pro–Gly–Lys–Asp–Gly–Val–*Arg*–Gly–Leu–Thr–Gly–Pro–Ile–Gly–Pro–Hyp–Gly–Pro–Ala–Gly–Ala–

600 605 610 615 620 625
Hyp–Gly–Asp–Lys–Gly–Glu–Ala–Gly–Pro–Ser–Gly–Pro–Ala–Gly–Thr–*Arg*–Gly–Ala–Hyp–Gly–Asp–*Arg*–Gly–Glu–Hyp–Gly–Pro–Hyp–Gly–Pro–

630 635 640 645 650 655
Ala–Gly–Phe–Ala–Gly–Pro–Hyp–Gly–Ala–Asp–Gly–Gln–Hyp–Gly–Ala–Lys–Gly–Glu–Hyp–Gly–Asp–Ala–Gly–Ala–Lys–Gly–Asp–Ala–Gly–Pro–

660 665 670 675 680 685
Hyp–Gly–Pro–Ala–Gly–Pro–Ala–Gly–Pro–Hyp–Gly–Pro–Ile–Gly–Asn–Val–Gly–Ala–Hyp–Gly–Pro–Hyl–Gly–Ala–*Arg*–Gly–Ser–Ala–Gly–Pro–

690 695 700 705 710 715
Hyp–Gly–Ala–Thr–Gly–Phe–Hyp–Gly–Ala–Ala–Gly–*Arg*–Val–Gly–Pro–Hyp–Gly–Pro–Ser–Gly–Asn–Ala–Gly–Pro–Hyp–Gly–Pro–Hyp–Gly–Pro–

720 725 730 735 740 745
Ala–Gly–Lys–Glu–Gly–Ser–Lys–*Gly–Pro–Arg*–Gly–Glu–Thr–Gly–Pro–Als–Gly–*Arg*–Hyp–Gly–Glu–Val–Gly–Pro–Hyp–Gly–Pro–Hyp–Gly–Pro–

750 755 760 765 770 775
Ala–Gly–Glu–Lys–Gly–Ala–Hyp–Gly–Ala–Asp–Gly–Pro–Ala–Gly–Ala–Hyp–Gly–Thr–Pro–Gly–Pro–Gln–Gly–Ile–Ala–Gly–Gln–*Arg*–Gly–Val–

780 785 790 795 800 805
Val–Gly–Leu–Hyp–Gly–Gln–*Arg*–Gly–Glu–*Arg*–Gly–Phe–Hyp–Gly–Leu–Hyp–Gly–Pro–Ser–Gly–Glu–Hyp–Gly–Lys–Gln–Gly–Pro–Ser–Gly–Ala–

810 815 820 825 830 835
Ser–Gly–Glu–*Arg*–Gly–Pro–Hyp–Gly–Pro– Met–Gly–Pro–Hyp–Gly–Leu–Ala–Gly–Pro–Hyp–Gly–Glu–Ser–Gly–*Arg*–Glu–Gly–Ala–Hyp–Gly–Ala–

840 845 850 855 860 865
Glu–Gly–Ser–Hyp–Gly–*Arg*–Asp–Gly–Ser–Hyp–Gly–Ala–Lys–Gly–Asp–*Arg*–Gly–Glu–Thr–Gly–Pro–Ala–Gly–Pro–Hyp–Gly–Ala–Hyp–Gly–Ala–

870 875 880 885 890 895
Hyp–Gly–Ala–Hyp–Gly–Pro–Val–Gly–Pro–Ala–Gly–Lys–Ser–Gly–Asp–*Arg*–Gly–Glu–Thr–Gly–Pro–Ala–Gly–Pro–Ile–Gly–Pro–Val–Gly–Pro–

900 905 910 915 920 925
Ala–Gly–Ala–*Arg*–Gly–Pro–Ala–Gly–Pro–Gln–*Gly–Pro–Arg*–Gly–Asx–Hyl–Gly–Glx–Thr–Gly–Glx–Glx–Gly–Asx–*Arg*–Gly–Ile–Hyl–Gly–His–

930 935 940 945 950 955
Arg–Gly–Phe–Ser–Gly–Leu–Gln–Gly–Pro–Hyp–Gly–Pro–Hyp–Gly–Ser–Hyp–Gly–Glu–Gln–Gly–Pro–Ser–Gly–Ala–Ser–Gly–Pro–Ala–*Gly–Pro*–

960 965 970 975 980 985
Arg–Gly–Pro–Hyp–Gly–Ser–Ala–Gly–Ser–Hyp–Gly–Lys–Asp–Gly–Leu–Asn–Gly–Leu–Hyp–Gly–Pro–Ile–Gly–Hyp–Hyp–*Gly–Pro–Arg*–Gly–*Arg*–

990 995 1000 1005 1010
Thr–Gly–Asp–Ala–Gly–Pro–Ala–Gly–Pro–Hyp–Gly–Pro–Hyp–Gly–Pro–Hyp–Gly–Pro–Hyp–Gly–Pro–Pro)

1974; Dabich and Andary 1974; Hoversland and Weitlauf 1978). Since only insignificant endopeptidase activity could be demonstrated in implanting mouse blastocysts using histochemical substrate film tests (Bergström 1970; Denker, unpublished) it was assumed that uterine secretion proteinase activity is physiologically more significant in this species (Mintz 1971). Species differences were also emphasized by Blandau (1949) who found gelatinolytic activity in late preimplantation blastocysts of the *guinea pig* (which show more invasive properties) but not in those of the *rat* (where the trophoblast seems to be less invasive, while the uterine epithelium degenerates more readily). The physiological function of the biochemically detected proteases of mouse blastocysts remains obscure in this context in spite of observations of an increase in activity of some of these enzymes toward implantation (Andary 1974; Andary and Dabich 1974).

Only histochemical investigations have been performed so far in the guinea pig and the cat. In the *cat*, considerable gelatinolytic activity is found in late preimplantation trophoblast and increases toward implantation. It is particularly high in the girdle-shaped invasion zone at 14 days post coitum, where it is located in the trophoblast, between it and the uterine epithelium and in the widened endometrial crypts (Denker et al. 1978). Experiments on the pH-dependence and on inhibition by various inhibitors revealed that this activity is predominantly due to an enzyme which is totally different from rabbit trophoblast-dependent (trypsinlike) proteinase, namely to an SH-dependent *cathepsin* with an acid pH optimum. It is in many respects similar to cathepsin B, but since no BANA-splitting activity has been demonstrated so far it may also be related to cathepsin L or the "collagenolytic cathepsin" (Barrett 1977) as is likewise discussed for rabbit entodermal cathepsin (see below). In the *guinea pig*, only limited experiments on pH-dependence and on inhibition by various inhibitors have been performed on post-implantation stages (Owers and Blandau 1968, 1971; Owers 1970, 1971). The results suggest that here, too, the predominant proteolytic activity is due to a cathepsin which might be of the same type. According to the acid pH optimum, an intracellular (intra-lysosomal) site of action has to be considered for cathepsins, thus differing from the assumed site of action of rabbit blastolemmase (i.e., at the interface between trophoblast and endometrium, where an alkaline pH is measured). In both the cat and the guinea pig, however, available data do not exclude the presence in addition, of a trypsinlike enzyme comparable to blastolemmase particularly in preattachment and attachment stages, where it might be involved in the dissolution of the zona pellucida (Denker 1977). This point will merit further investigation.

3.2 Entodermal Cathepsin

In the rabbit, considerable activity of an endopeptidase which is different from the described trophoblast-dependent enzyme becomes demonstrable in the entoderm after 7 days post coitum (Denker 1977). When the gelatin substrate film test is employed, first traces of activity are detected around 7 days post coitum, and high activity is found from 8 1/2 days post coitum on, which persists at least until 11 1/2 days post coitum (latest stage investigated).

Interestingly, the activity is not found in all entoderm cells but is restricted to a certain area of this germ layer. While at 7-7 1/2 days post coitum the first traces of activity are seen in the embryonic disk entoderm, the typical higher activity of later stages is restricted to parts of the extraembryonic entoderm lateral to the embryonic area, while the embryonic entoderm and the abembryonic-lateral parts are nearly nonreactive (Fig. 7). The highly active part of entoderm spreads more laterally until 11 1/2 days post coitum, but still the very abembryonic-lateral entoderm remains without any activity worthy of mention. As far as can be judged from present observations, the high activity is only found in those regions of extraembryonic entoderm where a well-developed mesoderm is also present. It should be interesting to see whether this correlation can be confirmed by more detailed future investigations, and what it might have to do with functional and structural relationships between the two germ layers.

Preliminary attempts at a *biochemical characterization* of this entodermal proteinase have clearly shown that it is different from the trypsinlike proteinases of trophoblast and uterine secretion of the rabbit (Denker 1977). Nearly all properties observed suggest that is is closely related to *cathepsin B*, it is an SH enzyme because it is inhibited by iodoacetamide and activated by cysteine, it is inhibited by antipain and chymostatin, and its pH optimum is in the acid region (in the range of pH 5.0 with the gelatin substrate film test, although this allows only a rough estimation of optimum pH values) (for references and discussion of inhibitor specificity, etc., see Denker 1977). However, it has not been possible so far to demonstrate any hydrolysis of BANA. The enzyme could, therefore, be related to *cathepsin L* and the *"collagenolytic cathepsin"* (see Barrett 1977). Further experiments including recently developed synthetic cathepsin substrates are needed before final classification can be done. The enzyme is definitely not related to cathepsin D, e.g., because it is an SH proteinase, and because it is not inhibited by pepstatin. Furthermore, cathepsin D does not have any significant gelatinolytic (or collagenolytic) activity (Burleigh 1977), while cathepsin B has.

The most remarkable aspect of the entodermal cathepsin is its high level of activity. It seems reasonable to assume that this enzyme plays a significant role in the physiological functions of the yolk sac entoderm as related to the "nutritional" requirements of the embryo. High rates of protein degradation may well be needed in this context. While the enzyme can be assumed to act intracellularly, its compartmentalization is not known so far (lysosomes?). Rabbit entoderm cells of these stages are particularly rich in cytoplasmatic granules which show extremely strong toluidine blue uptake in semithin sections (Denker 1977). The nature of these granules and their possible relation to entodermal cathepsin activity have to be studied further.

A cathepsin-B-type endopeptidase of quite similar properties is also found in certain rabbit *endometrial stroma cells* (Denker 1977), and in implantation-stage *trophoblast* and *endomatrial crypts* of the *cat* (see above). The cathepsin activity of cat implantation stage trophoblast is particularly high and will merit further attention. At the present time it is not possible yet to decide whether it is in any way directly involved in the process of attachment, or whether its function is merely related to the high nutritional requirements of the blastocyst in this phase of elevated metabolic activity.

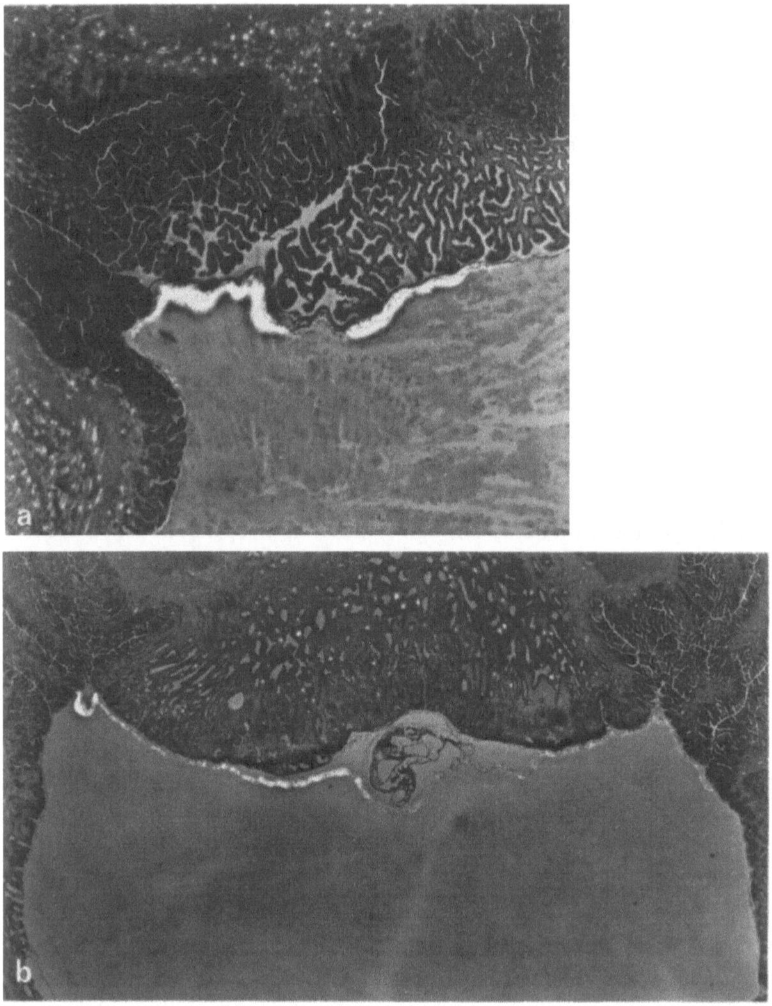

Fig. 7 a,b. Entodermal cathepsin activity in the rabbit. Cryostat sections through normal implantation sites, mesometrial region (embryonic pole), gelatin substrate film test. **a** 8 1/2 d p.c., x 16; **b** 9 1/2 d p.c., x 9. In **b** the differentiation of the embryonic shield is already considerably more advanced than in **a**. The embryonic body is folding off, and the amnion has developed. Invasion of the trophoblast into the endometrium of the placental folds is under way. In both stages, proteinase activity (*bright lysis zones*) is found in those parts of extraembryonic entoderm which are close to the embryonic anlage. In contrast, the lateral and abembryonic parts of extraembryonic entoderm and the embryonic entoderm show no or only very little activity

Table 2. Inhibition of rabbit entoderm proteinase (cathepsin) in vitro. (gelatin substrate film test)

Inhibitor	Concentration (mol/l)	Inhibitory effect
Iodoacetamide	10^{-2}	+
Antipain	10^{-3}	+
Chymostatin	10^{-3}	+
Pepstatin	10^{-3}	∅
Aprotinin (Trasylol)	10^{-3}	∅
SBTI	10^{-4}	∅
Ovomucoid (chicken)	10^{-3}	∅
EDTA	10^{-2}	∅

For references on the specificity of inhibitors see Denker (1976a, 1977)

3.3 Hormonal Control of Trophoblast-Dependent Proteinase Activity and of Implantation

Initiation of implantation is well known to depend on maternal sex steroid hormone levels, as shown by numerous investigations in a number of species. In the rat and mouse an "*estrogen surge*" plays a central role in triggering events which lead to receptivity of the endometrium and which allow the blastocyst to implant (reviewed by Psychoyos 1973). A similar role for a maternal estrogen surge has been postulated for the Alaskan fur seal (Daniel 1974). It seems possible that a maternal estrogen surge is involved in implantation initiation particularly in such species in which the blastocyst is able to enter, in certain physiological situations, a state of dormancy (*diapause*) (McLaren 1972) during which the metabolism, further development and implantation are blocked by *inhibiting factors derived from the uterus.* In fact there is good evidence that the estrogen effect is mediated via removal of a so far unidentified inhibitory factor present in progesterone-dominated uteri, rather than via induction of any new specific activity of the endometrium or the embryo.

In a number of species, on the other hand, there is no indication for the existence and/or any physiologically significant role of a rise in maternal estrogens around the time of implantation. It should be kept in mind, however, that in various cases only ovariectomy (and hormone replacement therapy) experiments have been performed in these species, so that the possibility is not completely excluded that estrogens of extraovarian origin (e.g., adrenal) may be involved. Nevertheless, progesterone replacement therapy experiments performed after ovariectomy strongly suggest a major role for *progesterone* but not for maternal estrogens in some species, including the rabbit (Allen and Corner 1929; Wu and Allen 1959; Lutwak-Mann et al. 1962; Hafez 1963; Rennie and Davies 1965; Pickworth and Lamming 1967; Denker 1972; Kwun and Emmens 1974; a peak of maternal plasma estradiol-17β has been reported, however, to occur on day 6 by Challis et al. 1973), the Syrian hamster (Orsini and Psychoyos 1965, Harper et al. 1966), the ferret (Wu and Chang 1973; Murphy and Mead 1976), the sheep (Bindon 1971;

Moore 1975) and the rhesus monkey (Bosu and Johansson 1975). In the guinea pig, implantation occurs after ovariectomy even without progesterone replacement therapy; here it seems to be needed more for early phases of preimplantation development (Deanesly 1960).

The idea that the *blastocyst* may produce and secrete estrogens (and other hormones) could provide an explanation for the observed independence from maternal estrogens if we assume that blastocyst-derived estrogens are sufficient for induction of implantation in these species, while in others maternal estrogens are needed (in addition?). However, good evidence for an actual production of estrogens by the preimplanted blastocyst has only been shown in the pig (but which is barely comparable to most other species because it has only a superficial, epitheliochorial "implantation") (see Heap et al., this volume) while for other species this is still being questioned (Bhatt and Bullock 1974; Dickmann and Dey 1974; Dickmann et al. 1975, 1977; Sherman and Salomon 1975). However, recently published data (George and Wilson 1978) indicate that estrogens may in fact be also produced by implantation stage rabbit embryos.

In the rabbit, implantation does not occur when the animals are ovariectomized at 6 days post coitum, i.e., 1 day before attachment begins (Lutwak-Mann et al. 1962). Interestingly, a correlation was found between this inability to implant and a failure to develop a regular blastocyst proteinase activity (Denker 1972). Instead, the latter enzyme activity (as estimated using the gelatin substrate film test) remains at a low level, the blastocyst coverings are not dissolved, and attachment of the trophoblast to the endometrium does not occur: the blastocyst remain free in the uterine lumen (Fig. 8). By appropriate progesterone replacement therapy, however, it was possible to get complete normalization of blastocyst proteinase activity, regular dissolution of blastocyst coverings, and implantation. It seemed to be very important, in these experiments, to choose an appropriate dose schedule and way of application of the hormone, in particular to place a subcutaneous depot of hormone 1 day before ovariectomy, and to use very large volumes of vehicle in order to have a slow resorption, resulting in fairly constant hormone levels. The following schedule was found suitable for subcutaneous injections: 1 mg progesterone in 1 ml sesame oil at 5 days post coitum, 4 mg in 2 ml at ovariectomy (6 days post coitum) and every 24 h thereafter (Denker 1972). The importance of sticking to this type of schedule is demonstrated by the fact that Lutwak-Mann et al. (1962) using a different schedule did not succeed with progesterone replacement therapy after ovariectomy at the same stage (which, being so close to implantation, might be a particularly delicate stage for ovariectomy and hormone replacement therapy experiments).

The hormonal regulation of implantation does not seem to involve direct effects of maternal hormones on the blastocysts. Only a few reports claim to have shown direct influences of steroid hormones on biochemical parameters of blastocysts in vitro which are thought to mimick in vivo effects, e.g., "induction" of mouse blastocyst chymotrypsinlike protease by estradiol-17β (or cAMP or PGE_1) (Andary 1974). However, numerous physiological experiments performed by various authors indicate that steroid hormones generally show more detrimental effects, if any, on blastocysts when directly applied in vitro, and there is good reason to believe that metabolic activity and invasive properties of the blastocysts are regulated via changes in the metabolism of the *endometrium* and in composition of the *uterine secretion* (for discussion see Denker 1977).

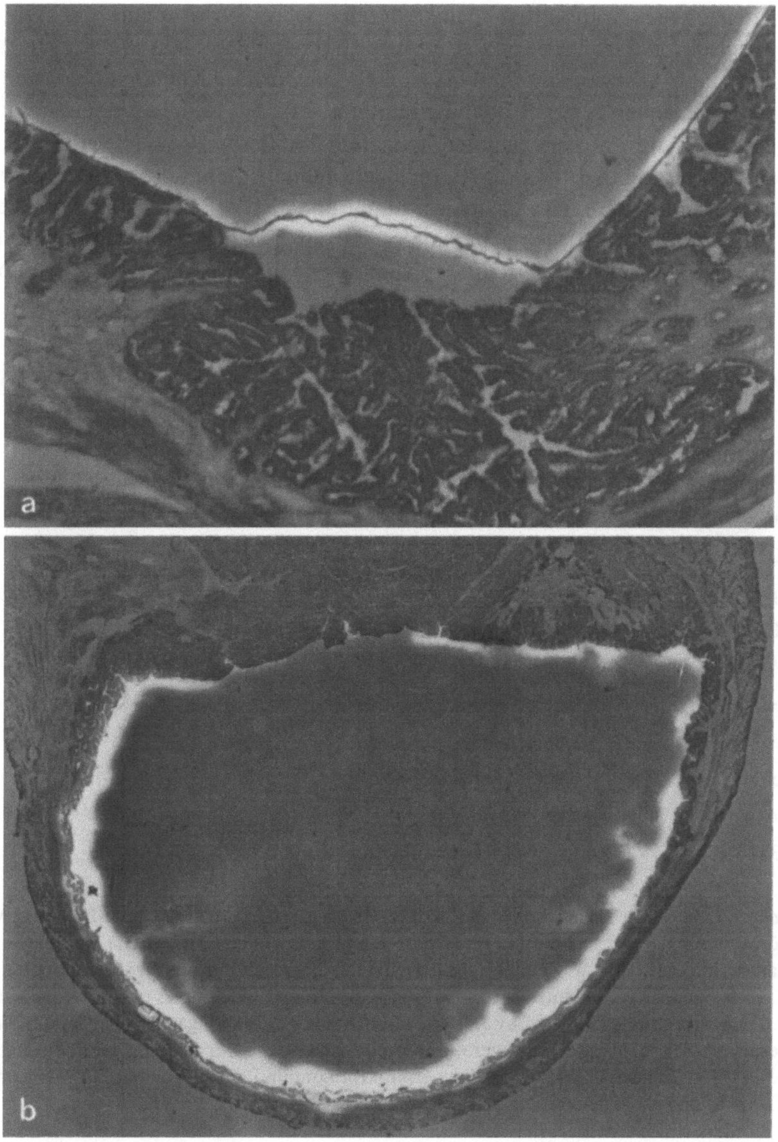

Fig. 8 a,b. Hormonal regulation of blastocyst proteinase activity and of implantation in the rabbit. Pregnant animals were ovariectomized at 6 d p.c., and proteinase activity was studied on cryostat sections of blastocyst sites at 7 1/3 d p.c. using the gelatin substrate film test. **a** No hormonal replacement therapy; detail of abembryonic blastocyst wall and antimesometrial endometrium, x 35. Proteinase activity is reduced as compared to the controls (cf. Fig. 4); lysis zones are predominantly found around those parts of the blastocyst wall which lie more remote from the endometrium, while those which approach the uterine tissues show little activity (due to endometrial proteinase inhibitors?). Blastocyst coverings have not been dissolved (*dark band* represents the complex of blastocyst coverings plus trophoblast). **b** Progesterone replacement therapy has normalized the proteinase activity (cf. Fig. 4), blastocyst coverings are dissolved and attachment of the trophoblast is under way. Overview, x 14.5

There is some indication that changes in rabbit blastocyst proteinase activity after ovariectomy and its restitution after progesterone replacement therapy are associated with changes in *proteinase inhibitor* activity present in the endometrium and uterine secretion (Denker 1972). Proteinase inhibitor activities in the uterine secretion have in fact been shown to be regulated by maternal hormone levels (Beier 1970), and it seems interesting that the progesterone-dependent protein fraction uteroglobin has recently been found to have proteinase inhibitor activity (Beier 1977). Possible changes in proenzyme synthesis (or activation of activating enzymes) are also involved in the regulation of proteinase activity. Much more experimental work has to be done before it can be decided how complex the system really is.

Abbreviations

AlaNA	alanine-β-naphthylamide
BANA	benzoyl-arginine-β-naphthylamide
d.p.c.	days post coitum
EDTA	ethylene diamine tetraacetate
h p.c.	hours post coitum
GPNA	glutaryl-phenylalanine-β-naphthylamide
LAP	leucine aminopeptidase
LeuNA	leucine-β-naphthylamide
$LeuNH_2$	leucineamide
LeuPA	leucine-p-nitroanilide
NA	β-naphthylamide
SBTI	soybean trypsin inhibitor (Kunitz)

Acknowledgments. The author wishes to express his gratitude to Mrs. Gerda Bohr and Mrs. Edith Höricht for excellent technical assistance, to Mrs. Ria Becht and Mrs. Gisela Mathieu for typing the manuscript and to Mr. W. Graulich for drawing the diagrams. Sincere thanks are due to Prof. Dr. H. Fritz (München, Germany) for valuable suggestions and for samples of inhibitors. Trasylol and antipain were kindly provided by Dr. E. Truscheit and Dr. W. Wingender (Bayer AG, Wuppertal, Germany). These investigations were supported by Deutsche Forschungsgemeinschaft grants No. De 181/1-4, 7 and 8 as part of the Schwerpunktprogramm Physiologie und Pathologie der Fortpflanzung and Biologie und Klinik der Reproduktion.

References

Allen WM, Corner GW (1929) Physiology of the corpus luteum. III. Normal growth and implantation of embryos after very early ablation of the ovaries, under the influence of extracts of the corpus luteum. Am J Physiol 88:340-346

Ardary TJ (1974) Trypsin – and chymotrypsin – like enzymes in preimplantation mouse embryos. Medical disserations, Wayne State University, Detroit

Barret MC (ed) (1977)Proteinases in mammalian cells and tissues. North-Holland, Amsterdam New York Oxford (Research monographs in cell and tissue physiology, vol. 2)

Beier HM (1970) Hormonal stimulation of protease inhibitor activity in endometrial secretion during early pregnancy. Acta Endocrinol (Kbh) 63:141-149

Beier HM (1974) Oviducal and uterine fluids. J Reprod Fertil 37:221-237
Beier HM (1977) Immunologische und biochemische Analysen am Uteroglobin und dem Uteroglobin-ähnlichen Antigen der Lunge. Med Welt 28:788-792
Beier HM, Kühnel W, Petry G (1972) Morphologische und biochemische Befunde am pseudograviden Kaninchenendometrium nach gonadotroper Stimulierung. Verh Anat Ges 66:445-457
Bergström S (1970) Estimation of proteolytic activity at mouse implantation sites by the gelatin digestion method. J Reprod Fertil 23:481-485
Bhatt BM, Bullock DW (1974) Binding of oestradiol to rabbit blastocysts and its possible role in implantation. J Reprod Fertil 39:65-70
Bindon BM (1971) Role of progesterone in implantation in the sheep. J Reprod Fertil 24:146
Blandau RJ (1949) Embryo-endometrial interrelationship in the rat and guinea pig. Anat Rec 104: 331-359
Böving BG (1963) Implantation mechanisms. In: Hartman CG (ed) Conference on physiological mechanisms concerned with conception. Pergamon, Oxford London New York Paris, pp 321-396
Bornstein P, Traub W (1979) The chemistry and biology of collagen. In: Neurath H, Hill RL (eds) The proteins, 3rd edn vol IV. Academic Press, New York San Francisco London, pp 411-632
Bosu WTK, Johansson EDB (1975) Implantation and maintenance of pregnancy in mated rhesus monkeys following bilateral oophorectomy or luteectomy with and without progesterone replacement. Acta Endocrinol (Kbh) 79:598-609
Burleigh MC (1977) Degradation of collagen by non-specific proteinases. In: Barrett AJ (ed) Proteinases in mammalian cells and tissues. North-Holland, Amsterdam New York Oxford (Research monographs in cell and tissue physiology) pp 285-309
Challis JRG, Davies IJ, Ryan KJ (1973) The concentrations of progesterone, estrone and estradiol-17ß in the plasma of pregnant rabbits. Endocrinology 93:971-976
Dabich D, Andary TJ (1974) Prevention of blastocyst implantation in mice with proteinase inhibitors. Fertil Steril 25:954-957
Dabich D, Andary TJ (1976) Tryptic- and chymotryptic-like proteinases in early and late preimplantation mouse blastocysts. Biochim Biophys Acta 444:147-153
Daniel JC Jr (1974) Circulating levels of oestradiol-17ß during early pregnancy in the Alaskan fur seal showing an oestrogen surge preceding implantation. J Reprod Fertil 37:425-428
Daniel JC Jr, Krishnan RS (1967) Amino acid requirements for growth of the rabbit blastocyst in vitro. J Cell Physiol 70:155-160
Deanesly R (1960) Implantation and early pregnancy in ovariectomized guinea-pigs. J Reprod Fertil 1:242-248
Denker H-W (1969) Zur Enzym-Topochemie von Frühentwicklung und Implantation des Kaninchens. Medical dissertation, University of Marburg
Denker H-W (1970a) Topochemie hochmolekularer Kohlenhydratsubstanzen in Frühentwicklung und Implantation des Kaninchens. I. Allgemeine Lokalisierung und Charakterisierung hochmolekularer Kohlenhydratsubstanzen in frühen Embryonalstadien. Zool Jahrb Abt Allgem Zool Physiol 75:141-245
Denker H-W (1970b) Topochemie hochmolekularer Kohlenhydratsubstanzen in Frühentwicklung und Implantation des Kaninchens. II. Beiträge zu entwiclungsphysiologischen Fragestellungen. Zool Jahrb Abt Allgem Zool Physiol 75:246-308
Denker H-W (1971) Substratfilmtest für den Proteasennachweis. Acta Histochem [Suppl] (Jena) 10:303-305
Denker H-W (1972) Blastocyst protease and implantation: Effect of ovariectomy and progesterone substitution in the rabbit. Acta Endocrinol (Kbh) 70:591-602
Denker H-W (1974a) Protease substrate film test. Histochemistry 38:331-338; Histochemistry 39: 193
Denker H-W (1974b) Trophoblastic factors involved in lysis of the blastocyst coverings and in implantation in the rabbit: Observations on inversely orientated blastocysts. J Embryol Exp Morphol 32:739-748
Denker H-W (1975) Implantation des Kaninchenkeims im Uterus. Bedeutung von Enzymen des Embryos bzw. des Endometriums. Verh Anat Ges 69:281-289

Denker H-W (1976a) Interaction of proteinase inhibitors with blastocyst proteinases involved in implantation. In: Peeters H (ed) Protides of the biological fluids. Proceedings of the *XXIII*rd Colloquium, Brugge 1975. Pergamon, Oxford, pp 63-68
Denker H-W (1976b) Cooper IUD-induced loss of endometrial arylamidase activity in the rabbit. Biol Reprod 15:519-522
Denker H-W (1976c) Wechselbeziehungen zwischen Blastozyste und Endometrium bei der Implantation: Beeinflussung der endometrialen Aminosäure-Arylamidase-Aktivität durch die Blastozyste. Verh Anat Ges 70:140:839-847
Denker H-W (1977) Implantation: The role of proteinases, and blockage of implantation by proteinase inhibitors. Springer, Berlin Heidelberg New York, Advances in anatomy, embryology and cell biology, vol 53/5
Denker H-W (1978a) The role of trophoblast-dependent and of uterine proteases in initiation of implantation. In: Ludwig H, Tauber PF (eds) Human fertilization. Proceedings of an international workshop, Essen 1976. Thieme, Stuttgart: pp 204-214
Denker H-W (1978b) The role of trophoblastic factors in implantation. In: Spilman CH (ed) Novel aspects of reproductive physiology. Seventh Brook Lodge workshop on problems of reproductive biology, 1977. Spectrum, New York, pp 181-212
Denker H-W, Eng LA, Hamner CE (1978) Studies on the early development and implantation in the cat. II. Implantation: proteinases. Anat Embryol 154:39-54
Denker H-W, Fritz H (1979) Enzymic characterization of rabbit blastocyst proteinase with synthetic substrates of trypsin-like enzymes. Hoppe Seylers Z Physiol Chem 360:107-113
Denker H-W, Gerdes H-J (1979) The dynamic structure of rabbit blastocyst coverings. I. Transformation during regular preimplantation development. Anat Embryol 157:15-34
Denker H-W, Hafez ESE (1975) Proteases and implantation in the rabbit: Role of trophoblast vs. uterine secretion. Cytobiologie 11:101-109
Denker H-W, Petzoldt U (1977) Proteinases involved in implantation initiation in the rabbit: microdisc electrophoretic studies. Cytobiologie 15:363-371
Denker H-W, Stangl R (1976) Versuche zur Lokalisierung und Abgrenzung verschiedener Aminosäure-Arylamidasen in Uterus und Blastozyste des Kaninchens. XVII. Symposium der Gesellschaft für Histochemie, Bozen 1974. Acta Histochem [Suppl] (Jena) 16:249-257
Denker H-W, Van Hoorn G (1974) Peptidases related to implantation in the rabbit: Local stimulation of endometrial arylamidase secretion by the blastocyst immediately preceding implantation. Abstract, VIIth Annual Meeting of the Society for the Study of Reproduction, Ottawa, Canada
Dickmann Z, Dey SK (1974) Steroidogenesis in the preimplantation rat embryo and its possible influence on morula-blastocyst transformation and implantation. J Reprod Fertil 37:91-93
Dickmann Z, Dey SK, Gupta JS (1975) Steroidogenesis in rabbit preimplantation embryos. Proc Natl Acad Sci USA 72:298-300
Dickmann Z, Gupta JS, Dey SK (1977) Does "blastocyst estrogen" initiate implantation? Science 195:687-688
George FW, Wilson JD (1978) Estrogen formation in the early rabbit embryo. Science 199:200-201
Gerdes H-J, Denker H-W (1979) Die strukturelle Entwicklung der Keimhüllen des Kaninchens vom Stadium der Blastozystenbildung bis zur Implantation. Verh Anat Ges 73, Anat. Anz. Suppl. 146:427-434
Gwatkin RBL (1966) Amino acid requirements for attachment and outgrowth of the mouse blastocyst in vitro. J Cell Physiol 68:335-343
Hafez ESE (1963) Gestation in salpingectomized-ovariectomized progesterone-treated rabbits. Proc Soc Exp Biol Med 114:604-607
Harper MJK, Prostkoff B, Reeve RJ (1966) Implantation and embryonic development in the ovariectomized hamster. Acta Endocrinol (Kbh) 52:465-470
Hoorn G van, Denker H-W (1975) Effect of the blastocyst on a uterine amino acid arylamidase in the rabbit. J Reprod Fertil 45:359-362
Hoversland RC, Weitlauf HM (1978) The effect of estrogen and progesterone on the level of amidase activity in fluid flushed from the uteri of ovariectomized mice. Biol Reprod 19:908-912
Jaszczak S, Choroszewska A, Bentyn K (1970) α-Amino nitrogen concentrations in endometrial secretion and in blastocyst fluid between fertilization and implantation in rabbit. Pol Med J 9:1308-1315

Kirchner C (1972) Uterine protease activities and lysis of the blastocyst covering in the rabbit. J Embryol Exp Morphol 28:177-183
Kirchner C, Mootz U (1974) Untersuchungen über die Auflösung der Keimhüllen beim Kaninchen: Der Einfluß des uterinen Milieus. Wilhelm Roux Arch Entwicklungsmech Org 174:172-180
Kirchner C, Hirschhäuser C, Kionke M (1971) Protease activity in rabbit uterine secretion 24 hours before implantation. J Reprod Fertil 27:259-260
Kühnel W, Beier HM, Petry G (1971) Untersuchungen zur hormonellen Regulation der Praeimplantationsphase der Gravidität. II. Histologische, topochemische und biochemische Analysen am hormonbehandelten, pseudograviden Kaninchenuterus. Cytobiologie 4:9-40
Kwun JK, Emmens CW (1974) Implantation and pregnancy in the ovariectomized rabbit (Abstract). J Reprod Fertil 36:465-466
Leroy F, Finn CA, Psychoyos A, Hubinont PO (1980) Blastocyst endometrium interrelationships. Prog Reprod Biol 7
Lutwak-Mann C, Hay MF, Adams CE (1962) The effect of ovariectomy on rabbit blastocysts. J Endocrinol (Kbh) 24:185-197
Maurer RR, Onuma H, Foote RH (1970) Viability of cultured and transferred rabbit embryos. J Reprod Fertil 21:417-422
McLaren A (1972) Endocrinology of implantation. Abstract, 1st South-East Asia and Oceana med. sci. Congr. Int. Plann, Parent, Fedn., Sydney, p 16
Mintz B (1971) Control of embryo implantation and survival. In: Raspe G (ed) Schering symposium on intrinsic and extrinsic factors in early mammalian development, Venice 1970. Pergamon, Oxford; Vieweg, Braunschweig (Advances in the biosciences, vol. 6, pp 317-342)
Moore NW (1975) Progesterone requirements for the maintenance of pregnancy in the ovariectomized ewe. J Reprod Fertil 43:386-387
Murphy BD, Mead RA (1976) Effects of antibodies to oestrogens on implantation in ferrets. J Reprod Fertil 46:261-263
Orsini MW, Psychoyos A (1965) Implantation of blastocysts transferred into progesterone treated virgin hamsters previously ovariectomized (Abstract) J Reprod Fertil 10:300-301
Owers NO (1970) Comparison of the proteolytic activity of the implanting rat and guinea-pig blastocyst (Abstract). Anat Rec 166:358
Owers NO (1971) Ingestive properties of guinea pig trophoblast grown in tissue culture: A possible lysosomal mechanism. In: Blandau RJ (ed) The biology of the blastocyst. The University of Chicago Press, Chicago London, pp 225-241
Owers NO, Blandau RJ (1968) Enzymatic activities of implanting guinea pig embryos. Anat Rec 160:404
Owers NO, Blandau RJ (1971) Proteolytic activity of the rat and guinea pig blastocyst in vitro. In: Blandau RJ (ed) The biology of the blastocyst. The University of Chicago Press, Chicago London, pp 207-223
Petry G, Kühnel W, Beier HM (1970) Untersuchungen zur hormonellen Regulation der Praeimplantationsphase der Gravidität. I. Histologische, topochemische und biochemische Analysen am normalen Kaninchenuterus. Cytobiologie 2:1-32
Pickworth S, Lamming GE (1967) Observations on pregnancy in intact rabbits and in ovariectomized rabbits treated with 17α-hydroxy-progesterone caproate. J Reprod Fertil 13:457-468
Pinsker MC, Sacco AG, Mintz B (1974) Implantation-associated proteinase in mouse uterine fluid. Dev Biol 38:285-290
Psychoyos A (1973) Endocrine control of egg implantation. In: Greep RO (ed) Female reproductive system. American Physiological Society, Washington (Handbook of physiology, vol II/2, sect 7, pp 187-215
Rennie P, Davies J (1965) Implantation in the rabbit following administration of 20α-hydroxypregnen-3-one and 20ß-hydroxypregnen-3-one. Endocrinology 76:535-537
Ribbons DW, Brew K (eds) (1976) Proteolysis and physiological regulation. Academic Press, New York San Francisco London (Miami Winter symposia, vol 11)
Rosenfeld MG, Joshi MS (1977) A possible role of a specific uterine fluid peptidase in implantation in the rat. J Reprod Fert 51:137-139

Schumacher GFB, Swartwout JR, Zuspan FP (1971) Fertility experiments in mice and rabbits with the trypsin-kallikrein inhibitor from bovine lung. In: Fritz H, Tschesche H (eds) Proceedings of the international research conference on proteinase inhibitors, Munich 1970. de Gruyter, Berlin New York, pp 247-252
Sherman MI, Salomon DS (1975) The relationships between the early mouse embryo and its environment. In: Markert CL, Papaconstantinou J (eds) The developmental biology of reproduction. Academic Press, New York San Francisco London, pp 277-309
Spindle AI, Pedersen RA (1973) Hatching, attachment, and outgrowth of mouse blastocysts in vitro; fixed nitrogen requirements. J Exp Zool 186:305-318
Strickland S, Reich E, Sherman MI (1976) Plasminogen activator in early embryogenesis: enzyme production by trophoblast and parietal endoderm. Cell 9:231-240
Tauber PF, Wolf AS, Herting W, Zaneveld LJD (1977) Hemorrhage induced by intrauterine devices: control by local proteinase inhibition. Fertil Steril 28:1375-1377
Weitlauf HM (1971) Influence of ovarian hormones in the incorporation of amino acids by blastocysts in vivo. In: Blandau RJ (ed) The biology of the blastocyst. University of Chicago Press, Chicago London, pp 277-290
Winkle LJ van, Dabich D, Andary TJ (1972) Amino acid incorporation into normal and delayed implantation blastocysts (Abstract). J Cell Biol 55:266 A
Wu DH, Allen WM (1959) Maintenance of pregnancy in castrated rabbits by 17-alpha-hydroxy-progesterone caproate and by progesterone. Fertil Steril 10:439-460
Wu JT, Chang MC (1973) Hormonal requirement for implantation and embryonic development in the ferret. Biol Reprod 9: 350-355

Embryo-Associated Plasminogen Activator Prior to and During Implantation in the Mouse

Michael I. Sherman

Roche Institute of Molecular Biology, Nutley, New Jersey, USA

1 Protease Activities During Implantation

A number of investigators have studied the presence of proteases in the mammalian genital tract prior to, or during, implantation. Roles for these enzymes have been proposed in various aspects of implantation, including hatching from the zona pellucida (e.g. McLaren 1970; Mintz 1971, 1972; Kirchner et al. 1971; Kirchner 1972; Denker 1974), adhesion of the blastocyst to the uterine epithelium (e.g., Kirchner et al. 1971; Bergström 1972; Pinsker et al. 1974) and invasion of trophoblast cells into the uterine wall (e.g. Blackwood et al. 1968; Owers and Blandau 1971; Strickland et al. 1976; Sherman et al. 1976). Certain synthetic protease inhibitors can block implantation when administered intraluminally (Dabich and Andary 1974), although naturally occurring protease inhibitors can be found in the uterus throughout the implantation period (Blackwood et al. 1968; Beier 1970). In contrast to the above studies, it has been reported that some protease activities drop at the time of implantation (e.g., Van Hoorn and Denker 1975; Liedholm and Åstedt 1975), and in one of these reports, it has been postulated that the *absence* of certain proteolytic activities may facilitate implantation (Liedholm and Åstedt 1975).

Although there is general agreement that proteases are in one way or another involved in implantation, a great deal of confusion has been generated concerning details, because different investigators have studied proteases with different specificities in different tissues of different mammals. For example, proteases have been monitored with substrates as diverse as synthetic compounds susceptible to trypsinlike, chymotrypsinlike, or exopeptidase activity (Blackwood et al. 1968; Bergström 1972; Van Hoorn and Denker 1975; Dabich and Andary 1976), naturally occurring macromolecules such as fibrin or plasminogen, which are cleaved by relatively specific groups of proteases (Albrechtsen 1959; Liedholm and Åstedt 1975; Strickland et al. 1976; Sherman et al. 1976), or other natural substrates, such as gelatin, which are subject to hydrolysis by a more general class of enzymes (Blandau 1949; Owers and Blandau 1971; Bergström 1970; Denker 1971; Kirchner et al. 1971). Studies have been carried out on fluids and tissues in the genital tract of rabbits, rats, mice, guinea pigs, and humans; direct comparisons have shown that differences in proteases content do exist from one species to the next, e.g., guinea pig trophoblast shows gelatinase activity that is not evident in rat trophoblast of equivalent age under the same assay conditons (Owers and Blandau 1971), and while dog placentas contain plasminogen activator activity (Lewis and Ferguson 1950), efforts to find this enzyme in human placentas have been unsuccessful (Page et al. 1951; Albrechtsen 1959; Beller et al. 1962).

Although proteases have been detected in the tissues of the reproductive tract, in genital tract secretions, and in parts of the conceptus, the actual source of the enzymes is not always evident. For instance, the site of synthesis of proteases present in oviducal or uterine fluid or in menstrual blood is often unclear. Furthermore, the association of a protease with the conceptus does not assure that the origin is fetal. As an example, Denker (1971) proposed that a gelatinase activity present in rabbit embryos on the 7th day post coitum originated in the trophoblast, whereas Kirchner (1972) reported that the activity was in actuality associated with blastocyst coverings and that the trophoblast itself showed no sign of gelatinase activity. Denker (1974) has subsequently accepted this interpretation, but the origin of the enzyme is still uncertain. In other cases, investigators have been able to locate the site of synthesis of certain proteases with some confidence. For example, enzymes of uterine origin can sometimes be found in uterine cells distant from the implantation site (e.g., Bergström 1970; Van Hoorn and Denker 1975) or in uterine tissue of pseudopregnant animals (Albers et al. 1961), precluding the possibility that they were produced and secreted by the conceptus. Alternatively, production of proteases by embryos or embryonic tissues in culture can provide conclusive evidence of the fetal origin of these enzymes (Owers and Blandau 1971; Dabich and Andary 1976; Strickland et al. 1976; Sherman et al. 1976; Sherman 1980).

2 Plasminogen Activators in the Reproductive Tract

To begin our investigations on the role of proteases in implantation (Strickland et al. 1976; Sherman et al. 1976), we chose to study plasminogen activator production by the mouse conceptus, for a number of reasons. First, as already mentioned, proteases with this activity have a well defined and naturally occurring substrate, the zymogen plasminogen. The product of the reaction, plasmin, can be conveniently measured by its action on one of its natural substrates, fibrin. Thus, appropriate coupled assay systems can be set up for quantitative measurements or for highly sensitive qualitative determinations (see Beers et al. 1975). Further, the observation that cultured cells acquire plasminogen activator activity coincident with oncogenic transformation (Ossowski et al. 1973; Unkeless et al. 1973) led us to consider the possibility that trophoblast cells might also produce this enzyme, since they resemble neoplastic cells when they are in the invasive phase of implantation that is seen in many mammals including the mouse (see Amoroso 1952).

Plasminogen activators have been found in numerous tissues and body fluids (see Lewis and Ferguson 1950; Albrechtsen 1957, 1959), including various parts of the reproductive system: ovarian granulosa cells in the rat (Beers et al. 1975), preimplantation rat embryos (Liedholm and Åstedt 1975), postimplantation mouse and dog extraembryonic tissues (Strickland et al. 1976; Sherman et al. 1976; Bode and Dziadek 1979; Lewis and Ferguson 1950), oviducal musosa (Tympanidis and Astrup 1968), and uterine cells including endometrium and myometrium (see Albrechtsen 1959). Uterine fluid, cervical mucus and menstrual blood are all rich in plasminogen activity (Albrechtsen 1959; Harpel et al. 1966; Weiss and Beller 1969). Because these proteases are ubiquitous and are often secreted, we have not felt it adequate for our purposes merely to show enzyme activity in the conceptus; rather, we have endeavored to determine the site of origin of plasminogen activators associated with the conceptus and to attempt to learn more about the function of these enzymes by drawing correlations between their appearance and events occurring during implantation.

3 Embryo-Associated Plasminogen Activator Prior to Implantation

Liedholm and Åstedt (1975) reported that rat embryos possess plasminogen activator activity at all stages prior to implantation. In our previous studies using ^{125}I-labeled fibrin as a substrate for plasmin formed from plasminogen by activator, we failed to detect evidence for the release of plasminogen activator into the culture medium by preimplantation-stage mouse embryos (Strickland et al. 1976). In more recent analyses (Sherman 1980) with a sensitive fibrin-agar assay (Beers et al. 1975), similar observations to those of Liedholm and Åstedt (1975) have been made: plasminogen activator activity is associated with embryos between the two-cell and blastocyst stages. This activity is observed whether the embryos have developed in the genital tract or in culture. Furthermore, the pattern of activity both in vivo and in vitro resembles that associated with rat embryos (Liedholm and Åstedt 1975): the size of the lysis zone diminishes progressively as the embryos pass through preimplantation stages. However, it has been shown conclusively that the plasminogen activator acitvity detected on preimplantation mouse embryos is, in fact, associated with the zona pellucida (Sherman 1980): when zonae and embryos are separated, all detectable plasminogen activator activity remains with the zona and the denuded embryos do not produce any activity during subsequent development to the blastocyst stage; furthermore, zonae surrounding embryos which have been arrested in development, and even zonae containing completely degenerated embryos, still have plasminogen activator associated with them. The indication from these studies that plasminogen activator remains tightly adherent to the zona pellucida (Sherman 1980), explains why we originally failed to detect this activity in our assays of the culture medium from preimplantation embryos (Strickland et al. 1976).

An obvious conclusion from these studies is that zona-associated plasminogen activator is not produced by the embryo. What, then, is the source of the enzyme? It is possible that the enzyme is taken up from genital tract fluids, as these are known to contain plasminogen activator (Harpel et al. 1966, 1967; M.I. Sherman and S. Strickland, unpublished observations). On the other hand, the observation that activity decreases steadily as preimplantation development proceeds, both in vivo and in vitro, raises the possibility that the zona pellucida contains residual plasminogen activator that was originally produced and secreted by granulosa cells during ovulation (Beers et al. 1975). It is well known that processes from these cells permeate the zona pellucida (see, e.g., Baker 1972, Fig. 2.7) and thus deposition and trapping of the enzyme in the zona matrix could easily take place.

4 Embryo-Associated Plasminogen Activator at the Onset of Implantation

After the mouse blastocyst has hatched from its zona and become properly aligned in a uterine crypt, it attaches to the uterine epithelium. We have attempted to test for the production of plasminogen activator by mouse embryos early on the 5th day of gestation. Blastocysts so obtained were of three types: expanded and rounded blastocysts, as yet unhatched; expanded and elongated blastocysts which had already hatched from their zonae; and elongated and flattened blastocysts, also hatched. The latter class of embryos often contained a few trophoblast cells which appeared to be damaged. It is likely that these blastocysts had already begun to adhere to the uterine epithelium but had been detached by the vigorous flushing of the uteri.

When tested for plasminogen activator activity over a subsequent period of 24 h, most unhatched blastocysts produced small lysis zones; these were not observed when the zonae were removed by pronase treatment prior to assay. Blastocysts which had hatched spontaneously prior to collection were negative in almost every case, whether they were expanded or flattened.

We have provided evidence elsewhere that mouse blastocysts in culture undergo a series of changes which are analogous to those which occur during implantation in utero (Sherman and Salomon 1975; Sherman and Wudl 1976; Sherman and Atienza-Samols 1978). One of these is an alteration of the blastocyst surface from nonadhesive to "sticky", an event which could be necessary for attachment of the trophoblast to the uterine wall. We have, therefore, assayed blastocysts cultured from the fourth day of gestation for plasminogen activator production over the 24-h period during which most would be expected to acquire surface adhesiveness, i.e., from the 5th to the 6th equivalent gestation day (EGD–the age of the embryo had it been left in utero). Two such experiments were carried out, and the results were quite different. In the first, less than 15% of the 5th EGD blastocysts had hatched spontaneously in culture by the beginning of the assay period (hatching will not take place subsequently under the fibrin-agar overlay; in fact, by the end of the 24-h incubation period, the blastocoele cavities have collapsed and many of the cells have degenerated). Of those embryos which had hatched, none possessed plasminogen activator. The unhatched embryos all produced small lysis zones, but this was due to residual zona-associated plasminogen activator described in the previous section, since activity was abolished when the zona was dissolved.

In the second culture experiment, the embryos were more advanced by the 5th EGD, probably due either to their being further developed when removed from the uteri or to some subtle difference in culture conditions: in this case, almost half of the embryos had hatched prior to the start of the assay. Some of the unhatched blastocysts appeared to have lost the zona-associated plasminogen activator activity by this time. Conversely, more than half of the spontaneously hatched embryos produced lysis zones, suggesting that they had begun to synthesize plasminogen activator.

In summary, 5th day blastocysts assayed for plasminogen activator directly upon removal from the uterus appear not to produce this protease over the next several hours (after which they start to degenerate under the fibrin-agar overlay). However, some cultured blastocysts which are relatively advanced in their development do begin to secrete plasminogen activator at the equivalent time. Meanwhile, the zona-associated plasminogen activator has dropped to very low levels or is absent by the time the blastocyst begins to hatch.

5 Plasminogen Activator Production by Trophoblast Cells During the the Invasive Phase

An account of plasminogen activator production by blastocysts that have "implanted" and outgrown on tissue culture dishes has already been published (Strickland et al. 1976; Sherman et al. 1976; Sherman 1980). Regardless of the time at which preimplantation embryos are removed from the genital tract and placed in culture, they will produce plasminogen activator during the period of trophoblast outgrowth along the substratum, a phenomenon which we believe to be analogous to trophoblast invasion of the uterine wall (Sherman and Wudl 1976).

We have found previously that some cultured blastocysts fail to hatch, but continue to develop inside their zonae for several days, as shown by their acquisition of markers of trophoblast differentiation (Barlow and Sherman 1972; Sherman 1972 a,b). Furthermore, when the zonae are removed with pronase these "unhatched" blastocysts quickly attach to the culture dish and give rise to outgrowths which are normal in appearance (Barlow and Sherman 1972). The inability of these blastocysts, which had been in vitro from the two-cell stage, to hatch spontaneously within 5 days of culture is not due to their failure to produce plasminogen activator, since most of the embryos are positive, even after treatment with pronase to remove any residual zona-associated enzyme (Sherman 1980). Thus, attachment and outgrowth are not necessary stimuli for production of plasminogen activator by blastocysts.

We have recently demonstrated that plasminogen activator is produced by single trophoblast cells developing from blastomeres isolated at the 16-cell stage (Sherman and Atienza-Samols 1979). Therefore, neither cell interaction nor cell division is required for the expression of this enzyme by trophoblast cells.

By the quantitative ^{125}I-fibrin lysis assay, we found that plasminogen activator can first be detected in blastocyst culture medium between the 6th and 7th EGD (Strickland et al. 1976); the secretion of small amounts of protease between the 5th and 6th EGD would probably have escaped detection by this assay, which is less sensitive than the fibrin-agar overlay procedure. With increasing time in culture, activity characteristically rises to a peak, falls, and then continues to rise once again. When lysates of the cultured blastocysts were tested for plasminogen activator activity, none could be detected prior to the 6th EGD and at subsequent stages the pattern of enzyme activity was similar to that observed in the culture media (Strickland et al. 1976). The amounts of plasminogen activator activity within the cells were small relative to that of enzyme secreted (S. Strickland and M.I. Sherman, unpublished observations).

In a series of experiments described by Strickland et al. (1976) and Sherman et al. (1976), it was demonstrated that both trophoblast and parietal endoderm cells produce and secrete plasminogen activator. To determine the fraction of activity that is due to trophoblast cells, we prepared cultures containing only trophoblast by making and further culturing trophectodermal vesicles or by killing the inner cell mass derivatives of blastocysts with bromodeoxyuridine. We found that trophoblast cells are responsible for a substantial proportion of plasminogen activator activity up to the 9th EGD. Thereafter, much of the activity is due to secretion by parietal endoderm cells. These studies also indicated that the fibrinolysis is strictly plasminogen-dependent, confirming that the protease secreted by trophoblast and parietal endoderm cells is indeed a plasminogen activator and cannot act directly upon the fibrin substrate (Strickland et al. 1976).

6 Is Plasminogen Activator Production by the Conceptus Under Hormonal Control?

Both proteases (e.g., Albers et al. 1961; Mintz 1971, 1972) and protease inhibitors (Beier 1970) found in the reproductive tract and shown to be of maternal origin have been reported to be under control of estrogens, progestins, or both. In 1951, Page et al. reported that endometrial fibrinolytic activity was stimulated by estrogen withdrawal and more recently Harpel et al. (1966, 1967) found that plasminogen activator levels in the uterine fluid of the mouse were depressed by administration of estrogens, progestins, or their synthetic counterparts. Beers et al. (1975) demonstrated that plasminogen activator production by rat ovarian granulosa cells was stimulated by luteinizing hormones or dibutyryl cyclic AMP.

There have been very few direct studies on hormonal control over proteases of fetal origin. Denker (1972) reported that the gelatinase activity associated with the 7th day rabbit conceptus was partially suppressed by prior ovariectomy, but as mentioned above, the origin of the enzyme(s) with this activity is uncertain. Since Dabich and Andary (1976) found that the trypsinlike and chymotrypsinlike enzyme activities in mouse blastocysts appeared in embryos developing from the two-cell stage in a hormone-free medium; it is likely that these enzymes are not under direct hormonal control.

When blastocysts that had been delayed from implanting for 1 week in ovariectomized mothers were removed and placed into culture, they were found initially to produce low levels of plasminogen activator (Sellens and Sherman 1980). Thereafter activity rose and the characteristic biphasic curve was seen as the embryos attached to the culture dish and trophoblast and parietal endoderm cells developed (M.I. Sherman and S. Strickland, unpublished observations). In fact, the profile of plasminogen activator activity was qualitatively the same as that of control blastocysts removed from untreated mothers on the fourth day of gestation. It might be interpreted from these results that plasminogen activator production or secretion by the conceptus is under direct hormonal control, and that the ultimate appearance of the protease was due to exposure of the embryos to hormones in the serum added to the culture medium. This is unlikely to be the case, however, since plasminogen activator activity appears, and rises with time, when blastocysts are cultured in serum-free media devoid of steroid hormones and containing albumin and/or fetuin as the only macromolecules (Rizzino and Sherman 1979; Sellens and Sherman 1980). Taken together with the results described in the previous section, the evidence suggests that the production of plasminogen activator by trophoblast cells is not hormone-dependent but is intrinsically programmed.

7 Possible Roles for Plasminogen Activator in Implantation

When considering the functions of proteases in vivo, one must keep in mind that we are defining these enzymes in an operational sense. Thus, we have used the term "plasminogen activator" to denote the ability of a particular protease or group of proteases to cleave the zymogen plasminogen to the active form, plasmin. A plasminogen activator may just as effectively activate other zymogens (Werb et al. 1977); although plasminogen activators cannot degrade fibrin, they may nevertheless act directly upon other structural proteins. Therefore, when discussing the possible roles of plasminogen activators in implantation, we must also consider those events which appear not to involve the activation of plasminogen.

The role of the plasminogen activator associated with the zona of preimplantation embryos might well be clarified if its origin could be established. If it represents the residual activity from granulosa cells during ovulation (see Sect. 3), then it may no longer have any useful purpose, and the experiments described above may merely be documenting the slow decay of the protease. On the other hand, even though the pattern of enzyme activity (i.e., a decrease up to the time of hatching) tends to speak against it, the protease may be involved in dissociation of the zona pellucida. It is very unlikely that the enzyme acts upon the zona either directly or via the activation of plasminogen because the zona remains intact following the enzyme assay period, even though there is substantial fibrinolysis (see Sherman 1980). Furthermore, blastocysts which have failed to hatch after several days in culture do produce plasminogen activator in most instances (Sherman 1980). However, the protease that we are studying may activate some other enzyme which participates in hatching, e.g., the zona lysin (Mintz 1971, 1972). Kirchner (1972) has speculated that one or more proteases possibly involved in dissolution of the coverings of the rabbit blastocyst may have to be activated before they carry out their function.

Plasminogen activator does not seem to participate in the attachment of the blastocyst to the uterine epithelium. Blastocysts which appear to have succeeded in attaching to the uterine wall are not necessarily secreting detectable levels of plasminogen activator (see Sect. 4). Since blastocysts at this stage degenerate under the conditions of the fibrin-agar assay and many cells lyse before the end of the 24-h incubation period, plasminogen activator produced but not secreted when the cells were intact should ultimately have been released. On the other hand, these blastocysts may have adsorbed from the uterine milieu, or may have been producing, inhibitors of plasmin or plasminogen activator which could have obliterated the detection of any activator produced. However, other experiments suggest that if plasminogen activator is not produced by attachment-stage blastocysts, it begins to be synthesized very shortly thereafter (Sherman 1980).

Finally, we have already commented that the profile of plasminogen activator production and secretion by trophoblast correlates closely with the period of invasiveness of these cells into the uterine wall (Strickland et al. 1976; Sherman et al. 1976). The enzyme may be functioning over the invasive period to dissolve fibrin via activation of plasminogen. Other roles may be envisaged. For example, a zymogen of collagenase might be activated (Werb et al. 1977) to break down the thick basement membrane of the uterine epithelium (see Potts 1969). Also, the trophoblast plasminogen activator may help to digest the debris remaining from uterine epithelial cells which appear to degenerate by autolysis (El-Shershaby and Hinchliffe 1974, 1975). We have noted elsewhere that we consider it unlikely that the plasminogen activator secreted by the parietal endoderm cells functions in implantation (Sherman et al. 1976).

Our previous studies have indicated that plasminogen activator activity falls at the time when trophoblast cells would be expected to lose their invasive properties in utero (Strickland et al. 1976; Sherman et al. 1976). We have also demonstrated a decreased ability of trophoblast cells to displace cultured uterine cells during this period (Sherman and Salomon 1975). Other factors may also be involved in combating the invasiveness of trophoblast cells in utero; for example, the observation that trophoblast cells fail to invade actively into healthy decidual tissue (see Billington 1971) may be due to the production of protease inhibitors by the latter.

References

Albers HJ, Bedford JM, Chang MC (1961) Uterine peptidase activity in the rat and rabbit during pseudopregnancy. Am J Physiol 201:554-556

Albrechtsen OK (1957) The fibrinolytic activity of animal tissues. Acta Physiol Scand 39:284-290

Albrechtsen OK (1959) Fibrinolytic activity in the organism. Acta Physiol Scand [Suppl] 165:1-112

Amoroso EC (1952) Placentation. In: Parkes AS (ed) Marshall's physiology of reproduction, vol II. Longman, London, pp 127-311

Baker TG (1972) Oogenesis and ovulation. In: Austin CR, Short RV (eds) Germ cells and fertilization. Cambridge University Press, London, pp 14-45

Barlow PW, Sherman MI (1972) The biochemistry of differentiation of mouse trophoblast: Studies on polyploidy. J Embryol Exp Morphol 27:447-465

Beers WH, Strickland S, Reich E (1975) Ovarian plasminogen activator: Relationship to ovulation and hormonal regulation. Cell 6:387-394

Beier HM (1970) Hormonal stimulation of protease inhibitor activity in endometrial secretion during early pregnancy. Acta Endocrinol (Kbh) 63:141-149

Beller FK, Goessner W, Herrschlein HJ (1962) Tissue activator of the fibrinolytic system in placental tissue. Obstet Gynecol 20:117-119

Bergström S (1970) Estimation of proteolytic activity at mouse implantation sites by the gelatin digestion method. J Reprod Fertil 23:481-485

Bergström S (1972) Histochemical locatilization of acid uterine aminoacylnaphthylamidases in early pregnancy and in different hormonal states of the mouse. J Reprod Fertil 30:177-183

Billington WD (1971) Biology of the trophoblast. Adv Reprod Physiol 5:27-66

Blackwood CE, Hosannah Y, Mandl I (1968) Proteolytic enzyme studies in developing rat tissues. J Reprod Fertil 17:19-33

Blandau RJ (1949) Embryo-endometrial interrelationships in the rat and guinea pig. Anat Rec 104: 331-359

Bode V, Dziadek MA (1979) Plasminogen activator secretion during mouse embryogenesis. Dev Biol 73:272-289

Dabich D, Andary TJ (1974) Prevention of blastocyst implantation in mice with proteinase inhibitors. Fertil Steril 25:954-957

Dabich D, Andary TJ (1976) Tryptic- and chymotryptic-like proteinases in early and late preimplantation mouse blastocysts. Biochim Biophys Acta 444:147-153

Denker HW (1971) Enzym-Topochemie von Frühentwicklung und Implantation des Kaninchens. III. Proteasen. Histochemie 25: 344-360

Denker HW (1972) Blastocyst protease and implantation: Effect of ovariectomy and progesterone substitution in the rabbit. Acta Endocrinol (Kbh) 70:591-602

Denker HW (1974) Trophoblastic factors involved in lysis of the blastocyst coverings and in implantation in the rabbit: Observations on inversely oriented blastocysts. J Embryol Exp Morphol 32: 739-748

El-Shershaby AM, Hinchliffe JR (1974) Cell redundancy in the zona-intact preimplantation mouse blastocyst: A light and electron microscope study of dead cells and their fate. J Embryol Exp Morphol 31:643-654

El-Shershaby AM, Hinchliffe JR (1975) Epithelial autolysis during implantation of the mouse blastocyst: An ultrastructural study. J Embryol Exp Morphol 33:1067-1080

Harpel P, Bang NU, Homburger F, Treger A (1966) Plasminogen activator in the mouse uterine fluid: Its suppression by estradiol and progesterone. Proc Soc Exp Biol Med 122:1192-1195

Harpel P, Homburger F, Treger A (1967) Mouse uterine fluid plasminogen activator, acid phosphatase and contraceptive hormones. Am J Physiol 215:928-931
Kirchner C (1972) Uterine protease activities and lysis of the blastocyst covering in the rabbit. J Embryol Exp Morphol 28:177-183
Kirchner C, Hirschhäuser C, Kionke M (1971) Protease activity in rabbit uterine secretion 24 hours before implantation. J Reprod Fertil 27:259-260
Lewis JH, Ferguson JH (1950) Studies on a proteolytic enzyme system of the blood. II. Fibrinolysokinase activators for profibrinolysin. J Clin Invest 29:1059-1068
Liedholm P, Åstedt B (1975) Fibrinolytic activity of the rat ovum, appearance during tubal passage and disappearance at implantation. Int J Fertil 20:24-26
McLaren A (1970) The fate of the zona pellucida in mice. J Embryol Exp Morphol 23:1-19
Mintz B (1971) Control of embryo implantation and survival. Adv Biosci 6:317-342
Mintz B (1972) Implantation-initiating factor from mouse uterus. In: Moghissi KS, Hafez ESE (eds) Biology of mammalian fertilization and implantation. Thomas, Springfield, pp 343-356
Ossowski L, Unkeless JC, Tobia A, Quigley JP, Rifkin DB, Reich E (1973) An enzymatic function associated with transformation of fibroblasts by oncogenic viruses. II. Mammalian fibroblast cultures transformed by DNA and RNA tumor viruses. J Exp Med 137:112-126
Owers NO, Blandau RJ (1971) Proteolytic activity of the rat and guinea pig blastocyst *in vitro*. In: Blandau RJ (ed) The biology of the blastocyst. University of Chicago Press, Chicago, pp 207-223
Page EW, Glendening MB, Parkinson D (1951) Cyclic biochemical changes in the human endometrium. Am J Obstet Gynecol 62:1100-1105
Pinsker MC, Sacco AG, Mintz B (1974) Implantation-associated proteinase in mouse uterine fluid. Dev Biol 38:285-290
Potts M (1969) The ultrastructure of egg implantation. Adv Reprod Physiol 4:241-267
Rizzino A, Sherman MI (1979) Development and differentiation of mouse blastocysts in serum-free medium. Exp Cell Res 121:221-233
Sellens MH, Sherman MI (1980) Effects of culture conditions on the developmental programme of mouse blastocysts. J Embryol Exp Morphol 56:1-22
Sherman MI (1972a) The biochemistry of differentiation of mouse trophoblast: Alkaline phosphatase. Dev Biol 27:337-349
Sherman MI (1972b) The biochemistry of differentiation of mouse trophoblast: Esterase. Exp Cell Res 75:449-459
Sherman MI (1980) Studies on the temporal correlation between secretion of plasminogen activator and stages of early mouse embryogenesis. Oncodev Biol Med 1:7-17
Sherman MI, Atienza-Samols SB (1978) *In vitro* studies on the surface adhesiveness of mouse blastocysts. In: Ludwig H, Tauber PF (eds) Human fertilization. Thieme, Stuttgart, pp 179-184
Sherman MI, Atienza-Samols SB (1979) Differentiation of mouse trophoblast does not require cell-cell interaction. Exp Cell Res 123:73-77
Sherman MI, Salomon DS (1975) The relationship between the early mouse embryo and its environment. In: Markert CL, Papaconstantinou J (eds) The developmental biology of reproduction. Academic Press, New York, pp 277-309
Sherman MI, Wudl LR (1976) The implanting mouse blastocyst. In: Poste G, Nicolson GL (eds) Cell surface interactions in embryogenesis. North Holland, Amsterdam, pp 81-125
Sherman MI, Strickland S, Reich E (1976) Differentiation of early mouse embryonic and teratocarcinoma cells *in vitro*: Plasminogen activator production. Cancer Res 36:4208-4216
Strickland S, Reich E, Sherman MI (1976) Plasminogen activator in early embryogenesis: Enzyme production by trophoblast and parietal endoderm. Cell 9:231-240
Tympanidis K, Astrup T (1968) Localization of fibrinolytic activity in the rat oviduct. Obstet Gynecol 31:727-731
Unkeless JC, Tobia A, Ossowski L, Quigley JP, Rifkin DB, Reich E (1973) An enzymatic function associated with transformation of fibroblasts by oncogenic viruses: 1. Chick embryo fibroblast cultures transformed by avian RNA tumor viruses. J Exp Med 137:85-111
Van Hoorn G, Denker H-W (1975) Effect of the blastocyst on a uterine amino acid arylamidase in the rabbit. J Reprod Fertil 45:359-362

Weiss G, Beller FK (1969) Tissue activator of the fibrinolytic enzyme in the female reproductive system. Obstet Gynecol 34:809-819

Werb Z, Mainardi CL, Vater CA, Harris ED (1977) Endogenous activation of latent collagenase by rheumatoid synovial cells. Evidence for a role of plasminogen activator. N Engl J Med 296:1017-1023

Postfertilization Activation of Stored Maternal mRNA in the One-Cell Mouse Egg

Nancy K. Love

Department of Zoology, University of Maryland, College Park, Maryland, USA

1 Introduction

The mechanisms which regulate genetic expression during early development of the mammalian embryo remain essentially unknown. Expression of the recombined embryonic genome in the mouse embryo has been documented as early as the two-cell stage by RNA precursor studies (Monesi and Salfi 1967; Knowland and Graham 1972; Siracusa 1973), and at the four- to eight-cell stage by investigations of the time of appearance of paternally determined isoenzyme patterns (Chapman et al. 1976). DNA-dependent RNA-polymerase activity can be assayed at the two-cell stage, but not in the earlier one-cell pronuclear embryo (Moore 1975). Thus, it appears that the diploid genome of the mouse embryo may become transcriptionally active after the embryonic cleavage divisions have begun, but not prior to that time.

The information derived from this cleavage-stage transcription is clearly required for further development of the mammalian embryo, for if transcription is inhibited in the two-cell embryo and later stages, developmental arrest occurs (Golbus et al. 1973; Warner and Versteegh 1974). Thus, the mammalian embryo appears to be precocious in a developmental sense, when compared to the embryos of non-mammalian species. Non-mammalian embryos exhibit developmental sensitivity to transcription inhibition only much later in development, in the blastula and gastrula stages, indicating that embryonic transcription required for further development is not initiated until the much later developmental stages.

However, if the absolute time frame of mammalian development is considered, this apparent precocity disappears. It is important to note that the first embryonic cleavage of the mouse embryo occurs approximately 20 h after fertilization. This period is quite protracted compared to the postfertilization time of onset of cleavage in non-mammalian species. Non-mammalian embryos initiate cleavage much more quickly after fertilization, and with a much shorter cell cycle than mammalian embryos, and will achieve blastulation in a time period very similar to that required for the first cleavage of the mouse embryo. Thus, in real time, the period between fertilization of the egg and the onset of transcription of the embryo's genome, which is required for further normal development, may be quite similar in both mammalian and non-mammalian species. This is an important point to keep in mind when considering the molecular biology of the initiation of development, many features of which may cross species lines.

The molecular aspects of development in amphibian and echinoderm embryos may bear great similarity to the early events of mammalian development. In the embryos of many non-mammalian species, there exist stable messenger RNA templates which are transcribed from the maternal genome during oogenesis and then are stored in the egg's cytoplasm for postfertilization translation (Brachet and Denis 1963; Gross and Cousineau 1964; Humphreys 1971). As part of the complex response of the egg to the fertilization stimulus, these stored maternal mRNA templates enter active translation and total cellular protein synthesis is stimulated (Gross 1967). This postfertilization protein synthesis is required for embryonic development (Hultin 1961), and proceeds at a high rate in the face of transcription inhibition (Gross and Cousineau 1964). Transcription of the embryo's own genome, although it occurs during early cleavage development (Kedes and Gross 1969; Slater and Spiegelman 1970), is not required for completion of the early developmental events. Rather the products of maternal gene expression, synthesized prior to ovulation, regulate the early developmental events in embryos of these species.

We may ask then, how many of these events occur in the mammalian embryo? And, secondarily, what are the regulatory mechanisms which direct mammalian embryonic development during the hours after fertilization, prior to the initiation of embryonic genetic expression?

It has been hypothesized in recent years (Epstein and Smith 1974; Van Blerkom and Brockway 1975) that the mammalian egg, similar to echinoderm and amphibian eggs, may contain stable mRNA templates, products of maternal gene expression. In the following discussion, evidence for the existence of such stored maternal mRNAs in the mouse egg and for postfertilization alterations in protein synthesis which depend on these stored mRNAs will be discussed.

2 Translation in the Fertilized Mouse Egg

Upon fertilization of the mouse egg, there are dramatic alterations in the regulation of protein synthesis within the one-cell egg---alterations both quantitative and qualitative in nature. Correlations between incorporation studies and electrophoretic analyses suggest that these alterations in translation occur in a sequence specific to the progression of the embryo toward the first embryonic cleavage.

2.1 Quantitative Alterations in Protein Synthesis After Fertilization

The quantitative changes were monitored by measurements of the in vitro uptake and incorporation of exogenous ^{3}H-labelled amino acids by each stage of preimplantation embryonic development (Fig. 1). As has been demonstrated in other laboratories (Monesi and Salfi 1967; Brinster et al. 1976), there are increases in both uptake and incorporation of amino acids as the embryo progresses from the four-cell to the blastocyst stage. The increase in incorporation correlates to increased cellular machinery for protein synthesis: appearance of new polysomes and greater abundance of cytoplasmic ribosomal subunits (Calarco and Brown 1969; Hillman and Tasca 1969) and the transcription of mRNA, rRNA, and tRNA from the embryonic genome (Ellem and Gwatkin 1968; Woodland and Graham 1969). It can be concluded that the increases in protein-synthetic activity during cleavage are related to the activation of the embryonic genome (cf. Johnson, Handyside and Braude 1977).

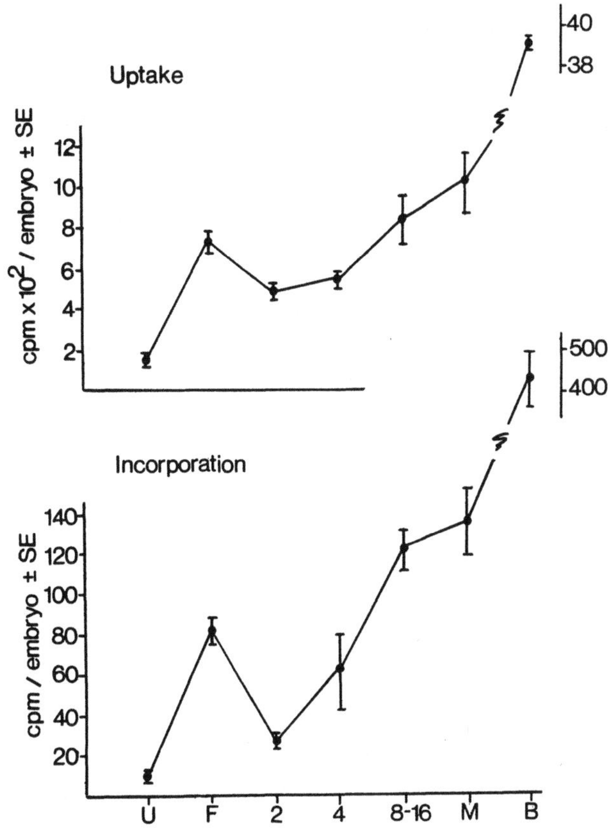

Fig. 1. The in vitro uptake and incorporation of exogenous ^{3}H-amino acids (^{3}H-Reconstituted Protein Hydrolysate, Schwarz-Mann) by preimplantation stages of the mouse embryo. Uptake was measured in washed eggs after 3 h in culture; incorporation was determined by an acid-precipitation filter assay. Developmental stages: *U* unfertilized egg, *F* midcycle, fertilized egg, *2* 2-cell embryo, *4* 4-cell embryo, *8-16* 8-16 cell embryo, *M* morula, *B* blastocyst

Unique to the present study however is the finding that there is a sharp postfertilization increase in incorporation, and a decline after the first cleavage division. This increase in protein-synthetic activity occurs at a time prior to the initiation of genetic expression by the embryo's own genome; the source of the mRNA templates which support this postfertilization activation of protein synthesis is explored below.

Since there is also a postfertilization increase in amino acid uptake, an interpretation of the increase in incorporation as an increase in protein synthetic rate must be made cautiously. In order to take the effect of varying uptake into account, the incorporation data are expressed as percent incorporation in Fig. 2. Percent incorporation (the ratio of ^{3}H-precursor incorporated to the total ^{3}H-precursor taken up by the embryo) is a more valid way to compare rates of macromolecular synthesis under conditions of varying uptake, than is the determination of incorporation alone (Berg 1970). Berg and coworkers have shown that when both the absolute incorporation and the percent incorporation alter in the same direction, one can conclude that a real change in the rate of protein synthesis has occurred.

The difficulty in measuring actual endogenous amino acid pool size in mammalian embryos was surmounted at first for the amino acid leucine (Brinster et al. 1976). In that case, it was found that the size of the leucine pool does not alter after fertilization, nor does the apparent rate of leucine incorporation change after fertilization. The endogenous pool sizes of 17 amino acids were measured recently directly in samples of mouse eggs, 8-cell embryos and blastocysts (Schultz et al. 1981). In addition, Sellens et al. (1981) described a rapid and sensitive procedure for the simultaneous measurement of total protein and free amino acid contents in early mouse embryos (in-vivo and in-vitro under various culture conditions).It is difficult to compare the present results with those reported from leucine studies, because in the present case, a mixture of amino acids has been used. Thus, the present observation of a post-fertilization spike in protein synthesis reflects a summation of the incorporation of this group of amino acids rather than a single one. It is not improbable that study of a single amino acid might produce a synthetic pattern not wholly representative of the total cellular activity.

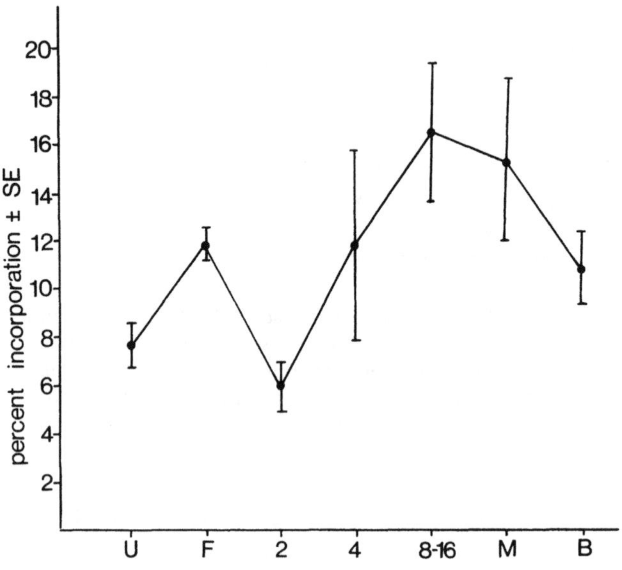

Fig. 2. Percent incorporation (incorporation/uptake x 100) of exogenous ^{3}H-amino acids. Developmental stages as in Fig. 1

Figure 2 shows the percent incorporation at each developmental stage prior to implantation. There is a significant rise in percent incorporation after fertilization, and a decline after the first cleavage with both alterations paralleling the changes in incorporation. Thus, it can be concluded, since absolute incorporation plus percent incorporation alter in the same direction, that after fertilization, a stimulation in the rate of protein synthesis occurs. This stimulation of translation rate is limited to the one-cell stage.

The percent incorporation in the later cleavage stages (four-cell to blastocyst) does not vary significantly, although there is a dramatic increase in absolute incorporation during this period (cf. Monesi and Salfi 1967; Brinster et al. 1976; Abreu and Brinster 1978; Braude 1979; Schultz et al. 1979). It is concluded that the rate of protein synthesis is relatively constant in these developmental stages. The observed increase in incorporation is due solely to increase in amino acid uptake as the embryonic cell number increases, and not to an altered rate of amino acid incorporation into protein within these cells.

The postfertilization spike in protein synthesis is examined in greater detail in Table 1, which compares the protein-synthetic activity of unfertilized eggs to that of fertilized eggs, labelled at three periods after fertilization. By 3-6 h after fertilization of the egg, there is a fourfold increase in uptake, and an eightfold increase in incorporation, with the percent incorporation approximately doubled. There is a secondary small increase in uptake at 10-13 h, but more significantly, the incorporation shows a dramatic second increase at 15-18 h after fertilization. At this time, the percent incorporation is also increased, compared to the immediate postfertilization value, and is approximately three times that of the unfertilized egg.

Thus, there are two points of stimulation of the rate of translation in the mouse egg: immediately after fertilization, and again just prior to the first cleavage. Overall, there is a threefold stimulation of the translation rate after fertilization, a value comparable to the activation of protein synthesis after fertilization of the sea urchin egg (Epel 1967; Fry and Gross 1970).

Table 1. Determination of uptake and incorporation of ^{3}H-amino acids by unfertilized eggs, and fertilized eggs at three periods after fertilization. The unfertilized eggs were collected at a time equivalent to the mid-cycle fertilized eggs (10-13 h after fertilization)

	Uptake	Incorporation	Percent incorporation
	mean cpm/embryo/3 h ± s.e.		% ± s.e.
Unfertilized	138 ± 22	10 ± 2	7.6 ± 1.0
Fertilized, 3-6 h after fertilization	588 ± 47	81 ± 19	13.3 ± 2.3
10-13 h after fertilization	715 ± 56	82 ± 7	11.9 ± 0.8
15-18 h after fertilization	779 ± 77	181 ± 18	24.0 ± 3.0

2.2 *Qualitative Alterations in Protein Synthesis After Fertilization*

If the qualitative patterns of proteins being synthesized in unfertilized and fertilized eggs, at two of the three periods after fertilization, are examined, dramatic differences can be observed.

The proteins synthesized at each stage and labelled by the in vitro incorporation of ^{35}S-methionine are separated according to molecular weight, by SDS-polyacrylamide slab gel electrophoresis. Figures 3-5 show the separation patterns of proteins synthesized in the unfertilized egg, the early fertilized egg (at 3-6 h after fertilization), and the mid-cycle fertilized egg (at 10-13 h after fertilization), respectively. In all cases, most of the incorporated label is found in proteins of molecular weight greater than 25 000 daltons.

The unfertilized egg and the early fertilized egg (at 3-6 h) show very similar qualitative patterns of protein synthesis, with no clear differences. In contrast, the mid-cycle fertilized egg, at 10-13 h after fertilization, is synthesizing new proteins which are not synthesized in the unfertilized or the recently fertilized egg. By mid-cycle, the fertilized egg is synthesizing new species of proteins, of apparent molecular weight approximately 37 000-39 000 daltons, with 5.2% of total radioactivity being incorporated into proteins in this region of the gel. (This region is indicated by the arrow in Figs. 3-5). By late first cell cycle, at 15-18 h after fertilization, the percent of total counts being incorporated into these new proteins has increased to 7.1%, which represents a significant increase in the synthetic rate of the new proteins, relative to the entire spectrum of proteins under active synthesis, as the embryo approaches the first cleavage.

After the first embryonic cleavage, the percent of counts in this 37 000-39 000 molecular weight region declines sharply, to 2.5%, representing in turn a relative decrease in synthetic rate of the new proteins. As cleavage proceeds, the figure declines to that observed in the unfertilized egg. Thus, the new proteins are stage-specific to the one cell egg in their synthesis and appear only after fertilization of the egg.

The complexity of the population of proteins under active synthesis is seen most clearly by autoradiography of the slab gels, for fractionation of the gels, as in Figs. 3-5, is limited in its resolving power. Figure 6 shows an autoradiogram of a gel on which samples of proteins from the unfertilized egg and the fertilized egg (at 10-13 h after fertilization) were run in parallel. The identity of the qualitative synthetic patterns is quite remarkable for the majority of over 100 bands on the patterns. There are three striking differences between the patterns of unfertilized and fertilized eggs however: most important, the appearance of the new proteins in the fertilized egg at molecular weight 37 000-39 000 (Fig. 6, *arrow*); and two minor alterations, a decrease in synthesis of a band of slightly higher molecular weight, and the conversion of a triplet, of molecular weight approximately 28 000, to the emphasis of the lower band in the triplet.

2.3 Correlation of the Qualitative and Quantitative Changes in Translation

The timing of postfertilization alterations in the quantitative pattern of protein synthesis may be related in a significant manner to the time of the qualitative alterations. The initial postfertilization increase in the rate of translation, reflected in the doubling of percent incorporation by 3-6 h after fertilization (Table 1), is interpreted as a stimulation of the synthetic rate of those proteins already under synthesis in the unfertilized egg. That is, there is a population of mRNA templates being actively translated in the unfertilized egg; immediately upon fertilization, there is a stimulation of the rate of translation of this same template population, with no significant changes in the type of proteins being synthesized (see Figs. 3 and 4).

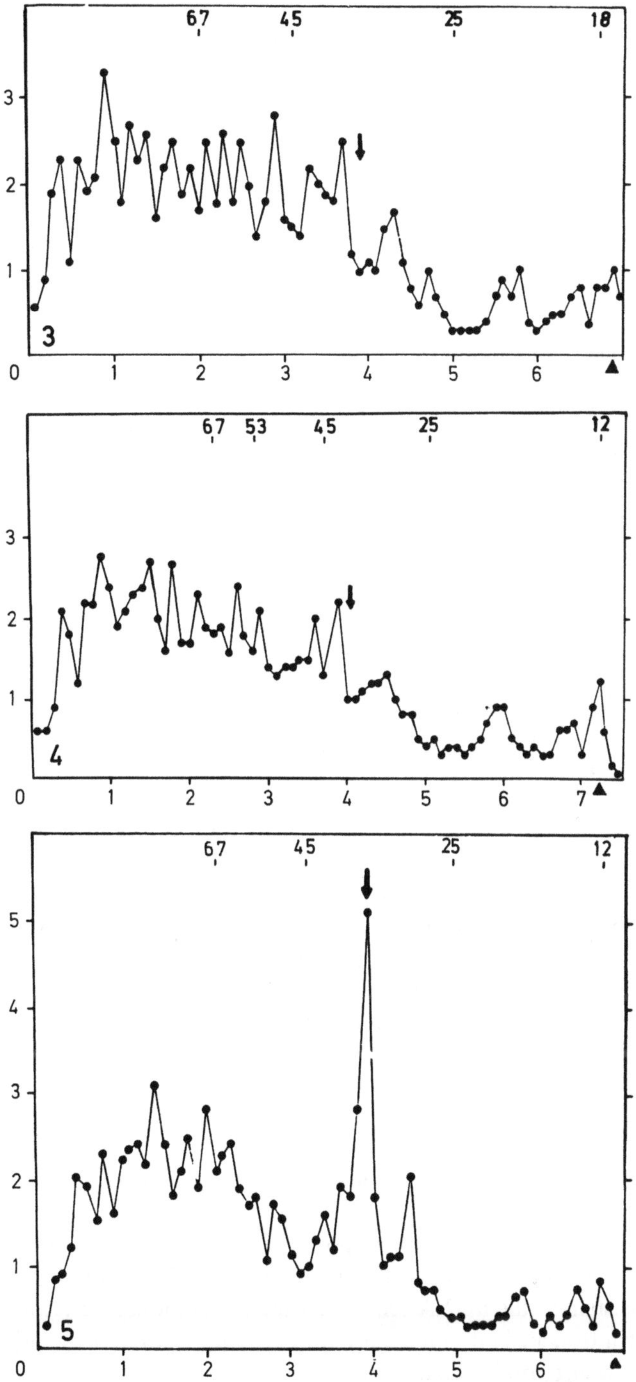

Fig. 3-5. Separation of ^{35}S-methionine-labelled proteins synthesized during 3-h labelling of one-cell mouse eggs. Proteins were separated on SDS-polyacrylamide slab gels (10% acrylamide). The gels were sectioned into 1-mm fractions, and the radioactivity in each fraction determined. The percent of total counts located in each fraction (*ordinate*) is plotted against centimeters (*abscissa*) moved from the origin (top of gel to left). In the upper portion of each panel is shown the electrophoretic position of molecular weight marker proteins (molecular weight x 10^3): *67* bovine serum albumin, *45* ovalbumin, *25* chymotrypsinogen A, *18* myoglobin, *12* cytochrome C

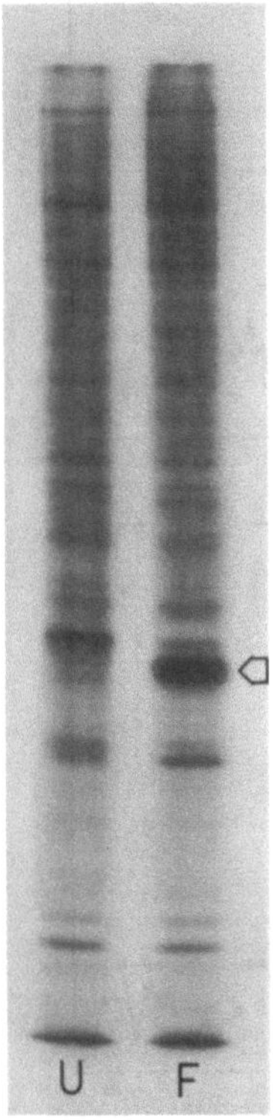

Fig. 6. Autoradiogram of a 10% acrylamide slab gel on which samples of ^{35}S-methionine-labelled proteins from unfertilized and fertilized eggs were separated. The arrow marks the position of approximately 37 000-39 000 molecular weight. *U* unfertilized, *F* fertilized eggs

The second increase in translation rate, somewhat later in the first cell cycle (15-18 h after fertilization, Table 1), is associated firstly with the appearance of new, stage-specific proteins (37 000-39 000 molecular weight). These new proteins then show a selective stimulation of their translation rate, with an increase in synthesis relative to the remainder of the protein population. This selective stimulation may well be sufficient to account for the second translation rate increase, as reflected in total cellular incorporation.

3 Source of the mRNA in the One-cell Mouse Egg

The source of the mRNA supporting these postfertilization alterations in protein synthesis is a consideration of prime importance. There are three possible sources to be considered: the mRNA templates may be synthesized after fertilization; the templates are introduced into the egg by the fertilizing sperm; or the mRNA templates are stored in the egg prior to fertilization.

3.1 Absence of a Requirement for RNA Transcription

The possibility of postfertilization synthesis of RNA was examined by measurements of the in vitro uptake and incorporation of exogenous ^{3}H-uridine. Although a clear postfertilization stimulation (six-to tenfold) of uridine uptake was observed, no acid-precipitable uridine incorporation could be detected, even with pooled samples of 100 or more fertilized eggs. The conclusion that there may be no RNA synthesis in the fertilized egg is made conditionally, for there remains the chance of postfertilization mRNA synthesis, but at a level below detection by incorporation and acid-precipitation methods. However, the conclusion is strenthened by the lack of detectable RNA polymerase activity in the one-cell egg (Moore 1975).

To test whether the postfertilization stimulations of protein synthesis require simultaneous RNA synthesis, fertilized eggs were incubated in the presence of inhibitors of transcription, at concentrations known in other cell types to depress all RNA synthesis. The transcription inhibitors used were actinomycin D, at 10 ng–5 μg/ml; alpha-amanitin, at 1–10 μg/ml; and MPB (mercapto-pyridethyl-benzidimazole), at 10–50 μg/ml. In the presence of each of these inhibitors, the in vitro uptake and incorporation of amino acids by the fertilized egg were monitored. No effect of the transcription inhibitors could be observed on either translation rate, the postfertilization alterations observed in translation rate, or the postfertilization appearance of the new proteins in the egg. The qualitative pattern of proteins synthesized in the fertilized egg (at 10-13 h after fertilization), cultured 3 h in the simultaneous presence of actinomycin D (5 μg/ml) plus radiolabelled amino acid is shown in Fig. 7. There is no significant variation from the pattern shown in Fig. 5, for the untreated, mid-cycle fertilized egg.

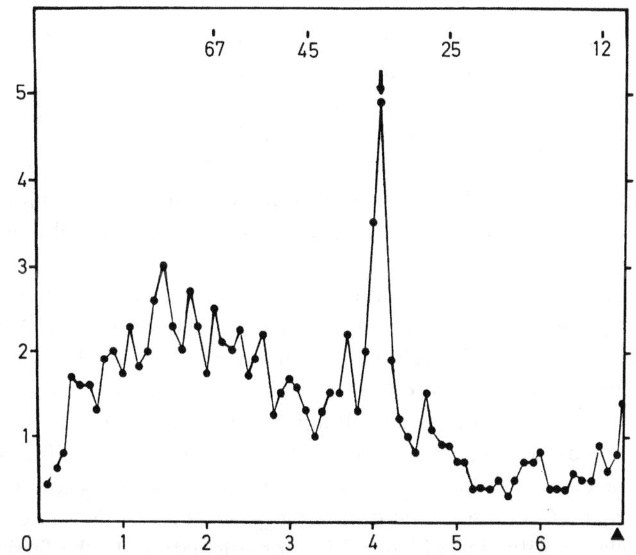

Fig. 7. Separation pattern of ^{35}S-methionine-labelled proteins from fertilized eggs, incubated in actinomycin D (5μg/ml). Ordinate and abscissa and molecular weight markers as in Fig. 3-5

An identical result is obtained if the transcription inhibitor is present from the time of fertilization until the introduction of the radiolabelled amino acid at mid-cycle. That is, preincubation of the eggs in the inhibitor from time zero to 10 h after fertilization had no depressive effect on either translation rate or qualitative synthetic pattern during the period of labelling (10-13 h after fertilization).

Thus, the possible requirement for postfertilization RNA synthesis to support the postfertilization alterations in translation is ruled out by both the absence of detectable uridine incorporation and the demonstration that inhibitors of transcription have no effect on protein synthesis. Therefore, even if there is transcription in the one-cell egg (Clegg and Piko 1977), it is not required for the phenomena related to translation alterations.

3.2 Irrelevance of the Sperm

As will be clearly demonstrated in subsequent paragraphs, the fertilizing sperm cannot be the source of the mRNA templates which support postfertilization protein synthesis, since the presence of a fertilizing sperm is not required for the egg to undergo the described changes in translation. Whether some component of the sperm may, in the normal scheme of embryonic development, trigger the sequence of events leading to the activation of translational alterations remains to be investigated.

3.3 Storage of Maternal mRNA in the One-cell Mouse Egg

Unfertilized mouse eggs may be stimulated to initiate parthenogenetic development by a number of artificial methods, one of which is brief exposure to calcium-free culture medium (personal communication, D.G. Whittingham). Successful initiation of development, or activation, is indicated by the formation of a visible pronucleus within the egg. Thus, the egg can initiate development in the absence of any influence from the sperm. This system of induction of activation by exposure to calcium-free medium was utilized to examine the egg for maternal control of the early developmental changes in translation.

Recently ovulated, unfertilized one-cell mouse eggs are recovered from the oviducts, and placed in calcium-free medium for 1 h. After the initial 1-h period, the eggs are transferred (cumulus cells having been removed by hyaluronidase) to calcium-containing culture medium; culture is continued for a further 5 h, when the eggs are examined for the appearance of pronuclei. Radioactive amino acids may be added during the last 2 h of this procedure, or the culture period may be extended for 2-3 h for radiolabelling.

The quantitative and qualitative patterns of protein synthesis in the activated egg were compared to those of the fertilized egg. In all cases, the successfully activated egg (pronucleus positive) was identical in synthetic pattern to the fertilized egg, while the nonresponsive egg (pronucleus negative) was identical to the unfertilized egg, as would be expected. Fig. 8 shows an autoradiograph of proteins synthesized in activated and fertilized eggs of equivalent developmental stage. The patterns are identical, with the activated egg successfully initiating the synthesis of the new, postfertilization proteins (arrow) of approximately 37 000 molecular weight. These new proteins dominate the synthetic pattern in the activated egg in a manner similar to the fertilized egg.

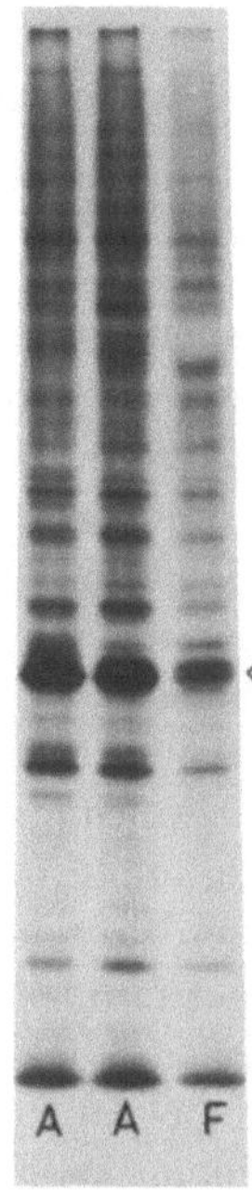

Fig. 8. Autoradiogram of a 10% acrylamide slab gel showing the separation patterns of proteins from two samples of activated eggs, and one of fertilized eggs. The arrow indicates the position of the newly appearing proteins in both activated and fertilized eggs. *A* activated eggs, *F* fertilized eggs

Fig. 8

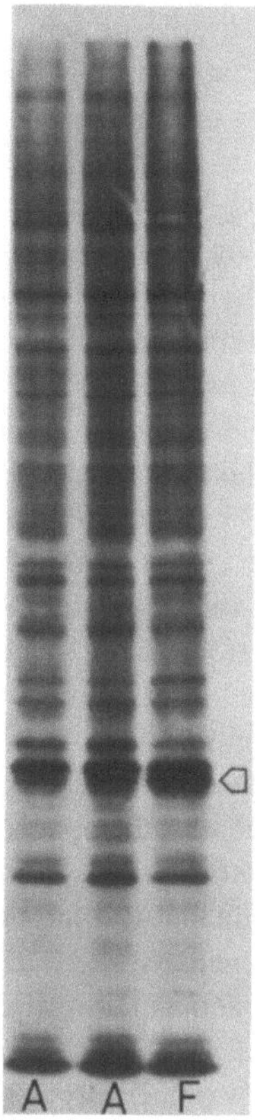

Fig. 9. Autoradiogram of a 10% acrylamide slab gel showing the separation patterns of proteins from eggs activated in the presence of 10 μg alpha-amanitin/ml, compared to proteins from fertilized eggs; see Fig. 8

Fig. 9

In Fig. 8, the protein pattern for the fertilized egg (on the right) is underexposed, to demonstrate that the newly appearing species of proteins can be resolved into a doublet of bands in this one-dimensional electrophoresis. The bands, very close in molecular weight, can be detected as a pair by autoradiography, but appear as a single peak when the gel is fractionated for scintillation counting, as in Figs. 3-5 and 7. This single peak is due to low resolution by the gel fractionation apparatus, and not to lack of resolution in the gel itself.

The appearance of these one-cell-specific, new proteins is associated with the initiation of development, whether that is achieved by fertilization or activation. Clearly, the sperm has no direct role in these translation alterations, but there remained the possibility that certain mRNAs could be synthesized after activation. This possibility seemed remote, because it would require transcription from the egg's haploid genome---an unlikely event. However, to test for the requirement of RNA synthesis after activation, eggs were activated in calcium-free medium containing an inhibitor of transcription, alpha-amanitin at 10 μg/ml. The inhibitor was continuously present from the time of initiation of activation, through the period of labelling (at the time equivalent to mid-cycle of the fertilized egg). Figure 9 shows the autoradiographic pattern of proteins synthesized in eggs activated in alpha-amanitin, compared to mid-cycle fertilized eggs. No difference in the qualitative protein synthetic patterns can be detected. Thus, it can be concluded that postactivation RNA synthesis is not required to support the onset of synthesis of the new proteins.

The question of penetration by the transcription inhibitors is part of all inhibitor studies. The mammalian embryo is not considered to present the problems of low membrane permeability to exogenous precursors, etc., as has been described for embryos of other species. In the present investigations, uptake of ^{3}H-actinomycin D was examined; radiolabelled alpha-amanitin was not available. Uptake of the inhibitor was clearly demonstrable, and a small postfertilization increase was observed. Therefore, it seems probable that the transcription inhibitors were certainly penetrating the egg but having no effect on translation.

One is led then to the conclusion that in the mouse egg, as in amphibian and echinoderm eggs, the postfertilization changes observed in protein synthesis depend on the utilization of maternal mRNA, which has been synthesized prior to ovulation. The mRNA is stored in an inactive state, and is translated only after embryonic development is initiated. It is an important difference to note, when comparing mammalian embryos to non-mammalian, that the postfertilization activation of stored templates in non-mammalian embryos does not give rise to new species of proteins (Brandhorst 1976). Consequently, the appearance of the new species in the population of proteins under stimulated synthesis in the developing mouse embryo may be a feature unique to mammals.

Finally, it is noteworthy to mention that the very recent elegant experiments of Illmensee (1981) and of Petzold et al. (1980) demonstrate that the enucleation of fertilized and unfertilized mouse eggs leads to remaining cytoplasts which are able to synthesize proteins for several days. After two days of egg culture the appearance of some new proteins typical for two-cell embryos could be observed in enucleated fertilized eggs only. These findings suggest clearly that maternal RNA stored during oogenesis is utilized during development of the preimplantation period.

Acknowledgments. Part of this work was carried out in the Department of Cell Biology, Baylor College of Medicine, Houston, Texas, USA, and was supported by NIH grant number HD 07495 to the Center for Population Research. Additional studies were conducted at the MRC Mammalian Development Unit, University College, London, UK, while the author held a Rockefeller Foundation postdoctoral fellowship.

References

Abreu SL, Brinster RL (1978) Synthesis of tubulin and actin during the preimplantation development of the mouse. Exp Cell Res 114:135-141

Berg WE (1970) Further studies on the kinetics of incorporation of valine in the sea urchin embryo. Exp Cell Res 60:210-217

Brachet J, Denis H (1963) Effects of actinomycin D on morphogenesis. Nature 198:205-206

Brandhorst BP (1976) Two-dimensional gel patterns of protein synthesis before and after fertilization of sea urchin eggs. Dev Biol 52:310-317

Braude PR (1979) Control of protein synthesis during blastocyst formation in the mouse. Dev Biol 68:440-452

Brinster RL, Wiebold JL, Brunner S (1976) Protein metabolism in preimplanted mouse ova. Dev Biol 51:215-224

Calarco PG, Brown EH (1969) An ultrastructural and cytological study of preimplantation development of the mouse. J Exp Zool 171:253-283

Chapman VM, Adler D, Labarca C, Wudl L (1976) Genetic variation of ß-glucuronidase expression during early embryogenesis. Ciba Found Symp 40:115-124
Clegg KB, Piko L (1977) Size and specific activity of the UTP pool and overall rates of RNA synthesis in early mouse embryos. Dev Biol 58:76-95
Ellem KAO, Gwatkin RBL (1968) Patterns of nucleic acid synthesis in the early mouse embryo. Dev Biol 18:311-330
Epel D (1967) Protein synthesis in sea urchin eggs: A "late" response to fertilization. Proc Natl Acad Sci USA 57:899-906
Epstein CJ, Smith SA (1974) Electrophoretic analysis of proteins synthesized by preimplantation mouse embryos. Dev Biol 40: 233-244
Fry BJ, Gross PR (1970) Patterns and rates of protein synthesis in sea urchin embryos. Dev Biol 21:125-146
Golbus MS, Calarco PG, Epstein CJ (1973) The effects of inhibitors of RNA synthesis (α-amanitin and actinomycin D) on preimplantation mouse embryogenesis. J Exp Zool 186:207-216
Gross PR (1967) The control of protein synthesis in embryonic development and differentiation. In: Moscona AA, Monroy A (eds) Current topics in developmental biology, vol 2. Academic Press, New York, pp 1-46
Gross PR, Cousineau GH (1964) Macromolecular synthesis and the influence of actinomycin on early development. Exp Cell Res 33:368-395
Hillman N, Tasca RJ (1969) Ultrastructural and autoradiographic studies of mouse cleavage stages. Am J Anat 126:151-173
Hultin T (1961) The effect of puromycin on protein metabolism and cell division in fertilized sea urchin eggs. Experientia 17:410-411
Humphreys T (1971) Measurements of messenger RNA entering polysomes upon fertilization of sea urchin eggs. Dev Biol 26:201-208
Illmensee K (1981) Experimental manipulation of the mouse embryo: Genetic and developmental consequences. Abstract-Vol. of III. World Congr Human Reproduction, Berlin, p 220
Johnson MH, Handyside AH, Braude PR (1977) Control mechanisms in early mammalian development. In: Johnson MH (ed) Development in mammals, vol 2, North-Holland Publ Comp, Amsterdam New York Oxford, pp 67-97
Kedes LH, Gross PR (1969) Identification in cleaving embryos of three RNA species serving as templates for the synthesis of nuclear proteins. Nature 223:1335-1339
Knowland J, Graham C (1972) RNA synthesis at the two-cell stage of mouse development. J Embryol Exp Morphol 27:167-176
Monesi V, Salfi V (1967) Macromolecular synthesis during early development in the mouse embryo. Exp Cell Res 46:632-635
Moore GPM (1975) The RNA polymerase activity of the preimplantation mouse embryo. J Embryol Exp Morphol 34:291-298
Petzold U, Hoppe PC, Illmensee K (1980) Protein synthesis in enucleated fertilized and unfertilized mouse eggs. Wilhelm Roux's Archives 189:215-219
Schultz GA, Kaye PL, McKay DJ, Johnson MH (1981) Endogenous amino acid pool sizes in mouse eggs and preimplantation embryos. J Reprod Fertil 61:387-393
Schultz RM, Letourneau GE, Wassarman PM (1979) Program of early development in the mammal. Changes in patterns and absolute rates of tubulin and total protein synthesis during oogenesis and early embryogenesis in the mouse. Dev Biol 68:341-359
Sellens MH, Stein S, Sherman MI (1981) Protein and free amino acid content in preimplantation mouse embryos and in blastocyst under various culture conditions. J Reprod Fertil 61:307-315
Siracusa G (1973) RNA polymerase during early development in mouse embryo. Exp Cell Res 78: 460-462
Slater DW, Spiegelman S (1970) Transcriptive expression during sea urchin embryogenesis. Biochim Biophys Acta 213:194-207
Van Blerkom J, Brockway GO (1975) Qualitative patterns of protein synthesis in the preimplantation mouse embryo. Dev Biol 44:148-157
Warner CM, Versteegh LR (1974) In vivo and in vitro effect of alpha-amanitin on preimplantation mouse embryo RNA polymerase. Nature 248:678-680
Woodland HR, Graham CF (1969) RNA synthesis during early development of the mouse. Nature 221:327-332

The Control of Blastocyst Activity

R.J. Aitken

Medical Research Council, Department of Reproductive Biology, Centre for Reproductive Biology, Edinburgh, U.K.

1 Introduction

It is perhaps fortuitous that the phenomenon of embryonic diapause is exhibited by laboratory species such as the rat and mouse since it has prvided us with an extremely effective tool for elucidating the endometrial-blastocyst interactions leading to implantation. Much of the early work with these species concentrated on determining the hormonal requirements for implantation (Canivenc and Laffargue 1956; McLaren 1970; Psychoyos 1973). It was discovered that the administration of progesterone to animals ovariectomised in early pregnancy resulted in a state of diapause during which gestation was indefinitely suspended at the immediate preimplantation phase. The subsequent administration of a minute dose of oestrogen, however, resulted in the rapid precipitation of an implantation response (Nalbandov 1971). The way in which the presence or absence of oestrogen led to the stimulation or inhibition of implantation has been the focus of much current research on the endocrine control of endometrial function. As a consequence of these investigations, two hypotheses have been advanced concerning the role of this hormone, (a) that oestrogen induces the release of embryotrophic endometrial secretions which are either not synthesized or not released during the delay phase, and (b) that oestrogen induces the inactivation or withdrawal of a blastocyst inhibitor which is manufactured by the progesterone-dominated uterus during diapause. The evidence for and against each of these alternative but not mutually exclusive hypotheses is presented in the following discussion.

2 Endometrial Secretory Activity and Implantation

Evidence for the retention or release of endometrial secretions by species exhibiting delayed implantation has come from ultrastructural studies of the uterine mucosa and biochemical analyses of washings recovered from the uterine lumen.

Ultrastructural evidence for a lack of endometrial secretory activity during the delay phase has been obtained in such species as the roe deer (Aitken et al. 1973, Aitken 1975), mink (Enders et al. 1963) and mouse (Nilsson 1970, 1974). The roe deer uterus is characterized by the progressive accumulation of clear supranuclear vesicles in the ducts and glands during delayed implantation, suggesting the possible storage of an endometrial secretion at this time. Clear vesicles have also been observed in the luminal epithelium of the mouse uterus during diapause, while large numbers of electron-dense granules collect in the glandular epithelium of the mink uterus during this phase. In all three species the termination of diapause is associated with the sudden disappearance of these vesicles and granules, presumably as a result of their release into the uterine lumen (Enders et al. 1963, Enders and Enders 1963, Aitken et al. 1973, Aitken 1974a, b,1975, Nilsson 1974). Other secretory changes occurring in the roe deer endometrium at the end of diapause include an increase in the number and diameter of endometrial duct openings, the appearance of apical protrusions on the surface of the luminal epithelium, and the hypertrophy of the granular endoplasmic reticulum in the endometrial duct cells (Aitken 1974a,b, 1975). An increase in the quantity of granular endoplasmic reticulum has also been observed in the luminal epithelium of the mouse (Nilsson 1974) and the glandular epithelium of the mink uterus (Enders et al. 1963) at the time of blastocyst activation and implantation.

Biochemical analysis of flushings recovered from the roe deer uterine lumen during diapause revealed extremely low levels of α-amino nitrogen, protein, total hexose, and calcium, confirming the ultrastructural evidence for a lack of endometrial secretory activity at this time. In contrast, the termination of diapause (signified by a rapid elongation of the blastocyst in this species) was associated with a highly significant ($P < 0.001$) increase in the luminal concentration of all these factors (Aitken 1974a,b). A similar increase with respect to the protein content of the uterine flushings has been observed in the wallaby (Renfree 1973) and fur seal (Daniel 1971) during the transition from diapause to implantation. The progesterone-dominated rat uterus is also associated with a lack of luminal protein (Surani 1975), and recent analyses of uterine flushings recovered during lactational and experimental delayed implantation in the mouse has produced identical findings (Aitken 1977a, Pratt 1977, Fishel 1979, Figs. 1 and 2). When an implantation-inducing dose of oestradiol-17β was administered to rats or mice during diapause, however, a significant increase in luminal protein concentration was observed (Surani 1975; Aitken 1977a; Figs. 3 and 4). Biochemical analysis of uterine flushings recovered during early pregnancy in the mouse also revealed a highly significant increase ($P < 0.001$) in luminal protein concentration on day 4, the day of implantation (Fig. 1, Aitken 1977a). This change was not a consequence of implantation because a similar increase was observed on day 4 of pseudopregnancy (Aitken 1977b; Fig. 5).

Acrylamide gel electrophoresis was used to investigate the qualitative changes accompanying these shifts in protein concentration. On day 4 of pregnancy and pseudopregnancy, flushings of the mouse uterine lumen contained 36-38 detectable proteins of which 14 were of serum and the remainder of non-serum origin (Fig. 2). The major non-serum components were observed at the origin of the gels and in the slow α-globulin (Ra 0.06, 0.10, 0.25, 0.32, 0.34; Ra albumin= 1), fast α-globulin (Ra 0.82, 0.89), and prealbumin (Ra 1.19, 1.32) regions (Fig. 2). During experimental and lactational delayed implantation the only proteins present in significant quantities were serum transferrin and albumin. Traces of another 30-35 proteins were observed in the delay flushings, however, and a majority of these corresponded to proteins detected in larger quantities at the time of implantation (Aitken 1977a). This finding suggested that the induction of blastocyst activity was associated with a quantitative rather than a marked qualitative change in the protein content of the uterine lumen. This concept was supported by the results of a study involving the in vitro culture of blastocysts in media inoculated with the luminal contents of mice autopsied during diapause or at the time of implantation. In the first series of experiments, each millilitre of culture medium (Bowman and McLaren 1970) was flushed through the uterine lumina of three mice in rapid succession,

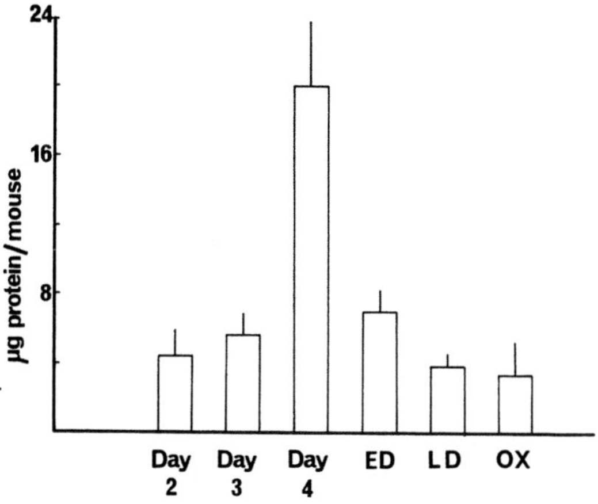

Fig. 1. Luminal protein concentration in mice during early pregnancy and delayed implantation. *Day 2, 3 and 4,* day of pregnancy; *ED* experimental delay; *LD* lactational delay; *OX* ovariectomized animals; *vertical bars,* standard deviation

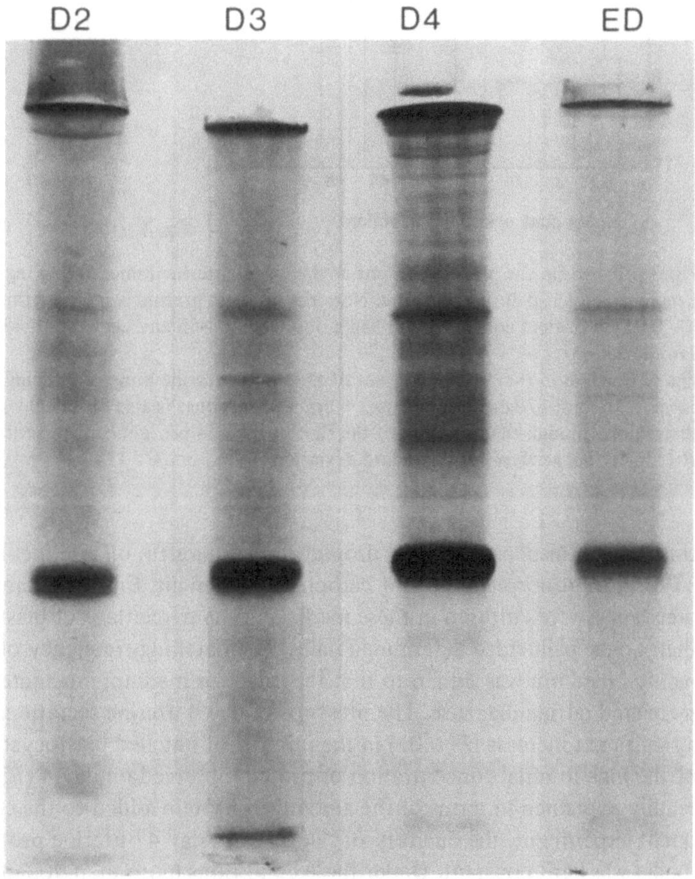

Fig. 2. Polyacrylamide gel electrophoresis of uterine flushings recovered on days 2, 3, and 4 of pregnancy (*D2, D3, D4*) and during experimental delayed implantation (*ED*) in the mouse. Note the absence of protein during diapause

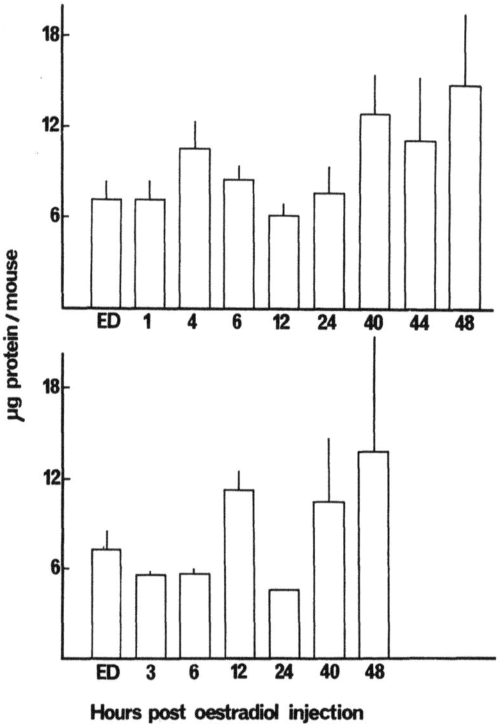

Fig. 3

Fig. 4

Fig. 3. Changes in the protein content of the mouse uterine lumen following the administration of 1 µg oestradiol-17β during diapause. Note the biphasic uterine response with significant increases ($P < 0.01$) in protein concentration being observed at 4-6h and again at 40-48h; *vertical bars,* standard deviation

Fig. 4. Changes in the protein content of the mouse uterine lumen following the administration of 50 ng oestradiol-17β during diapause. A biphasic response is also observed with this dose level of oestradiol although the initiation of the first increase in protein concentration is delayed until 12 h after injection; *vertical bars,* standard deviation

centrifuged, dispersed in 50-µl droplets under paraffin oil and incubated for 4 days at 37°C in an atmosphere of 10% carbon dioxide in air. Embryos harvested on day 3 of pregnancy were cultured in these media and the percentage of blastocysts hatching from their zonae pellucidae determined daily. The hatching frequency observed in unsupplemented medium was similar to that recorded for medium inoculated with uterine fluids recovered during diapause. The presence of day 4 uterine secretions, however, induced a significant increase ($P < 0.01$) in the number of hatched blastocysts (Table 1). In view of the high luminal concentration of protein observed on day 4 (Fig. 1), this result was readily explained in terms of the amount of protein added to the cultures. In a subsequent experiment, the quantity of "delay" or "day 4" uterine protein added to the media was kept constant. Under these conditions more than 90% of the blastocysts hatched from their zonae regardless of the source of supplementary protein (Table 2). These results confirmed the biochemical data in suggesting that blastocyst activation in the mouse depends upon an oestrogen-induced change in the concentration rather than the species of protein in the uterine lumen (Aitken 1977c).

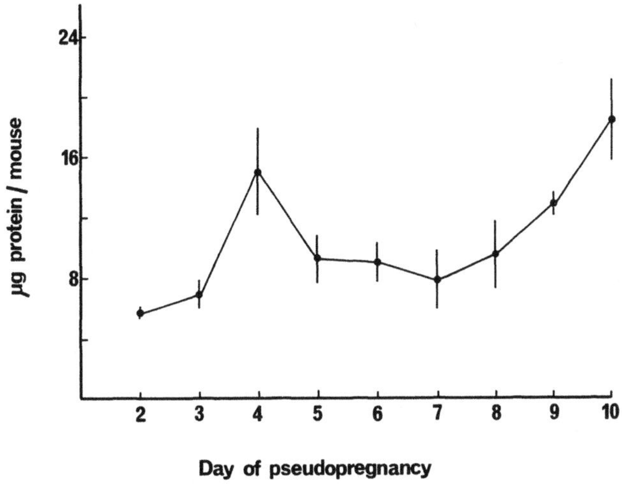

Fig. 5. Changes in the protein content of the mouse uterine lumen during pseudopregnancy. Note the increase in protein concentration observed on day 4, the day on which implantation would have occurred in pregnant animals. The subsequent increase in protein concentration observed on days 8-10 of pseudopregnancy is associated with the return of the mice into oestrus; *vertical bars*, standard deviation

Table 1. Hatching of mouse blastocysts in media supplemented with the uterine fluids of 3 mice/ml

Supplement	Hatching frequency (%) on day 4 of culture ± s.e.	No. blastocysts	No. trials
No supplement	51.0 ± 4.9	164	8
Delayed implantation	46.3 ± 7.9	116	6
Day 4	82.8 ± 4.7	184	7

Table 2. Hatching of mouse blastocysts in media supplemented with 30 μg/ml luminal protein

Supplement	Hatching frequency (%) on day 4 of culture ± s.e.	No. blastocysts	No. trials
Bovine serum albumin	58.3 ± 5.4	90	4
Delayed implantation	94.4 ± 5.6	49	2
Day 4	96.6 ± 3.5	53	2

Further evidence for the way in which oestrogen induces implantation in the rat and mouse has been obtained by the careful analysis of uterine flushings recovered at various time intervals after the administration of this hormone (Surani 1975, Aitken 1977a). In the mouse this response to oestradiol was found to be biphasic (Figs. 3 and 4), significant increases in protein content being observed within 12 h and again at 40-48 h after injection. The elevated protein levels observed during the first 12 h were associated with a marked increase in the concentration of both non-serum and, particularly, serum proteins in the uterine flushings, a majority of these bands migrating as high-molecular-weight slow α-globulins (Figs. 6 and 7). Confirmation of this early secretory phase has been obtained by Bergstrom and Nilsson (1975) who detected the appearance of an electron-dense secretion in the mouse uterine lumen 8 h after oestrogen administration. Studies on the time course of blastocyst activation following oestradiol injection clearly suggest that this initial period of endometrial secretory activity is responsible for the termination of diapause in the mouse. Hence, ultrastructural signs of blastocyst activation have been observed within 8 h of oestrogen administration (Nilsson 1974), carbon dioxide and protein synthesis by the blastocysts are elevated within 12 h (Weitlauf 1971, Torbit and Weitlauf 1974), and a pontamine blue response may be detected in the uterine stroma within 17 h (McLaren 1970). Similar conclusions have been reached by Surani (1975) who detected an influx of high-molecular-weight proteins into the rat lumen 13-20 h after the administration of oestradiol-17β.

The second part of the uterine response to oestradiol in the mouse was observed 40-48 h after injection and involved an increase in the concentration of several non-serum proteins, most serum proteins being present in trace amounts only (Aitken 1977a, Figs. 6 and 7). The dominant non-serum component was a slow α-globulin (Ra 0.32; Figs. 6 and 7) which closely resembled proteins observed in the uterine fluids of rats (Surani 1975), women (Wolf and Mastroianni 1975; Maathuis and Aitken 1978), baboons (Peplow et al. 1973), and roe deer (Aitken 1974a). The identity and functional significance of this protein have yet to be established.

Circumstantial biochemical and ultrastructural evidence has therefore been obtained in a number of diverse species to suggest a relationship between the induction and termination of embryonic diapause and the inhibition or stimulation of endometrial secretory activity. Experimental evidence also exists to support this hypothesis. Hence, if mouse blastocysts are cultured in a defined medium lacking either whole serum or certain amino acids they enter a state of diapause similar to that observed in vivo (Gwatkin 1966a,b, McLaren 1973, Aitken 1979, Van Blerkom et al. 1979); the subsequent addition of the missing constituents, however, results in a rapid resumption of blastocyst activity (McLaren 1973, Aitken 1979, Van Blerkom et al. 1979). The termination of diapause, whether induced by the addition of serum in vitro or the administration of oestrogen in vivo, certainly appears to involve similar increases in the rate of RNA and DNA synthesis (Ellem and Gwatkin 1968, Prasad et al. 1968, Psychoyos and Bitton-Casimiri 1969), carbon dioxide production (McLaren and Menke 1971), and cell division (Gwatkin and Meckley 1966, Bowman and McLaren 1970).

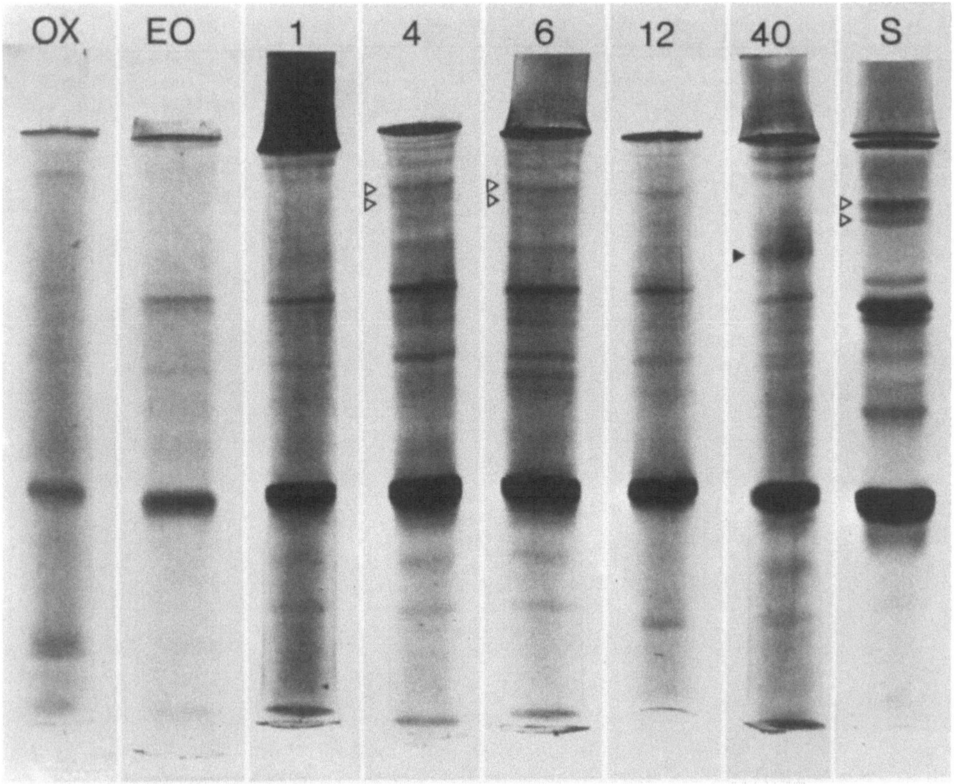

Fig. 6. Polyacrylamide gel electrophoresis of mouse flushings following the administration of 1 μg oestradiol-17β to pregnant animals during diapause. *OX* ovariectomized animals not receiving hormone treatment; *EO* experimental delayed implantation; *S* serum; *1, 4, 6, 12,* and *40* h after oestradiol administration. Note the pronounced increase in the number and intensity of bands 4-6 h after injection, the general decline in protein concentration at 12 h and the subsequent increase in non-serum proteins at 40 h. At *S* and 4 and 6 h *open arrows,* serum slow α-globulins, Ra 0.16 and 0.20; at 40 h *closed arrow,* major non-serum slow α-globulin at Ra 0.32

3 Endometrial Inhibitors and Implantation

Recent studies in the rat (Psychoyos 1973) and mouse (Finn 1974; Weitlauf 1976, 1978) suggest that embryonic diapause may be associated with the presence of an inhibitor of blastocyst activity in the uterine lumen rather than the simple absence of embryotrophic factors. The principal evidence for this concept is that flushings recovered from the uterine lumina of rats (Psychoyos 1973) and mice (Weitlauf 1976, 1978) during diapause possess one or more component(s) capable of suppressing the incorporation of ^{3}H-uridine by blastocysts in vitro. In the mouse fractionation of uterine flushings by Sephadex G25 chromatography (Weitlauf 1978) has succeeded in identifying 12 separate inhibitory fractions. Some non-specific inhibitory activity was detected in this study although one of the fractions, isolated in the void volume of the G25 column, was present in greater amounts in "delayed implanting" than "non-delayed implanting" uteri and was

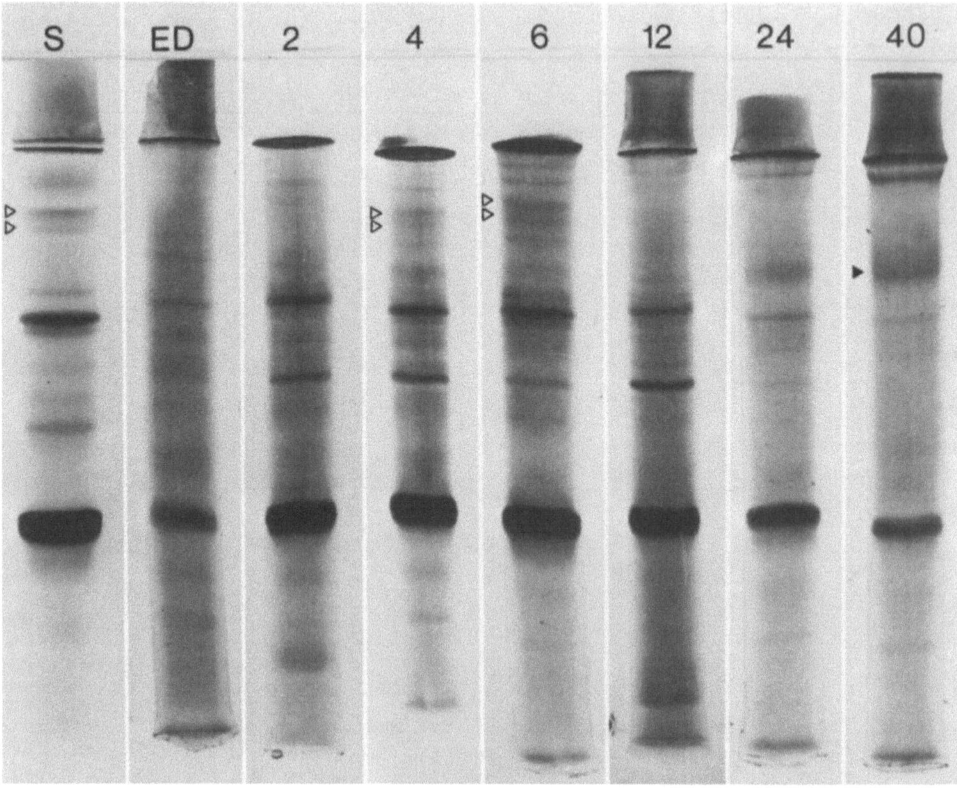

Fig. 7. Polyacrylamide gel electrophoresis of uterine flushings following the administration of 1 μg oestradiol-17β to pseudopregnant mice during diapause. *S* serum; *ED* experimental delayed implantation; *2, 4, 6, 12, 24,* and *40* h after oestradiol administration. Note the pronounced increase in the number and intensity of protein bands 4-6 h after injection, the decline in protein concentration at 12 h and the subsequent increase in non-serum proteins at 24 and 40 h. The similarity of this pattern to that observed in pregnant animals (Fig. 6) indicates that the qualitative aspects of the uterine response to oestradiol-17-β are independent of the presence of blastocysts. At *S* and 4 and 6 *open arrows,* serum slow α-globulins, Ra 0.16 and 0.20; at 40 h *closed arrow,* major non-serum protein at Ra 0.32

absent from the uterine fluids of control mice treated with oil vehicle alone. These results suggest that the termination of delayed implantation in the mouse may be associated with both a decline in the concentration of an 'inhibitor' and the simultaneous neutralisation of its activity by the influx of a protein-rich secretion into the uterine lumen from the endometrial glands and vasculature.

Psychoyos (1973) has proposed that the rat inhibitor may be a lipid since alcohol-ether or methanol extracts of the uterine flushings retain the inhibitory properties exhibited by the original material. Endometrial lipid deposits are, indeed, a conspicuous feature of several species during delayed implantation, e.g. the rat (Warren and Enders 1964), mouse (Nilsson 1970), roe deer, (Aitken et al. 1973), black bear (Wimsatt 1963), and armadillo (Enders 1961). This may be a consequence rather than a cause of delayed implantation however, since mustelids such as the mink (Enders and Enders 1963), long-tailed weasel (Wright 1963), and stoat (Enders 1967) accumulate an alternative storage material, glycogen, in the uterus during the delay phase.

Further credence was given to the concept of an inhibitor by Finn (1974) who demonstrated that the intraperitoneal injection of actinomycin D during diapause induced implantation in the mouse. Evidence for the mode of action of this compound was obtained by Pollard et al. (1976) who found that actinomycin D administration was followed by the disapperance of a protein with a molecular weight of about 70 000, from the luminal epithelium of the mouse uterus.

4 Conclusions

A combination of biochemical, ultrastructural, and experimental evidence has been obtained to suggest that delayed implantation is normally associated with a lack of secretory material within the uterine lumen. Although the recovery of uterine flushings is subject to the risk of contamination by serum or stromal proteins, it is difficult to envisage how a "false negative" can be obtained with this system. A simple relationship between the absence of endometrial secretory activity and the inhibition of blastocyst development may not exist, however, because actinomycin D is capable of simultaneously inducing implantation and a highly significant ($P < 0.001$) fall in luminal protein concentration (Aitken 1977d, Fig. 8). It is therefore possible that the induction of embryonic diapause depends upon the combined effects of a lack of embryotrophic substrate and the presence of an inhibitor. Such an inhibitor may directly suppress blastocyst activity, as suggested by the data of Psychoyos (1973) and Weitlauf (1976), and/or prevent blastocyst attachment, as suggested by the results obtained by Pollard et al. (1976). The precise nature, functional significance, and mode of action of these putative inhibitors urgently require clarification.

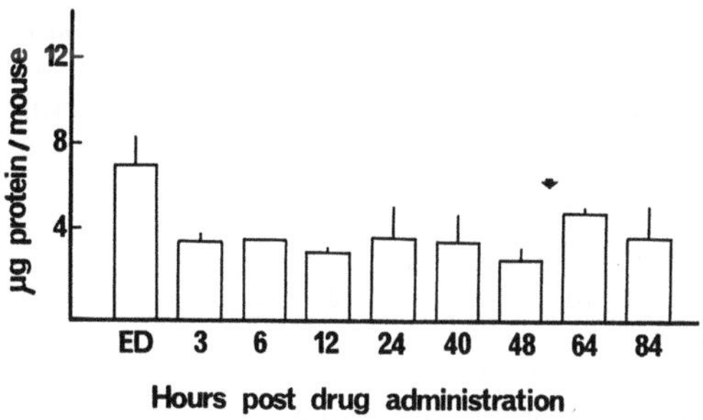

Fig. 8. Changes in the protein content of the mouse uterine lumen following the administration of 15 µg actinomycin D during delayed implantation. *ED* experimental delayed implantation; *vertical bars,* standard deviation; *arrow,* approximate time of implantation

References

Aitken RJ (1974a) Delayed implantation in the roe deer (*Capreolus capreolus).* PhD dissertation, University of Cambridge

Aitken RJ (1974b) Delayed implantation in roe deer (*Capreolus capreolus).* J Reprod Fertil 39: 225-233

Aitken RJ (1975) Ultrastructure of the blastocyst and endometrium of the roe deer (*Capreolus capreolus)* during delayed implantation. J Anat 119:369-384

Aitken RJ (1977a) Changes in the protein content of mouse uterine flushings during normal pregnancy and delayed implantation and following ovariectomy and oestradiol administration. J Reprod Fertil 50:29-36

Aitken RJ (1977b) Changes in the protein content of mouse uterine flushings during pseudo-pregnancy. J Reprod Fertil 50:191-192

Aitken RJ (1977c) The culture of mouse blastocysts in the presence of uterine flushings collected during normal pregnancy, delayed implantation and pro-oestrus. J Embryol Exp Morphol 41:295-300

Aitken RJ (1977d) The influence of actinomycin D on the protein composition of mouse uterine flushings. J Reprod Fertil 50:193-195

Aitken RJ (1979) The hormonal control of implantation. In: Whelan J (ed) Maternal recognition of pregnancy. Ciba Foundation Symposium 64. Excerpta Medica, Amsterdam, pp 53-73

Aitken RJ, Burton J, Hawkins J, Kerr-Wilson R, Short RV, Steven DH (1973) Histological and ultrastructural changes in the blastocyst and reproductive tract of the roe deer, *Capreolus capreolus,* during delayed implantation. J Reprod Fertil 34:481-493

Bergstrom S, Nilsson O (1975) Embryo-endometrial relationship in the mouse during activation of the blastocyst by oestradiol. J Reprod Fertil 44:117-120

Bowman P, McLaren A (1970) Cleavage rate of mouse ova *in vivo* and *in vitro.* J Embryol Exp Morphol 24:203-207

Canivenc R, Laffargue M (1956) Survie prolongee d'oeufs fecondes non implantes dans l'uterus de rattes castrees et injectees de progesterone. Cr Acad Sci (Paris) 242:2857-2860

Daniel JC Jr (1971) Growth of the preimplantation embryo of the northern fur seal and its correlation with changes in uterine protein. Dev Biol 26:316-332

Ellem KAO, Gwatkin RBL (1968) Patterns of nucleic acid synthesis in the early mouse embryo. Dev Biol 18:311-330

Enders AC (1961) Comparative studies on the endometrium of delayed implantation. Anat Rec 139:483-497

Enders AC (1967) Uterus in delayed implantation. In: Wynn RM (ed) Cellular biology of the uterus. Meredith, Amsterdam, pp 151-190

Enders RK, Enders AC (1963) Morphology of the female reproductive tract during delayed implantation in the mink. In: Enders AC (ed) Delayed implantation. University of Chicago Press, Chicago, pp 129-139

Enders AC, Enders RK, Schlafke SJ (1963) An electron microscope study of the gland cells of the mink endometrium. J Cell Biol 18:405-418

Finn CA (1974) The induction of implantation in mice by actinomycin D. J Endocrinol 60:199-200

Fishel SB (1979) Analysis of mouse uterine proteins at pro-oestrus during early pregnancy and after administration of exogenous steroids. J Reprod Fertil 55:91-100

Gwatkin RBL (1966a) Defined media and development of mammalian eggs *in vitro.* Ann NY Acad Sci 139:79-90

Gwatkin RBL (1966b) Amino acid requirements for attachment and outgrowth of the mouse blastocyst *in vitro.* J Cell Physiol 68:335-343

Gwatkin RBL, Meckley DE (1966) Chromosomes of the mouse blastocyst following its attachment and outgrowth *in vitro.* Ann Med Exp Biol Fenn 44:125-127

Maathuis JB, Aitken RJ (1978) Protein patterns of human uterine flushings collected at various stages of the menstrual cycle. J Reprod Fertil 53:343-348

McLaren A (1970) Early embryo-endometrial relationships. In: Hubinot PO, Leroy F, Robyn C, Leleux P (eds) Ovo implantation, human gonadotrophins and prolactin. Karger, Basel, pp 18-37

McLaren A (1973) Blastocyst activation. In: Segal SJ, Crozier R, Corfman PA, Condliffe PG (eds) The regulation of mammalian reproduction. Thomas, Springfield, pp 321-328

McLaren A, Menke TM (1971) CO_2 output of mouse blastocysts *in vitro,* in normal pregnancy and in delay. J Reprod Fertil [Suppl] 14:23-29

Nalbandov AV (1971) Endocrine control of implantation. In: Blandau RJ (ed) The biology of the blastocyst. University of Chicago Press, Chicago, pp 383-392

Nilsson O (1970) Some ultrastructural aspects of ovo-implantatation. In: Hubinot PO, Leroy F, Robyn C, Leleux P (eds) Ovo implantation, human gonadotrophins and prolactin. Karger, Basel, pp 52-72
Nilsson O (1974) The morphology of blastocyst implantation. J Reprod Fertil 39:187-194
Peplow V, Breed WG, Jones CMJ, Eckstein MD (1973) Studies on uterine flushings in the baboon. 1. Method of collection, cellular composition and protein electrophoretic profiles in animals with and without intrauterine contraceptive devices. Am J Obstet Gynec 116:771-779
Pollard JW, Finn CA, Martin L (1976) Actinomycin D and uterine epithelial protein synthesis. J Endocrinol 69:161-162
Prasad MRN, Dass CMS, Mohla S (1968) Action of oestrogen on the blastocyst and uterus in delayed implantation – autoradiographic study. J Reprod Fertil 16:97-103
Pratt HPM (1977) Uterine proteins and the activation of embryos from mice during delayed implantation. J Reprod Fertil 50:1-8
Psychoyos A (1973) Hormonal control of ovo implantation. Vitam Horm 31:201-256
Psychoyos A, Bitton-Casimiri V (1969) Caption *in vitro* d'un precurseur d'acide ribonucleique (ARN) (uridine-5-^{3}H) par le blastocyste du rat: difference entre balstocyste normaux et blastocyste en diapause. Cr Acad Sci (Paris) 268:188-192
Renfree MB (1973) Proteins in the uterine secretions of the marsupial, *Macropus eugenii.* Dev Biol 32:41-49
Surani MAH (1975) Hormonal regulation of proteins in the uterine secretion of ovariectomized rats and the implications for implantation and embryonic diapause. J Reprod Fertil 43:411-417
Torbit CA, Weitlauf HM (1974) The effect of oestrogen and progesterone on CO_2 production by 'delayed implanting' mouse embryos. J Reprod Fertil 39:379-382
Van Blerkom J, Chavez DJ, Bell H (1979) Molecular and cellular aspects of facultative delayed implantation in the mouse. In: Whelan J (ed) Maternal recognition of pregnancy. Ciba Foundation Symposium 64. Excerpta Medica, Amsterdam, pp 141-162
Warren RH, Enders AC (1964) An electron microscope study of the rat endometrium during delayed implantation. Anat Rec 148:177-195
Weitlauf HM (1971) Influence of ovarian hormones on the incorporation of amino acids by blastocysts *in vivo.* In: Blandau RJ (ed) The biology of the blastocyst. University of Chicago Press, Chicago, pp 277-290
Weitlauf HM (1976) Effect of uterine flushings on RNA synthesis by 'implanting' and 'delayed implanting' mouse blastocysts *in vitro.* Biol Reprod 14:566-571
Weitlauf HM (1978) Factors in mouse uterine fluid that inhibit the incorporation of ^{3}H uridine by blastocysts *in vitro.* J Reprod Fertil 52:321-325
Wimsatt WA (1963) Delayed implantation in the Ursidae with particular reference to the black bear *(Ursus americanus).* In: Enders AC (ed) Delayed implantation. University of Chicago Press, Chicago, pp 49-76
Wolf DP, Mastroianni L (1975) Protein composition of human uterine fluid. Fertil Steril 26:240-247
Wright PL (1963) Variations in reproductive cycles in North American mustelids. In: Enders AC (ed) Delayed implantation. University of Chicago Press, Chicago, pp 77-98

Cellular and Molecular Aspects of Decidualization and Implantation

Stanley R. Glasser and Shirley A. McCormack

Department of Cell Biology, Baylor College of Medicine, Houston, Texas, USA

1 Introduction

A comprehensive analysis of the molecular and cellular mechanisms underlying decidualization and implantation is not possible today. This is because the three concurrent lines of research, i.e., molecular endocrinology, cell cycle kinetics and biochemistry, which have contributed the recent advances to these studies, have progressed with almost no interaction in parallel fashion. An initial attempt at integrating these data as they apply to the mechanisms of decidualization and implantation will be the task of this contribution.

2 Mechanism of Steroid Hormone Action

2.1 Dynamics of Steroid Receptor Complexes

The developmental processes by which the "neutral" or nonreceptive uterus is transformed to a uterus receptive to implantation of a blastocyst are initiated and ordered by progesterone and/or estrogen. The patterns of DNA, RNA, and subsequently protein synthesis which characterize these changes are induced or regulated by these hormones.

The biochemical mechanisms by which steroid hormones act on target tissues to alter their responses are set forth as a generally accepted theoretical framework which, if not dogma, has at least achieved the status of a paradigm. Almost all the modern work on steroid hormone action takes place within this conceptual framework supported by a mass of evidence (O'Malley and Means 1973; King and Mainwaring 1974; Yamamoto and Alberts 1976; Gorski and Gannon 1976).

Our current state of knowledge regarding the regulatory aspects of hormone action considers that a steroid hormone, upon entering a steroid responsive or target cell, combines with a specific cytoplasmic macromolecule. It is this specific binding which is the first step in a sequence of steps which results in the production of hormone specific responses. The idea that cells which respond preferentially to a hormone (target cells) contain specific recognition factors or receptors is not new. For many years the idea was not testable. However, the synthesis of ^{3}H-estradiol with a high specific radioactivity (Jensen and Jacobson 1962) provided the methodology that made it possible to measure the differential accumulation, distribution and retention of estradiol by its target cells. Steroid-target organ interractions have since been described for all steroid hormones. These relationships are fundamental to our understanding of hormone action.

2.1.1 Steroid Binding Macromolecules in Blood

The quantity of free hormone which is available for entry into a target cell will modify steroid hormone-receptor interaction. Amongst the factors which determine the availability of free hormone are the specific and non-specific binding macromolecules found in the blood. The affinity of these blood binding agents for steroid hormones varies from K_d 10^{-3}M (very weak) to K_d 10^{-10}M (very strong). Thus, these proteins can limit the amount of hormone which is free and available for receptor binding thereby exercising an important control over hormone action.

Simple measurements of steroid concentrations in blood may give rise to false conclusions regarding the level of hormone stimulation at the target tissue site. Biologically important blood binders such as TEBG (testosterone-estrogen binding globulin) and α-fetoprotein which bind testosterone and estradiol with high affinity can serve as effective regulators (Murphy 1968; Soloff et al. 1971). Of particular interest to the developmental endocrinologist is α-fetoprotein. The titer of this binder falls from high neonatal levels to low prepubescent levels (Raynaud 1973) thereby making increasing amounts of free steroid available to the target cells of the maturing animal.

Nonspecific binders such as serum albumin may also play an important role in establishing the level of free steroid. Serum albumin serves as an ubiquitous, relatively weak affinity binder (K_d 10^{-4}-10^{-5}M). A specific example which may be cited (Clark et al. 1977a) considers the importance of serum albumin in altering the relative potencies of hormones. The K_d values for estradiol (E_2) and estriol (E_3) complexed with their E receptor at 4°C are 10^{-10}M and 10^{-9}M respectively. E_2 should be ten times more potent than E_3. However, in terms of their ability to provoke the early events related to the uterotrophic response, e.g., water imbibition, RNA and protein synthesis, etc., which occur within the first 6 h following injection, the two hormones are of equal effectiveness. This lack of agreement between the theoretical and observed potencies of E_2 and E_3 results from their differential binding to plasma components. As noted, the affinity of a binder such as serum albumin for E_2 is 10-100 fold greater than that for E_3. Therefore the amount of E_3 which is free and available after injection is much greater than free E_2. These elevated levels of E_3 result in levels of receptor binding and stimulation of the early (0-6 h) uterotrophic events that are approximately equal to E_2.

E_2 plays a role in the amount of serum albumin available for binding (Peterson and Spaziani 1971). In comparison to nontarget tissues, E_2 enhances the accumulation and retention of serum albumin in target tissues. In regard to the E binding characteristics of serum albumin, this particular E_2 action may be involved in establishing levels of total estrogen binding in the uterus that exceed titers in the general circulation and are maintained at these levels for longer periods of time (Peck and Clark 1974).

2.1.2 Entry of Steroid into Cells

Free steroids enter target and nontarget tissues with equal facility at a rate proportional to the steroid concentration over a range of 0.2-5000 ng/ml (Gurpide and Welch 1969; Gurpide et al. 1971). These results are most easily interpretable in terms of simple diffusion. They do not disallow consideration of a carrier mediated process. Peck et al. (1973a,b) have stringently demonstrated that E_2 initially partitions between medium and tissue (target and nontarget) in a nonspecific passive manner.

2.1.3 Receptor-Steroid Binding in Cytoplasm

Upon entry into the target cell estrogen may bind to either the cytoplasmic receptor (Rc) to form the estrogen receptor complex (R_cE) or with a variety of low affinity nonspecific binding components (NS) that are to be found in the cytoplasmic compartment (Peck et al. 1973a; Rochefort and Baulieu 1969). Thus, there exist two classes or species of binding sites in the cytosol: (1) high affinity, specific, low capacity and (2) low affinity, nonspecific, high capacity. Any investigation of the role of the receptor in hormone action must distinguish between these two modes of binding (Fig. 1).

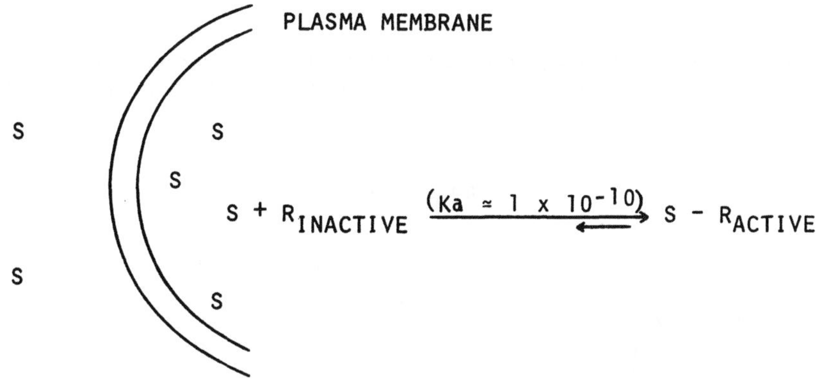

Fig. 1. Steroid (*S*) binding to the specific macromolecular receptor (*R*) in the target cell. *R* is believed to exist in the cytoplasm in an "inactive form". Steroid binding to the receptor is thought to produce a conformational change which converts the receptor to an "active" form. (Glasser et al. 1977)

2.1.4 Translocation and Accumulation in the Nucleus

R_cE is a high affinity but reversible complex. As a consequence of binding, the receptor undergoes a conformational transformation to an "active" state. The activated R_cE may now be translocated through membrane pores into the nuclear compartment of the target cell. Nuclear accumulation of the hormone receptor complex concomitant with the depletion of the receptor complex in the cytoplasmic compartment (Anderson et al. 1972a; Katzenellenbogen and Gorski 1975) serves as evidence that the "active" R_cE is translocated from the cytoplasm to the nucleus (R_nE).

R_cE and R_nE share similar characteristics, i.e., dissociation constants, hormone specificities and sedimentation coefficients. This may be taken as proof of the concept that R_cE, per se, is translocated to the nucleus. Specific thermodynamic evidence in support of the translocation process has been provided by Williams and Gorski (1974) (Figs. 2 and 3).

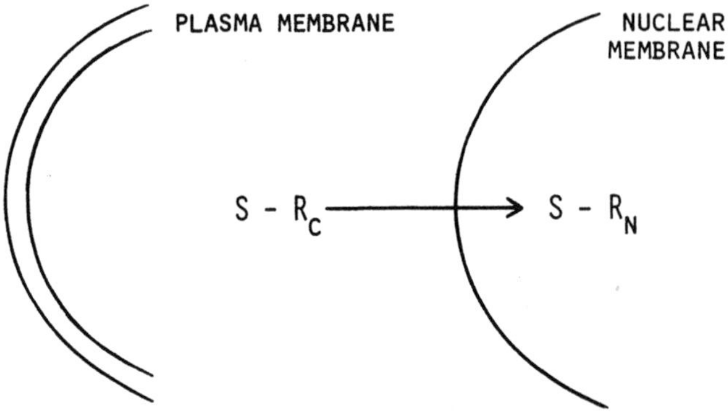

Fig. 2. The active hormone receptor complex (*S-Rc*) is translocated from the cytoplasm to the nucleus (*S-Rn*). The exact mechanism by which translocation is accomplished is still unclear. (Glasser et al. 1977)

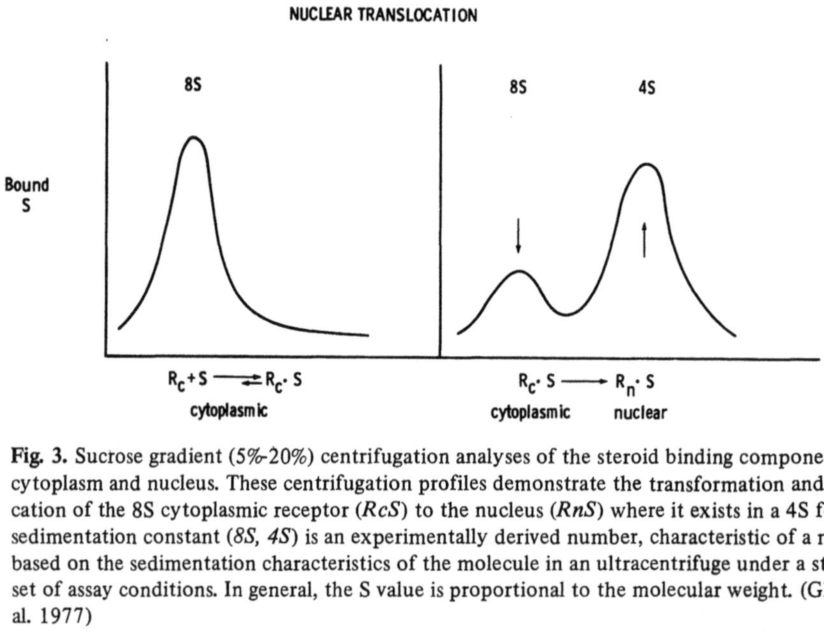

Fig. 3. Sucrose gradient (5%-20%) centrifugation analyses of the steroid binding components of the cytoplasm and nucleus. These centrifugation profiles demonstrate the transformation and translocation of the 8S cytoplasmic receptor (*RcS*) to the nucleus (*RnS*) where it exists in a 4S form. The sedimentation constant (*8S, 4S*) is an experimentally derived number, characteristic of a molecule, based on the sedimentation characteristics of the molecule in an ultracentrifuge under a standard set of assay conditions. In general, the S value is proportional to the molecular weight. (Glasser et al. 1977)

2.1.5 Nuclear Binding Mechanisms

The translocated hormone receptor complex (R_nE) participates in a high affinity binding reaction with the interphase chromosomes of the nucleus. The binding of R_nE onto chromatin is associated with the nonhistone protein (NHP) components of chromatin (O'Malley et al. 1972). Further studies have identified a subclass of chromosomal NHP which, when complexed to DNA, constitute the chromatin "acceptor" sites for R_nE binding (Spelsberg et al. 1975) (Fig. 4).

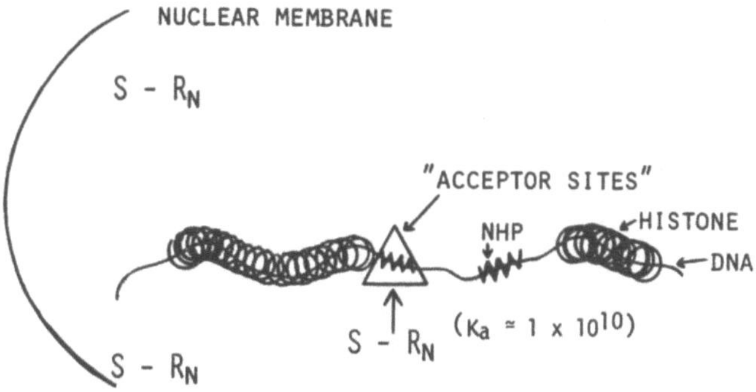

Fig. 4. Binding of the nuclear receptor hormone complex (*S-Rn*) to nuclear receptor sites. The absolute number of binding sites available for SRn is thought to be determined by the nonhistone (*NHP*) or acidic chromosomal proteins. DNA appears to be a requisite component for these binding reactions. (Glasser et al. 1977)

Many details of the mechanism of gene modulation as the result of steroid hormone action remain unclear, but it would appear that the presence of steroid hormone-receptor complexes such as R_nE, bound to NHP-DNA complexes at "acceptor" sites on chromatin, appear to alter these sites in a manner which promotes transcriptional activity by nuclear DNA dependent classes of RNA polymerase. Thus, the binding reaction with the nuclear "acceptor" sites is an obligatory event for subsequent steps in hormone action. It has been shown that the E_2 induced increase in chick oviduct chromatin transcriptional activity arises from an increase in the number of polymerase binding and initiation sites in chromatin. This yields a threefold increase in the number of RNA chains initiated per picogram DNA template (Tsai et al. 1975; Schwartz et al. 1975).

The criteria for establishing the identity of a steroid hormone receptor and the theoretical and practical considerations involved in the selection and in the execution of any particular method of analysis have been comprehensively reviewed (Schrader and O'Malley 1977; King and Mainwaring 1974). There are many methods for the assay of cytosol and nuclear receptors. All offer advantages which are offset with particular pitfalls and limitations. A decision or "trade-off" between the pluses and minuses of available methods must be made. The investigator must be aware that there are no "cookbook" methods. Assay techniques must be worked out and modified for each new target cell receptor population.

The interaction of the hormone-receptor complex and the genome generates new hormone specific messenger RNAs. It is not known if there is any nuclear posttranscriptional modification of these new messages before they are exported to the cytoplasm for translation into new proteins capable of executing the hormone mediated functions of the specific target cell. With the completion of this sequence of events, it is believed that the hormone-receptor complex dissociates from the chromatin acceptor site as part of the "off" mechanism of hormone action (Fig. 5).

It is speculated that the receptor proteins are not destroyed during these reactions but may return to the cytoplasmic compartment where they would be available to bind to another steroid hormone molecule and reinitiate the cycle as part of the process of continuous hormonal stimulation. Following its dissociation from the nucleus the

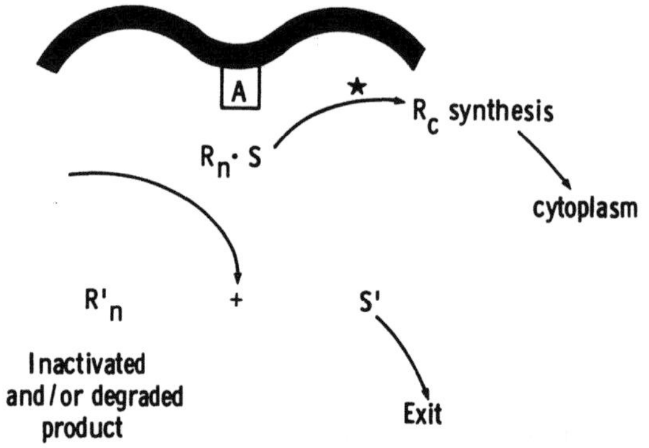

Fig. 5. Nuclear processing of Rn.S. Possible pathways in the nucleus. The binding of RnS to acceptor sites (*A*) on chromatin of the target cell stimulates the synthesis of Rc-mRNA which is responsible for the replenishment of Rc in the cytoplasm. RnS is also inactivated or degraded by "processing factors" that produce an ineffective form of the receptor, Rn. The steroid (*S*) is also metabolized to an inactive metabolite (*S*') that subsequently leaves the cell (Glasser et al. 1977)

steroid hormone may be altered to a biologically inactive molecule and diffuse out of the target cell.

2.2 Biological Responses

2.2.1 Early Uterotrophic Responses (Figs. 6 and 7)

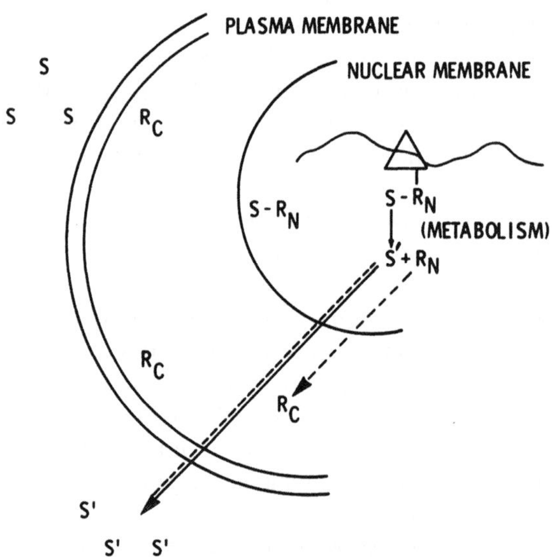

Fig. 6. Deinduction or "turn-off" of the specific hormone inductive process, possibly a consequence of metabolic alteration of steroid hormone binding to chromatin. Such changes would promote the dissociation of the receptor complex (SRn→S+Rn) from the acceptor site thereby diminishing the inductive reactions. The factors and mechanisms by which the "off" reaction is accomplished are not known. (Glasser et al. 1977)

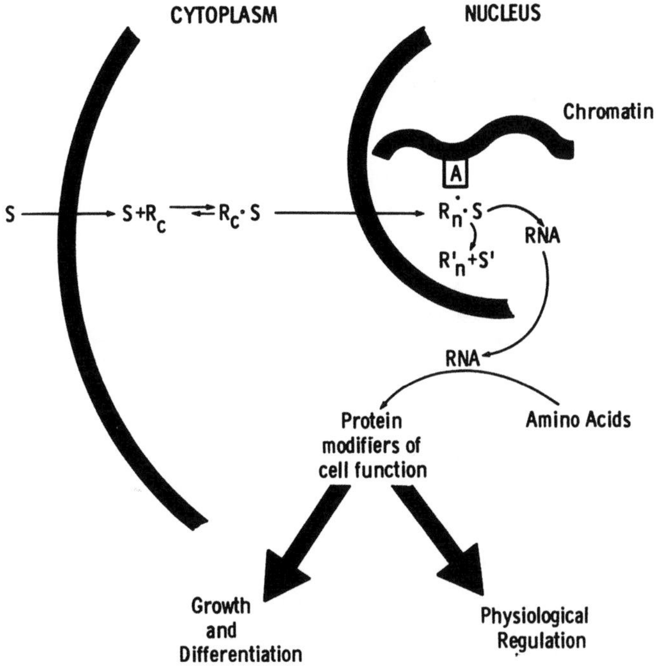

Fig. 7. Summary of the dynamics of steroid hormone-receptor interaction. The steroid hormone (*S*) enters the cell freely to interact with the cytoplasmic receptor (*Rc*). The "active" complex (*RcS*) is translocated to the nucleus where the nuclear complex (*RnS*) binds to the acceptor site (*A*) on chromatin. This reaction alters transcription of RNA directing the synthesis of specific mRNA species which translate hormone specific proteins in the cytoplasm. Some of these proteins are essential to the physiologic regulation of the target cell. Other proteins, reflecting the unique nature of their specific hormone induction regulate the differentiation and growth of the cell. (Glasser et al. 1977)

The biological effectiveness of any hormone in modulating the biochemical maturation of the uterus for ovoimplantation could be regulated at each of the focal points noted above. Thus, in the case of estrogens, the ability of an individual estrogen to stimulate growth and differentiation of the uterine endometrium is dependent not only on the initial accumulation of the hormone-receptor complex by the target cell nucleus but, in a more important sense, on the characteristics of the binding reaction between the R_nE complex and nuclear acceptor sites. The importance of these binding processes in the qualitative and quantitative regulation of uterine responses can be appreciated by the study of the relative biological effectiveness of estradiol and estriol. It has been suggested and proven that the failure of estriol (E_3) to produce a significant increase in true uterine growth within 24 h relates to that binding reaction. Thus, the R_nE_3 complex dissociates readily and is not retained by the nucleus. The response of the uterus to estradiol (E_2) is entirely different. The RE_2 complex binds to and is retained by the nucleus for significantly longer periods of time. Within 24 h E_2 has stimulated uterine cell division and growth (Anderson et al. 1972b) (Fig. 8).

Comparative analysis of the response of immature rats to either 1 μg E_2 or E_3 was

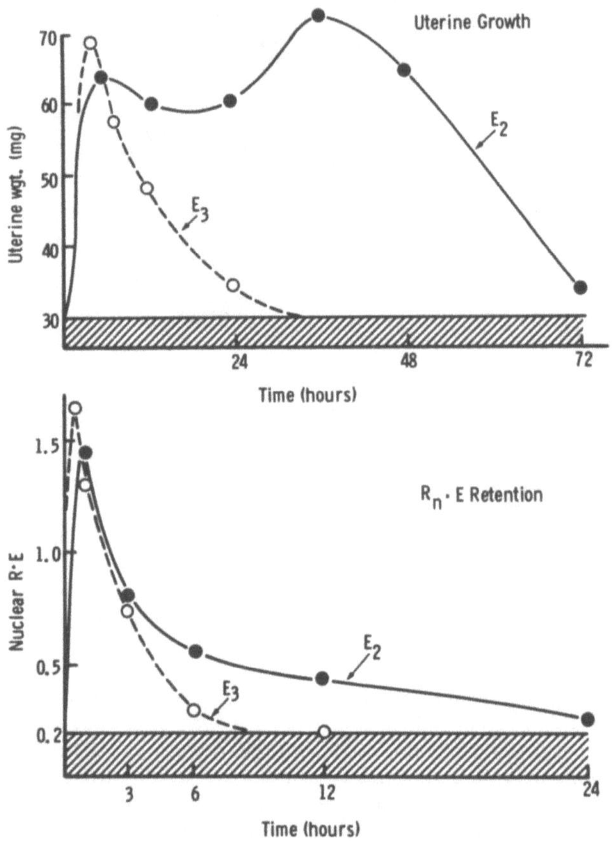

Fig. 8. Comparison of uterine growth curves and retention of the nuclear steroid hormone receptor complex at various times following the intraperitoneal injection of 1 μg of either estradiol-17β *(E_2/●)*, or estriol *(E_3/o)* into 21-day-old immature rats. The failure of E_3 to stimulate uterine growth correlates with its short retention time in the nucleus. E_2 is retained for a significantly longer period (check 6-12 h time points) and stimulates true uterine growth. Cross-hatched areas represent data collected from saline injection control rats. (Anderson et al. 1972b)

performed by the [^{3}H]estradiol exchange method. Uterine weights were measured at the same time periods. At 3 h following hormone injection the concentration of nuclear hormone injection the concentration of nuclear receptor complexes, R_nE_2 or R_nE_3, were equivalent. At 6 h the concentration of R_nE_3 had declined to near control levels whereas R_nE_2 remained significantly elevated.

The rapid decline of nuclear R_nE_3 is paralleled by the inability to maintain uterine weight in the estriol treated animal. True uterine growth may then be considered as a long-term response to estrogen. In order to stimulate those nuclear events which produce such long-term responses the receptor estrogen complex should be retained by the nucleus for at least 6 h. These conditions are met by the binding of R_nE_2, but not R_nE_3, to the nucleus. Nuclear RE_3 is not retained and fails to induce those nuclear events which promote true uterine growth (Fig. 9).

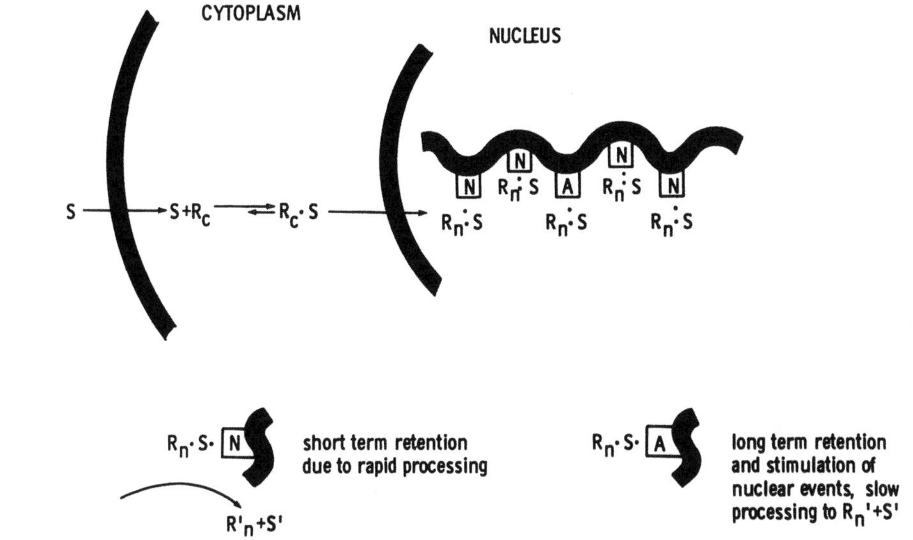

Fig. 9. The differential uterine growth response to E_2 versus E_3 may be a reflection of retention and processing of the nuclear steroid receptor complex. *N* and *A* represent different qualities of the chromatin binding site. The characteristics of the binding reactions may be determined by both the specific steroid hormone and the nature of the site. (Glasser et al. 1977)

2.2.2 RNA Polymerase and Uterotrophic Responses

Further biochemical characterization of the dependency of true uterine growth on the retention of the hormone-receptor complex by the nucleus has been advanced by studies in both mature castrate (Glasser et al. 1972a, 1974) and immature rats (Hardin et al. 1976). It was demonstrated that the initial response to E_2 is a rapid but transient increase in the activity of nuclear DNA dependent RNA polymerase II (Glasser et al. 1972). Translocation of the R_cE_2 complex to the nucleus is being completed during this period. The relationship between these two events may be critical to the initial action of estrogen since polymerase II codes for DNA-like RNA and may include some messenger species. After 30 min, enzyme activity begins to fall. This occurs as the capacity of uterine chromatin template to transcribe and the activity of polymerase I (codes for rRNA) begin to rise. Between 60 and 120 min after the injection of E_2, these latter changes are significant, as is the secondary rise in the activity of polymerase II. A change in the composition of uterine chromatin due to elevation of NHP/DNA ratios accompanies these changes (Fig. 10).

An extensive literature supports the contention that on an equal weight basis, E_2 will produce more uterine growth than E_3. Analysis of short-term data (0-6 h) would seem to repudiate this premise. It has been proven that short-term effects are not particularly reliable in predicting the regulatory effectiveness of an estrogen. Injection of E_3 to adult castrate females stimulates the early peak of RNA polymerase II activity. The exchange assay which measures nuclear concentrations of R_nE indicates that, within 30 min following injection, the R_nE_3 complex is equal to or greater than the concentration of

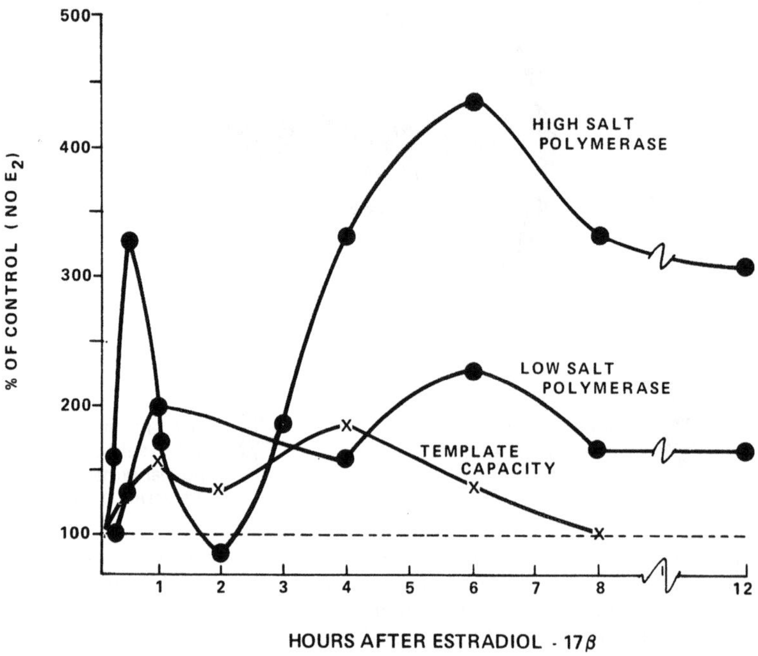

Fig. 10. The temporal response of RNA polymerase II (high salt), RNA polymerase I (low salt) and template activity of uterine chromatin of castrate adult rats injected intravenously with 1 μg estradiol-17β. The primary response of the DNA dependent nuclear RNA polymerase II can be elicited by E_2 and E_3 but the secondary increase in activity, along with the increase in polymerase I can be produced by E_2 but not E_3. For this reason (Glasser et al. 1972a; Hardin et al. 1976) the secondary rise in polymerase II may be regarded as an obligatory biochemical step in the pathway to produce true uterine growth

R_nE_2. However, E_3 (unlike E_2) fails to stimulate any significant changes in the subsequent nuclear biochemical events (which occur after the second postinjection hour) that relate to true uterine growth (Glasser et al. 1974). Thus, the increase in uterine weight observed between 4 and 6 h after E_3 injection (similar to E_2) reflects only a transient water imbibition by uterine tissue. E_3 (unlike E_2) fails to stimulate the later increase in uterine protein and dry weight that begins at approximately the 12th postinjection hour preceding cell division (20-24 h), nor do the early biochemical changes observed in the uterus display the hormone specificity characteristic of the later growth-related alterations. The biologically inactive E_2-17α (which cannot produce a uterotrophic response) will elicit the full spectrum of early effects including the stimulation of "I.P." synthesis (Katzenellenbogen and Gorski 1975).

2.2.3 Relationship to Nidation

This prompted our initial measurements of the quantities of uterine nuclear RE during the period from conception to implantation. R_nE was measured using the [^{3}H]estradiol exchange assay (Anderson et al. 1972a) in pregnant animals of proven cyclicity and fertility. The presence of a sperm-positive vaginal smear and a copulatory plug designates day 0 of the pregnancy. A more comprehensive discussion of the nature and the role of preimplantation estrogen is presented elsewhere (Glasser and Clark 1975).

The quantity of nuclear RE is low on day 0 (which essentially reflects the metestrus levels of the estrous cycle). Nuclear accumulation increases significantly on days 2 and 3 following conception. The peak on day 3 is followed by a decline on day 4 to levels that are not significantly different from those measured on day 0. The elevated levels of nuclear RE on days 2 and 3 were attributed to the increase in estrogen blood titers reported to occur on these days (Yoshinaga et al. 1969; Nimrod et al. 1972). The increase in the concentration of the nuclear estrogen receptor complex is probably involved in the initiation of transcriptive events which characterize the biochemical maturation of the uterus preparatory to implantation (Fig. 11).

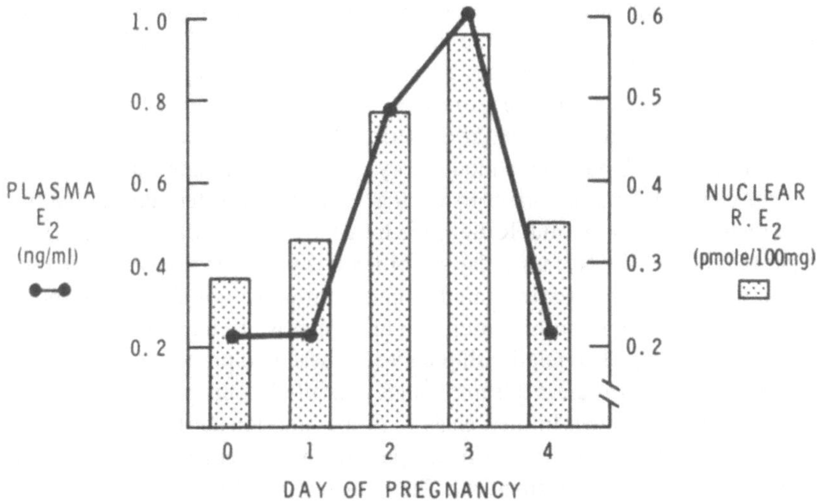

Fig. 11. The concentration of the receptor estrogen complex in the nucleus of uterine cells during the early days of pregnancy. The amount of nuclear RE was measured by the [^{3}H]estradiol exchange method and compared to blood estrogen levels. (Glasser and Clark 1975)

The decline in nuclear RE that is noted between days 3 and 4 is more complex and difficult to interpret. One of the contributing factors may be the reported decline in blood levels of estrogen at this time. Additionally, it has been reported by Baylor colleagues that progesterone reduces the quantity of cytoplasmic estrogen receptor by interfering with the replenishment process (Hsueh et al. 1976). This effectively reduces the capacity of the uterus to respond to available estrogen. Thus, the rising progesterone of this period might play a role in three important processes related to uterine receptivity. First, P may influence the dynamics of its own receptor protein. This is not well understood, particularly in the rat and mouse, because we do not have a reliable assay method for measuring the P receptor in these animals. Second, because of its putative action on replenishment, P may decrease the availability of the R_cE receptor thus eventually reducing nuclear RE. Third, and perhaps most important in terms of regulation, rising titers of P may influence the gene expression induced by the binding of R_nE to DNA chromatin. In addition to modifications of E-induced transcription are translation relevant alterations or R_cP and R_nP, per se, which may also perform a regulatory role by directing the transcription and/or translation of P-specific regulatory factors. The net result of these changes would be the progressive differentiation of cells, with or without cell division, whose functional capacity, evolving during the period immediately surrounding day 4, would permit them to respond to the presence of the blastocyst.

It is evident from these studies that a molecular role for prenidatory E remains difficult to define. We must deal with the constant observation that although E is released in sufficient amounts by the ovary it is unable to stimulate the specific endometrial events leading to egg implantation (Psychoyos 1973a). We have established a determinant role for P, independent of E, in the development of uterine sensitivity to decidualization and subsequent receptivity to implantation (Glasser 1972; Glasser and Clark 1975). Integrating the data developed from studies of the E receptor in a variety of animal models together with the correlates deriving from studies of the estrous cycle (Clark et al. 1972) and preimplantation period (Glasser and Clark 1975), we come to the conclusion that E has only a modulatory influence on P-determined gene expression. Thus, we feel that it is not correct to say that E is without effect during the first 72 h of the preimplantation period. Rather, the influence of E is not rate limiting or determinant (Glasser 1972). Not until day 4, for reasons yet to be resolved, can a specific recognizable set of E responses related to implantation be noted. These responses include: (1) the evolution of the sensitive uterus to one receptive to implantation and (2) the loss of uterine sensitivity. The relationship between these two processes is not known.

Another attempt to address this problem is presented in the provocative work of Martel and Psychoyos (1976). In their analysis of the functions of E during the preimplantation period they describe two different patterns of concentration of nuclear E binding sites in the uterus of the pregnant rat. They note an initial period, up to day 4, during which the amount of R_nE increases periodically every night. During the second period, from day 4 onward, the concentration of R_nE maintains a plateau at its minimal level. The pivot between these two periods coincides with the onset of estrogen-dependent functional changes of the uterus related to its receptivity for nidation. The paper includes a number of observations and assumptions which remain to be confirmed.

A singular finding by these authors is a circadianlike rhythmicity in the concentration of both soluble and insoluble forms of R_nE which peaks (2400 molecules/cell) between 10 and 12 P.M. with minimal values (800 molecules/cell) noted at noon. The figures cited represent the high values at the end of the first period and they certify that the excursion in values, from noon until midnight, could exceed 200%. The authors contend that the time interval between maximal E secretion (Watson et al. 1975) and maximal R_nE concentration can not be explained by prolonged retention and gradual accumulation of the hormone receptor complex in the nucleus. The model proposed to explain these interesting dynamics is that a sudden nuclear charge of the complex follows a gradual cytoplasmic accumulation of R_cE. When a threshold cytoplasmic concentration of R_cE is reached, translocation is precipitated. This charging of E, apparently bound at the level of uterine nuclear chromatin, takes place each night. It is the last of these charges (midnight, day 3) that seems to be the one involved in the induction of E-dependent events. All elements for uterine competency, according to the authors, are present by day 2 but initiation of the response requires additional factors.

These data pose interesting and important questions regarding the kinetics of translocation, the synthesis, degradation and replenishment of the receptor protein, and the influence of varying plasma E titers on these events. At this time there simply are not enough data of the species necessary to fix these points.

On the basis of the fact that the nuclear content of R_nE is lower during the preimplantation period than that observed during the cycle, Martel and Psychoyos (1976) have suggested that the number of R_nE complexes implicated in the nuclear events which regulate the induction of implantation is low. This assumption, although of considerable interest, is without foundation or confirmation. As a conclusion it must remain speculative until we can competently measure the efficiency of transcription stimulated by a hormone in any specific situation.

Technical refinements developed in our laboratory for the assay of R_nE_2 prompted us to consider a reinvestigation of these interrelationships between nuclear RE and the maturation of the preimplantation uterus. The detailed methodology has been published as part of our study of the receptor dynamics of the rat placental trophoblast (McCormack and Glasser 1976, 1978). Important features of those studies include the use of clean but not purified nuclei (*vs.* a rather crude nuclear pellet) which permit rigorous kinetic studies. Data are then expressed in terms of a 12-point Scatchard analysis rather than a 1-point charcoal exchange assay.

Our present data confirm those we have reported before (Glasser and Clark 1975). The concentrations of both R_cE and R_nE on day 0 reflect the residuum of proestrus plasma estrogen titers and the values on day 1 represent diestrus plasma estrogen (Fig. 12). Plasma progesterone concentrations are very low on both days (Fig. 13).

Both plasma estrogen and progesterone rise markedly between days 2 and 3 of gestation. The change in steroid levels is paralleled by a 10-fold increase in R_cE and a 50-fold increase in R_nE. Presently we are assigning this rise in RE to the influence of estrogen on the synthesis, translocation, and replenishment of its own receptor. There is a paucity of data regarding progesterone so we have made the assumption that the primary locus of progesterone action is its depression of RE replenishment (Hsueh et al. 1976) in which case one would not expect to see an obvious progesterone effect on the other aspects of RE dynamics at this early stage.

The concentration of R_nE falls steadily from its day 2 peak to plateau on day 4. This precedes the fall in R_cE which begins on day 3 after a continued rise from day 2 levels. Provisionally we suggest the fall in R_cE reflects the interaction between rising progesterone (diminished replenishment) and falling estrogen (depressed synthesis). We cannot account for the fall in R_nE while R_cE is still rising. If R_nE values on days 2 and 3 of pregnancy are indeed different we do not believe the fall is due to depressed synthesis and replenishment of RE. Hypothetically we would advance the idea of a direct effect of rising progesterone on the translocation process and/or a direct influence of RE itself.

It is obvious that the story of receptor dynamics in early pregnancy is unresolved. None of our data supports, or really contradicts, the thesis advanced by Martel and Psychoyos

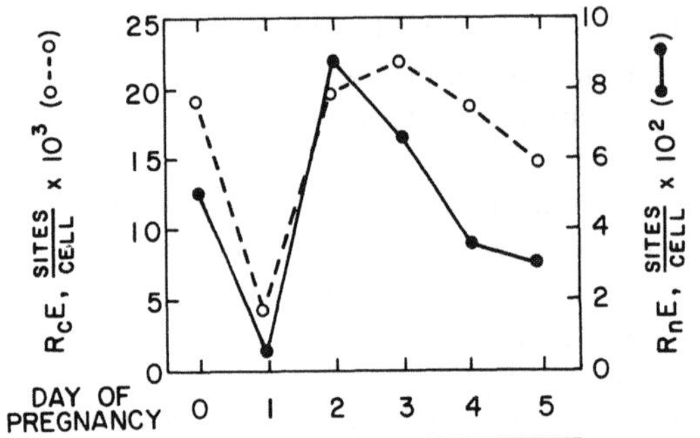

Fig. 12

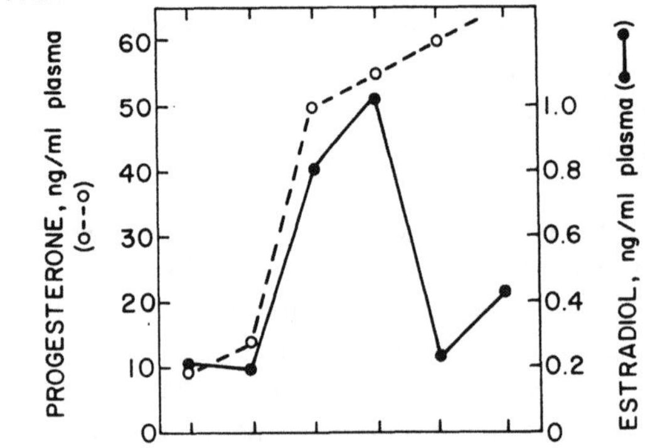

Fig. 13

Fig. 12. Comparative populations of cytoplasmic (*RcE)* and nuclear (*RnE)* hormone receptor complexes in the uterus expressed as sites per cell. Day 0 of pregnancy is the day of vaginal sperm. Compare with Figs. 11 and 13

Fig. 13. Comparative plasma titers of progesterone and estradiol from the same pregnant rats as used in Fig. 12. Compare with Figs. 11 and 12

(1976). More stringent and comprehensive protocols will be required to produce the data necessary to clarify this issue.

3 Cell Proliferation

3.1 Patterns in the Endometrium of Pregnant and Pseudopregnant Rats

Althought it is generally accepted that the uterus is a complex yet formally organized community of diverse cell types only a very few of the investigators in the field of implantation and decidualization have cogently integrated the role of cell division with that of the regulatory endocrine biology of early pregnancy. There has been a paucity of systematic investigations into the kinetic analysis of uterine cell proliferation as it relates to blastocyst-endometrial interactions. Most recently Martin et al. (1973a,b,c, 1976), Martin and Finn (1968, 1969, 1971) Heald (1977), and O'Grady and Bell (1977) have initiated such rigorous analyses of hormonally directed epithelial and stromal proliferation. As the fundamentals of uterine cell division become clarified, it is important that these data be regarded in perspective to their facultative and/or permissive role in decidualization and nidation.

Some of the analytical difficulties derive from the fact that preparation for ovum implantation requires the conjoined action of progesterone and estrogen (Shelesnyak 1960; DeFeo 1967; Finn 1971). However, much of the data arising from studies of cell division during the estrous cycle is of restricted utility because the prolonged luteal phase which characterizes pregnancy occurs outside the limits of the normal reproductive cycle of the rodent. This is a pivotal point, for only after a period of progesterone secretion is estrogen able to render the rat uterus responsive to the blastocyst (Psychoyos 1973b; Glasser 1972) or sensitize the mouse uterus to deciduogenic stimuli (Finn and Martin 1970). The capacity for decidualization in the rat, however, is determined by progesterone, the role of estrogen being a modulating one that generally (but not always) enhances the response of the progesterone-treated uterus (Glasser 1972; Glasser and Clark 1975).

A marked pattern of cellular proliferation characterizes the natural history of the preimplantation period of rodent pregnancy. In both pregnant and pseudopregnant rodents there is a spontaneous shift of mitotic acitvity from the epithelial to the stromal components of the endometrium (Tachi et al. 1972; Tachi and Tachi 1975) which occurs between noon of day 2 and noon of day 3 in the rat. This pattern of cell division may be studied by analysis of autoradiographic patterns of incorporated ^{3}H-uridine and thymidine. This type of data accents a fundamental that should be more generally appreciated. There are three cell types which comprise the uterine endometrium (luminal epithelium, LE; glandular epithelium, GE; and stroma, S). Each of these cell types has an individual response to growth stimulators and information regulators that is distinct for that cell type and defines its specific role in pre- and peri-implantation processes.

3.1.1 Epithelial Cells

Mitotic activity in LE and GE of pregnant and pseudopregnant rats peaks by 24 h after conception. The mitotic index in the LE and GE of pregnant rats is 75 and 50 respectively, while in the pseudopregnant rat the corresponding values are 40 (LE) and 20 (GE). These differences between pregnant and pseudopregnant animals may reflect the differential plasma steroid patterns that have been observed between these two groups. These data may be part of a growing literature which indicates that, even at the beginning of the prolonged luteal phase, the pseudopregnant animal may be an imperfect model for the study of pregnancy. This data also prompts the speculation that such differences may result from the presence or absence of an embryonic signal or maternal pregnancy recognition factor as has been suggested by Haour and Saxena (1974). This hCG-like substance has been the object of multiple investigations and discussions (Holt et al. 1976; Asch et al. 1978, 1979; Channing et al. 1978, Browning et al. 1980). Other molecules have been introduced into this discussion, preferably steroids (Perry et al. 1976, Singh and Adams 1978, Flint et al. 1979, Heap et al., this volume p. 157; Gadsby, this volume p. 173).

3.1.2 Stroma

Qualitatively, pregnant and pseudopregnant rat uteri are similar until midnight of day 3. Epithelial mitoses may be marked until midnight of day 2 and have ceased by noon of day 3. By midnight of day 3 there is brisk mitotic activity in the subepithelial antimesometrial (SEAM) stroma. Stromal mitoses are confined to the SEAM region through day 7. Mitosis then ceases in the pseudopregnant rat, but in the pregnant rat there is evidence that cell division spreads to deeper regions in the vicinity of the implanting blastocyst. This constitutes evidence that the attachment of the blastocyst on day 4 becomes a factor which stimulates a further increase in the number of stromal cells undergoing mitosis. Quantitative and qualitative differences between implantation site and intersite stroma become increasingly pronounced.

3.1.3 Significance of Stromal Mitosis

Analytically there are two facets to the control of endometrial cell division. The first, the preimplantation phase, is regulated by the ovarian hormones which determine the pattern and extent of cell division. At first reading this seems rather simple and amenable to analysis. However, cell proliferation is such a striking phenomenon that too little attention has been directed to questions which consider the mechanisms by which each hormone modulates cell division, the essentiality of cell division to preimplantation events as well as the extent and importance of biochemical differentiation in the absence of cell proliferation. Our understanding of regulatory reproductive biology is, in fact, dependent on qualifying the relative importance of the role of preovulatory estrogen compared to the role of prenidatory estrogen. This comparison must be made not solely in terms of the action of a specific hormone on cell division but with respect to all the sequelae which constitute the response of a target cell to a hormone. Additionally, a local factor or signal related to or coincident with blastocyst attachment must be considered. This "local" factor modulates endometrial cell division during implantation and the period immediately following. This is an important facet of endometrial response for which there is too little data. Transformation of stromal cells to decidual cells may, of course, also be stimulated by artificial stimuli (DeFeo 1963a) in the absence of appreciable cell division (Glasser 1972).

As we proceed with our analysis of model systems for the study of hormonally directed uterine proliferation we should recall that the prenidatory wave of mitosis which occurs in the stroma of rodents and in humans is thought to be associated with the preparation of stromal cells for decidualization. Yet transformation can occur in the absence of proliferation (Glasser 1972), and it is unclear whether such cell divisions occur in animals which do not exhibit decidualization. Thus an essential question to be asked in any study of the hormonal control of uterine cell proliferation is whether the purpose of the prenidatory shift in mitotic activity that occurs on day 3 after conception (rat) is to prepare a cadre of stromal cells that could respond to a deciduogenic stimulus in sufficient numbers to ensure the success of implantation. This speaks to the necessity of some threshold population of stromal cells but does not include any qualitative considerations.

In this particular area we are without adequate guidelines to assess the quality of information being generated. In advancing their concept of quantal mitosis, Stockdale and Topper (1966) spoke of the necessity for cells to divide before they could differentiate ("clean the genes"). Yet there is no proven necessity for these events to be linked in the uterine stroma. The progressive progesterone induced biochemical differentiation of nonproliferating stromal cells which renders them sensitive to deciduogenic stimuli has been documented (Glasser 1972; Glasser and Clark 1975). Prenidatory estrogen (best demonstrated in the delayed implantation model) evokes yet another state of differentiation which converts the sensitive uterus to one which is receptive to the implanting blastocyst. The functional relationships between implantation, the transcriptive events characteristic of uterine sensitivity and receptivity, and the stimulation of stromal cell division is unclear. The peak of mitosis in such cells does not occur until almost 24 h after exposure to estrogen at which time the uterus has already become receptive and implantation has been initiated. In fact, by 30-36 h after estrogen the endometrium is no longer receptive and has rapidly begun to lose its sensitivity (Glasser and McCormack 1979a, 1979b, 1980).

3.2 Hormonal Effects on Cell Cycle

3.2.1 General Principles

The fundamentals by which the hormonal effects on the various aspects of the cell cycle may be analyzed have been elaborated by Epifanova (1971). Animal models should be simple, allowing for the exclusive discrimination of the action of the individual hormone on the target cell. Studies of cell cycle cannot be considered separately from the general principles of hormone regulation. However, it is important to note that regulation of cell division is only one particular manifestation of the general regulatory action of hormones.

Modern scientists define a hormonal effect on a target organ so that tissue function is included along with increases in cell size and numbers. These are, of course, interdigitated processes, but there are provisional gains to be made if, for the moment, cell proliferation is considered separately by an analysis of its specific and necessarily more restricted literature and methodology (Quastler 1960, 1963; Mendelsohn 1962; Baserga 1965, 1968; Gelfant 1966). Thus, in spite of its pitfalls, the development and extended use of autoradiographic techniques, in vivo and in vitro, along with other special labeling methods have allowed investigators to elaborate stages of the cell cycle and permitted the kinetic analysis of dynamic cell populations. Together with the judicious use of inhibitors of biosynthetic pathways it has been possible to interpret cytological data resulting from hormone action in the idiom of biochemistry and molecular biology.

Initially the analysis of cell division proceeded according to concepts developed to find a key enzymatic system sensitive to hormonal regulation. Bullough (1955) advanced the thesis that endocrine control of the rate limiting glucokinase reaction regulated target cell proliferation. Subsequently, Swann (1958) rationally displaced Bullough's hypothesis with his own which proposed that hormones act as inducers of new patterns of nucleic acid and protein synthesis (including the mitotic apparatus). These in turn determine the further metabolic activity of the cell. It is evident that Swann's hypothesis is similar to current concepts of regulatory hormone action.

Preliminary characterization of the cell cycle is requisite to the study of mitogenic hormones. Epifanova (1971) recognizes three phases of the cycle. The proliferative or mitotic cycle includes mitosis (M), a period of DNA synthesis (S) and G_2, the premitotic period. There are also the null or resting states, R_1 and R_2. R_1 is the equivalent of what was termed G_0. Finally, there is the state of specific function. It is assumed that a cell is able to pass into a resting period not only after completing mitosis (R_1) but also after the completing of DNA synthesis (R_2). It may remain in this state for long periods of time retaining its potential for proliferation and functioning until an adequate stimulus directs it back into the mitotic cycle (this latter move is age-dependent).

The earliest work with tritiated thymidine recognized the advantage of using the female reproductive system as a model for the study of mitogenic hormones. Early studies were beset with technical problems of all sorts. Legitimate questions were posed about labeling techniques, sampling times, principles of cell kinetic analysis, choice of hormone, dose, and schedule. Nevertheless, it was learned from these early experiments that estrogen-induced uterine cell division was associated with stimulation of DNA synthesis (Perotta 1962, 1966); that estrogen significantly reduced the duration of the mitotic cycle largely at the expense of S and G phases (Peckham and Kiekhofer 1962); that the increase in mitotic index was not due to a reduction in generation time but to accelerated passage through the mitotic cycle (Epifanova 1966; Bullough 1967; Leroy and Galand 1969); and that estrogen significantly increased the number of new cells becoming competent for DNA replication and capable of entering the mitotic cycle (Tables 6.1 and 6.2 in Epifanova 1971).

It has also been learned and confirmed that mitogenic hormones can reduce the length of the S period, but that there is a minimum duration of DNA synthesis characteristic of a cell (Bresciani 1965), thereby invalidating the concept of a constant S period in mammals (Pilgrim and Maurer 1965). Nevertheless, because of technical problems, we are still without accurate data regarding generation time, the size of the proliferating pool and the homogeneity of uterine cell populations.

3.2.2 Simulation of Pregnancy by Steroid Hormones

In an effort to assign specific roles to each of the hormones which regulate cell growth and proliferation during the preimplantation period, any number of investigators have attempted to simulate the patterns of pregnancy by preparing ovariectomized rodents, supported on either estrogen or estrogen and progesterone, as models for study. Few investigators have reported data from animals supported on progesterone alone. The studies of Tachi et al. (1972) are exemplary in that their rat preparations were both ovariectomized and adrenalectomized so that their animals were completely steroidoprivic prior to experimentation.

Simulated patterns in the castrate female show that estradiol-17β (E_2) stimulates intense cell proliferation in the epithelial components of the endometrium. The response of LE, which peaks at 24 h, can be initiated by a single dose (Allen et al. 1937) and is preceded by an increase in DNA synthesis (Epifanova 1966; Galand et al. 1967). The same dose of E_2 elicits a later response in GE (peak at 48 h) which is less intense. At their respective peaks the mitotic index for LE is 250 and for GE 30. There are no stromal mitoses.

In the castrate female mouse the generation time of the uterine epithelial cell is estimated at 276 h with an average S period of 10.5 h (Das 1972). Therefore most cells are in G_1 at the time of treatment. E_2 (50 ng) produced marked hyperplasia in uterine LE of previously E_2-primed castrates. There was a fourfold increase in cells (20% of the uterine population) within 34-40 h (Martin et al. 1973a). This increase in cell proliferation was rapid, highly synchronized, short lived, and attributable to a shortening of the cell cycle. Approximately 60% of LE were in S at 10-12 h after E_2; the G_1 phase of the mitotic cycle was therefore the most markedly influenced, but S and possibly R_1 were also shortened (Epifanova 1966). Analysis of the labeling index curves show a "hump" 25 h after the initial dose of E_2 which suggests that some of the original stimulated cells are entering a second round of DNA synthesis (Das 1972; Martin et al. 1973b).

Progesterone inhibits the mitogenic action of E_2 on LE if it is given at the same time as the E_2. Forty percent of the cells are prevented from entering DNA synthesis but the G_1 period of the remaining cells (60%) was not lengthened. The reduction of cells entering the cycle is reflected in the decrease but not the delay of the labeling and mitotic indices (Martin et al. 1973b). The 60% of the cells which do enter the mitotic cycle have their passage accelerated, in the main, by a shortening of the G_1 phase. Pretreatment with progesterone did not prevent the shortening of the S phase produced by E_2 (Das 1972) or significantly slow the entry of cells into M, i.e., prolong G_2. Progesterone pretreatment did block the cells from entering a second round of DNA synthesis. Epithelial cells are sensitive to progesterone only during a limited period in the G_1 period (Martin et al. 1973c). Cells in this particular phase of the period are blocked, but others are not affected by progesterone and pass through one cycle to be blocked as they enter the sensitive phase of the subsequent G_1.

Progesterone has been found to be effective for at least 10 h. Because of the extended G_1 period in untreated animals, if P is given up to 10 h prior to E_2, it will block the actions of that hormone, but, if it is given after E_2, it will be ineffective since all LE and GE cells will have already been accelerated into the progesterone-insensitive part of the G_1 period.

Although the E_2 response of the progesterone pretreated castrate rat has not been as extensively investigated or analyzed, the same principle holds true. Three characteristic estrogen responses of rat uterine LE cells are altered or abolished by prior exposure to progesterone for a least 12 h prior to estrogen. These are: (1) increased rate of DNA synthesis, mitosis; (2) formation of large prominent nucleoli, and (3) increased ^{3}H-uridine incorporation. If progesterone is given for at least 30 h prior to estrogen, these estrogen-stimulated responses are not diminished but are translocated to the stromal cells.

Having established mitosis as a consequence of hormone action, we may appropriately inquire to what extent are the progesterone influences on estrogen responses due to alterations in the specific binding capacities of the different uterine cells. The selective effect of progesterone on LE and GE could be exercised on (1) E receptor replenishment, (2) the quality and degree of expression elicited by binding of the nuclear receptor-E_2 complex to the genome, (3) P-receptor synthesis, and (4) the influence of the nuclear receptor progesterone complex on gene expression. A study of "detached" uterine epithelium in ovariectomized mice (Smith et al. 1970) showed that progesterone did not alter the amount of E receptor protein but the amount of estrogen that bound to a 100 000 x *g* pellet was increased. Similar data were reported by Psychoyos et al. (1968) who noted an increase in the overall estrogen binding capacity of uteri from castrate animals treated with progesterone.

A cautionary note is due here. In reporting their data on ^{3}H-E_2 binding, which is decreased in a manner parallel to the suppression of E_2 responses of LE cells by progesterone, the authors operate with the misconception that ^3H-E_2 binding (specific and nonspecific) is an index of "responsiveness". It should be noted that the data reported in all these studies are essentially uptake data. There are no proofs of specific binding to a receptor protein. Uptake, specific binding, and mitosis are not required to be correlatively linked.

In consideration of what we do know about hormone action and mitotic cycle kinetics, the peak of LE mitosis observed 24 h after conception is likely to have been stimulated by the high estrogen titers at proestrus. It is difficult, however, to identify the specific cyclic event which provokes GE mitosis. The peak proliferative activity of these cells is delayed relative to LE and is qualitatively and quantitatively different (Finn and Martin 1973).

There is relatively little mitotic activity as the result of a single injection to an unprimed castrate animal. Continued injections of E_2 produce a mitotic peak on the 3rd day which falls away with continued injection. A second rise in proliferative activity can be noted 72 h after cessation of a treatment schedule which requires that at least two injections of E_2 have been given initially before withdrawing the hormone. No immediate rationalization for these events is at hand, but considerations of the interactions between mitotic cycles and secretory cycles might provide some insights.

Progesterone inhibits both the primary and secondary responses of GE to E_2 (Martin et al. 1970; Martin and Finn 1971). If progesterone is given on the same schedule as are the three injections of E_2, both the first and second mitotic waves will be arrested. If progesterone is delayed until 24 h after the last of the three E_2 injections, progesterone will stop the second mitotic wave. This is difficult to explain. By 24 h after E_2 injection, the surviving E_2 fraction should be nuclear bound and the sequelae of that binding to the genome should have been or be in the process of being expressed. Is the secondary wave of mitosis so late an effect of E_2 that progesterone can still interfere with that particular expression of R_nE?

Of all the estrogen responses elicited in GE, only the effect on mitosis is influenced by progesterone. Neither the uptake of ^3H-E_2 or ^3H-uridine are depressed or delayed by progesterone. Does this represent a form of differential response analogous to that observed in the chick oviduct (O'Malley and Means 1974)? Relative to the uterus, is estrogen required for proliferation while the combination of progesterone and estrogen is required for secretory activity of the glandular epithelium?

In terms of cell proliferation the stromal cells do not respond to estrogen alone. Only when modeling conditions are optimized and the animal is estrogen primed before progesterone is introduced into the treatment schedule will stromal cells divide (Finn and Martin 1970). Effectively progesterone redirects the mitogenic activity of E_2 to the stroma. The essential proestruslike pattern of the epithelium is suppressed by 12 h after progesterone. A predominantly stromal pattern of proliferation appears (Tachi et al. 1972). Mitosis is most marked on the 3rd day of progesterone treatment. This is a complex process to analyse. Progesterone alone has no effect on stromal mitosis (O'Grady and Bell 1977); however, a single "implanting" dose of E_2 (0.2 μg in the rat) will cause a marked increase in stromal mitosis. This is a response to estrogen that occurs independently of E priming (Martin and Finn 1968). These data are supported by observations in the rat maintained in diapause (Tachi et al. 1972). Animals placed in delay by ovariectomy (60-64 h after conception) and maintained on progesterone show no evidence of mitotic activity in either LE or GE. The same implanting dose of E_2 noted above will, within 24 h, initiate implantation of the diapaused blastocyst and also evoke marked stromal mitosis. Characteristic alterations in nucleolar appearance and function and autoradiographic evidence of DNA synthesis support these data. The mitotic index of the stroma continues to rise and represents a decidual contribution. No mitotic figures are noted in the epithelial cells.

The basic pattern of the stromal cell mitotic response to estrogen varies with the interval between E_2 priming and the start of the progesterone regimen (Finn and Martin 1970). Without priming there are no stromal mitoses, but after E_2 priming the number of cells dividing increases. The maximum number of stromal divisions occurs when progesterone treatment is initiated 3-4 days after E_2 priming.

Tachi et al. (1972) have demonstrated the complete absence of E_2-stimulated mitosis in the same population of stromal cells which take up ^3H-E_2. This suggests that a different mechanism underlies the permissive action of progesterone relative to estrogen action on the stroma and the suppressive action of progesterone relative to the action of estrogen on LE. However, this may not necessarily be the case. Alternatively, a similar mechanism can exist in both stromal and LE cells although their responses are different. For instance, the processes of glucocorticoid receptor interaction with HTC and thymocyte nuclei are similar, but the responses of these cells to the hormone are opposite (Cake and Litwack 1975).

There is, relative to the hormonal response of stromal cells, some controversy as to which stages of the cell and mitotic cycles are the most sensitive. The maximum rate of ^3H-thymidine incorporation coincides (Tachi et al. 1972) with the pattern of maximal uterine sensitivity (day 4). This is later than the peak of stromal cell proliferation which has peaked (day 3) and is declining sharply on the morning of day 4. This apparent schism between the two parameters of cell division could signify a prolongation of the G_2 phase in stromal cells immediately prior to implantation. This debatable position has been advanced by Galassi (1968) who argues that it is those stromal cells, arrested in G_2, which are responsible for the initial decidual cell reaction. This idea has been challenged by LeRoy and Galand (1969). As noted previously (Martin et al. 1973c), progesterone stimulated E_2-primed stromal cells from quiescent state (G_0) to an active state (G_1). After an accelerated passage through the cycle (G_1 shortening) the cells then withdraw from the mitotic cycle. What is the nature and function of these cells? Are they to become the primary decidual cells or do the changing titers of progesterone and estrogen alter thresholds so that a new cycle of cell arises from a base population?

It has also been noted (Tachi et al. 1972) that the implanting blastocyst initiates a further rise in DNA not seen in the pseudopregnant rats. This is interpreted to indicate that early decidual proliferation does not rely entirely on preformed DNA. These data do not rule out a significant role for a metastable stromal population lingering in G_2 (Galassi 1968; Marcus 1974). The response of these cells may precipitate recruitment of other deep and quiescent cells (attachment stimulus).

These considerations do not resolve the question as to whether cell division is to be considered one of many forms of gene expression induced by a hormone or whether cell division is a prerequisite to hormone induced differentiation. There have been few studies in mammalian physiology bearing on this question. Except for the data regarding the response of GE to the combination of progesterone and estrogen, the directive not to consider hormonally stimulated mitotic activity independent of other functions of the target cell has been set aside (Epifanova 1971). Questions then present themselves regarding nidation which are all too obvious. How closely is the LE cell which has been through two mitotic cycles related to the LE cell which has divided only once? Relative to ovum attachment which is the operative mitotic cycle? Are the cell products of the first more important than the products of the second mitotic cycle? What proportion of the epithelial population must that critical cell represent? Or are both cells from both mitotic cycles functional? Could it be the action of progesterone on the postmitotic LE cell, overlying a substratum of differentiating stromal cells, that is of importance for blastocyst-endometrial interaction? The answers are not immediately at hand.

3.2.3 Weak Versus Strong Estrogens

Additional relevant information about the capacitites of the LE cell may be forthcoming from studies of the response of the cell to E_2 compared to estriol (E_3), a weak or impeded estrogen (Clark et al. 1977b). E_2 will produce growth (hypertrophy, hyperplasia, significant net increases in weight and protein on a dry weight basis) within 24 h of a single injection. Comparatively, E_3 will not produce true uterine growth. Measured 24 h after a single injection, E_3 behaves as a typical impeded estrogen with low potency and a flat dose-response curve.

Our understanding of how a hormone acts is limited by the fact that almost all relevant information derives from models perturbated by a single or pulse injection of a hormone. Thus the rudiments of the mechanism of estrogen action in almost all mammalian systems are the products of analysis of the rodent uterotrophic model (intact immature or castrate) stimulated by a single injection of E_2. Figure 18 represents the sequelae of such an injection and its chronology. Convenience has classified them into "early" and "late" events of the uterotrophic response when, in fact, this is really a sequence, or perhaps a series of sequences, of interdependent events some of which represent the obligative pathway to true uterine growth. We cannot at the present time recognize which of these events are obligate nor do we really have a good grasp on the mechanism of response of a target to multiple or chronic dosages with a single hormone. Neither do we understand how different hormones interact on the varied targets of the uterus.

Generally, all estrogens stimulate those uterine events we would classify as "early". Thus E_2 and diethylstilbestrol (DES) (strong, long-acting estrogens) as well as E_3 and dimethylstilbestrol (DMS) (weak, short-lived estrogens) produce a transient early increase in uterine wet weight, an early synthesis of DNA, a specific "induced protein", all classes of nonhistone chromosomal protein, specific glucose utilization, and early (15-30 min) activation of uterine nuclear RNA polymerase II. However, true uterine growth is not produced by weak impeded estrogens. Weak estrogens fail to elicit later secondary rises in RNA polymerase II activity that occur at the same time as do the initial E_2 stimulated changes in template capacity, RNA polymerase I activity, and a compositional change in chromatin (significant increases in nonhistone proteins). Studies with E_3 indicate that, in the absence of this second rise in polymerase II, there are no alterations in the transcriptional capacity of chromatin, its composition, or in RNA polymerase I activity and no uterine growth occurs (Hardin et al. 1976). Correlative studies indicate that a significant concentration of nuclear receptor steroid hormone complex must remain bound to the genome for a period in excess of 6 h in order that those events (late?) which result in true uterine growth may occur. There has been much dwelling on the number 6 h. This is not a mystical number but a recurrent data point that is representative of the idea that, if these events which produce true uterine growth are to occur, the binding of the hormone receptor complex to the genome must have certain specific physicochemical characteristics. This binding is manifested in our experiments by a prolonged nuclear residency or occupancy of the receptor-hormone complex. Our present methodology does not allow us to determine whether an ordered sequence of obligatory events is specified and initiated by the nuclear interaction or whether certain of these nuclear events trigger a later process integral to uterine growth.

Since it is the pharmacodynamics of the single injection of E_3 which renders it unable to cause true uterine growth, altering the pharmacodynamics by giving E_3 more frequently or chronically increases its bioassayable potency. The dose-response curve becomes steep and the uterotrophic effect is comparable to that of E_2.

A most elegant report on multiple E_3 injections affords an excellent model for analysis (Martin et al. 1976). In this study one to five injections of 50 ng E_3 were given at 5-h intervals. Successive injections produced successive increases in uterine weight which declined within 5 h after the last injection was given. At least four injections (0,5,10, 15 h) were required in order to produce an increase in uterine weight comparable to that of E_2. No studies were done on the nuclear RNA polymerases, chromatin template capacity or chromatin composition. The number of epithelial cells in the S phase was increased by maintaining organ levels of E_3 with repeated doses but the lag between injection and response was not shortened once the cells were stimulated. Similarly, one injection of E_3 increased the mitotic index, a second injection (5 h) produced an added effect, but subsequent injections at 10, 15, and 20 h were without effect. It should be noted that whereas the second injection produced no increase in the number of cells in S phase (labeling index) it did increase the mitotic index. This analysis suggests that the second injection significantly decreased the number of dead cells which could be counted at 15 h, probably by preventing cell death in late S and G_2.

The unstimulated uterus is not necessarily quiescent. We find active RNA synthesis; as many as 4000 initiation sites/pg uterine chromatin DNA can be measured (Glasser and McCormack 1979a). There are cells in DNA synthesis, cells in mitosis and cells dying E_2 inhibits cell death by karyorrhexis long before it stimulates DNA synthesis and cell division (Martin et al. 1973b). E_3 is as effective as E_2 in reducing epithelial cell death to zero by 3 h. After E_2, LE cell death rate remains low for 21 h or more, whereas after E_3 it starts to increase at 12 h. This parallels the mitotic index so that there is no net gain in cell number. Many cells die prematurely during S or shortly after without dividing.

Estrogens, as a class, increase the rate at which cells enter the S phase. Once a cell has entered the S phase, the sequence of replication and division goes to completion unless cell death intervenes. The results of Martin et al. (1976) indicate that cell death will intervene unless certain early events are maintained. These events are considered "early" because they prevent cell death within 3-5 h of estrogen stimulation which is an earlier response than the introduction (late?) of DNA synthesis. Attempts to correlate these events do accent a possible obligatory role for a net increase in cell numbers in estrogen action; they also emphasize that insistence on categorizing events as early or late may be parochial. Furthermore, we may have to appreciate the fact that although a particular step, e.g., early DNA synthesis, arresting cell death, may be less specific in terms of a response to estrogens as a class (E_2, E_3, etc.) the response itself may be none the less essential (McCormack and Glasser 1980).

Biochemistry

4.1 DNA Synthesis

4.1.1 Introduction

It is not difficult to follow the natural history of the postconception uterus as it matures biochemically to the state at which it is receptive to a blastocyst. This progression would seem to be most efficiently ordered by a limited number of hormonally induced regulatory proteins (enzymes?). It is the differential analysis of the role of each modulating hormone that is complex. Biochemical analysis of the changes occurring during the preimplantation period must be consistent with our knowledge of the mechanisms of hormone action as they influence the course and efficiency of transçripiton and translation. It should be obvious that even tentative conclusions must factor the alteration of the target populations by cell division, endoreduplication, cell aging, cell death, and the interaction between the regulatory hormones.

In an effort to define the precise biochemical mechanisms of progesterone and estrogen action we have used two mammalian models. The uterotrophic responses of the castrate mature female have yielded much data concerning estrogen action. In the absence of specific endpoints this model has not been useful as yet for the study of progesterone. The second model has been the decidual cell response system. As we have defined this system (Glasser 1972; Glasser and Clark 1975) this has been most productive in differentiating between the individual actions and interaction of estrogen and progesterone.

The generalized protocol for such studies is presented in Fig. 14. Tissue may be harvested at any time during the preinductive period. After a properly sensitized uterine horn is stimualted with a deciduogenic agent, that tissue is analyzed separately from tissue taken from the unstimulated or control horn. It is critically important that the control tissue for any particular study be taken from either the unstimulated contralateral horn or the interimplantation sites of the same animal. Nonpregnant animals or animals not subjected to exactly the same hormonal regimen are not appropriate primary controls for any of these studies.

Historically, biochemical analyses of uterine tissue have concentrated on peri- and postimplantation events. There has been less work on the preimplantation uterus and those general observations that have been recorded, have been obtained from the whole unfractionated uterus. The absence of data relating biochemical events to individual populations of endometrial cells makes the interpretation of hormone action in terms of cell cycle phenomena very difficult. Most recently with the development of reason-

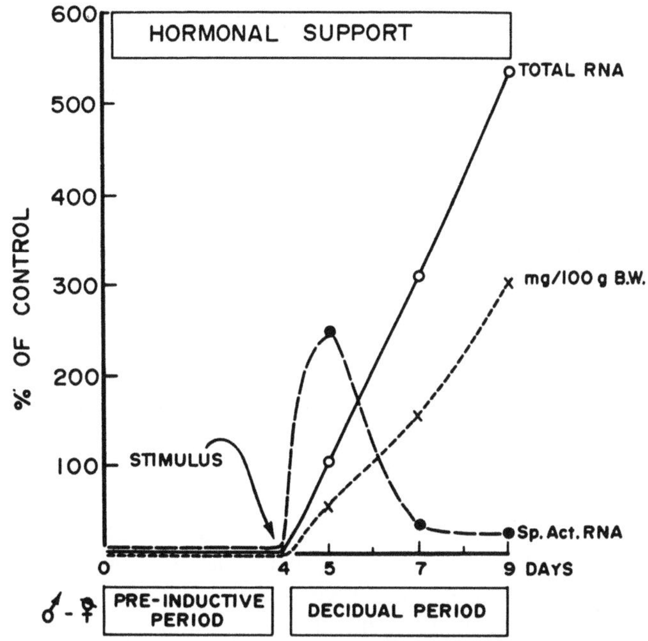

Fig. 14. A generalized protocol developed for studies on the mechanism of progesterone action. These data represent the response of the progesterone (2 mg/day) sensitized uterus to a deciduogenic stimulus. They are prototypic in a qualitative sense for other active progestogens with or without estrogen. See Fig. 15. (Glasser 1972; Glasser and Clark 1975)

ably rapid and reliable methods for fractionating endometrial cell populations there have been systematic studies of the protein and nucleic acid metabolism of these individual cell types (Heald 1976; O'Grady and Bell 1977; McCormack and Glasser 1980; Glasser and McCormack 1981).

4.1.2 DNA and Hormone Interaction

Hormonally induced alterations in target cell nucleic acids transduce changes in the functions of these cells. Such regulated changes which may include functional differentiation of the cell and/or cell proliferation are directed by specific messenger RNAs transcribed of DNA templates. Compared with our knowledge of the role of various RNA species, relatively little has been done to elucidate the influence estrogen and/or progesterone may have on DNA composition, structure and function. The fundamentals of hormone-receptor binding to DNA, hormone effects on chromatin composition and conformation, and hormone receptor interaction with chromosomal proteins have been reviewed by O'Malley and Means (1974).

The extent to which DNA participates in the binding of hormone receptor complexes to chromatin varies somewhat depending on the hormone and has yet to be adequately explained. From studies with the chick oviduct we may expect that the region of the genome to which estrogen receptors bind is apparently more susceptible to digestion by DNase than corresponding regions to which progesterone receptors bind (Buller et al. 1975). It has also been demonstrated that chick oviduct progesterone receptors do not bind to single strand DNA or double strand regions accessible to digestion by DNase I. These data are in contradiction to Crick's general model for gene control which requires that regulatory macromolecules interact with single strand regions of the genome (Crick 1971).

It seems difficult for DNA per se to be the sole determinant of receptor binding to the genome. Evidence in support of the concept that chromosomal proteins, particularly nonhistone (acidic) proteins, play an active role in receptor interactions with the genome is discussed by Buller and O'Malley (1976). They advance a rationale to support the suggestion that nonhistone proteins (NHPs) are a logical choice to confer acceptor site specificity. The selection in favor of nonhistone proteins (*vs* histones, chromosomal RNA) is based on these criteria: (1) NHPs represent a heterogeneous class resolved by size, charge, and immunological properties, (2) NHPs turn over rapidly, (3) they possess tissue specificity, (4) NHPs demonstrate specific changes in response to hormonal perturbation (Teng and Hamilton 1968; Shelton and Allfrey 1970; Glasser et al. 1972a), and (5) binding capacity and specificity of target tissue chromatin can be transferred to nontarget chromatin by the substitution of a particular fraction of target tissue NHP (Spelsberg et al. 1972). Thus, it would appear that actual hormone-receptor acceptor sites on the genome consist of a DNA backbone structurally modified by chromatin-associated NHP.

After $^3H\text{-}E_2$ is taken up by the uterine cell and is translocated in conjunction with its specific receptor binding protein to the nucleus, the amount of $^3H\text{-}E_2$ recovered per endometrial nucleus is 40% greater than the amount bound to myometrial nuclei (Leake et al. 1975). This difference is reflected not only in short-term responses of the endometrium to E_2 (increased activity of RNA polymerase, synthesis of "induced protein" and enhanced glucose transport) but also in the longer-term response of DNA synthesis (Bennett and Leake,cited according to Leake et al. 1975). These data suggest a direct connection between entry of the E_2 receptor complex into the nucleus and the extent of subsequent DNA synthesis.

4.1.3 Hormone Influence on DNA Synthesis

The first peak of DNA synthesis in the immature rat is noted 24 h after exposure to E_2 (Kaye et al. 1972) and occurs simultaneously in both endometrium and myometrium (Bennett and Leake, cited according to Leake et al. 1975). Therefore, with respect to DNA synthesis there may be quantitative differences in the E_2 response of different target cells but the qualitative responses are similar.

The synthesis of DNA is viewed by Leake et al. (1975) as a "late" response relative to nuclear binding of E_2. Is the presence of a bound hormone receptor complex required over the entire interval between uptake and response? Are there any intermediates required for the response? Using the estradiol exchange assay to differentiate binding and response characteristics of E_2 and E_3, Bennett and Leake (cited according to Leake et al. 1975) concluded that, although the "late" (18-h) peak of glucose transport required the continued presence of the bound hormone-receptor complex, stimulation of DNA synthesis by estrogens did not require the continued occupancy of the complex. E_2 stimulation of DNA synthesis is initiated shortly after estrogen enters the nucleus. Maintenance of synthesis throughout the period is assigned to an intermediate that is synthesized early. The putative intermediate is supposedly a protein that is constantly turned over. The mechanism by which the intermediate initiates DNA synthesis is not known but studies involving estrogen influence on DNA polymerases are in progress (Leake et al. 1975).

How do estrogens stimulate DNA synthesis? Eukaryotic DNA molecules consist of replicative units arranged end to end, from the middle of which DNA synthesis proceeds bidirectionally at divergent fork growing points until fusion occurs with adjacent sections. Chain growth rate and the mean distance between initiation points prove to be cell specific. It can be inferred from this that the estrogen-stimulated shortening of the S phase in uterine epithelial cells is due to a speed up of replication along preexisting units or to multiplication of stimultaneous replicative sections or both.

Autoradiography of DNA fibers confirm that E_2 increases the rate of DNA fork replication (growing rate). The tentative estimate of a 50% increase in fork growth rate accounts for a 33.3% shortening of the S phase (Leroy et al. 1975). This value is to be compared to the proposals of other investigators who estimate the shortening of the S phase in E_2-stimulated epithelial cells at 30%-60% (Das 1972; Eide 1975). The authors propose that their corrected true fork growth rates of 0.3 μm/min for controls and 0.45 μm/min for E_2-treated animals fall within the range observed for other mammalian cells (Callan 1973).

4.1.4 Endometrial DNA Synthesis in Pregnancy and Pseudopregnancy

Autoradiographic analysis of DNA synthesis can be related to individual cell types and to the known changes in cell cycle occurring during the preimplantation period. In vivo incorporation of ^{3}H-thymidine into the total DNA of pregnant uteri disclosed patterns that correlated with mitotic activity of LE and GE (Tachi et al. 1972). Qualitative confirmation of these conclusions can be obtained from the work of Galassi (1968) and Leroy and Galand (1969). The technical features of their work do not permit a more critical correlation. Thus a peak in incorporation could be noted on day 1 which declined along with mitotic activity on day 2. A second peak in incorporation was related to increased mitotic activity in stromal cells on day 3 (Tachi et al. 1972; Marcus 1974; O'Grady and Bell 1977). Subsequent to implantation the concurrence of mitosis and polyploidy diminish the usefulness of this method. There is a rapid increase in DNA synthesis at implantation sites and in decidual tissue that is not observed in intersite samples, the unstimulated horn of the pseudopregnant animal, or the sterile horn of the pregnant rat.

A more detailed analysis of these events has been provided by O'Grady and Heald (1976), who fractionated the uterine cell population after the incorporation of a 1-h pulse of ^{3}H-thymidine in vivo. The isotope was given at 6-h intervals to pregnant rats which were unilaterally ovariectomized presumably to provide a sterile pseudopregnant horn as a control unit. (Why not simply ligate that oviduct, providing two functional ovaries rather than a single ovary undergoing compensatory hypertrophy?) The elevated incorporation in LE on days 1 and 2 was a reflection of the mitotic activity of these cells. The "stromal" fraction also incorporated increased levels of ^{3}H-thymidine on day 1 which was probably a reflection of the contamination of these cells with endometrial glands. In both pregnant and pseudopregnant horns elevated incorporation into stromal cells was noted at 0900 hours on day 3 and was maintained at the 1500 hours measurement. Synthesis in other fractions was low at this time as was the mitotic index. This is the generalized picture obtained by analysis of the total, unfractionated uterus (Tachi and Tachi 1975).

Stromal mitoses and ^{3}H-thymidine incorporation decrease and remain at low levels throughout day 4. The peak of stromal mitosis on day 3 is believed to be related to a regional increase in DNA synthesis in the primary decidual zone. It has been suggested (Leroy et al. 1974) that a proportion of the stromal cells pass through G_1 and S and are arrested in G_2 in a tetraploid state prior to the initiation of deciduoma-related DNA synthesis. The arrest of these cells in G_2 is proffered as an explanation for the low rate of DNA synthesis noted early (0300-0900 h) on day 4.

Following attachment of the blastocyst at about 1200 hours on day 4 (Tachi et al. 1972) and the onset of decidualization, there is an increase in DNA synthesis noted at 1500 h on day 4. This elevation of DNA synthesis is representative of the DNA synthesis associated with the differentiation of the SEAM stroma to decidual cells. It is characteristic of this particular process for there is no appreciable synthesis in the sterile horn.

The castrate animal maintained on progesterone alone allows certain insights into the manner in which that hormone modulates DNA. In mice (Martin et al. 1973b) autoradiographic analysis of uteri from progesterone-treated animals showed that DNA synthesis of GE cells was depressed whether the animal was primed with estrogen or not. DNA synthesis in LE cells was also depressed but the presence of estrogen enhanced reentry of some cells into S. Progesterone alters the nuclear morphology of stromal cells. An increased number of cells leave the R_1 state to reenter the cell cycle and the population synthesizing DNA is increased. If estrogen is given in addition to progesterone, 40% of the stromal cells will enter S within 10-15 h. A single dose of estrogen accelerates the passage of stromal cells through a single round of cell division, shortening the interval between M and S. The cells then withdraw from the cycle. There is no response to a second estrogen injection.

Castrate adult rats maintained on progesterone alone will respond to artificially stimulated decidualization with a 100% increase in total uterine DNA by 24 h. There is a gradual increase over the next 48 h to 150% which is significantly enhanced at the 72-h point (300%) if 1 μg estrone is combined with the 2 mg daily dose of progesterone (Fig. 15).

4.2 Uterine Chromatin Template Activity

4.2.1 General Principles

Changes in protein synthesis coincident with hormone mediated differentiation and growth appear to be mediated, to a large extent, by the nature and degree of DNA transcription. It has been noted that observed increases in hormone-induced protein synthesis are preceded by marked changes in the level of nuclear RNA synthesis (Means and Hamilton 1966), specific RNA polymerase activities (Glasser et al. 1972a), and chromatin template activity (Teng and Hamilton 1968; Glasser et al. 1972b).

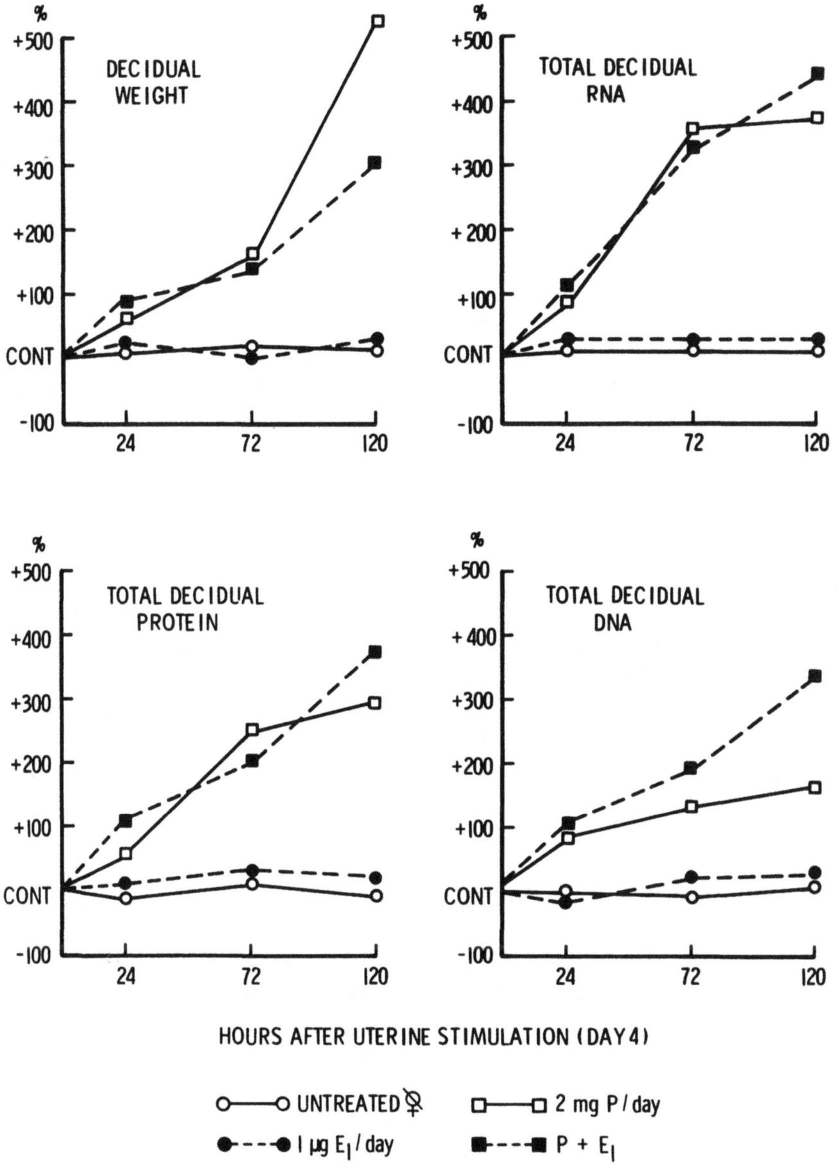

Fig. 15. Relative changes in some of the parameters used to monitor decidual growth in mature ovariectomized rats maintained on either 1 µg estrone/day (E_1), 2 mg progesterone/day (*P*) or a combination of these hormones ($P+E_1$). One group of rats was left untreated. One horn of the uterus was stimulated on day 4. The contralateral horn served as an internal nonstimulated control. These data demonstrate that when using appropriate controls, there are no significant differences in responses when P is compared to P+E during the first 72 postinductive hours. Thus both Fig. 14 and 15 show effective decidualization can be induced with P alone. There is a general growth response to E which occurs equally in both horns but does not contribute to decidual growth. (Glasser 1972; Glasser and Clark 1975)

The measurement of total template activity assesses the ability of isolated chromatin to serve as a template for the exogenous RNA polymerase used in the test system. The assay is designed to measure the percentage of DNA sequences in chromatin which are available for transcription in vivo. Practically the assay provides, at best, only a rough estimate of DNA availability due to the complexity of the reaction being measured. For this reason measurements of total template capacity should be used only for comparing the relative activities of different chromatin samples.

Exact quantitation of template activity is very sensitive to small changes in reaction conditions, i.e., ionic strength, purity of the RNA polymerase, etc. One must exercise extreme care in the control of all reaction conditions when comparing chromatin samples from different sources. Preferentially these measurements should be done simultaneously. Methods for the analysis of chromatin template activity are discussed comprehensively and critically by Tsai M-J et al. (1976). Specific modifications used in our own work with the uterus are cited in the following papers (Glasser et al. 1972a, Steggles et al. 1971).

The major problem in the determination of template activity is the limited solubility of chromatin. Preparation of chromatin of a quality which can be used in the assay, either directly from whole uterus or from isolated uterine nuclei, is difficult because of the myofibrillar matrix of the organ which interferes with yield, extractability, and purification. The assay for template capacity itself is complicated by the fact that the concentrations of metal ions and the ionic strengths that are optimal for RNA synthesis are very effective in precipitating chromatin. It is obvious that less RNA can be transcribed from insoluble than from soluble chromatin. While shearing chromatin by sonication does increase total RNA synthesis, the risk of altering the nature of the RNA product transcribed is great due to erroneous initiation and termination at points of DNA strand breakage.

4.2.2 Hormonal Effects on Chromatin

The first data regarding changes in uterine chromatin template capacity were developed by Teng and Hamilton (1968) who reported a 25% increase in transcriptive capacity within 30 min of treating adult ovariectomized rats with E_2. A similar response was noted for endometrial chromatin of castrate rabbits within 10 min (Church and McCarthy 1970). Because of the pitfalls regarding the measurement of total template capacity noted above, we decided to restudy the problem of estrogen influence on chromatin template capacity. The chemical composition of the chromatin used in the assay was carefully monitored.

The first significant elevation in total template activity occurs 1 h after E_2 is intravenously injected into castrate female rats (Glasser et al.1972b). The response curve is statistically bimodal (Fig. 16d), but there is no direct evidence that the peak differences between 1 and 4 h are of biological importance. It should be noted that the peak at 1 h occurs after the initial rise in RNA polymerase II activity has fallen and as the rise in RNA polymerase I activity begins. During this same period there are no compositional changes (*vs* untreated controls) in uterine chromatin (Fig. 16b). Template capacity is still elevated at 4 h after E_2. The possibility that products of transcription may be different at various times after estrogen stimulation, i.e., 30 min *vs* 4 h, is suggested by

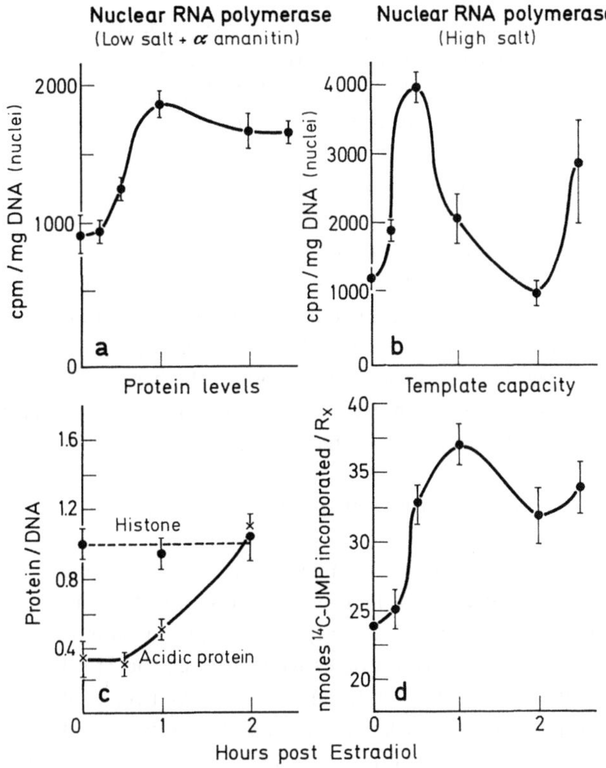

Fig. 16 a-d. The short-term response of ovariectomized mature rats to a single 1-μg intravenous injection of estradiol-17β. Panels **a, b,** and **d** restate Fig. 10. Of particular interest here is the increase, both absolute and relative, in nonhistone (acidic) chromosomal proteins. This significant response to E_2 occurs after the first hour and may be related to the increases in polymerase I, chromatin template activity, and the secondary rise in RNA polymerase II (high salt) of uterine nuclei. (Glasser et al. 1972a)

concurrent elevations in the activity of RNA polymerase I and II and a significant alteration in chromatin composition at the later time. At the 4-h postinjection period the ratio of nonhistone (acidic) protein (NHP) to DNA has shifted from the control value of 0.38 to 1.57. It is important to note the fact that both the hormone-stimulated chromatin template activity and the changes in chromatin composition are not sustained but have returned to near control levels by 12 h after E_2. At this time we have no data regarding these responses to a second injection of E_2 given either at 12 or at 24 h after the first dose.

SDS gel electrophoretic analysis of ovariectomized mouse epithelium and stroma showed that all classes of NHP were stimulated by estrogen (Pollard and Martin 1975). NHP are considered to play important roles in hormone receptor complex binding to chromatin and in cell division (Tsuboi and Baserga 1972). Increased synthesis of NHP precedes entry of cells into S but NHP synthesis can also be substantially enhanced in cell populations where proliferative activity is not altered (Enea and Allfrey 1973). Subsequent analysis by isoelectric focusing showed that pretreatment of the castrate mouse with progesterone increased the proportion of basic NHP in uterine epithelium within 4 h (Pollard and Martin 1975). This effect was unaltered after estrogen treatment. Estrogen increased synthesis of all classes of NHP. The level of basic NHPs is higher in the stroma than in the epithelium of untreated castrate mice. Progesterone pretreatment did not increase stromal synthesis of basic NHP per se but arrested the synthesis of the more acidic NHPs which are produced following estrogen stimulation. These changes do not correlate with the proliferative status of the tissue since they also occur in epithelium tissue in which progesterone suppresses the mitogenic action of estrogen as well as in stroma where cell division is enhanced by the same combination of hormones (Pollard and Martin 1975). The important concept that emerges is that, by altering the structure of chromatin through the NHPs, estrogen can redirect gene expression. Furthermore, progesterone, through its action on the basic NHPs, can redirect estrogen action by changing the accessibility of different regions on chromatin to the estrogen receptor complex or the progesterone receptor complex.

4.2.3 Template Activity During Early Pregnancy

These general considerations might make it productive to extend the analysis of chromatin template capacity to the uterine chromatin of animals during the early stages of pregnancy. The transcriptive capacity of chromatin isolated from uterine nuclei of the pregnant rat uterus increases progressively through the 3rd day of pregnancy (Fig. 17). There is a twofold increase between days 3 and 4 (Glasser et al. 1972b; Glasser and Clark 1975). Following attachment (day 4), the template activity of implantation site chromatin remains elevated whereas intersite tissue is markedly repressed (O'Grady et al. 1975). The greater mass of intersite tissue accounts for the drop to basal levels in the measurement of template capacity of chromatin prepared from the whole pregnant uterus on day 5. It is noteworthy that the pattern of template activity of chromatin from ovariectomized pregnant rats maintained on progesterone alone is identical to that of the intact pregnant rat with the exception that, even in the absence of uterine stimulation on day 4, the capacity to transcribe remains elevated. Artificial stimulation to decidualize on day 4 would evoke even greated template activity (*vs* nonstimulated) but the important point to note is that the template activity of the "diapaused" uterus (maintained on progesterone alone) remains unrestricted or "open". Template activity in these animals remains significantly elevated over other treatment groups for at least 10 days.

These events should be contrasted with the response of rats maintained on estrogen only (a continuous depression of activity after the initial transient rise!) and with that of animals maintained on 1 μg estrone and 2000 μg progesterone per day. This combination of hormones will maintain pregnancy, allow the maximum decidual cell response to traumatization and the normal temporal and spatial expression of uterine cell proliferation. Yet the pattern of template activity in these animals is not a faithful copy of that of the intact pregnant animal. The peak of template activity is greater and is shifted 24 h so that it occurs on day 3 rather than day 4. Template activity falls between days 3 and 4 and plateaus on day 5 at a level that indicates that a greater proportion of the template of this uterus is available for transcription than in the day 5 pregnant uterus. Regionally, the implantation site chromatin of intact pregnant rats has much greater template compared to intersite tissue so that it is difficult to compare the models. The

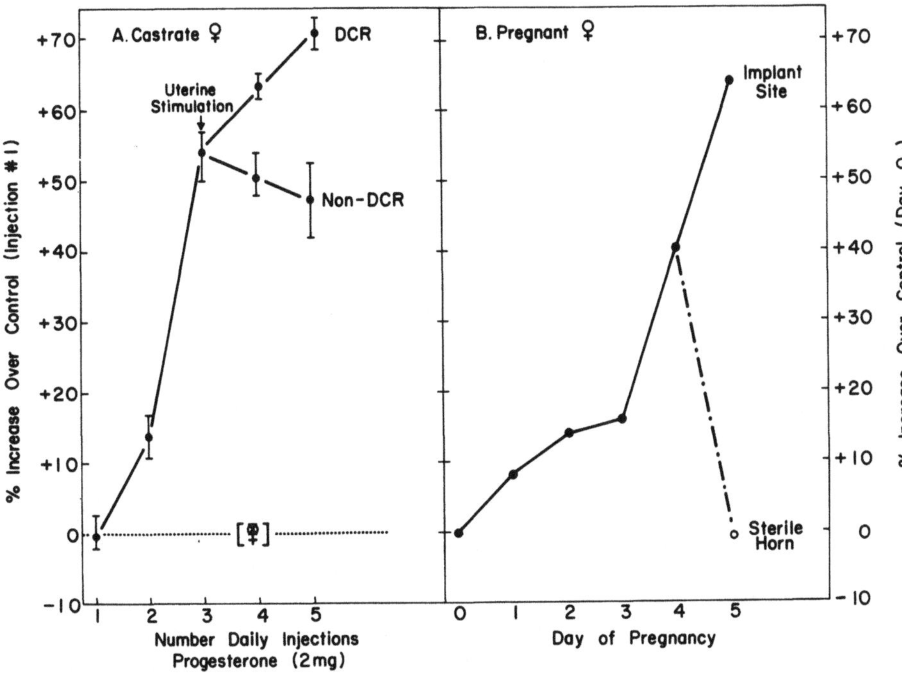

Fig. 17. A. The proportion of the uterine genome available for transcription increases with increasing numbers of progesterone injection to an ovariectomized rat. The template capacity of uterine chromatin increases through the third injection. If the uterus is not stimulated at that time the transcriptional capacity of the chromatin decreases (*non-DCR*). However, if the uterus is stimulated (*DCR*) that tissue (not mesometrial) yields chromatin of even greater capacity for transcription. **B.** Template capacity of uterine chromatin isolated from pregnant uteri. (Day 0 is the day of vaginal sperm). Capacity for transcription increases steadily through day 3. Note precipitous increase between days 3 and 4. If implantation does not occur the capacity for transcription of chromatin from sterile (horn) sites is diminished sharply. Template capacity was assayed daily with purified *E. coli* RNA polymerase. Under conditions of the assay template (chromatin DNA) is rate limiting and the incorporation of radioactivity into and acid-soluble product is linear. (Glasser et al. 1972b)

dissimilarity between the pregnant animal and its model (castrate plus progesterone and estrone) should be remarked. These simulated conditions are not then really analogous to the hormonal environment of the preimplantation stages of pregnancy. Furthermore, the differential responses of estrogen and progesterone suggest that estrogen may, in these circumstances (*vs* the uterotrophic model), restrict gene expression altering and eventually terminating uterine sensitivity.

4.2.5 RNA Chain Initiation Site Assay

As we have noted above, while measurement of chromatin template activity generally reflects the amount of DNA sequences available to RNA polymerase, the components of the reaction are so complex that the interpretation of the data is very difficult and results must be generalized. The events that may be involved in determining changes in template activity are: (1) RNA polymerase binding to chromatin; (2) RNA polymerase searching for initiation sites; (3) the formation of stable RNA polymerase – DNA preinitiation complexes; (4) the initiation of RNA chain synthesis; (5) RNA chain propagation rate; (6) the termination and release of the RNA chain; and (7) the reinitiation of the transcription process. Thus the gross measurement of chromatin template activity can be influenced by a change in any one of these components thereby limiting the discriminitive use of this assay in the analysis of hormone action.

The inability to distinguish between the number of available RNA polymerase initiation sites and the rate of RNA chain elongation posed a major block to our understanding of steroid-directed gene expression. Techniques adapted from bacterial and bacteriophage systems have been successfully developed in laboratories of this department which allow the direct quantitative measurement of both the polymerase initiation sites and the RNA chain propagation rate (Tsai M-J. et al. 1975; Tsai SY et al. 1975; Schwartz et al. 1975). The method is based on the reversible and random binding of *E. coli* RNA polymerase to DNA. After the enzyme is bound to a promoter site, there is a transition to a stable complex which forms only under specific conditions. These complexes are probably specific in prokaryotic systems since they reportedly form at genuine promoter sites. Formation of the stable complex involves conformational changes in both the DNA promoter site and the enzyme.

Use of the antibiotic rifampicin has allowed the initiation process to be elucidated to the point where it could be used as an assay. Rifampicin inhibits RNA synthesis by binding to the RNA polymerase. The antibiotic is inhibiting to RNA polymerase only prior to RNA chain initiation; after the formation of the first phosphodiester bond, elongation of nascent RNA chains is insensitive to rifampicin. Therefore, it is possible to determine that fraction of RNA polymerase that exists in a stable binary complex with DNA. This stable complex can initiate RNA synthesis and can be determined by the assay (Fig. 18). That RNA polymerase which is free or randomly bound (nicks, single stranded) is inhibited by rifampicin as are all secondary initiations. The technique for the assay of RNA chain initiation sites in DNA or chromatin is outlined in Fig. 18a. Details may be obtained from the following references: Tsai M-J et al. (1975), Schwartz et al. (1975). The bottom panel of Fig. 18 is a schematic representation of the steps of RNA chain initiation from open preinitiation complexes.

The assay of RNA chain initiation sites was utilized to show that estrogen increased the number of RNA polymerase binding and initiation sites in chick oviduct chromatin without changing the elongation rate or the size of the average product (Schwartz et al. 1975). Withdrawal of estrogen caused a reduction in the number of initiation sites. This reduction in the transcriptive capacity of chromatin was correlated with a reduction in nuclear estrogen receptor complexes (Tsai et al. 1976). These parameters remained correlated during reinitiation of hormone treatment and suggested that the hormonal receptor complexes act on chromatin to mediate these changes in genetic transcription activity. Further analysis showed that chromatin proteins are involved in the selection of initiation sites on chromatin for RNA polymerase. The efficiency with which a specific gene is transcribed appears to depend on the interaction between RNA polymerase and the chromosomal elements in the region of initiation (Tsai M-J et al. 1975).

The working hypothesis constructed from such data stipulates that steroid hormone action involves the interaction of steroid hormone receptor complexes with acceptor

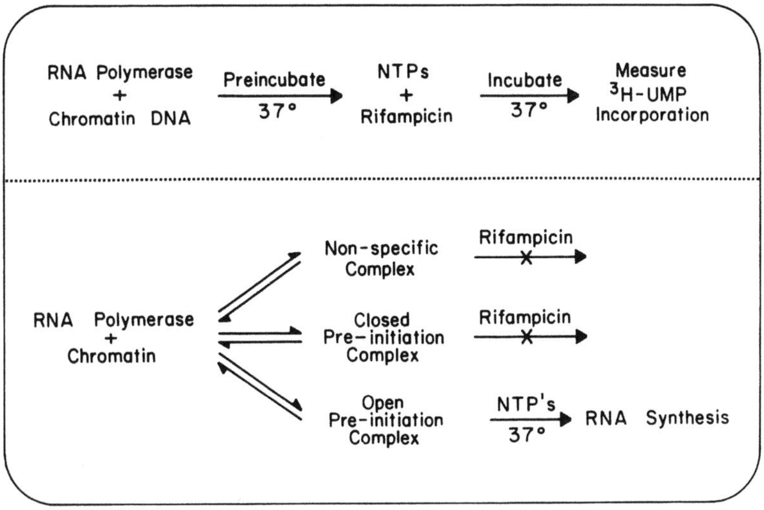

A

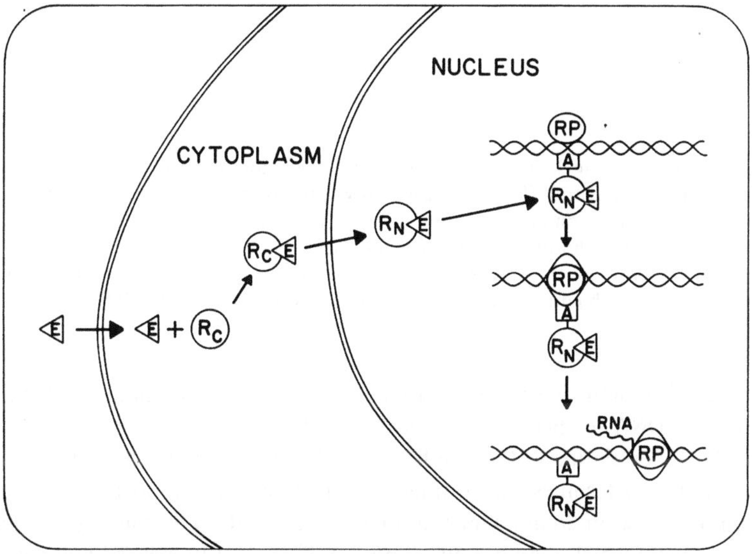

B

Fig. 18. A. Scheme of the rifampicin challenge transcription assay. RNA polymerase and template (uterine chromatin DNA) are preincubated for a period of time sufficient to allow formation of stable preinitiation complexes. RNA synthesis is initiated by the addition of 4 nucleotide triphosphates (NTPs) in the presence of rifampicin (40 μg/ml). The enzyme is titrated against a fixed amount of chromatin. Under conditions in which chromatin is saturated with enzyme the number of RNA chains synthesized proves equivalent to the available initiation sites. The synthesis of RNA from open stable preinitiation complexes is very efficient at this inhibitor concentration. No synthesis proceeds from nonspecific or closed preinitiation complexes. Chain elongation and secondary initiations are strongly inhibited. Incorporation of labeled NTP into the RNA product will allow the transcriptive activity to be measured and expressed as RNA initiation sites/pg chromatin DNA. (See Tsai M-J et al. 1975; Schwartz et al. 1975, for assay details and discussion). **B** Hypothetical scheme for gene activation by the receptor estrogen complex. The complex (*RnE*) binds to an acceptor protein (*A*) on DNA. The binding of the complex to a specific binding site on DNA results in the "melting out" of the DNA strands and opens the gene for transcription by RNA polymerase

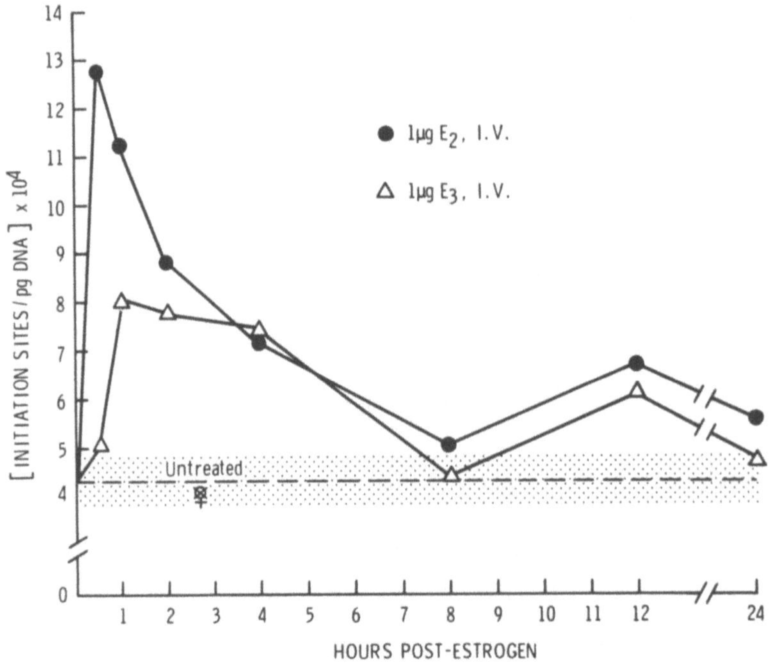

Fig. 19. Initiation site data obtained from adult ovariectomized rats which received 1 μg estradiol-17β *(E_2/●)* or 1 μg estriol *(E_3/△)* intravenously. Groups of animals were killed at various times after injection and chromatin prepared from their uteri. The data (0-2 h) is qualitatively similar to that for nuclear receptor hormone complex binding and may be accepted as evidence of hormonally induced transcription. Differences between initiation site, polymerase activity, and nuclear receptor hormone complex binding data may be resolved by further work with RNA product analysis

sites on chromatin. This leads to an increase in the number of available initiation sequences in the DNA template and the synthesis of new RNA sequences. The data describing the direct correlation in the number of nuclear bound receptor molecules and polymerase initiation sites in estrogen-stimulated chick oviduct supports this thesis. The steroid hormone receptor complex may influence transcription by causing a structural change in chromatin which then permits exposure of previously masked polymerase binding and initiation sites. Following destabilization of the DNA duplex at predetermined induction sites, RNA polymerases are then able to saturate the newly exposed sites resulting in the synthesis of steroid-induced gene transcripts. The presence of steroid receptor is required to maintain this increased number of RNA chain initiation sites on the chromatin.

4.2.5 Hormonal Effects on RNA Chain Initiation

The work on the chick oviduct must be regarded as prototypic. It is not known if E_2-stimulated uterine chromatin transcription is due to an increase in the number of available initiation sites on chromatin or to an increase in the rate of chain elongation. It is not adequate to extrapolate chick oviduct data. Understanding the mechanism of steroid hormone action in gene expression in uterine target tissue requires the quantitation of these processes. As is customary in our laboratory, our first attempt to resolve this question about E_2-induced transcription was to apply the rifampicin challenge assay to the study of the uterotrophic response.

Utilizing this method and calculating the number of available initiation sites on chromatin it was estimated that E_2 (1 μg, i.v.) stimulated RNA chain initiations almost 200% within 1 h (*vs* saline-injected castrate controls). Consistent with previously reported template activities, there was a progressive reduction to 50% between 4 and 24 h in the number of available initiation sites. E_3 (1 μg, i.v.) also stimulated the number of initiated chains (80%-110%) during the 1st hour before falling directly to a plateau (45%-55% of the peak) at 8 h. Subsequent fall to baseline levels was much more rapid in E_3-treated uteri (Glasser et al. 1976).

A consistent technical problem is the preparation of compositionally acceptable chromatin that, in the final preparative step, will undergo hydration, gelation, and swelling. For this reason we have validated the above data by alternative methods. Newly synthesized transcripts were labeled at their 5' ends with $\gamma^{32}P$ (Schwartz et al. 1975). Separation was accomplished on PEI chromatography plates. The new transcripts were scraped from the plates and counted. Estradiol-stimulated synthesis, measured at 1 h, was significantly greater (0.94 pmol $\gamma^{32}P$ incorporated/pg chromatin DNA) than saline (0.28 pmol $\gamma^{32}P$/pg DNA) and E_3-stimulated synthesis (0.61 pmoles/pg DNA). These reproducible increases are consistent with the data obtained with the rifampicin challenge assay. In no case could we obtain evidence of a hormonally stimulated increase in chain elongation. Validation, by two separate methods, of the increase in the number of RNA chain initiation sites during E_2 stimulation confirms the hypothesis that this hormone exercises its regulatory control over uterine target cells at the level of transcription (*vs* translation). These data are also consistent with the thesis that obligatory pathways to true uterine growth are closely correlated with prolonged nuclear retention of the receptor hormone complex.

4.2.6 RNA Chain Initiation and Uterine Sensitivity

The utility of the rifampicin assay recommended itself to us as a means of reinforcing the data establishing progesterone as a major determinant in the development of uterine sensitivity to decidualization (Glasser and Clark 1975). It has been shown that progesterone stimulates the synthesis of specific RNAs and subsequently of specific proteins in the uterus of a castrate pregnant rat maintained on progesterone alone. The products of these syntheses are thought to be responsible for the competence of the free blastocyst to differentiate embryonic derivatives after nidation, the developmental inhibition of the blastocyst prior to implantation and, after a minimum of 48 h, the development of uterine sensitivity to the deciduogenic stimuli. The dependency of uterine sensitivity on the progesterone-stimulated synthesis of unique or specifically stimulated RNAs and proteins has been proven. However, we were unable to identify in these studies a specific progesterone inducible inhibitor as had been suggested by Psychoyos (1969). At that time we felt that the idea of an inhibitor was still viable but we did not have an appropriate endpoint.

An environment dominated by progesterone alone will not sustain a blastocyst-endometrial interaction. Estrogen, as we have indicated, also stimulates the synthesis of specific RNAs and proteins. These estrogen-produced metabolites are directed toward uterine growth, but they, by themselves, are not compatible with decidualization or nidation. At first we pictured estrogen modulating the synthetic processes of the progesterone-dominated (sensitive) uterus thereby altering the threshold conditions required for implantation. Completing this hormonal sequence leads to the evolution of uterine receptivity to ovoimplantation and the related but delayed restriction and loss of uterine sensitivity (Glasser and Clark 1975).

Our subsequent experiments were oriented toward the elucidation of those biochemical processes by which the sensitive uterus becomes a receptive one. We addressed the question of what specific message or information was estrogen inducing that rendered nidation possible. After many unsuccessful attempts at isolating a unique biochemical message that specified attachment and implantation, we concluded that perhaps we were trapped by the semantics of hormone induction. We therefore altered the question to address the loss of uterine sensitivity rather than the attainment of ovum receptivity. We intended to apply our methodology for the study of gene expression, including the rifampicin challenge assay, to the analysis of loss of sensitivity of the uterus with the bias that this was an exercise in restrictive biology imposed by viviparity rather than a further estradiol-induced stimulation of the synthesis of new informational molecules.

The protocol we utilized was one suggested by Meyers (1970) and adapted by Psychoyos (1973b). Rats were ovariectomized 7-10 days prior to their first injection with progesterone (2 mg/day). We have continued this regimen as long as 20 days but at any time after the third injection (Px3) the uterus can be successfully decidualized by artificial stimuli. We have used a variety of deciduogenic agents but in these studies, for purposes of overstatement, uteri were traumatized by needle scratch. An untraumatized contralateral horn served as an internal control.

Loss of uterine sensitivity was initiated by injecting 0.2 μg E_2. Usually this was done at the same time as the third progesterone injection (0900 hours). From that point, one horn of the uterus was traumatized at 4-h intervals during the course of the experiment. Daily injections of progesterone were administered through the end of the experiment (day 7). The time course of the loss of uterine sensitivity, as measured by the decidual cell response (DCR), is depicted in Fig. 20. Data are expressed as a percentage function calculated as the weight of the traumatized or decidualized horn (*D*) less the weight of its contralateral control (*C*) divided by the weight of the control horn, i.e. $\frac{D-C}{C} \times 100 = \%$ DCR response or uterine sensitivity.

Following the injection of E_2, it is fully 24 h before there is any evidence of diminution in the decidual cell response. Uterine sensitivity is sustained for at least 24 h and then falls precipitously even though a fourth (Px4) injection of progesterone is given. Subsequent progesterone injections do not reverse this change.

An alternative method to produce loss of uterine sensitivity is to withdraw progesterone. When the uterus is challenged just prior to the fourth progesterone injection (24 h after Px3) it appears to be fully sensitive. However, if the fourth injection (Px4) is missed then uterine sensitivity is rapidly lost in the next 4-8 h (28-32 h after Px3). Our results with animals maintained continuously on daily injections of progesterone had shown that there was no significant decrease in sensitivity during the 24 h following an

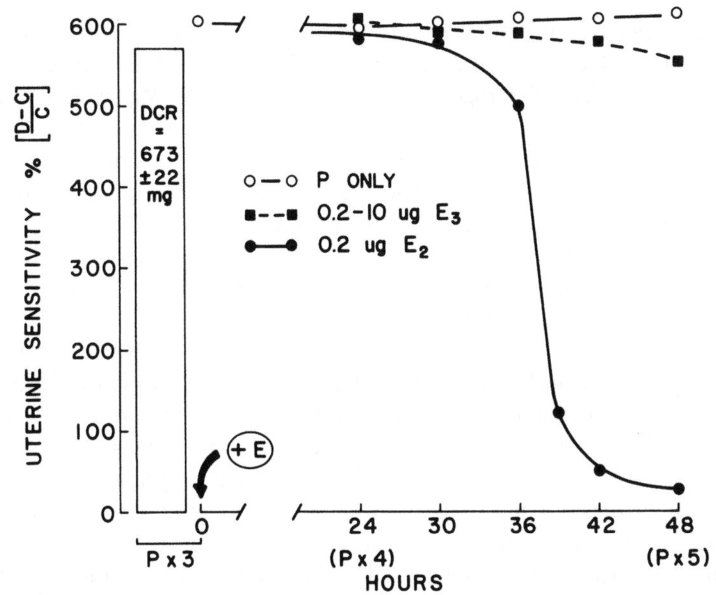

Fig. 20. Loss of uterine sensitivity (DCR). The progesterone sensitized uterus (*P x 3*) maintains its sensitivity to respond fully to deciduogenic stimuli through the 5th (*Px 5*) to at least the 10th day (P x 10, not shown). Estriol (E_3) in single doses of 0.2-10 μg will not alter this response pattern. Estradiol-17β (E_2) in doses as low as 0.2 μg, subcutaneously, will cause the loss of uterine sensitivity if given any time after P x 3. Regression of response is not apparent until the 30th hour after E_2 after which loss of sensitivity is rapid. Uterine sensitivity is expressed in terms of the weight of the decidual response (673 ± 22 mg or 585% greater than internal control at P x 3) which is measured 120 h after the uterus is sensitized. (Glasser and McCormack 1979a)

injection, i.e., during the interval between Px3 and Px4 or Px8 and Px9. Therefore, the extremely rapid loss of uterine sensitivity when the Px4 injection was withheld at 0900 hours on that day was surprising. A more gradual decline had been anticipated. The mechanisms underlying this change will have to be investigated.

The response to P withdrawal appears to be an earlier change, even with reference to the time of the Px3 injection, than does the response to E_2. The net result is the same loss of sensitivity to DCR but, whereas the progressive loss of uterine sensitivity in E_2-treated animals is at least coincident with, if not correlated with, ovoimplantation, the loss following progesterone withdrawal is associated with uterine alterations that will not support nidation.

4.3 RNA Synthesis

4.3.1 Early Studies

Whatever knowledge we now possess regarding the regulation of nucleic acid and protein metabolism of the pregnant uterus has its origins in the laboratory of Mueller et al. (1958) who almost 20 years ago demonstrated that estrogen stimulated the incorporation of radioactive precursors into total uterine cell RNA. This response was measured in hours and led to a controversy about whether the stimulation of RNA synthesis was a cause or effect of estrogen-modulated translation (O'Malley and Means 1974). With methodologic advances it was eventually shown that a 40% increase in the synthesis of rapidly labeled RNA occurred, not in hours, but within 2 min of estrogen injection (Means and Hamilton 1966). The response was biphasic, peaking initially at 20-30 min and followed by an abrupt decrease and a second rise at 2 h which was maintained for 24 h. The first phase of the response was blocked by actinomycin D but not by cycloheximide, leading to the idea that the primary response of the nucleus to estradiol was not dependent on continued protein synthesis.

Initial studies of base composition indicated that the RNA which was labeled rapidly was ribosomelike (Hamilton et al. 1968). Although quantitative E_2-induced changes are most marked in rRNA and begin within 4-6 h after hormone administration (Knowler and Smellie 1971), there is new evidence that this change is preceded by and perhaps dependent upon the synthesis of mRNA (Knowler and Smellie 1973). Sedimentation characteristics of rapidly labeled RNA and the kinetics of E_2-stimulated activity of uterine nuclear RNA polymerase II favor an early synthesis of a RNA of more DNA-like character (Knowler and Smellie 1973; Glasser et al. 1972a, Luck and Hamilton 1972). This high-molecular-weight RNA which appears in nuclei extracted from uteri of immature rats within 30-45 min of E_2 stimulation shows all the characteristics of heterogeneous mRNA. Within a few hours E_2 administration results in the stimulation of both ribosomal RNA, 28S, 18S, and 5S ribosomal RNA, and 4S transfer RNA (Knowler and Smellie 1973; Means and O'Malley 1972). All classes of uterine RNA are thus eventually increased by estrogen.

4.3.2 DNA-RNA Hybridization

The use of DNA-RNA hybridization for studying the effects of estrogen and progesterone on the transcriptional apparatus of the uterus was first successfully carried out by Church and McCarthy (1970). They examined the influence of E_2 on the appearance of new populations of nuclear RNA sequences and recorded the rapid induction of the synthesis of different populations of RNA. The technique as employed by Church and McCarthy was of limited use for hormone responsive tissues because the conditions of hybridization permitted analysis only of changes in the repetitive sequences of DNA. The portion of the genome containing unique sequences could not be assayed.

Methodological advances proposed by Britten and Kohne (1968) allowing the genome to be effectively subdivided into unique and repeating sequences were subsequently applied to the study of hormone action. The major work concerning estrogen action has been done with the chick oviduct (Liarakos et al. 1973). These data suggest that estrogen stimulates the synthesis of new proteins required for growth of oviductal cells via differential gene transcription rather than major gene duplication or restriction. The amount of RNA transcribed from unique sequence DNA that is processed into the cytoplasm does not appear to vary as a result of estrogen stimulation. There are qualitative differences that appear to exist in the cytoplasmic mRNA population at different stages of estrogen-induced differentiation.

Competitive DNA/RNA hybridization has been used to compare the nature of the RNA synthesized on different days of pregnancy (O'Grady et al. 1975). Although this method compared mostly repetitive sequences, certain modifications (exclusion of cytoplasmic RNA) increased the possibility of revealing any new species. Uterine segments from postconception uteri, day 1 and day 4, and from day 6 implantation site and intersite tissue were labeled in vitro with ^{3}H-uridine. Nuclear RNA was isolated from these preparations and similar unlabeled preparations. Chromatin was also prepared from these same tissues in order to synthesize RNA (labeled and unlabeled). From studies of competitive hybridization between labeled and unlabeled preparations derived from these various groups of tissues, it was shown that there are RNA species present on day 4 which are not present on the other days of pregnancy that were studied. Furthermore, the total RNA present on day 4 contains in addition to the specific RNAs all RNA species present at other days of pregnancy. These results are in agreement with existing studies which demonstrate that unique or specifically synthesized species of RNA characterize the period of uterine sensitivity (day 4). Hybridization of uterine preparations from uteri of ovariectomized or immature rabbits pretreated with estrogen before progesterone produced data which suggested that progesterone, independent of estrogen, opened a gene locus leading to specific or unique RNA transcription (Moriyama 1974). These data may also be considered to support the hypothesis that both estrogen and progesterone regulate cell differentiation by means of their individual influences on gene transcription.

4.3.3 Sites of RNA Synthesis

Two methods have been utilized to attempt to localize the sites of RNA synthesis in the preimplantation uterus of the rat. In vivo exposure of the uterus to ^{3}H-uridine is included in the autoradiographic studies by Tachi et al. (1972). The same isotope has been used in the biochemical analysis of RNA synthesis in separated luminal epithelium, stromal and myometrial fractions (Heald et al. 1975a). The dynamics of RNA synthesis recommend the latter experiments for future study.

Analytical studies of total uterine RNA are consistent in reporting an increase in RNA synthesis initially marked on day 2 of pregnancy. Utilizing the parameter of ^{3}H-uridine "uptake into uterine RNA", Heald et al. (1975b) report a second peak between 0900 and 1000 hours on day 4. Our own data on specific activity of either ^{3}H-uridine or ^{3}H-cytidine incorporated into total uterine and nuclear RNA show an early peak at day 2. We have also used a 1-h pulse and have done both in vivo and in vitro labeling. Although the values we note on day 4 are greater than on days 2 and 3, we cannot statistically identify a biphasic modality. Rather, what is unique about the day 4 RNA synthesis is the evidence obtained from double-labeled gels which indicate that either unique RNA species or selected populations of existing RNA species are being synthesized at this time.

4.3.4 Nature of the RNA Product

The exact nature of the RNA product which is synthesized at each stage of the preimplantation period is not known. Studies of this nature have, in the main, been rather perfunctory. Sucrose density gradient analysis of uterine RNA suggests that the rate of synthesis of rRNA is increased on pregnancy days 4 and 5. Processing of rRNA is stimulated on day 4. This is more pronounced in decidualizing tissue (day 5) but reduced in non-decidual tissue (Heald et al. 1972). This subject deserves more extensive and rigorous study. The data cited above should be validated and the extent of agreement (or disagreement) with the fundamentals of hormone action needs to be resolved. There is a preliminary report of an endometrial 10S RNA fraction in the rabbit (Vittorelli et al. 1967). This has, as yet, not been confirmed for the rat, but the possible presence of a 10S fraction which could contain putative mRNA requires reexamination. One would predict that more discriminating methods would identify hormone-induced messages which would specify those RNA and protein species that distinguish the processes of decidualization and implantation.

It is recognized that the differential activities of the endogenous nucleoplasmic and nucleolar DNA-dependent RNA polymerase participate in the regulation of the quality of the RNA being synthesized by the uterine cells. There have been a number of reports of the hormonal regulation of these activities. However, none has been addressed to study of the RNA synthesized at each stage in the preimplantation uterus.

The first observations regarding the activation of DNA-dependent RNA polymerase in the uterus recorded an estrogen-stimulated increase of a Mg^{2+} dependent polymerase within 1-h of a single injection of E_2 (Gorski 1964). It was subsequently demonstrated that the Mg^{2+} dependent enzyme (RNA polymerase I) was restricted to the nucleolus and directed the synthesis of ribosomal RNA. A second, high-salt Mn^{2+} dependent enzyme (RNA polymerase II) was located in the nucleus and synthesized an RNA product that was DNA-like.

From these data it appeared that the estrogen-stimulated synthesis of ribosomal RNA occurred before any apparent influence of DNA-like RNA formation: Subsequent studies (Barry and Gorski 1971) reported that within 1 h after estrogen there was an increase in the rate of chain elongation. The number of growing chains was unaltered by the hormone. The thesis was advanced that estrogen stimulated the activity but not the number of RNA polymerase molecules. The effect of estrogen at this level is believed to be only minimally due to transcription of additional template.

For several years the rapid but transient stimulation of uterine RNA synthesis could not be explained because neither chromatin template activity nor RNA polymerase activity seemed to increase at the same time as RNA (Hamilton 1968; O'Malley and Means 1974). Stringent kinetic analysis of the polymerase response to estrogen has shown a rapid but transient increase in the activity of RNA polymerase II occurring within 15 min of hormone injection (Glasser et al. 1972a). The chronology of this activation was consistent with the uptake of RNA precursors and their incorporation into rapidly labeled nuclear RNA. It is of interest to note that the initial enhancement of polymerase II occurs prior to a detectable increase in the activity of the chromatin template or the activity of RNA polymerase I. This study confirmed the time of polymerase I activity as reported by others (Gorski 1964; Hamilton et al. 1975) and recorded, for the first time, a secondary significant increase in polymerase II activity.

A second set of preliminary studies readdressed the question as to whether some or all of the effects of E_2 on transcription required preliminary or continued protein synthesis. As noted, the pattern of RNA polymerase II activity evoked by E_2 was biphasic. The first peak of 15-30 min was followed by a drop to baseline and a second sustained rise beginning after the first hour. When actinomycin D was administered (i.p.), some 30 min before E_2, both peaks were abolished. Cycloheximide did not affect the first peak but the second peak was deleted (Glasser and Spelsberg 1973). These data which accent the role of continuing RNA synthesis in estrogen action also provide an explanation for the findings of Gorski et al. (1965) and Nicolette and Babler (1972). In their experiments, stopping uterine protein synthesis prevented enhancement of RNA synthesis. Primary responsibility for estrogen action was placed on continuing protein synthesis rather than on the transcriptional process. However, these data were obtained 2-6 h after estrogen administration, long after the first peak of RNA polymerase II activity was over. Additionally, DeAngelo and Gorski (1970) found that synthesis of the estrogen-specific induced protein "IP" was blocked by actinomycin D but not by cycloheximide. From data such as these the suggestion was derived that the early increase in RNA polymerase II activity could be necessary for subsequent events. Activation of enzyme would not commit the uterine cells to a determinative plan but would permit a programmed sequence of "later" events to occur.

4.3.5 Transcription Versus Enzyme Activation

The data presented above are by no means uncontested. It is the opinion of certain investigators (Bouton 1976; Courvalin et al. 1976) that no clear conclusion may be drawn at the present time from in vivo studies. They view the results of early quantitative and qualitative changes in uterine RNA studied by the incorporation of radioactive RNA precursors as conflicting. It is not clear to these investigators whether the noted alterations are related to the incorporation of precursors into all RNA species or to an increase which is limited to either preribosomal or heterogeneous high-molecular-weight nuclear RNA. Experiments with RNA polymerase solubilized from whole uterine homogenates (Courvalin et al. 1976) showed that the activities of α-amanitin sensitive (polymerase II) and α-amanitin insensitive (polymerase I) enzymes were constant during the first 6 h after estradiol. The number of polymerase II molecules evolved at 24 h paralleled the 40% increase in activity suggesting that any increase in polymerase activity was due to the synthesis of new enzyme molecules.

In a second series of experiments Bouton (1976) and Courvalin et al. (1976) studied the enzyme solubilized from disrupted nuclei under conditions (0.15M $(NH_4)_2SO_4$) which presumptively release polymerase molecules not firmly bound to DNA in transcription complexes. Under these conditions α-amanitin sensitive molecules which remain firmly bound do not alter activity for the first 6 h after estrogen treatment. From 6 through 24 h there is a doubling of activity. Enzyme insensitive to α-amanitin which remained firmly bound to DNA was stimulated by estrogen. The earliest differences (polymerase I) were detectable at 1-2 h. At 12 and 24 h the increase was 50% and 100%, respectively.

The interpretation offered by these investigators is that the progressive increase in polymerase I agrees with those reports of estrogen elevated ribosomal RNA synthesis (Hamilton1968) notable at 6 h. The delayed enhancement of tightly bound polymerase II correlates with those studies reporting a delayed increase in the synthesis of mRNA. It is of interest that the citation used to validate these reluts is that of Billing et al. (1969) but no note is made of later work from the same laboratory which supports the concept of a very early estrogen stimulation of uterine nuclear transcription (Knowler and Smellie

1973; Borthwick and Smellie 1975). The latter paper presents data gathered from the immature rabbit uterus which strongly confirm our own data from the rat uterus (Glasser et al. 1972a). Observed differences in polymerase I and II activities are attributed to changes in the structure and function of chromatin rather than modifications in the RNA polymerases themselves.

5 A Model for Stromal Cell Transformation

5.1 Historical Perspective

More than 10 years ago, in a series of incisive experiments, Psychoyos (1969) developed the data from which this laboratory derived the concept that progesterone was a major determinant in the processes of decidualization and implantation (Glasser 1972). Until that time the compelling thesis that influenced studies of these processes emphasized the related and sequential roles of preovulatory estrogen and prenidatory estrogen (Shelesnyak and Marcus 1971). It is surprising how relatively little insight the long, still-unresolved controversy regarding the nature of the prenidatory estrogen has provided to our understanding of uterine sensitivity.

A second idea which had great bearing on the earlier work on the decidual cell reaction (DCR) came from the laboratory of DeFeo (1967). In defining the hormonal and temporal parameters of the DCR, DeFeo and Yochim evaluated the effectiveness of their procedures by measuring the decidual growth at a point 120 h after the deciduogenic stimulus (DeFeo 1963a, b; Yochim and DeFeo 1963). Their optimum ratio of estrone: progesterone was 1 μg: 2000 μg, the optimal day of stimulation was day 4, and the optimal time for induction was between 1000 and 1400 h. This produced the greatest response and the idea of "maximum uterine sensitivity" became a useful tool in the analysis of deciduogenesis.

The concept of maximum uterine sensitivity was, however, of limited utility in trying to analyze the inductive process, the role of each specific hormone, and the response of individual uterine cell types. As elaborated in previous publications (Glasser 1972; Glasser and Clark 1975), the end point of a 120-h growth period is too far removed in time and therefore too nondiscriminatory to be useful in the genre of experiments concerned with the molecular basis of hormone action.

Our initial investigations on the role of progesterone were not intended to challenge either the Shelesnyak or the DeFeo interpretation. Rather they were designed to gather data which would define the nature of progesterone action per se. We were quite prepared at that time to accept a modulatory role for progesterone. Therefore, the observation that the DCR response was fully inducible with progesterone alone in the castrate rat, unprimed with estrogen, was exciting. Under these conditions the structural and biochemical characterization of decidualization was qualitatively identical with deciduoma raised on progesterone and estrogen. Compared with the unstimulated contralateral control horn in the same animal, the relative quantitative responses (weight, DNA, RNA, protein) of castrate animals maintained on P were identical with those of rats on P + E for the first 72 h after DCR induction; only after that period is the response enhanced by E (Glasser 1972).

The usefulness of the endpoint "maximal uterine response" seems increasingly restricted particularly in the evaluation of critical postinductive changes in stimulated uteri. Even in considering more general responses this endpoint is of questionable utility. In a pregnant animal where the total amount of decidual tissue is approximately 300 mg what would be the value and purpose of the 1000-1200 mg of deciduoma produced with artificial stimuli ? In pilot studies where we studied the effect of varying doses of actinomycin D on implantation (patent horn) and decidualization (ligated horn) in the same animal we arrived at the operational conclusion that the hormonal environment which would support a decidual cell response of 250 $\pm$ 23 mg/100 g body weight would also support a normal complement of implanting embryos (5.2 $\pm$ 0.7/horn). There was no greater implantation success in animals supporting deciduoma in the range of 600-740 mg, but, when the DCR fell to 200 mg/100 g body weight and below, the success of implantation was seriously impaired (0-3 embryos/horn).

With the recognition that the understanding of decidualization and implantation would be productively advanced by even the limited study of inductive rather than growth phenomena and that effective decidualization is inducible with progesterone alone, these investigations created an entirely new generation of questions.

5.2 Progesterone as a Determinant

The first version of our concept that progesterone played a primary determinant role in the induction of the decidual cell response was derived from studies in which castrate mature and immature rats maintained on 2 mg progesterone/day only were among the important experimental groups. Initially we set no criteria but as the data accumulated we recognized that there occurred a preferential increase in the specific activity of newly synthesized RNA (Glasser 1975) and the subsequent appearance of unique or preferentially synthesized species of RNA and protein were also characteristic of the induction of deciduoma. These changes could also be closely related to the role of progesterone in these processes. Additional studies on the template capacity of chromatin (Glasser and Clark 1975) and the demonstration that the inductive process was dependent on continuing RNA synthesis (Glasser and Spelsberg 1973) integrated these specific progesterone actions with the growing literature of molecular endocrinology.

In 1975 we presented a further modification of this model for the induction of decidualization to the Society for Developmental Biology (Glasser and Clark 1975). That proposition indicted progesterone as the determinant regulatory factor that induced, in nonproliferating populations of stromal cells, specific species of RNA and subsequently proteins. This increase in unique transcriptive activity is effected by the binding of the progesterone-nuclear receptor complex to the nucleus of the stromal cell. Among the translational products that result are those factors which, after 48 h exposure to progesterone, transform the neutral or nonresponsive uterus to one that is sensitive to deciduogenic stimuli. It must be remarked that, although estrogen itself induces populations of unique RNAs and proteins, this type of stimulation leaves the uterus unresponsive to even the crudest type of deciduogenic agents. The progesterone-dominated uterus can remain sensitive to deciduogenic stimuli for periods reportedly in excess of 21 days (Nutting and Meyer 1963) and provides an environment in which the blastocyst can be maintained in diapause. Yet implantation cannot occur in such a uterus. A further modification of gene expression is necessary to activate the blastocyst and convert the uterus from a sensitive to a receptive state in order to initiate implantation. The role for these latter actions falls to estrogen.

5.3 Criteria for Modeling

Organization of the data presented here into an operational model of the hormonal regulation of the cellular interactions involved in the processes of decidualization and implantation should avoid overinterpretation of experimental results. Currently there are an increasing number of models, either original or revised, being advanced. They are too often faulted by being overextended in terms of speculative inflation of data. Models should be inclusive and consistent. A successful model should integrate all contributory disciplines in such a way that they are not contradictory. Although the roles played by progesterone and estrogen in linking and synchronizing the embryo-uterine endometrium systems are well recognized, too little attention has been paid to the advances made in molecular endocrinology and molecular biology. Relevant data from both these disciplines has been either neglected or only superficially integrated into the analysis of decidualization and nidation. As an example, the principles of cell kinetics and proliferation have been adequately studied but only a very few investigators (Martin et al. 1976; Epifanova 1971) have viewed these responses as an integral part of hormone action. Failure to do so has led to unfounded and misleading biases.

Models for the development of endometrial sensitivity to decidualization have been developed which assign primary determinative roles to mitotically linked differentiation of the epithelium (O'Grady and Bell 1977). Three key integral systems are poposed:

1. The transient production of a metastable "sensitive" luminal epithelium characterized by the competence of the luminal epithelium to (a) activate the free blastocyst, (b) receive and transmit a signal/stimulus from/to the stroma, (c) activate "programmed cell death" in response to a stimulus from the blastocyst.

There is no confirmed data in support of items (a) and (b). Cell death is an integral response of luminal epithelium to estrogen. There is no direct evidence for the presence of a blastocyst signal or its intervention with any differentiated form of the epithelium.

2. The transient production of a metastable "sensitive" stromal cell population (presumptive decidual cells, PD cells). PD cells are competent to a signal/stimulus transmitted from the "sensitive" epithelium.

Decidualization can be induced in progesterone differentiated stromal cells. This condition leaves the epithelial cell apparently unaltered. Under these conditions, if there is a signal from the luminal epithelium, it would not seem to require only mitotically linked differentiation to produce the signal. Decidualization as an integral process in implantation occurs in a hormonal environment that will stimulate cell proliferation and differentiation of the luminal epithelium as well as the stroma. At the present time there is no evidence for an epithelial signal to the stroma in this system although the possibility of such a signal should not be discounted.

3. Defined postinductive growth and regression of decidual cell systems. Decidual growth is almost exclusively a stromal cell phenomenon. The work of Noyes et al. (1963) has

shown that a temporal interlock operates by means of the dependency of the component systems (epithelium stroma, blastocyst) on the same two regulatory hormones. Theoretically this concept of synchrony is most plausible when applied to inductive processes. There is little or no data concerning the refractoriness of decidual growth or its subsequent regression to continued hormonal stimulation.

Thus, in the absence of any data regarding communication between endometrial cell types during the various phases of postinductive growth, there are only didactic reasons to suggest that mitotically linked differentiation of epithelial cells can regulate the course of stromal cell growth. It would seem no less plausible to suggest that the refractoriness and regression of the stromal cells are modulated by the products of their own mitotically linked differentiation.

In spite of this negative introduction, a role for the luminal and glandular epithelium must be provided for in the analysis of the decidualization-implantation process until they are specifically discounted. However, the analysis of the morphological and biochemical changes in these cell populations provides only the seeds for a most speculative type of model construction.

5.4 Possible Role of Epithelial Cells

The single most positive regulator of LE and GE populations is estrogen. This is evident from changes noted during the onset of pregnancy in the intact, cycling rat and in the castrate animal injected with estradiol-17β. In the intact pregnant rat the proliferation noted in the luminal epithelium might well be a late response of that cell population to the extremely high titers of E at proestrus. The first wave of LE mitosis is probably an actual response to estrogen secreted at the time of the proestrus peak, there is probably a sufficient amount of E still present during estrus (as titers are falling from the proestrus peak) for this purpose. If the data from studies on the castrate animal maintained on E can be extrapolated to the intact pregnant rat, then there is sufficient E secreted during day 0 and 1 of pregnancy to sustain the proliferative activity in the LE and extend it to the GE. It will remain difficult to assign this role to any specific period of estrogen secretion in the cyclic animal until we can, with fair accuracy, determine the effectiveness of specific E titers as well as their duration of effect.

The E-stimulated castrate animal may therefore be the more productive system to analyze. The untreated castrate, even if it has been previously estrogen primed (Finn and Martin 1973), is a hormone deficient animal with an unstimulated but not atrophic epithelium with a basal concentration of cytoplasmic E receptors. The unprimed animal possesses atrophic luminal epithelium along with a vanishing population of cytoplasmic receptors. The first daily E injection will serve to translocate the cytoplasmic receptor hormone complex to the nucleus. The binding of the hormone receptor complex to the nuclear acceptor site will initiate transcriptive events resulting in changes in RNA, DNA, and protein, which includes the synthesis of additional receptors as well as the entry of a proportion of the LE cells into the mitotic cycle.

Progressively, as the titers of E reach effective levels, organ specific proportions of uncommitted cells (UC) enter die G_1 phase of the mitotic cycle. According to the most comprehensive study available (Martin et al. 1976), this proliferative response to E is manifest at 15 h after injection and is accompanied by a decrease in cell death. Thus, E has produced at this time, from the formerly uncommitted matrix, a population of endometrial cells which is larger and increasingly complex. This population includes still uncommitted cells (UC), UC cells that are growing, cells entering the mitotic cycle, cells which have divided (M_1), as well as a mixed population of cells which have not died but are being maintained in the endometrial population.

With more time the proportion of M_1 increases as cells pass from G_0 to G_1 and enter the first mitotic cycle. A proportion of these M_1 cells enter and complete a second mitotic cycle (M_2) adding heterogeneity to the types of cells which comprise the endometrium, i.e., UC, UC which are growing, UC which are nondead, M1, M2, and M type cells that are nondead. Continued exposure to estrogen will not stimulate epithelial cells to enter a third mitotic cycle.

At the end of a 24-h period of estrogen exposure the endometrium is composed of a complex mixed population of biochemically altered cells. The proportion of M type cells has increased significantly, but it should be noted that, with the cessation of marked proliferative activity, the rate of cell death increases significantly. At that time in early pregnancy when progesterone titers begin to increase significantly (days 1-2), the epithelial population has been totally committed to the estrogen directed program. There is no further cell division in response to estrogen (Martin et al. 1976) and the effective presence of P would markedly alter any further regulatory response to estrogen. If there are further changes in the epithelial cells of the pregnant animal (days 3 and 4) they are, at present, undocumented.

If the epithelial cell or any of its variants is involved in the uterine response to a deciduogenic stimulus, either natural or artificial, it seems appropriate to ask if this ability to respond is characteristic of any one epithelial cell type. Is any particular cell "sensitized" to respond? If so, which, if any, of these single cell types plays an obligative role in the processes of decidualization and implantation? Is there a critical population of that cell type required for the response, or does participation require a specific mix of cell types from this heterogeneous epithelial population? Evidence, either pro or con, implicating the epithelial population is sparse. There is the report of Fainstat (1963) that unless the epithelium per se is involved in the physical stimulation of the uterus, there is no response. On the other hand there are the more comprehensive studies of Sherman and Salomon (1975) and Salomon and Sherman (1975) demonstrating implantation on monolayers of decidual cells. In these experiments nidation, as it often is in extrauterine sites, seems to be independent of the time restrictive program imposed by the uterus. Finally, we are left, as always, with unsatisfactory rationalizations of ectopic implantation.

Nevertheless we would not presume, at this time, a role for the unstimulated epithelial cell. A functional role, if any, is most logically assigned to those cells which have survived one if not two mitotic cycles. Does completion of the first mitotic cycle represent an altered differentiated state that is competent to carry out the functions of the epithelial cells that are involved in implantation? If M1 and M2 cells are different, is the competent cell that luminal epithelial cell which has completed the second mitotic cycle? Or is there a critical ratio of M1 and M2 cells that is required?

We must also consider that the role of the luminal epithelial cell is secondary. Very little has been recorded concerning the specific functions of the glandular epithelium. Significant changes in the glandular epithelium do not occur until relatively late in the response of the epithelium to E and numerically these cells do not represent a significant population of the endometrium. However a number of secretory products have been assigned to the glandular epithelial cells, e.g., a zonalysin (Pinsker et al. 1974), proteins including uteroglobinlike proteins (Surani 1975; Bullock 1977), implantation inhibitors (Psychoyos 1969), and almost metaphysically positive implantation signals. If these secretory products do play a regulatory role then it follows that the glandular epithelial cells may play an important and direct role in signal generation or secondarily in significantly modulating the biochemical differentiation and response of the luminal epithelium. In a recent study of Lejeune, Van Hoeck and Leroy (1981) it is postulated that the function of the luminal uterine epithelium in the induction of decidualization in rats may be an obligatory transmitter role. Release of an appropriate epithelial message to the stroma should require the local preservation of membrane relationships between both tissues.

Finally (and at this moment the least attractive) is the possibility that the putative role of the uterine epithelium is exercised via a lytic product produced by the death of these cells.

5.5 Nonproliferative Stromal Cell Transformation

A detailed analysis of stromal cell transformation promises to be more fruitful in delineating the factors involved in decidualization and nidation. It should be reemphasized that any model for the analysis of implantation related decidualization must explain, in terms of regulatory hormone action, the following: (1) differentiation of the individual cell types of the endometrium, (2) production of a transient period of uterine sensitivity, (3) induction of the decidual cell response, (4) growth, and (5) regression.

It has become increasingly evident, on the basis of data generated in our laboratory on the mechanism of hormone action, that the functional capacity of the stromal cell reflects not only the degree of gene restriction of that cell but the qualitative nature of gene expression. Thus, although the mitotic activity of the stromal cell may be a functional correlate of stromal cell function, cell proliferation is not required for hormonally induced stromal cell transformation to decidual cells. The induction of decidualization is not a consequence of cell division.

There has been undue confusion and contention regarding the obligatory role of cell division in the process of decidualization. In the course of the natural history of early pregnancy the two phenomena are very closely correlated. However, the judicious use of the rat as a model demonstrates that the two processes are indeed dissociable. The utility of the mouse as a model is not diminished by the fact that, in this system, cell proliferation and decidualization are apparently nondissociable. Rather, comparison of the two animal models indicates that each provides for a different experimental strategy. The rat uterine model has allowed us to catalog the functional capacity of each stromal cell state, to dissociate the processes of functional and biochemical differentiation from cell division, and to identify the exclusive roles of each of the regulatory steroid hormones; in short, to demonstrate the determinant role of progesterone in preimplantation pregnancy.

The stromal cell of what Psychoyos (1969) has called the "neutral uterus" is basically a genetically restricted cell with basal biochemical functions and a life-death cell replenishment program. Compared to the stromal cells of the atrophied untreated castrate rat, the neutral uterus is markedly both transcriptional and translationally active. The information being produced in the neutral uterus is akin to that produced in the diestrus stages of the reproductive cycle and provides for a maintenance physiology. It is a more elaborate environment than is produced in the more restrictive condition of the untreated castrate.

5.5.1 "A" State Cells (An)

On a theoretical basis it has been proposed by Smith and Martin (1973, 1974) that the cells of the epithelium and stroma are arrested following mitosis in a state (the A state) which requires an additional stimulus to the target cell, i.e., the stromal cell of the neutral uterus, for additional synthesis of RNA and DNA. These changes are at least coincident if not correlated with transformation of that cell, e.g., stromal cell ⇒ decidual cell. It remains to be proven if the stimulus in question is required to be mitogenic. Basically the A state cell may be considered a highly restricted cell with a high selective threshold to stimuli. At one time it was argued that estrogen was the primary stimulus to the A-state stromal cell. This is not true, but it is highly probable that estrogen, which is mitogenic, serves to condition the threshold setpoint of the A-state stromal cell.

Although the neutral (A state) uterus (An cell; Fig. 21a) will respond to both estrogen and progesterone, the history of estrogen as the determinant of uterine sensitivity can no longer be sustained. The response of the uterus to continued exposure to estrogen alone is most marked in the epithelial cell populations of the endometrium. A-state (An) epithelial cells are stimulated by estrogen. The uterus responds trophically with the correlated induction of unique RNA and protein species, which is the subject of an extensive literature (O'Malley and Means 1974). Two things should be noted about the uterine response to estrogens. First, growth is not limitless; uterine growth will reach a plateau. Second, none of the products of this extensive hormonal response will result in either epithelial or stromal cell populations sensitive to decidualization or receptive to nidation.

5.5.2 Progesterone Induced Sensitive Cells (As)

The neutral (A state) uterus will respond to progesterone. We designate the stromal cells of the neutral uterus as An. As we mark it, the uterine response (decidual cell response) requires at least 48 h preexposure to progesterone (three daily injections). It should be noted that even after the first progesterone injection (Px1) the stromal cell is altered even though there is no cellular proliferation, the biochemical changes are not significant, and the cell cannot be stimulated by deciduogenic agents. However, the cell is no longer neutral. If estrogen is injected, or if the next progesterone is missed, at least 48 h of a hormone-free environment are required before the neutral state can be reattained and the cycle reinitiated. If progesterone is injected on a daily basis before this hormone-free 48-h period is completed, the uterus and its stromal cells will be retained in a refractory nonresponsive state. The same basic finding and responses hold for the cell exposed to two injections of progesterone (Px2) although the biochemical characteristics of these cells are somewhat different from those found in the Px1 situation.

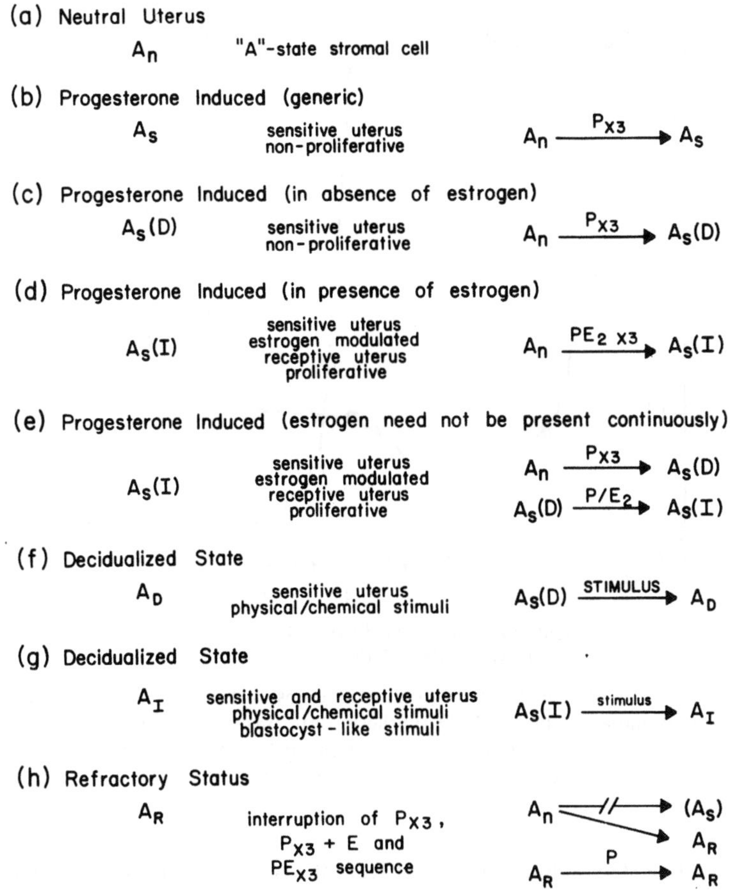

Fig. 21. A scheme to explain the various stages of stromal cell transformation and the hormonal conditions which induce and/or modulate these programs of differentiation.

The stromal cell of the previously untreated castrate rat exposed to at least three injections of progesterone (Px3) has been the subject of a great deal of the data set forth in this chapter. Analysis of the template capacity of uterine chromatin isolated from progesterone-treated castrate rats shows that a progressively increased percentage of the genome is available for transcription as the series of progesterone injections is extended (Fig. 22). The average number of RNA chains initiated in the neutral uterus is 4000/pg chromatin DNA. At Px3 the number has increased to almost 20 000/pg chromatin DNA. Significant elevation in the concentration of total RNA, the induction of new RNA, and subsequently protein species characterize these essentially nonproliferative As cells (Fig. 21b).

This As cell, the product of 3 progesterone injections, is different not only from the neutral or An stromal cell but from the transitional Px1 and Px2 stromal cells. Not only is there the marked increase in the rate of transcription but the informational molecules being produced specify that biochemically, if not structurally, this Px3 cell has differen-

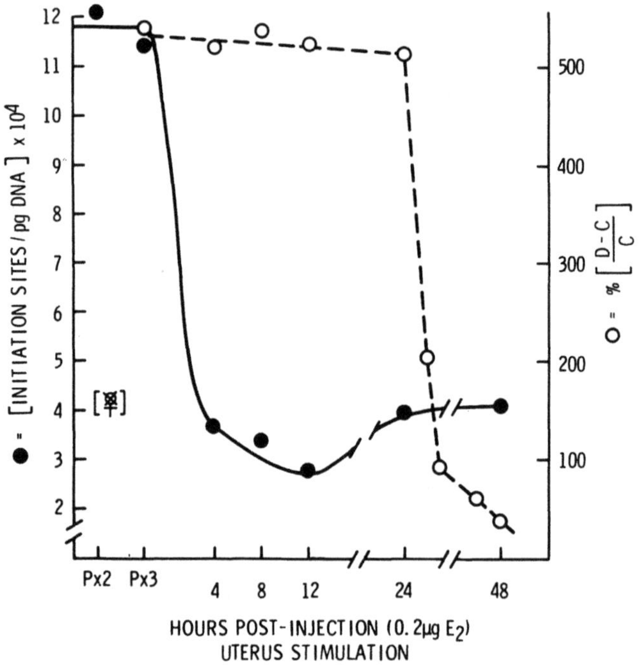

Fig. 22. Loss of uterine sensitivity (o) plotted against restriction of the uterine genome as expressed in terms of initiation sites for RNA synthesis. Both are estrogen stimulated events and the data suggest that E_2 induced qualitative gene restriction regulates the development of uterine receptivity, initiation of implantation and loss of uterine sensitivity

tiated to a state uniquely different from the An, Px1, and Px2 cells. The program to which this As cell is committed may be sustained for periods in excess of 21 days by continuous exposure to progesterone (Fig. 21b) or redirected into one of three other modes of expression.

The progesterone (Px3) induced differentiated stromal cell (As), committed to a program with multiple second stage potentials, is the cell we have previously designated at the "sensitive" cell. Challenging the sensitive stroma (Px3) with a deciduogenic stimulus will produce a fully developed decidual cell response. The endocrinology and biochemistry of this response have been comprehensively detailed in previous publications (Glasser 1972; Glasser and Clark 1975). One of the signal features established by our earlier work was the demonstration that the biochemical differentiation of the neutral uterus to a sensitive uterus competent to respond to deciduogenic agents is determined by progesterone; the role of estrogen in the induction of the decidual cell response and its early growth is presumably a modulating one (Glasser 1972).

In emphasizing that progesterone-directed biochemical differentiation of the uncommitted stromal cell (An) is not dependent in an obligatory way on stromal cell mitosis, we are insisting on the importance of identifying for each hormone its unique spectrum of actions and responses. We are also restating the importance of cataloging the individual cell types and the potential fate of each of these cell types as it is expressed in a specific hormone environment. Parenthetically, it should be noted that, compared with estrogen, very little data has been gathered relative to the action of progesterone itself. This type of analytical exercise does not fail to recognize the responses of individual endometrial cell types in the natural history of pregnancy. Rather it makes it possible to understand how and why the cells respond and interact. Furthermore, this information may be used to redirect cellular responses experimentally and therapeutically and alter the course of blastocyst-endometrial interaction.

We are also questioning the current fashion of calling the cells of the sensitive uterus "presumptive decidual cells". This term is too inclusive for the definition that it covers. O'Grady and Bell (1977) have defined presumptive decidual (PD) stromal cells as being formed from and inherently dependent on cellular division of the stromal stem cell. According to their hypothesis, it is during this essentially quantal mitotic division that cells are reprogrammed. The cells are said to be committed to this as yet unexpressed program. Expression takes the form of transformation of the stromal cell to the decidual cell. This sensitive PD cell is further described as transitory. If the cell program is not activated within a limited (?) period of time the cells "pass" into an insensitive state.

This description of the biological fate of the stromal cell is adequate during the preimplantation stages of a normal pregnancy. However, the definition of the PD cell makes no provision for recognizing the roles of each of the regulatory hormones, where and when each hormone acts, the nature and chronology of each response, and the relative importance of each of the responses in the transformation of the stromal cell to that type of cell which is receptive to the activated blastocyst as opposed to transformation to a stromal cell which is sensitive to deciduogenesis only.

In order to understand the early stages of pregnancy it is important to recognize that the important first stages of stromal cell differentiation can be accomplished by progesterone induction alone and in the absence of cell division (Fig. 21a-c). This cell (As), without further modification in gene expression, is competent (sensitive) to respond to deciduogenic stimuli and decidualize.

The study of this type of cell response provides the insight to phrase appropriate questions about the action of estrogen secreted in early pregnancy and its interactions with progesterone. Does estrogen have an effect on the stroma during the period it is stimulating epithelial cell proliferation? If so, how can it be measured? The response of the castrate rat to progesterone would seem to obviate any obligatory role for estrogen on the stroma or for biochemical signals derived from the dividing epithelium. It is possible, even in the context of molecular endocrinology, that the Psychoyos "paradox" is operable and that until the stroma has been exposed to 48 h of progesterone, it, in effect, does not "see" estrogen? We do not have receptor data on these particular cell populations to enable us to determine if, during the first days of pregnancy (days 0, 1, perhaps 2), the estrogen receptor complex binds to stromal nuclei with physical/chemical characteristics and occupancy time consistent with effective estrogen action.

As we will note in more detail below, it is only after the progesterone induced stage of differentiation is completed (Px3) that the stromal cell (As) becomes uniquely responsive to estrogen. In pregnant animals the source of the hormone is probably nidatory estrogen. Cell division must also be included as one of the potential responses of the (As) cell to estrogen. Estrogen induced mitosis of the progesterone differentiated stromal cell (As) occurs in a population of cells in which the hormone has already produced severe restriction of the genome (Fig. 22). Other important features of estrogen modulation of the progesterone induced cell are the competence to respond to the presence of the blastocyst as a stimulus (within 24 h) and the loss of uterine sensitivity (within 36-40 h).

The action of progesterone and the modulating effect of estrogen on the stromal cell [AS (D); AS (I)]: Because the action of progesterone and the modulating effect of estrogen occur sequentially, we now recognize two forms of the AS cell: (1) the progesterone-induced stromal cell which has developed sensitivity to deciduogenic stimuli, the D-form of the As cell noted as As (D) (Fig. 21c), and (2) the progesterone-induced stromal cell (As) which has been exposed to estrogen and has further developed receptivity to implantation. This cell is noted as an As (I) cell (Fig. 21d, e). Analysis of these predecidualization stages provides sufficient experimental evidence to characterize the As (D) cell as different from the As (I) cell and to suggest, albeit with less assurance, that the I form of the As cell is a hormonally stimulated mitotic derivative of the D form of the As cell (Fig. 21). It remains to be determined if both the loss of uterine sensitivity and blastocyst receptivity are properties of the pre- or postmitotic forms of the estrogen modulated progesterone-induced cell.

5.5.3 Response to Deciduogenic Stimuli

5.5.3.1 Progesterone (AD). These progesterone sensitized cells, As (D) and As (I), can be induced to a second stage of differentiation. These programs are diagrammed in Fig. 21f, g. Stimulating the sensitized uterus, maintained on progesterone, with a deciduogenic agent will transform the sensitized stromal cell, As (D), to a decidual cell (A_D) (Fig. 21f). This decidual cell (A_D) must be presumed to be different from the decidual cell (A_I) which is induced after the completely progesterone sensitized cell (As) has been exposed to even a single E_2. Although induction of decidualization in both situations is followed by derepression of the genome, there are differences between A_D and A_I which reflect the quality of the cellular information being transcribed at the time of decidual cell induction.

Induction of the A_D cell takes place in a cellular matrix in which the genome is open, markedly derepressed (*vs* An) by progesterone. The deciduogenic stimulus produces a further derepression of the progesterone stimulated stromal cell while, at the same time, the mesometrial stroma is increasingly restricted. Induction of the A_I cell from the same An progenitor requires not only 48 h of progesterone, but at least the transient presence of estradiol (Fig. 21e, g). This hormone produces marked restriction of stromal cell transcription as expressed in quantitative terms (Fig. 22). Thus, the deciduogenic stimulus is inducing transformation of a population of cells, As (I), in which the genome is being rapidly and significantly restricted. The deciduogenic process reverses this repression of the genome, but the gene expression of the mesometrial endometrium continues to be severely restricted.

5.5.3.2 Progesterone Plus Estrogen (AsI). We do have some preliminary and indirect evidence that there are important qualitative shifts in the products of transcription in the estrogen-treated animal. Our analysis shows that, within the first 4 h after estrogen injection to a Px3 animal, there are relevant shifts in the nonhistone protein/DNA ratio suggestive of both derepression of the genome and a shift in the types of informational molecules being produced. This increase in acidic protein occurs in both epithelial and stromal fractions and is similar to that which occurs within 4 h of estradiol injection in an uterotrophic animal model (Glasser et al. 1972 a,b; Pollard and Martin 1975).

This modulation of progesterone-induced transcription by the action of the estrogen receptor complex on the genome also results in the initiation of stromal cell proliferation. This specific response to estrogen hormone action must condition the As (D) → As (I) → AI transition and the biological fate of the A_I cell compared to the nonproliferative population in which A_D transformation occurs (Fig. 21e). Another product of the putative qualitative difference in transcription is the difference in the threshold for induction. After E_2 treatment the inductive stimulus can be more attenuated and selective, such as that given by the blastocyst. The nature and magnitude of the inductive stimulus is a very complex issue to interpret for there is still considerable ambivalence over what constitutes a valid end point; the use of maximal uterine sensitivity is not appropriate.

In spite of these differences in the environment in which decidualization is induced, in the level of the induction threshold, and in their eventual biological fate, the A_D and A_I cells share many characteristics. In both cells there is an early postinductive increase in the specific activity of decidual cell RNA (Glasser 1975) and the appearance of unique or selectively synthesized RNA species and subsequently proteins (Glasser 1972; Denari et al. 1976). To date none of these parameters has disclosed any differences between the A_D and A_I uteri during the first 72 h after induction (Glasser 1972) (Fig. 15).

None of this casts any additional light on the role of the epithelium in the inductive process. There is the reserved suggestion that the LE cells are involved in the acceptance (nonrejection) of the attaching blastocyst. Among other undocumented suggestions is the assignment to the cells of the luminal epithelium of the role of signal generator or signal transducer. These cells would function in the final synchronization between blastocyst and endometrium. There is not a great deal of data bearing on this aspect of epithelial cell function. We are in need of more studies of the genre of the provocative paper by Kleinfeld, Morrow and DeFeo (1976) which presents ultrastructural evidence bearing on the problem of cell-cell communication in the decidualizing uterus. Considering the remarkable influence that estrogen exerts on the epithelial cells compared to the essential lack of response on the part of these cells to progesterone, we might inquire if the cells of the epithelium contribute in any way at all to differences in A_I and A_D cells.

A signal lack of similarity between the A_D and A_I cells is that the former is representative of a biochemical environment nonreceptive to implantation whereas the latter (A_I) will support nidation. In the presence of the conceptus the life of the A_I cell may be prolonged beyond the 16th postconception day, which is the life span characteristic of the A_D cell or the A_I cell in the pseudopregnant animal. It is worth noting again that another factor which may contribute to the differences between the A_D and A_I cell is that the changes noted in the As → A_I transformation are occurring in a proliferating cell population whereas the As → A_D induction is nonproliferative. The qualitative importance of cell division in this process cannot be evaluated at present.

The relationship of what we must assume to be nuclear bound progesterone receptor complex to gene transcription is neither simple or unilateral. From our study of uterine sensitivity we can recognize, to date, two separate actions of progesterone. The first and more obvious effect of the hormone is its role in inducing stromal cell differentiation. There is an obvious increase in gene expression during the progesterone-induced transformation of An → As (Fig. 21b, c). The second and less obvious action of the hormone is the stabilization of existing programs of differentiated gene expression.

The genome of the As (D) cell (Fig. 21c) is stabilized and remains open and sensitive to subsequent reexpression for at least 21 days if the animal continues to receive progesterone alone. At any point after 48 h of progesterone (Px3) the As cell can be further transformed to either an A_D or an A_I cell (Fig. 21f, g). The direction of the second stage program is determined by the hormonal environment at the time of induction and the character of the deciduogenic stimulus. Artificial agents induce decidual cells in either a progesterone (A_D) or progesterone + estradiol (A_I) environment with life spans which do not extend beyond postconception day 16. In this case the presence of the estrogen will not qualitatively alter the response to the artificial stimulus. Under these conditions neither of the decidual cell types will support implantation or survive beyond day 16.

5.5.4 Withdrawal of Hormonal Support

5.5.4.1 Transformation and Maintenance of the "A_R" Cell. Not as well considered is the role that progesterone will play in stabilizing conditions of gene restriction. There are three experimental conditions which will alter the stromal cell so that it will become and remain refractory to hormonal or deciduogenic stimulation. We have designated these stromal cells, A_R (Fig. 21h).

1. Failure to complete without interference or interruption those hormonal programs (Px3, PE_2x3) which induce the transformation of stromal An → As cells of either the D or I form (Fig. 21c, d, e)
2. Failure to stimulate the sensitive uterus deciduogenically after completion of the basic P + E sequence (PE_2x3, $PPPE_2$) which was described by Psychoyos (1973b) and is diagrammed in Fig. 21d, e
3. Withdrawal of progesterone for a single injection, interrupting the schedule maintaining the diapaused uterus without stimulation of the uterus within 24 h.

In each of these cases the genome is restricted and the uterus loses its sensitivity. The uterus can be maintained in this repressed state by continued progesterone treatment. We have maintained these cells in A_R status (Fig. 21h) for at least 11 days. The failure of the A_R cell to respond can be reversed by withdrawing progesterone (all hormonal support) for a minimum of 48 h. At this point the cells of the stroma may be regarded as having returned to the An state. The various parameters we have used to describe other stages of stromal cell gene expression have not been applied to the A_R cell so it can not be compared, except in terms of its lack of responsiveness, to the other differentiated forms of the stromal cell. It is also possible that the A_R cell may not be only a product of the three stages described but may represent the stromal cell population of the mesometrial wall of the decidualized uterus. From the little preliminary data we do have, we can only speculate that the $A_R \rightarrow A_N$ conversion is a form of "off" reaction. The result of progesterone withdrawal for 48 h would be a qualitative shift in the transcripts being produced. Both A_R and A_N states are highly repressed states.

5.6 Proliferative Stromal Cell Transformation

5.6.1 Transformation During Preimplantation Stages of Pregnancy

The fundamentals regarding the progesterone induced differentiation of the stromal cell derived from various models can be introduced with consistency to the condition of pregnancy. The rules we have developed for the action of a steroid hormone must apply, without exclusion, to the interaction of progesterone and estrogen on the cells of the pregnant uterine endometrium. They must explain the prolonged luteal phase as it is expressed at the uterine level and provide some basis for the establishment and maintenance of pregnancy.

When progesterone and estradiol are used in continuous combination, either in the optimal experimental ratio of 2000 μg progesterone: 1 μg estrone or 2000 μg progesterone: 0.1 μg estradiol-17β or as they occur naturally during preimplantation stages, the developmental biology of the stromal cell is altered. Not only is the progesterone plus estradiol treatment of the castrate rat analagous to pregnancy, but it mimics the mouse model. When the uterus is treated in this manner the transformation of the stromal cell progresses directly to the As (I) cell, e.g., An → As → As (I) (Fig. 21e). When the uterus is stimulated, this decidual cell characterizes uterine receptivity to ovoimplantation.

A possible but as yet inexplicable difference in the developmental fate of the As cell depending on the nature of its hormonal induction must be considered. Following Px3 the As, As (D) cell is characterized by an open genome and remains sensitive as long as progesterone is continued. On the other hand, the transcriptive capacity of the As, As (I) cell induced by PE_2x3 is markedly repressed within 4 h following the third injection and the sensitivity of the uterus to any physical or hormonal stimulation is lost completely within 36-42 h. Artificial deciduogenic stimulation of the nonproliferative Px3 uterus will redirect the program of differentiation to produce the A_D cell which will not support implantation. The proliferative endometrium of the PE_2x3 uterus will respond to natural stimuli such as the blastocyst. The newly differentiated A_I cell is compatible with the establishment and maintenance of nidation. That the As (D) (Px3) cell has the same determinative potential as the As (I) (PE_2x3) cell can be demonstrated by the following experiment. At any time after the As cell has differentiated to an As (D) cell, i.e., after three, five, ten, etc. (Fig. 21c) injections of progesterone, the addition of a small "implanting" dose of E_2 (0.2 μg) will allow that cell to respond to a blastocyst and to support its implantation. Thus, the As cell in question is an As (D) cell with its developmental fate committed to that program until it is exposed to estrogen after, not before, the third progesterone injection. The programmatic future of that cell, As (D), is altered to that of As (I) (Fig. 21e). Again, within 4 h after E_2, initiation of new RNA chains on uterine chromatin is severely restricted (85%) and the uterus becomes insensitive to all forms of stimulation by 36-42 h. These data emphasize that the basic determinant of stromal cell transformation is progesterone. Progesterone, with or without estrogen, induces the transformation of the generic As cell. This is consistent with the paradox noted by Psychoyos (1969) that, in terms of the response of the uterus, there was no identifiable effect of estrogen on that organ. Not until the 48-h progesterone program is completed is estrogen "visible" (Meyers 1970; Psyhoyos 1973b).

The program of differentiation developed following the induction by progesterone in the absence of estrogen is designated by the As (D) cell (Fig. 21c). However, in the presence of estrogen, the program of differentiation is characterized by the As (I) cell (Fig. 21d) with the important exception that E does not have to be present continuously (Fig. 21e). Although we do not have any cellular or molecular evidence to support the idea of two variants of the As cell, e.g., As (D) and As (I), the circumstantial evidence we do possess favors that situation. As a working hypothesis it could be put forth that in the pregnant animal the transformation of An → As (I) is a multistep process the components of which can be identified by the manner in which the transition to As is induced. The apparent lack of effectiveness of the estrogen which accompanies progesterone during the first days of pregnancy or the first two progesterone injections in the PE_2x3 model is enigmatic. In the pregnant animal the existence of the As (D) may be transitional with the cell passing through that state of differentiation so rapidly that it does not seem to exist at all (mouse model).

The unique responses of the As (I) stroma, i.e., loss of transcriptive capacity, receptivity to the blastocyst, and the loss of uterine sensitivity, suggest qualitative differences in the information being transcribed as a result of estrogen modulation of the progesterone induced genome. Finally, another strong possibility is that the As (I) cell with its singular potential is the progeny of the As (D) cell. Thus, the pathway to stromal cell differentiation which results in the transformation of the sensitive uterus to one receptive to ovo-implantation is intimately tied with that facet of estrogen hormone action which stimulates stromal cell proliferation (Fig. 23).

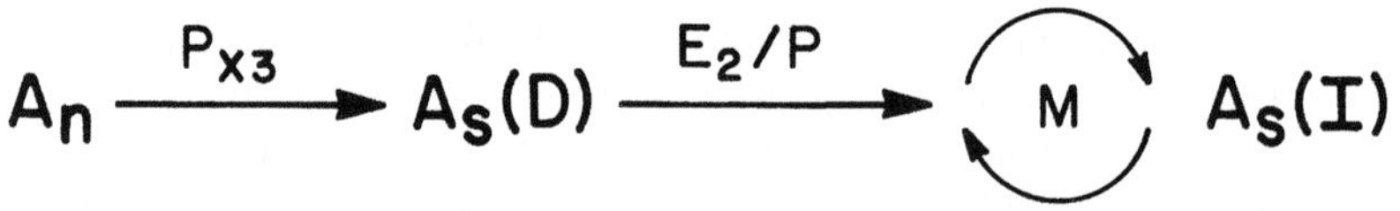

Fig. 23. A restatement of Fig. 21. The ultimate differentiation of the stromal cell which participates in implantation, As (I), is a function of E_2 stimulated mitosis. The I form of the As cell is a daughter of the D form

In terms of one definition of differentiation (Tsanev 1975), these cells are "reprogrammed" at each of these steps. Less mechanistically, although we are really deficient in receptor binding data, we should talk about these cells in terms of gene restriction or expression. Thus, among the RNA species being increasingly synthesized, a species of RNA appears on day 4 of pregnancy while the blastocyst is free in the uterus which is not present in the decidual cells on day 6. The transcription of this putatively unique RNA (mRNA?) is the product of progesterone-induced derepressed chromatin. This increased transcriptive capacity appears early on day 4. Chromatin from days 3 and 6 will apparently not transcribe this product. The identity of this new RNA species remains to be proven and its role in the induction of decidualization and/or implantation must be validated.

Further selectivity with respect to gene expression may be exerted by differences in qualitative and quantitative characteristics of the translation process. Differential hormonal induction of gene expression modulated through the dynamics of specific steroid hormone receptor complexes could provide the necessary quality of sequential control.

This ordering of genetic information could specify the target cell (receptor synthesis) and move that target cell through the required transformations (neutral → sensitive → receptive).

Failing the necessary external stimulus, the As (D) cell will continue to maintain the net synthesis required to maintain the sensitive state (diapause, delayed implantation). Should the hormonal states of the cell be shifted (by estradiol) the genetic expression of the cell will be altered. The new form As (I) cell will be a landmark for ovoreceptivity. In the absence of threshold deciduogenic stimulation the As (I) cell, unlike As (D), will lose its sensitivity and regress, not passively but as a consequence of an "off" reaction, to the refractory A_R state before returning to neutrality (A_N). The A_N cell is responsive to progesterone or progesterone plus estrogen but the nature of the regressive and refractory mechanisms is not known.

These considerations would modify the more classic definitions of differentiation. The process is defined in terms of the mechanisms which make information not readily available in a particular mother cell readily available in its daughter cells (Beermann et al. 1975). This is a rather circumscribed definition in that it fails to consider altered gene function in a mother cell which is not mitotically linked. The appropriate definition must also consider cell division as a consequence of altered (hormonally) gene expression, rather than a cause, and must not only consider the new information available but what proportion of this information is expressed.

Not all hormonal systems which will permit decidualization to be induced will stimulate stromal mitosis, but hormonal systems that stimulate stromal mitosis will all support the induction of decidual cells. These are the same hormonal conditions that are involved in blastocyst induced decidualization so that, with the fortuitious selection of the appropriate animal model system, evidence can be obtained circumscribing the transformation of stromal cells to decidual cells under the aegis of estrogen action to a mitotically linked induction. There is more than adequate evidence that cell division is not a prerequisite for the induction of decidualization.

It is true, however, that the hormonal conditions that favor stromal cell division support the induction of a variant decidual cell differentiated to express unique properties. This cell, A_I(Fig. 21), is indeed dependent for its last stage induction on estrogen induced stromal mitosis. Doses of weak estrogens (antiestrogens, i.e., nafoxidine, tamoxifen) that delay implantation (20 h) will also delay stromal mitoses (noon day 3 to morning day 4), and doses that inhibit one process (implantation) will inhibit the other (stromal mitosis) (Major and Heald 1974; O'Grady et al. 1974; Watson et al. 1975). Cells are recruited for decidualization from definite regions of the subeptihelial antimesometrial stroma. Those hormonal conditions that stimulate stromal mitoses reputedly can recruit cells from deeper regions. The developmental fate of stromal cells derived from different regions (mesometrial *vs* antimesometrial) can be different. This may be a function of their proportionate abilities to respond to hormonal regulations. Except for this broad generalization not very much is known about the hormone responsiveness and developmental fate of different cells of the same region (subepithelial *vs* deep).

The differences noted between the As (D) and As (I) variants of the stromal cell reemphasize that in spite of their obvious utility the DCR model and its attendant biochemistry are imperfect analogs for the final events required for implantation. There are well-meaning investigators, trained in other disciplines, amongst whom there is much confusion so that they equate decidualization with nidation, who are beginning studies in this area. There is no doubt on the basis of present data that these processes are linked. We cannot as yet differentiate between the cells of decidua and deciduoma on a molecular basis. Yet it is equally obvious that receptivity to implantation is more than, or different from, sensitivity to a deciduogenic stimulus. Nevertheless, the study of decidualization remains a forceful and productive tool. It should not be summarily dismissed for it defines certain recognizable limits within which implantation occurs.

Acknowledgments. The progress of our own researches reported here would not have been possible but for the expert and unselfish technical assistance of Ms. Gail Brady and Mr. Alan Gunn. Our investigations were supported by NIH Grants HD 08671, HD 07495, and HD 12964. During the course of these studies S.A.M. was a post-doctoral fellow in this laboratory supported by U.S. Public Health Service Fellowship (CA 02742) from the National Cancer Institute.

References

Allen E, Smith GM, Gardner WU (1937) Accentuation of the growth effect of Theelin on genital tissues of the ovariectomized mouse by arrest of mitosis with colchicine. Am J Anat 61:321-341

Anderson JN, Clark JH, Peck EJ (1972a) Oestrogen and nuclear binding sites: Determination of specific sites by [^{3}H] oestradiol exchange. Biochem J 126:561-567

Anderson JN, Clark JH, Peck EJ (1972b) The relationship between nuclear receptor-estrogen binding and uterotrophic responses. Biochem Biophys Res Commun 48:1460-1468

Asch RH, Fernandez EO, Magnasco L, Pauerstein CJ (1978) Demonstration of a chorionic gonadotropin-like substance in rabbit morulae. Fertil Steril 29:444-446

Asch RH, Fernandez EO, Siler-Kohdr TM, Pauerstein CJ (1979) Evidence for a human chorionic gonadotropin-like material in the rabbit blastocyst. Fertil Steril 32:697-703

Barry J, Gorski J (1971) Uterine ribonucleic acid polymerase. Effect of estrogen on nucleotide incorporation into 3' chain termini. Biochemistry 10:2384-2390

Baserga R (1965) The relationship of the cell cycle to tumor growth and control of cell division: A review. Cancer Res 25:581-595

Baserga R (1968) Biochemistry of the cell cycle: A review. Cell Tissue Kinet 1:167-191

Beermann W, Holtzer H, Reinert J, Ursprung H (1975) Preface. In: Reinert J, Holtzer H (eds) Cell cycle and cell differentiation. Springer, Berlin Heidelberg New York, pp 1-

Billing RJ, Barbiroli B, Smellie RMS (1969) Changes in the patterns of synthesis of ribonucleic acid species in immature rat uterus in response to oestradiol-17β. Biochem J 112:563-569

Borthwick NM, Smellie RMS (1975) The effects of oestradiol-17β on the ribonucleic acid polymerase of immature rabbit uterus. Biochem J 147:91-101

Bouton MM (1976) Differential effects of estradiol treatment on the number and activity of rat uterine RNA polymerase molecules. Abstract Volume, A-39, Fifth Intl. Congress of Endocrinology, Hamburg, p 16

Bresciani F (1965) A comparison of the cell generative cycle in normal, hyperplatic and neoplastic mammary gland of the C3H mouse. In: Anderson Hosp. and Tumor Inst. (ed) Cellular-radiation biology. William & Wilkins, Baltimore, pp 547-557

Britten RJ, Kohne DE (1968) Repeated sequences in DNA. Science 161:529-540

Browning JY, Keyes PL, Wolf RC (1980) Comparison of serum progesterone, 20α-dihydro-progesterone, and estradiol-17β in pregnant and pseudopregnant rabbits: Evidence for postimplantation recognition of pregnancy. Biol Reprod 23:1014-1019

Buller RE, Toft DO, Schrader WT, O'Malley BW (1975) Progesterone-binding components of chick oviduct: VIII receptor activation and hormone-dependent binding to purified nuclei. J Biol Chem 250:801-808

Buller RE, O'Malley BW (1976) The biology and mechanism of steroid hormone receptor interaction with the eukaryotic nucleus. Biochem Pharmacol 25:1-12

Bullock DW (1977) Progesterone induction of messenger RNA and protein synthesis in rabbit uterus. Ann NY Acad Sci 286:260-272
Bullough WS (1955) Hormones and mitotic activity. Vitam Horm 13:261-292
Bullough WS (1967) The evolution of differentiation. Academic Press, New York
Cake MH, Litwack G (1975) The glucocorticord receptor. In: Litwack G (ed) Biochemical actions of hormones. vol 3. Academic Press, New York, pp 319-390
Callan HG (1973) DNA replication in the chromosomes of eukaryotes. Cold Spring Harbor Symp Quant Biol 38:195-203
Channing CP, Stone SL, Sakai CN, Haour F, Saxena BB (1978) A stimulatory effect of the fluid from preimplantation rabbit blastocysts upon luteinization of monkey granulosa cell cultures. J Reprod Fertil 54:215-220
Church RB, McCarthy BJ (1970) Unstable nuclear RNA synthesis following estrogen stimulation. Biochim Biophys Acta 199:103-114
Clark JH, Anderson J, Peck EJ (1972) Receptor-estrogen complex in the nuclear fraction of rat uterine cells during the estrous cycle. Science 176:528-530
Clark JH, Peck EJ, Glasser SR (1977a) Mechanism of action of sex steroid hormones in the female. In: Cole HH, Cupp PT (eds) Reproduction in domestic animals, 3rd edn. Academic Press, New York, pp 143-174
Clark JH, Anderson JW, Hsueh AJ, Eriksson HA, Hardin JW, Peck EJ (1977b) Chromatin and chromosomal protein research. In: Stein G and Stein J (eds) Methods in cell biology, Academic Press, New York, pp 335-348
Courvalin JC, Bouton MM, Baulieu EE, Nuret P, Chambon P (1976) Effect of estradiol on rat uterus DNA-dependent RNA polymerases (Studies on solubilized enzymes). J Biol Chem 251:4843-4849
Crick F (1971) General model for the chromosomes of higher organisms. Nature 234:25-27
Das RM (1972) The effects of estrogen on the cell cycle in epithelial and connective tissues of the mouse uterus. J Endocrinol 55:21-30
DeAngelo AB, Gorski J (1970) Role of RNA synthesis in the estrogen induction of a specific uterine protein. Proc Natl Acad Sci USA 66:693-700
DeFeo VJ (1963a) Temporal aspect of uterine sensitivity in the pseudopregnant or pregnant rat. Endocrinology 72:305-316
DeFeo VJ (1963b) Determination of the sensitive period for the induction of deciduomata in the rat by different inducing procedures. Endocrinology 73:488-497
DeFeo VJ (1967) Decidualization. In: Wynn RM (ed) Biology of the uterus, 1st edn. Appleton Century Crofts, New York, pp 192-290
Denari JH, Germino NI, Rosner JM (1976) Early synthesis of uterine proteins after a decidual stimulus in the pseudopregnant rat. Biol Reprod 15:1-8
Eide A (1975) The effect of estradiol on the cell kinetics in the uterine and cervical epithelium of neonatal mice. Cell Tissue Kinet 8:249-257
Enea A, Allfrey VG (1973) Selective synthesis of liver nuclear acidic proteins following glucagon administration *in vivo*. Nature 242:265-267
Epifanova OI (1966) Mitotic cycles in estrogen-treated mice: A radioautographic study. Exp Cell Res 42: 562-577
Epifanova OI (1971) Effects of hormones on the cell cycle. In: Baserga R (ed) Cell cycle and cancer. Dekker, New York, pp 143-190
Fainstat T (1963) Extracellular studies of the uterus. (1) Disappearance of the discrete collagen bundles in endometrial stroma during various reproductive states in the rat. Am J Anat 112:337-369
Finn CA (1971) The biology of decidual cells. In: Bishop MWH (ed) Advances in reproductive physiology, vol 5. Academic Press, New York, pp 1-27
Finn CA, Martin L (1970) The role of the oestrogen secreted before oestrus in the preparation of the uterus for implantation in the mouse. J Endocrinol 47:431-438
Finn CA, Martin L (1973) Endocrine control of the timing of endometrial sensitivity to a decidual stimulus. Biol Reprod 7:82-86
Flint APF, Burton RD, Gadsby JE, Saunders PTK, Heap RB (1979) Blastocyst oestrogen synthesis and the maternal recognition of pregnancy. In: Heap RB, Whelan J (eds) Maternal Recognition of Pregnancy. Ciba Foundation Symp. 64 (New series), Excerpta Medica, Amsterdam, Oxford, New York, pp 209-238

Galand P, Rodesch F, Leroy F, Chretien J (1967) Radioautographic evaluation of the oestrogen-dependent proliferative pool in the stem cell compartment of the mouse uterine and vaginal epithelia. Exp Cell Res 48:595-604

Galassi L (1968) Autoradiographic study of the decidual cell reaction in the rat. Dev Biol 17:75-84

Gelfant S (1966) Patterns of cell division: The demonstration of discrete cell populations. In: Prescott D (ed) Methods in cell physiology, vol 2. Academic Press, New York, pp 359-395

Glasser SR (1972) The uterine environment in implantation and decidualization. In: Balin HA, Glasser SR (eds) Reproductive biology. Excerpta Medica, Amsterdam, pp 776-883

Glasser SR (1975) A molecular bioassay for progesterone and related compounds. In: O'Malley BW, Hardman J (eds) Methods in enzymology, vol. 36. Academic Press, New York, pp 456-465

Glasser SR, Clark JH (1975) A determinant role for progesterone in the development of uterine sensitivity to decidualization and ovoimplantation.In: Markert C, Papaconstantinou J (eds) The developmental biology of reproduction. Academic Press, New York, pp 311-345

Glasser SR, McCormack SA (1979a) Estrogen-modulated uterine gene transcription in relation to decidualization. Endocrinology 104:1112-1118

Glasser SR, McCormack SA (1979b) Functional development of rat trophoblast and decidual cells during establishment of the hemochorial placenta. Adv. Biosciences 25:165-197

Glasser SR, McCormack SA (1980) Role of stromal cell function diversity in altering endometrial cell responses. In: Leroy F, Finn CA, Psychoyos A, Hubinont PO (eds) Blastocyst-endometrium relationships. Progr. reprod Biol, vol 7, Karger Basel, pp 102-114

Glasser SR, McCormack SA (1981) Separated cell types as analytical tools in the study of decidualization and implantation. In: Glasser SR, Bullock DW (eds) Cellular and molecular aspects of implantation. Plenum,New York, pp 217-239

Glasser SR, Spelsberg TC (1973) Differential modulation of estrogen induced activity of uterine nuclear RNA polymerases (I and II) by cyclohexamide and actinomycin D. Abstract vol. Endocrine Society (55th Annual Meeting, p 80)

Glasser SR, Chytil F, Spelsberg TC (1972a) Early effects of oestradiol-17β on the chromatin and activity of the deoxyribonucleic acid-dependent ribonucleic acid polymerases (I and II) of the rat uterus. Biochem J 130:947-957

Glasser SR, Spelsberg TC, Chytil F, O'Malley BW (1972b) Progesterone induced changes in template capacity of uterine chromatid related to uterine sensitivity (Abstr) Biol Reprod 7: p 141

Glasser SR, Clark JH, Peck EJ (1974) Uterine nuclear estrogen receptor complexes and RNA polymerase activity. J Steroid Biochem 5:341

Glasser SR, Clark JH, Hardin JW (1976) Differential effect of estradiol on changes in the number of uterine chromatin initiation sites. Abstract vol., A-40, Fifth Internat. Congr. of Endocrinol. Hamburg p 16

Glasser SR, Clark JH, Smith RG, O'Malley BW (1977) Mechanism of steroid hormone action. In: Fuchs F, Klopper A (eds) Endocrinology of pregnancy, 2nd ed. Harper & Row, New York, pp 15-40

Gorski J (1964) Early estrogen effects on the activity of uterine ribonucleic acid polymerase. J Biol Chem 239:889-892

Gorski J, Gannon F (1976) Current models of steroid hormone action: A critique. An. Rev Physiol 38:425-450

Gorski J, Noteboom WD, Nicolette JA (1965) Estrogen control of the synthesis of RNA and protein in the uterus. J Cell Comp Physiol [Suppl 1] 66:91-110

Gurpide E, Welch M (1969) Dynamics of uptake of estrogens and androgens by human endometrium: Application of a double isotope perfusion technique. J Biol Chem 244:5159-5169

Gurpide E, Stolee A, Tseng L (1971) Quantitative studies of tissue uptake and disposition of hormones. In: Diczfalusy E (ed) *In vitro* methods in reproductive cell biology. Karolinska symposium on research methods in reproductive endocrinology. Karolinska Institutet, Stockholm, pp 247-278

Hamilton TH (1968) Control by estrogen of genetic transcription and translation. Science 161:649-661

Hamilton TH, Widnell CC, Tata JR (1965) Sequential stimulations by oestrogen of nuclear RNA synthesis and DNA-dependent RNA polymerase activities in rat uterus. Biochim Biophys Acta 108: 168-172

Hamilton TH, Teng CS, Means AR (1968) Early estrogen action: Nuclear synthesis and accumulation of protein correlated with enhancement of two DNA-dependent RNA polymerase activities. Proc Natl Acad Sci USA 59:1265-1272

Haour F, Saxena BB (1974) Detection of a gonadotropin in rabbit blastocyst before implantation. Science 185:444-445

Hardin JW, Clark JH, Glasser SR, Peck EJ (1976) Estrogen receptor binding by uterine nuclei: relationship to endogenous nuclear RNA polymerase activity. Biochemistry 15:1370-1374

Heald PJ (1976) Biochemical aspects of implantation. J Reprod Fertil. [Suppl] 25:29-52

Heald PJ, O'Grady JE, O'Hare A, Vass MA (1972) Changes in uterine RNA during early pregnancy in the rat. Biochim Biophys Acta 262:66-74

Heald PJ, Govan ADT, O'Grady JE (1975a) A simple method for the preparation of luminal epithelium and stromal cells from the rat uterus. J Reprod Fertil 45:593-595

Heald PJ, O'Grady JE, O'Hare A, Vass M (1975b) Nucleic acid metabolism of cells of the luminal epithelium of the rat uterus in early pregnancy. J Reprod Fertil 45:124-138

Holt JA, Heise WF, Wilson SM, Keyes PL (1976) Lack of gonadotrophic activity in the rabbit blastocyst prior to implantation. Endocrinology 98:904-909

Hsueh AJW, Peck EJ, Clark JH (1976) Control of uterine estrogen receptor levels by progesterone. Endocrinology 98:438-444

Jensen EV, Jacobson HI (1962) Basic guides to the mechanism of estrogen action. Recent Prog Horm Res 18:387-414

Katzenellenbogen BS, Gorski J (1975) Estrogen action on synthesis of macromolecules as target cells. In: Litwack G (ed) Biochemical actions of hormones, vol 3. Academic Press, New York, pp 188-245

Kaye AM, Sheratzky D, Lindner HR (1972) Kinetics of DNA synthesis in immature rat uterus: Age dependence and estradiol stimulation. Biochim Biophys Acta 261:475-486

King RJB, Mainwaring WIP (1974) Steroid-cell interactions. Butterworth, London

Kleinfeld RG, Morrow HA, DeFeo VJ (1976) Intercellular junctions between decidual cells in the growing deciduoma of the pseudopregnant rat uterus. Biol Reprod 15:593-603

Knowler JT, Smellie RMS (1973) The oestrogen-stimulation synthesis of heterogeneous nuclear ribonucleic acid in the uterus of immature rats. Biochem J 131:689-697

Leake R, McNeil W, Black M (1975) Steroids and the control of deoxyribonuclease acid synthesis in the uterus. Biochem Soc Trans 3:1180-1183

Lejeune B, Van Hoeck J, Leroy F (1981) Transmitter role of the luminal uterine epithelium in the induction of decidualization in rats. J Reprod Fertil 61:235-240

Leroy F, Galand P (1969) Radioautographic evaluation of mitotic parameters in the endometrium during the uterine sensitivity of the pseudopregnant rat. Fertil Steril 20:980-992

Leroy F, Bogaert C, Van Hoeck J, Delcroix C (1974) Cytophotometric and autoradiographic evaluation of cell kinetics in decidual growth in rats. J Reprod Fertil 39:207-211

Leroy F, Bogaert C, Van Hoeck J (1975) DNA replication and accelerated chain growth in oestrogen-stimulated uterine cells. Nature 258:259-260

Liarakos CD, Rosen JM, O'Malley BW (1973) Effect of estrogen on gene expression in the chick oviduct. II. Transcription of chick tritiated unique deoxyribonucleic acid as measured by hybridization in ribonucleic acid excess. Biochemistry 12:2809-2816

Luck DN, Hamilton TH (1972) Early estrogen action: Stimulation of the metabolism of high molecular weight and ribosomal RNAs. Proc Natl Acad Sci USA 69:151-161

Major J, Heald PJ (1974) The effects of ICI 46, 474 on ovum transport and implantation in the rat. J Reprod Fertil 36:117-124

Marcus GJ (1974) Mitosis in the rat uterus during the estrous cycle. Early pregnancy and early pseudopregnancy. Biol Reprod 10:447-452

Martel D, Psychoyos A (1976) Endometrial content of nuclear estrogen receptor and receptivity for ovoimplantation in the rat. Endocrinology 99:470-475

Martin L, Finn CA (1968) Hormonal regulation of cell division in epithelial and connective tissues of the mouse uterus. J Endocrinol 41:363-371

Martin L, Finn CA (1969) Duration of progesterone treatment required for a stromal response to estradiol 17-β in the uterus of the mouse. J Endocrinol 44:279-280

Martin L, Finn CA (1971) Oestrogen-gestagen interactions on mitosis in target tissues. In: Hubinont PO, Leroy F, Galand P (eds) Basic actions of sex steroids on target organs. Karger, Basel, pp 172-188

Martin L, Finn CA, Carter J (1970) Effects of progesterone and oestradiol 17-β on the luminal epithelium of the mouse uterus. J Reprod Fertil 21:461-469

Martin L, Finn CA, Trinder G (1973a) Hypertrophy and hyperplasia in the mouse uterus after oestrogen treatment: An autoradiographic study. J Endocrinol 56:133-144
Martin L, Finn CA, Trinder G (1973b) DNA synthesis in the endometrium of progesterone-treated mice. J Endocrinol 56:303-307
Martin L, Das RM, Finn CA (1973c) The inhibition by progesterone of uterine epithelial proliferation in the mouse. J Endocrinol 57:549-554
Martin L, Pollard JW, Fagg B (1976) Oestriol, oestradiol-17β and the proliferation and death of uterine cells. J Endocrinol 69:103-115
McCormack SA, Glasser SR (1976) A high-affinity estrogen-binding protein in rat placental trophoblast. Endocrinology 99:701-712
McCormack SA, Glasser SR (1978) Ontogeny and regulation of a rat placental estrogen receptor. Endocrinology 102:273-280
McCormack SA, Glasser SR (1980) Diferential response of individual uterine cell types from immature rats treated with estradiol. Endocrinology 106:1634-1649
Means AR, O'Malley BW (1972) Mechanism of estrogen action: Early transcriptional and translational events. Metabolism 21:357-370
Means AR, Hamilton TH (1966) Early estrogen action: Concomitant stimulations within two minutes of nuclear RNA synthesis and uptake of RNA precursor by the uterus. Proc Natl Acad Sci USA 56:1594-1598
Mendelsohn ML (1962) Autoradiographic analysis of cell proliferation in spontaneous breast cancer of C3H mouse. III. The growth fraction. J Natl Cancer Inst 28:1015-1029
Meyers KP (1970) Hormonal requirements for the maintenance of oestradiol-induced inhibition of uterine sensitivity in the ovariectomized rat. J Endocrinol 46:341-346
Moriyama I (1974) Effect of estrogen and progesterone on the characteristics of newly synthesized RNA in uterine tissue. Endocrinol Jpn 21:39-49
Mueller GC, Herranen AM, Jervell KF (1958) Studies on the mechanism of action of estrogens. Recent Prog Horm Res 14:95-139
Murphy BEP (1968) Binding of testosterone and estradiol in plasma. Can J Biochem 46:299-302
Nicolette JA, Babler M (1972) The regulatory role of protein in the estrogen-stimulated change in the nearest-neighbor frequency of rat uterine RNA synthesized *in vitro*. Arch Biochem Biophys 149:183-188
Nimrod A, Ladany S, Lindner HR (1972) Prenidatory ovarian oestrogen secretion in the pregnant rat, determined by gas chromatography with electron capture detection. J Endocrinol 53:249-260
Noyes RW, Dickmann Z, Doyle LL, Gates AH (1963) Ovum transfers, synchronous and asynchronous, in the study of implantation. In: Enders AC (ed) Delayed implantation. University of Chicago Press, Chicago, pp 197-212
Nutting EF, Meyer RK (1963) Implantation delay, nidation and embryonal survival in rats treated with ovarian hormones. In: Enders AC (ed) Delayed implantation. University of Chicago Press, Chicago, pp 233-252
O'Grady JE, Bell SC (1977) The role of the endometrium in blastocyst implantation. In: Johnson MH (ed) Development in mammals, vol 1, North Holland, Amsterdam, pp 165-244
O'Grady JE, Heald PJ (1976) Uterine nucleic acid and phospholipid metabolism in the early stages of pregnancy. J Endocrinol 68:33-34P
O'Grady JE, Armstrong EM, Moore IAR, Vass MA (1974) Effect of tamoxifen (ICI-46, 474) on mitosis in the uterus of the rat during the early stages of pregnancy. J Endocrinol 63:19P
O'Grady JE, Moffat GE, McMinn L, Vass MA, O'Hare A, Heald PJ (1975) Uterine chromatin template activity during the early stages of pregnancy in the rat. Biochim Biophys Acta 407:125-132
O'Malley BW, Means AR (eds) (1973) Receptors for reproductive hormones., vol. 36: Advances in experimental medicine and biology. Plenum Press, New York
O'Malley BW, Means AR (1974) Female steroid hormones and target cell nuclei. Science 183:610-620
O'Malley BW, Spelsberg TC, Schrader WT, Chytil F, Steggles AW (1972) Mechanism of interaction of a hormone-receptor complex with the genome of a eukaryotic target cell. Nature 235:141-144
Peckham B, Kiekhofer W (1962) Cellular behavior in the vaginal epithelium of estrogen-treated rats. Am J Obstet Gynecol 83:1021-1027
Perry JS, Heap RB, Burton RD, Gadsby JE (1976) Endocrinology of the blastocyst and its role in the establishment of pregnancy. J Reprod Fertil (Suppl) 25:85-104

Peck EJ, Clark JH (1974) The effect of physical state on the quantitation of the estrogen receptor. J Steroid Biochem 5:327-328
Peck EJ, Burgner J, Clark JH (1973a) Estrophilic binding sites of the uterus. Relation to uptake and retention of estradiol *in vitro*. Biochemistry 12:4596-4603
Peck EJ, DeLibero J, Richards R, Clark JH (1973b) Instability of the uterine estrogen receptor under *in vitro* conditions. Biochemistry 12:4603-4608
Perrotta CA (1962) Initiation of cell proliferation in the vaginal and uterine epithelia of the mouse. Am J Anat 111:195-201
Perrotta CA (1966) Effect of x-irradiation on DNA synthesis in the uterine epithelium. Radiat Res 28:232-242
Peterson RP, Spaziani E (1971) Autoradiographic localization of albumin-^{131}I accumulation in the estrogen-stimulated rat uterus. Endocrinology 89:1280-1286
Pilgrim CH, Maurer W (1965) Autoradiographische Untersuchung über die Konstanz der DNS-Verdopplungs-Dauer bei Zellarten von Maus und Ratte durch Doppelmarkierung mit ^{3}H- und ^{14}C-thymidin. Exp Cell Res 183-199
Pinsker MC, Sacco AG, Mintz B (1974) Implantation-associated proteinase in mouse uterine fluid. Dev Biol 38:285-290
Pollard JW, Martin L (1975) Cytoplasmic and nuclear non-histone proteins and mouse uterine cell proliferation. Mol Cell Endocrinol 2:183-191
Psychoyos A (1969) Hormonal requirement for egg implantation. In: Raspe G (ed) Adv. Biosciences Vol IV: Mechanisms involved in conception. Pergamon, London, pp 275-290
Psychoyos A (1973a) Hormonal control of ovoimplantation. Vitam Horm 31:201-256
Psychoyos A (1973b) Endocrine control of egg implantation. In: Greep RO, Astwood EB (eds) Handbook of physiology, sect 7, vol II 2. American Physiological Society, Washington, pp 187-215
Psychoyos A, Alberga A, Baulieu EE (1968) Incorporation de l'oestradiol par l'endométrie de rattes castrées prétraitées par differentes séquences hormonales. CR Acad Sci (Paris) 266:1407-1409
Quastler H (1960) Cell population kinetics. Ann NY Acad Sci 90:580-591
Quastler H (1963) The analysis of cell population kinetics. In: Lamerton LF, Fry RJM (eds) Cell proliferation. Blackwell, Oxford, pp 18-36
Raynaud JP (1973) Influence of rat estradiol binding plasma protein (EBB) on uterotrophic activity. Steroids 21:249-259
Rochefort H, Baulieu EE (1969) New *in vitro* studies of estradiol binding in castrated rat uterus. Endocrinology 84:108-116
Salomon DS, Sherman MI (1975) Implantation and invasiveness of mouse blastocysts on uterine monolayers. Exp Cell Res 90:261-268
Schrader WT, O'Malley BW (eds) (1981) Laboratory methods manual for hormone action and molecular endocrinology. Houston Biological Associates, Houston
Schwartz RJ, Tsai M-J, Tsai SY, O'Malley BW (1975) Effect of estrogen on gene expression in the chick oviduct. V. Changes in the number of RNA polymerase binding and initiation sites in chromatin. J Biol Chem 250:5175-5182
Shelesnyak MC (1960) Nidation of the fertilized ovum. Endeavour 19:81-86
Shelesnyak MC, Marcus GJ (1971) Steroidal conditioning of the endometrium for nidation. In: Raspe G (ed) Schering symposium on intrinsic and extrinsic factors in early mammalian development. Pergamon, New York: Advances in biosciences, vol VI, pp 303-316
Shelton KR, Allfrey VG (1970) Selective synthesis of a nuclear acidic protein in liver cells stimulated by cortisol. Nature 282:132-134
Sherman MI, Salomon DS (1975) The relationship between the early mouse embryo and its environment. In:Markert C, Papconstantinou J (eds) Developmental biology of reproduction. Academic Press, New York, pp 277-310
Singh MM, Adams CE (1978) Luteotrophic effect of the rabbit blastocyst. J Reprod Fertil 53:331-333
Smith JA, Martin L (1973) Do cells cycle? Proc Natl Acad Sci USA 70:1263-1267
Smith JA, Martin L (1974) Regulation of cell proliferation. In: Padilla GM, Cameron TL, Zimmerman Z (eds) Cell cycle controls. Academic Press, New York, pp 43-60
Smith JA, Martin L, King LJ, Vertes M (1970) Effects of oestradiol 17-β and progesterone on total and nuclear protein synthesis in epithelial and stromal tissues of the mouse uterus and of progesterone on the ability of these tissues to bind oestradiol-17β. Biochem J 119:773-784

Soloff MS, Creanage JE, Potts GO (1971) Unique estrogen-binding properties of rat pregnancy plasma. Endocrinology 88:427-432

Spelsberg TC, Steggles AW, Chytil F, O'Malley BW (1972) Progesterone-binding components in chick oviduct. V. Exchange of progesterone binding capacity from target to non-target tissue chromatins. J Biol Chem 247:1368-1374

Spelsberg TC, Webster R, Pikler GM (1975) Multiple binding sites for progesterone in hen oviduct nucleus: Evidence that acidic proteins represent the acceptors. In: Stein GS, Kleinsmith LJ (eds) Chromosomal proteins and their role in the regulation of gene expression. Academic Press, New York, pp 153-186

Steggles AW, Spelsberg TC, Glasser SR, O'Malley BW (1971) Soluble complexes between steroid hormones and target-tissue receptors bind specifically to target tissue chromatin. Proc Natl Acad Sci USA 68:1479-1482

Stockdale FE, Topper YJ (1966) The role of DNA synthesis and mitosis in hormone-dependent differentiation. Proc Natl Acad Sci USA 56:1283-1289

Surani MAH (1975) Hormonal regulation of proteins in the uterine secretions of ovariectomized rats and the implications for implantation and embryonic diapause. J Reprod Fertil 43:411-417

Swann MM (1958) The control of cell division: A review. II. Special mechanisms. Cancer Res 18: 1118-1160

Tachi C, Tachi S (1974) Cellular aspects of ovum implantation and decidualization in the rat. In: Coutino EM, Fuchs F (eds) Physiology and genetics of reproduction, part B. Plenum, New York, pp 263-286

Tachi C, Tachi S, Lindner HR (1972) Modification by progesterone of oestradiol-induced cell proliferation, RNA synthesis and estradiol distribution in the rat uterus. J Reprod Fertil 31:59-76

Teng CS, Hamilton TH (1968) The role of chromatin in estrogen action in the uterus. I. The control of template capacity and chemical composition and the binding of ^{3}H-estradiol. Proc Natl Acad Sci USA 60:1410-1417

Tsai M-J, Schwartz RJ, Tsai SY, O'Malley BW (1975) Effects of estrogen on gene expression in the chick oviduct: IV. Initiation of RNA synthesis on DNA and chromatin. J Biol Chem 250:5165-5174

Tsai M-J, Towle HC, Harris SE, O'Malley BW (1976) Effect of estrogen on gene expression in chick oviduct. Comparative aspects of RNA chain initiation in chromatin using homologous versus *E.coli* RNA polymerase. J Biol Chem 251:1960-1968

Tsai SY, Tsai M-J, Schwartz R, Kalimi M, Clark JH, O'Malley BW (1975) Effect of estrogen gene expression in chick oviduct: Nuclear receptor levels and initiation of transcription. Proc Natl Acad Sci USA 72:4228-4232

Tsanev R (1975) Cell cycle and liver function. In: Reinert J, Holtzer H (eds) Cell cycle and cell differentiation. Springer, Berlin Heidelberg New York, pp 197-248

Tsuboi A, Baserga R (1972) Synthesis of nuclear acidic proteins in density-inhibited fibroblasts stimulated to proliferation. J Cell Physiol 80:107-118

Vittorelli ML, Harrison RAP, Lutwak-Mann C (1967) Metabolism of ribonucleic acid in the endometrium of the rabbit during early pregnancy. Nature 214:890-892

Watson J, Anderson FB, Alam M, O'Grady JE, Heald PJ (1975) Plasma hormones and pituitary luteinizing hormone in the rat during the early stages of pregnancy and after post-coital treatment with tamoxifen. J Endocrinol 65:7-17

Williams D, Gorski J (1974) Equilibrium binding of estradiol by uterine cell suspension- and whole uteri *in vitro*. Biochemistry 13:5537-5542

Yamamoto KR, Alberts BM (1976) Steroid receptors. Elements for modulation of eukaryotic transcription. Annu Rev Biochem 45:722-746

Yochim JM, DeFeo VJ (1963) Hormonal control of the onset magnitude and duration of uterine sensitivity in the rat by steroid hormones of the ovary. Endocrinology 72:317-326

Yoshinaga K, Hawkins RA, Stocker JF (1969) Estrogen secretion by the rat ovary *in vivo* during the estrous cycle and pregnancy. Endocrinology 85:103-112

Pharmacological Aspects of Early Pregnancy and Contraception

W. Elger[1] and H.M. Beier[2]

[1] Dept. Endocrine Pharmacology, Schering Research Laboratories, Berlin, Germany

[2] Dept. of Anatomy and Reproductive Biology, Rheinisch-Westfälische Technische Hochschule, Aachen, Germany

1 Regulatory Principles of Reproduction

In mammals, the reproductive process depends at all stages on the coordinated functioning of several physiologically closed systems. In the genital tract itself these are the mucous membranes and muscular structures. Their functioning is dependent on very complex hormonal and nervous controls and also on the blood supply, which changes according to the situation. There is a lot of evidence that various hormones and drugs can have a direct and significant influence on the structure and function of the uterus. Since the ovaries and their regulating systems are also complex, it is understandable that widely varying drugs can also indirectly influence processes in the uterus.

The broad spectrum of the functional elements integrated in the reproductive system explains why the use of pharmacologically defined substances in reproduction research is not confined to experiments with sex hormones and sex hormone antagonists. Experiments with neurotransmitters and neurotropic drugs belong to the routine methods of neuroendocrinology. For example, we owe the enormous increase in our knowledge of the role of prolactin in the ovarian cycle, pregnancy, and lactation to the use of the dopaminergic substance bromergocryptin (CB 154), which inhibits prolactin secretion from the anterior lobe of the pituitary (Meites 1977; Frantz 1978; del Pozo et al. 1974).

Pharmacological experiments with α- and β-adrenergic substances have revealed that the effects of sex hormones on oviduct and uterus motility occur at least in part through the mediation of the sympathetic nervous system of the genital tract (see review of Moawad et al. 1977). The induction of a dominance of α- adrenergic receptors (estrogens) or β-adrenergic receptors (progesterone) determines whether the release of adrenergic neurotransmitters from the nerve endings will lead to an increase of uterine motility (α-dominance) or a reduction of it (β-dominance). For example, an α-receptor blocking counteracts "tube locking" by high doses of estrogens (Pauerstein et al. 1970).

We also owe new knowledge of the regulation of endocrine and generative ovary functions, placenta functions, and physiology of birth to studies with prostaglandins and, in particular, to inhibitors of prostaglandin synthesis (Karim and Hillier 1974; Challis 1974).

These few examples should serve to remind us that defined drugs need not only be useful for practical purposes. As instruments at the molecular level they can be used for purposes of basic reproduction research too.

Inversely, there are only a few examples which indicate the importance of basic research in this field for applied research. This is no doubt partly due to the fact that marked differences exist between the various species as regards the regulation of reproductive processes. This means that information gained from animal experiments can often be applied only to a very limited extent to the human situation. This seems to be particularly valid for pharmacological problems relating to pregnancy. On this basis, the search for animal models relevant to human pregnancy plays a decisive part in the further progress of pharmacologically oriented research.

Methods of hormonal contraception in use at present act mainly via the inhibition of ovulation and/or fertilization. As an alternative, methods of fertility control are conceivable which intervene in processes of early pregnancy or pregnancy in general. An example of the latter is the IUD (intrauterine device). Certainly the most serious subjective and objective disadvantage of the hormone contraceptives used today, and also of the IUD, is the necessity of prophylactic, and that means chronic, treatment. On the other hand, antifertility drugs which affect early pregnancy could provide a means of using durgs only over short periods or even only when necessary.

2 Strategies of Postconceptional Fertility Control

2.1 Synchronization of Embryonic Development and Genital Tract

That the ovary in mammals not only produces ova but especially the corpus luteum (CL), which has the morphological characteristics of an endocrine gland, and that its secretions might be essential for the maintenance of pregnancy was first suspected by the embryologist Born (Fraenkel 1903). With his encouragement, Fraenkel confirmed this suggestion by ovariectomy or lutectomy experiments on women and rabbits, long before the chemical nature of sex hormones was explained (Fraenkel and Cohn 1901; Fraenkel 1903).

It has been proved that in man the loss of the CL before the 6th week after conception leads with certainty to abortion. At a later stage, however, the placenta is able to maintain pregnancy alone (approx. 20 references are mentioned in Fraenkel 1903; Froewis 1963; Csapo et al. 1974).

The functions of progesterone during pregnancy are complex. Some of the various conditions controlled by progesterone remain constant throughout the pregnancy. Thus the cyclical ovarian activity and, in most mammals, psychosexual receptivity is interrupted during this time.

However, the fact that progesterone is unable to maintain blastocysts which reach the uterus too late or too early reveals other dynamic activities of this hormone in early pregnancy. The transitory endometrial receptivity for blastocysts is based on the one hand on the remarkable slowness with which the female genital tract reacts to progesterone, and on the other hand on the inability of the endometrium to retain a functional condition achieved in spite of the continued presence of the inducing hormone.

This endocrine regulation can easily be interpreted as a strategy for the synchronization of uterus and embryo. Estrogen production in the ovary is not diffuse but restricted to the follicles, in which the ovocytes for the coming ovulation also mature. Luteinization of the follicles and the increase in progesterone production are coupled to ovulation. Thus ovarian functions, early embryonic development, and processes in the genital tract relating to pregnancy are accurately tuned to one another.

Progesterone treatment before ovulation or high-dose estrogen treatment after ovulation can prevent nidation in rabbits. With the successful "too early" or "too late" transplantations of blastocysts after corresponding treatment it was possible to confirm that disturbance of the *timing* of early pregnancy is the basis of the antifertile effects mentioned (see the contributions by Beier and by de Visser, this volume).

These temporal relationships of early pregnancy must be borne in mind in pharmacological investigations of sex hormones and their analogues in pregnant animals. The timing of the treatment and the choice of species is often more important for obtaining abortion than the antifertile activity of any particular drug tested (Elger and Haae 1974; Prasad and Kaira 1967; Emmens 1970; Shearman 1973).

Strategies of fertility control aimed at endometrial desynchronization have not been able to achieve great practical importance. In the case of progestogens this is due to the fact that, if used *before* ovulation, they would delay or inhibit ovulation. Postovulatory use of estrogens, except in emergencies after a single exposure, has the disadvantage that large doses are needed. Regular use is also of no avail, owing to the uncertainty of the time of ovulation in women. This prevents accurate use *after conception*, i.e., once in the menstrual cycle.

2.2 The Transition from the Corpus Luteum of the Cycle to the Corpus Luteum of Pregnancy

Another vulnerable point of the reproductive cycle seems to be the mechanisms for the regulation of the life span of the CL. The eutherian mammals which give birth to living young originate from egg-laying ancestors. Phylogenetically the transference of embryonic and fetal development into the uterus made it necessary to solve the problem of extending CL function over the duration of the cycle. On the other hand, in the case, for example, of a nonfertile cycle, it was advantageous to be able to eliminate the CL prematurely in order to make a new ovulation possible. In most investigated mammals there are in fact mechanisms which serve to prolong or shorten luteal life span appropriately. In only a few mammals (e.g., most marsupials, the dog) duration of the cycle and the duration of pregnancy do agree.

The regulation of luteal function differs in many aspects from species to species; however, if the mechanisms for regulating the life span of the CL are interpreted as "strategy", only two basically different types of control are recognizable:

Type 1: The CL degenerated at the end of an infertile cycle either spontaneously or due to a lack of stimulation. In the case of pregnancy, additional stimuli, e.g., by chorionic gonadotrophins, prevent the loss of the CL. This regulatory principle is most distinctly established in primates and in man (Hobson and Wide 1972). An elevated secretion of prolactin from the pituitary gland after coitus (Rabii and Kragt 1972) and the secretion of a prolactinlike hormone from the trophoblast in more advanced stages of gestation have a comparable luteotropic function in rats (Takayama and Greenwald 1973; Matthies and Lyons 1971) and similarly in other rodent species.

Type 2: The CL can survive after the duration of the cycle even without additional stimuli. Its life span is controlled by active destruction (luteolysis). Toward the end of the nonfertile cycle, luteolytically active prostaglandins are released from the uterus in these species.

This regulatory principle is dominant over luteotropic factors in most farm and laboratory animals, e.g., all ruminant species, the pig, horse, rat, guinea pig, rabbit, hamster, and mouse (Pharris 1971; Cooper and Furr 1974). After hysterectomy CL persistence occurs in most of these species, as was first discovered in guinea pigs (Loeb 1923, 1927).

In the latter group of species a single injection of prostaglandins can cause the immediate involution of the CL and thus induce abortion (see Fig. 1).

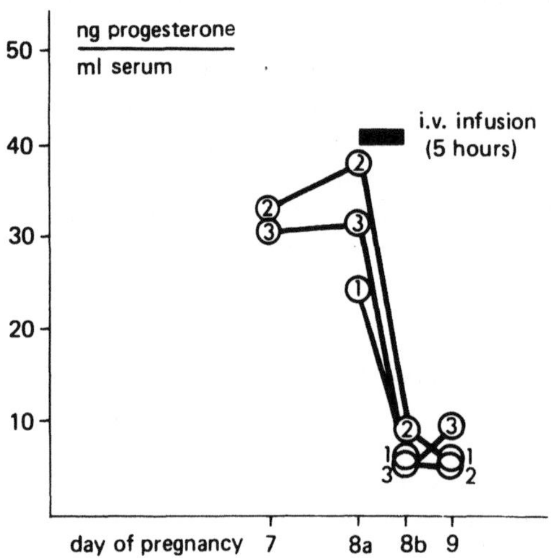

Fig. 1. Serum progesterone values of 3 rats before and after 0.01 mg w-tetranor-16-phenoxy-PGF2α-biphenyl-ester infused in 5 h on day 8 of gestation. Animals were killed 24 h after the end of infusion. Abortion bleeding was obvious first at that time

Artificial luteolysis can be applied for cycle synchronization in domestic farm animals.

Prostaglandins presumably cannot have a luteolytic effect on primates and man (Korda et al. 1975; Lyneham et al. 1975; Coudert et al. 1974). In man, in contrast to the above-named nonprimates, the ovarian cycle is maintained even after removal of the uterus (Doyle et al. 1971; Beling et al. 1970).

By immunologically eliminating the luteotropic factors of the trophoblast, progesterone secretion from the CL in early pregnancy can be excluded in monkeys (Prahalada et al. 1975; Sundaram 1976; Mougdal et al. 1971; Stevens and Crystle 1973) and in man (Stevens and Crystle 1973; Pala et al. 1976). In pregnant monkeys the loss of chorionic gonadotrophins after passive immunization manifests itself in an immediate drop in serum progesterone and subsequent menstruation (Prahalada et al. 1975; Sundaram 1976).

The idea of an "antipregnancy vaccine" seems tempting, since evidently immunization against human chorionic gonadotrophin (HCG) with corresponding preparation of β-subunits of HCG coupled to tetanus toxoid (Talwar et al. 1976; Salahuddin et al. 1976) is possible, without a cross-reaction with human luteinizing hormone occurring, which would result in disturbance of cycle and ovulation (Pala et al. 1976). First clinical trials started in India on the use of these immunological methods of fertility control for a while seemed to justify a certain optimism (Talwar et al. 1976a). However, according to newer reports, the first phase of the study did not proceed without serious difficulties and setbacks.

Large doses of progestogens, estrogens, and androgens (Johansson and Gemzell 1971; Lehmann et al. 1975; Henzl et al. 1973) cause markedly reduced serum progesterone values in man in the early luteal phase. A "luteolytic effect" has also been spoken of in this connection. The effect described seems to be based on an inhibition of luteinizing hormone secretion from the pituitary it can be counteracted by simultaneous administration of HCG. For this reason no "luteolytic" effect of this substance can be expected, particularly in the case of pregnancy (Henzl et al. 1973; Johansson 1971).

2.3 The Induction of Premature Uterine Contractions

Genital tract motility is subject to subtle control by hormonal and nervous regulatory mechanisms during all stages of pregnancy. At the time of nidation, muscular factors are *mainly* responsible for the determination of the nidation site known as "spacing" in polytocous species (Böving 1959). Clearly the disturbance of the sensory and muscular mechanisms involved in nidation can lead to the loss of blastocysts.

Nonmedicated IUDs perhaps take effect via interference in uterine motility. Apart from the difficulty of discovering (Böving 1959), and locating the blastocyst the release of prostaglandins from macrophages attached to the IUD surface should be discussed in this context (Myatt et al. 1975).

These phenomena are certainly not the only possibility for explaining the antifertile effect of nonmedicated and, especially, medicated IUDs (Hagenfeldt 1976).

The only drugs available at present for the induction of therapeutic abortion or menstrual regulation are prostaglandins of the E and F series. In this connection the effect of these drugs on the myometrium in primates, human beings and guinea pigs, but not in various other species is most important (Karim and Hillier 1974; Elger et al. 1976, 1977). This fact again confirms the significance of muscular mechanisms for the maintenance of pregnancy.

2.4 The Inhibition of Progesterone synthesis and Progesterone Activity

Theoretically, substances which inhibit the synthesis or action of progesterone provide optimal properties for contraceptive purposes in the postovulatory phase, since progesterone is undoubtedly the pivotal hormone in pregnancy of humans and animals.

Although there are various drugs which inhibit 3β-OH-steroid dehydrogenase (Goldman 1970) or other crucial steroidogenic enzymes, and thus inhibit progesterone synthesis, none of them seems to be suitable for clinical use. This is perhaps due to the difficulty of achieving a sufficiently high concentration of the substance in the CL. Some substances, the cyanoketone derivatives, must be excluded from clinical trials because of their extremely long half-life in target tissues and because of simultaneous impairment of adrenal steroidogenesis (Goldman 1970).

The search for suitable screening systems to evaluate compounds which are candidates for a substance that reveals an efficient antiprogesterone activity has led to an initially promising enzyme test using diamine oxidase (DAO) activity of rat and hamster uterine tissue as a marker for the progesterone effects (Spilman et al. 1980, 1981a). So far, however, the correlation of the DAO test results with progesterone receptor competition studies have not been satisfactory in experiments with rats, but better in studies with hamster uteri (Spilman et al. 1981b).

As far as we know, there is no compound at present which could be classified biochemically and pharmacologically as a competitive antagonist of progesterone. The main difficulties in the search for a competitive antagonist seem to be of a methodical nature. Thus it is difficult to explain why various 17α-ethyl-pregnane derivatives, with high affinity to the progesterone receptor in vitro, had no abortive effect in vivo (Kanojia et al. 1975). Even when close affinity exists to the progesterone receptor and an abortive effect can be shown as with the compound R 2323 (Raynaud et al. 1975) (see Fig. 2), this does not seem to be sufficient evidence for the assumption that the competitive inhibition of progesterone activity is the basis of the abortive effect.

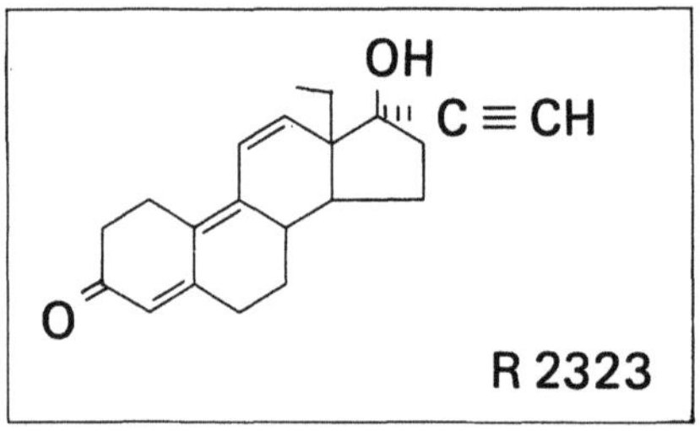

Fig. 2. Formula of R 2323

The same behavior could be expected, for example, of a progestogen with an inherent estrogenic activity (Elger and Haase 1974).

Since the effect of progesterone in the usual laboratory rodents is linked in a complicated manner to interaction with estrogens, it is easy to suppress certain progestogen effects noncompetively in these animals (e.g., rabbit, rat, mouse) by using estrogens. Such effects are often described as antiprogestogenic activity. However, even the very slight estrogenicity of a test substance suffices to obtain abortive effects in rats, for example, during the early stages of pregnancy, using only small doses. It can be very difficult to prove that the estrogenic properties are responsible for this effect, for instance, when the test substance also posesses progestogenic or androgenic properties. The latter activities are able to suppress the cornification of the vagina, which is otherwise a reliable parameter for estrogenicity in rats or mice in the Allen-Doisy bioassays.

Simple additional investigations often make possible the classification of the test substances as regards the abortive mechanism (see Fig. 3). Sometimes chemically and pharmacologically related substances cause abortion by different mechanisms. When, after treatment in early pregnancy, a uterus is found to be without implantation sites, estrogenicity can be excluded as the basis of the abortive effect when blastocysts are traced after rinsing. Inversely, the absence of blastocysts is an indication of estrogenic activity of the test substance. If the abortive effect of a "blastocyst-positive" substance can be counteracted by estrogen substitution, this substance is usually found to have an inhibitive effect on nidation, owing either to antiestrogenicity or to the inhibition of luteinizing hormone secretion. If the abortive effect cannot be neutralized by this treatment, then the possibility of a competitive progesterone antagonist should be considered. Figure 3 shows these possible analyses of abortive effects using three androgens as examples. If it is not recognized that estrogenicity is responsible for the abortive effect in rats, there is a danger of classifying a test substance falsely as "antiprogestogenic" or "antizygotic" and of testing this compound for abortive effects in humans too. In contrast to the usual experimental animals, however, pregnancies in primates are extremely resistant to estrogen after the first 3 days.

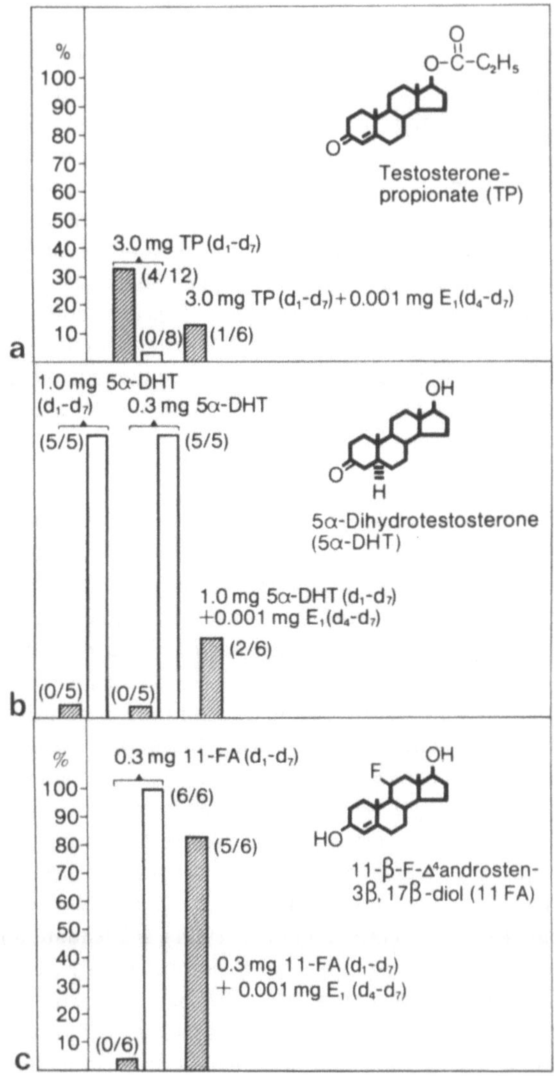

Fig. 3 a-c. Interference of androgenic steroids with early gestation in rats. (▨) Animals with implantation sites at autopsy on day 8 of gestation. (□) Blastocyst positive animals with no implantation sites at autopsy. (Numbers in brackets) Rate of blastocyst positive and implantation site-positive animals. (E_1) Estrone

For extensive reviews on this subject see Emmens (1970), Jacob and McLean Morris (1969), and Blye (1970).

3 Experience of Potential Abortion-Inducing Substances in Clinical Trials

3.1 General Considerations on Testing Induction of Abortion

Up to the present it has only been possible to induce abortion in humans irrespective of the stage of pregnancy with prostaglandins. This is in spite of the fact that there has certainly been no lack of relevant experiments with natural sex hormones and their steroid and nonsteroid analogues. When these (negative) investigations have been published it has often not been clear which consideration actually led to the start of clinical trails for abortion-inducing effects in human beings. It does, however, seem probably that results from animal experiments were the decisive factors. Thus, in cases which were obviously failures with regard to clinical trials it would seem sensible to look for possible errors in the choice of the test substance.

Another problem which must be taken into consideration here is that the completely inadequate abortive effect of a test subtance may be covered up, for example, by an unintentional inhibition of ovulation. This can easily occur when a test substance is used repeatedly in one cycle, e.g., as a "postcoital pill" or "weekly pill". The result could then be a supposed agreement between human and animal trials, which can only be refuted by evidence from other experimental models with human subjects. There are several examples of this in the literature.

3.2 The Antifertile Effect of "Postcoital" Progestogens

Various progestogens have been investigated in clinical trials as "postcoital pills". The question arises whether the available data can support the assumption that this form of application is at least partially *postconceptionally* contraceptive. Analysis of the experimental methods used in some of these clinical trials can clarify the situation.

The first progestogen postcoital pill used in humans was based on quingestanol acetate. This compound is a progestogen with marked estrogenicity and abortion-inducing power in rats (Giannina et al. 1969). There was also a good protection against pregnancies in women taking 0.8 mg p.o. after each coitus (Pearl index [1] of 1) (Rubio et al. 1970). Frequency of ingestion was on average 11 times per cycle in this study. The actual abortion-inducing potential in humans is certainly more realistically demonstrated by the finding that when the frequency of ingestion was lowered in other studies the Pearl index rose to 20 or 36 (Mischler et al. 1974). Inadequate fertility control (Pearl index 6.5) was also reported for *d*-norgestrel given postcoitally (dose 1.0 mg p.o.) (Larranaga et al. 1975). Detailed investigation of endocrine parameters of the cycle during postcoital use of *d*-norgestrel makes it seem likely that the postcoital administration of progestogens works mainly via inhibition of ovulation (Garmendia et al. 1976).

---

[1] Pearl index = number of pregnancies/100 years of use

Experiments which are even more informative than those quoted above have shown that the progestogens of the 19-nortestosterone series, which *in animals* are remarkably estrogenic and, as a result, also abortion-inducing, are hardly suitable for inducing abortion in humans. A Pearl index of 72 (!) was established after doses of 100 mg norethisterone p.o. per day on days 21 and 22 of the cycle, or 50 mg from day 19 to day 22, without any other methods of contraception (Nygren et al. 1972). When lower doses of 10 or 25 mg norethisterone p.o. per day were given earlier, from day 15 of the cycle (day 15 to day 22), on the other hand, a Pearl index of 29 was reached (Nygren et al. 1974).

The motive for these latter studies was admittedly not findings from animal experiments, but the already mentioned evidence of "luteolytic" effects of progestogens in monkeys and humans (Johansson 1971; Johansson and Gemzell 1971).

Experiments with R 2323 in human subjects, however, were obviously conducted because of its characteristics in rabbits, rats, and mice. This steroid has "slightly" progestogenic, estrogenic, and androgenic characteristics in animals, but "particularly potent antiprogestogenic activity" (Sakiz and Azadian-Boulanger 1971).Was the possibility underestimated of the androgenic and progestogenic activities of a test substance on the one hand, and the estrogenic activity on the other, each preventing the expression of the other with regard to some paramaters in rats and rabbits?

In doses of 5 mg p.o. R 2323 was investigated for its antifertile effect as a "weekly pill" in humans. A Pearl index of 3.7 was established by taking into consideration only the method failures (Mora et al. 1974). When specifically administered from day 15 to day 17 of the cycle in doses of 50 mg/day p.o. also a marked antifertile effect was found (Pearl index 4.6 drug failure + 5.5 patient failure) (Sakiz et al. 1974).

Hormonal cycle parameters indicate that neither the inhibition of ovulation nor the abortion-inducing effect are exclusively responsible for the antifertile activity of R 2323. When 5 mg R 2323 was given weekly, more than 50% of the cycles were in fact ovulatory (Tredway and Mishell 1974). Even with double this dose, ovulation occurred in some of the cycles observed (Victor and Johansson 1975). Thus it would seem reasonable to consider that R 2323 also has a postovulatory antifertile action in man. However, even doses of up to 100 mg R 2323 per day were not certain to induce abortion, when given at the time when menstruation should have started. Treatment at a later stage of pregnancy, i.e., when pregnancy was confirmed, was always a failure (Mora et al. 1975). In the light of these hard facts the antiprogestational effect of R 2323 in humans was doubted and its antifertile effect in the above-mentioned trials was considered mainly due to unintentional ovulation suppression (Mora et al. 1975).

3.3 The Antifertile Effect of Estrogens and of Estrogen/Progestogen Combinations Used Postconceptionally

It has already been mentioned that in contrast to most of the rodents usually used in research on reproduction, pregnancy in rhesus monkeys and humans is extremely resistant to estrogen after the first few days. In rhesus monkeys 250 mg diethylstilbestrol (DES)/day i.m. on about the 25th day of pregnancy fail to induce abortion. Neither was abortion induced by estradiol dipropionate, given at the same stage or later in the pregnancy in doses of 4-25 mg/day i.m. (McLean Morris and van Wagenen 1966).

There is no doubt, however, that estrogens taken in large doses very early in a pregnancy ("postcoitally") can prevent its development (for details and references see Haspels 1976; Haspels et al. 1977; Blye 1973; Kuchera 1974; McLean Morris and van Wagenen 1967; Döring 1971; Shearman 1973; Van Santen and Haspels 1980). The doses considered necessary for this therapy are between 25.0-50.0 mg DES p.o. per day for 5 days. Doses of 0.5-2.0 mg ethinylestradiol p.o. per day proved to be on the borderline as regards effectivity (McLean Morris and van Wagenen 1967). Doses of 5.0 mg ethinylestradiol p.o. per day are recommended today for greater safety (McLean Morris and van Wagenen 1967; Haspels and Andriesse 1973). The various authors agree that the time of treatment is crucial and as a rule should not be started later than 48 h after unprotected coitus.

Trials by Bacic et al. (1970) with high doses of ethinylestradiol at a somewhat later stage also document that this abortive effect of estrogens in humans is limited strictly to the first few days after ovulation. When used in the last week of the cycle, doses of up to 5.0 mg ethinylestradiol p.o. per day did not induce abortion at all (Bacic et al. 1970). We know from experiments with DES, which have now achieved sad fame, that the human pregnancy even tolerates much higher doses (Karnaky 1947). In order to test the estrogen-resistance of human pregnancy, doses of 100-200 mg DES/day p.o. or i.m. were administered. The earliest time of treatment was about day 14 of pregnancy. The surprising result was a suspicion that DES has an abortion-protecting effect. This work has been confirmed, at least as regards the lack of abortive effect in humans, by several other authors since the use of DES for the therapy of threatened and habitual abortion became popular among gynecologists in the United States.

It is assumed that the antifertile effect of postcoitally administered estrogens in rodents is due to effects on the genital tract rather than to luteal insufficiency. In any case, simultaneous treatment with progesterone only partially counteracts the effect of postcoital estrogens in rabbits (McLean Morris 1973). The suggestion that in man a similar situation exists is supported by evidence that a postcoitally administered combination of estrogens and progestogens (doses: 0.1 mg ethinylestradiol + 0.1 mg norgestrel, or 0.2 mg ethinylestradiol + 2 mg norgestrel, divided into two doses within 24 h), given in the first 5 days of pregnancy, prevents implantation (Yuzpe et al. 1974; Yuzpe and Lance 1977).

Obviously good results were also obtained with a combination of dienestrol and ethynodiol diacetate (2.5 mg + 0.2 mg p.o.). This dose was given three times within 36 h (Szontagh and Kovacs 1969).

The question arises whether the estrogen component in these combinations is dominant, or whether perhaps the combination has a stronger or different effect to the individual substances it contains.

4 The Influence of Synthetic Sex Hormones on the Uterine Milieu and Maintenance of Gestation Before Implantation

4.1 Significance of Research on Preimplantational Uterine Physiology

The biochemical changes in the uterine milieu under the influence of the CL have been only incompletely investigated until now and in only a few mammalian species, although there can be no doubt that the earliest stage of pregnancy, in particular, is subject to complex, and therefore vulnerable, regulatory mechanisms. Knowledge of these mechanisms is equally important for applied and for basic research.

Research in this connection has been concentrated in recent years mainly on the composition of the uterine secretion in relation to the reproductive cycle. The secretory protein from the rabbit endometrium, uteroglobin (Beier 1966) has been developed as a useful indicator as to whether the uterus is prepared for implantation (Beier 1970, 1973, 1976, 1980; see this volume pp. 39-72; Beier, Mootz and Fischer 1980).

The interest in the significance of this protein for practical considerations was also stimulated by experiments which seem to demonstrate that some potent progestogens are not able to induce the secretion of uteroglobin in the rabbit (Arthur and Chang 1974). These authors concluded that steroids without the C 19 atom cannot induce uteroglobin. Arthur and Chang stress the importance of the presence or absence of uteroglobin-inducing properties for the antifertile effects of progestogens (premature or no induction of uteroglobin). Before this conclusion is discussed, a report will be given of experiments of our own laboratory which were designed to test the reliability of Arthur and Chang's observations.

4.2 Maintenance of Blastocysts by Progestogens in the Rat

If certain progestogens are not able to induce the functional changes in the endometrium which are typical for early pregnancy, it might be expected that they cannot maintain pregnancy in the absence of progesterone. This hypothesis was tested on pregnant rats which were ovariectomized on the 3rd day of pregnancy (1st day = sperm in the vagina at the day following proestrus). Care was taken not to damage the Fallopian tubes during the ovariectomy, since the ova have not reached the uterus at this stage. The survival of the blastocysts in the female genital tract was checked by rinsing the uteri at a later time or by inducing implantation later by means of subsequent progesterone/estrogen treatment (see Fig. 4).

Figure 4 shows that in the absence of progesterone the number of blastocysts in the uterus decreases continuously, and tends toward zero on day 12 of ovariectomy. A similar trend was found after the induction of implantation. In this experimental model, progesterone has a positive effect on the survival of the blastocysts which are free in the uterus under treatment (see Fig. 5).

In this experiment with graded doses of progesterone for 8 days after castration the survival of blastocysts was estimated after induction of nidation, as indicated in Fig. 5. Even doses as low as 125 μg progesterone/day s.c. had statistically significant effects on blastocysts survival. This progesterone dose is much lower than is needed for the maintenance of pregnancy after ovariectomy (3.0-10 mg/day s.c.) in more advanced stages of gestation.

In extensive studies with various progestogens we found analoguous blastocyst-maintaining effects with all the progestogens investigated, irrespective of whether a C19 atom was present (cyproterone acetate, chlormadinone acetate, medroxyprogesterone acetate, etc.) or not present (norethindrone, norethindrone acetate, *d*-norgestrel, etc.). For examples see Figs. 6 and 7.

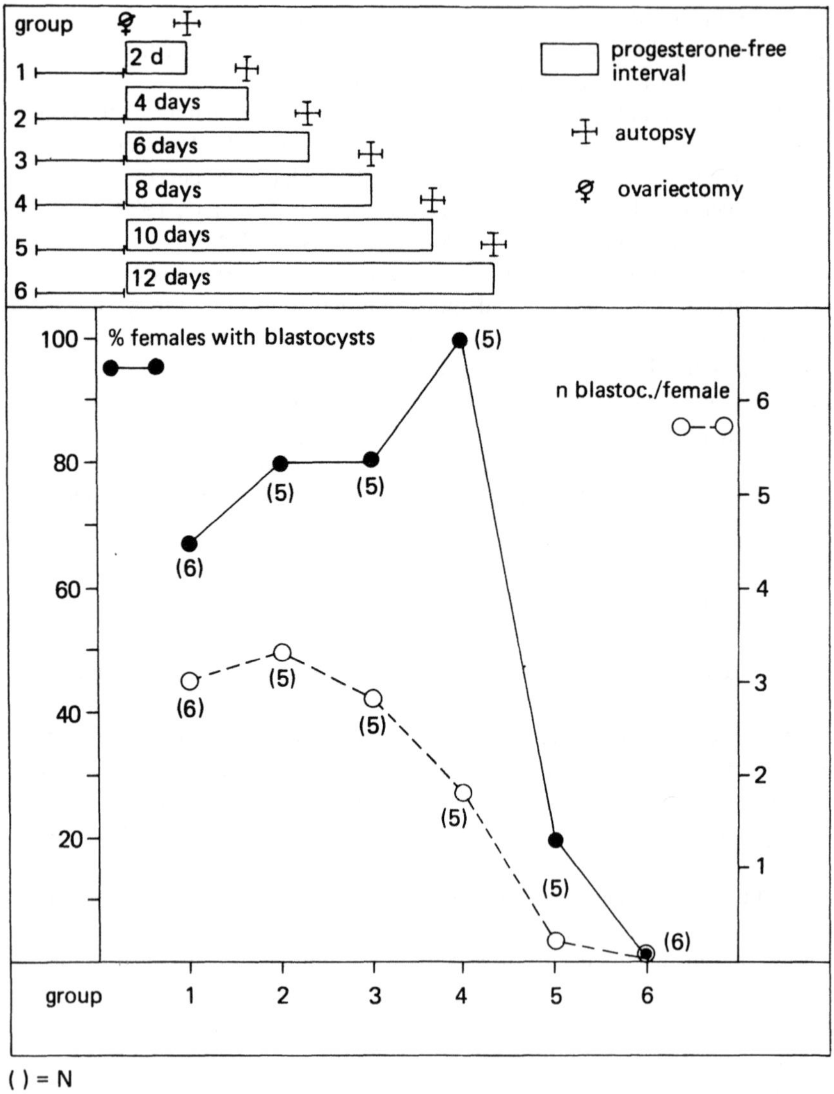

Fig. 4. Survival of blastocysts in the uteri of rats after ovariectomy on day 3 of gestation

In all cases investigated so far, the blastocyst-protecting effect correlated well with other parameters of progestational activity of the tested compounds in rats. Progestogenic compounds with estrogenic side effects induce nidation before estrogen substitution; this was found for higher doses of norethisterone acetate. In the case of norethisterone enanthate, which is relatively more estrogenic, a high proportion of pregnant animals but a very low number of surviving embryos per animal was found (see Fig. 7).

Using progesterone as an example, transplantation experiments were conducted to investigate whether improved implantation conditions would contribute to the "blastocyst-maintaining effect" of progesterone (Noack, Mühlenberg, Elger, unpublished data). Day-5 blastocysts were transferred to the uteri of rats which were receiving progesterone

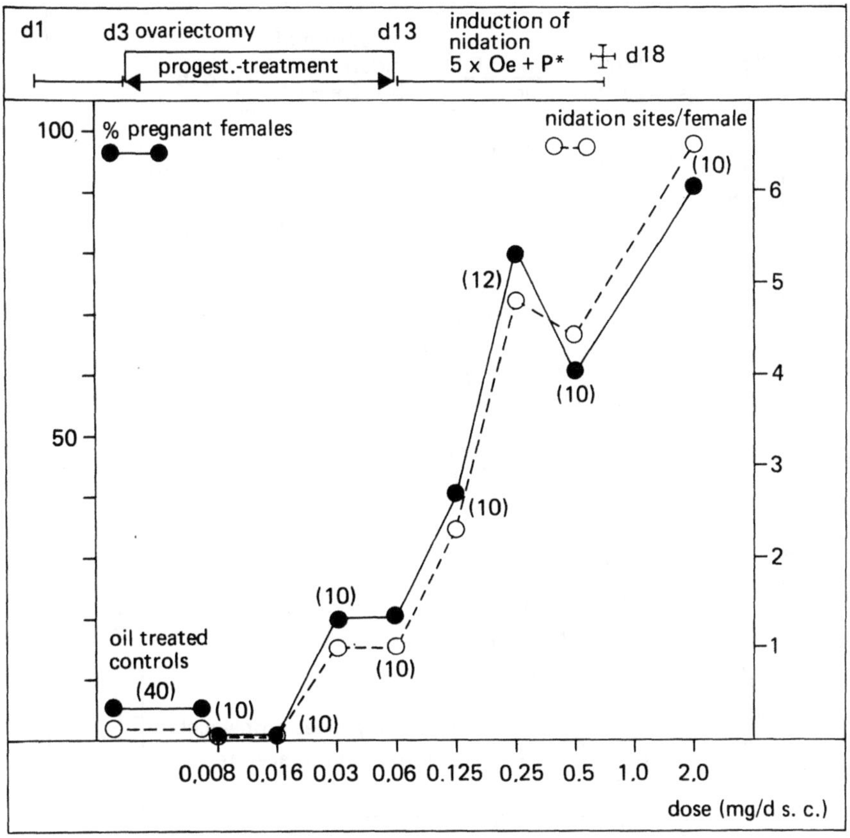

* 2.0 mg progesterone + 0.001 mg oestrone/day s. c.

( ) = N

Fig. 5. The blastocyst-maintaining effect of progesterone after ovariectomy on day 3 of gestation in rats. (Oe + P*) 2.0 mg progesterone + 0.001 mg oestrone/day s.c.

after ovariectomy and others which remained untreated at the same time. Nidation was induced by combined estrogen/progesterone treatment immediately after transfer (see Fig. 8). It was found that the uterus provides suitable conditions for nidation, independently of the prior progesterone treatment. We thus assume that progesterone in delayed nidation does not mainly act via mechanisms involved in nidation, but probably provides a uterine milieu that is suitable for the survival of nonimplanted blastocysts.

4.3 The Correlation of Biological and Morphological Parameters of Progestogenic Action in the Rabbit

In order to obtain in ovariectomized animals those structural changes in the endometrium which are typical for early pregnancy, it is necessary to administer estrogen before starting progesterone treatment. In rabbits the combined action of estrogens and progesterone is necessary for the initiation and maintenance of the traumatic decidual reaction or implantation (Chambon 1949a, b).

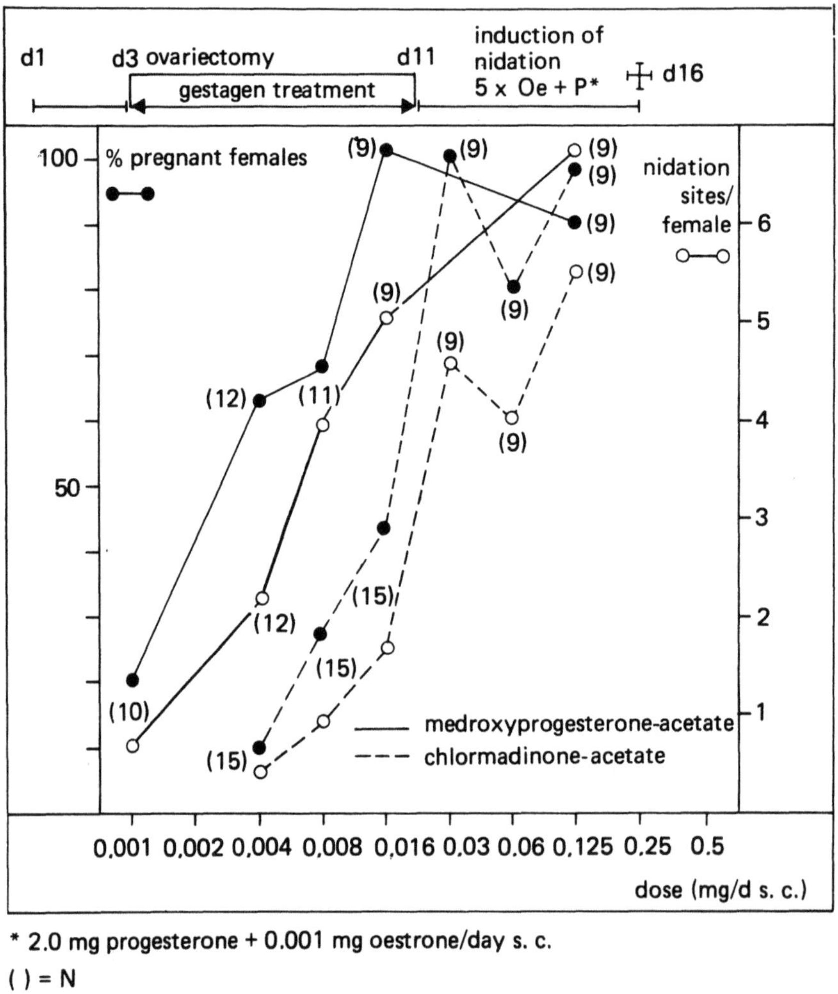

Fig. 6. The blastocyst-maintaining effect of artificial gestagens (medroxyprogesterone acetate, chlormadinone acetate) after ovariectomy on day 3 of gestation in rats. (Oe + P*) 2.0 mg progesterone + 0.001 mg oestrone/day s.c.

As far as we know no data exist which show whether uteroglobin synthesis and secretion are under the same hormonal control to that to which the structural and other functional changes of the uterus are subject.

Experimental results are available indicating that uteroglobin secretion is possible in ovariectomized animals by administration of progesterone even without accompanying or preceding estrogen treatment (Beier 1974, 1976; see also Fig. 9). Initially, when the progesterone treatment of freshly ovariectomized rabbits led to considerable uteroglobin secretion, we could not exclude the synergistic or even the "priming" effect of estrogens from sources of the body other than the ovaries (e.g., accumulated estrogens from adipose tissue depots or quickly metabolized progesterone molecules which could have

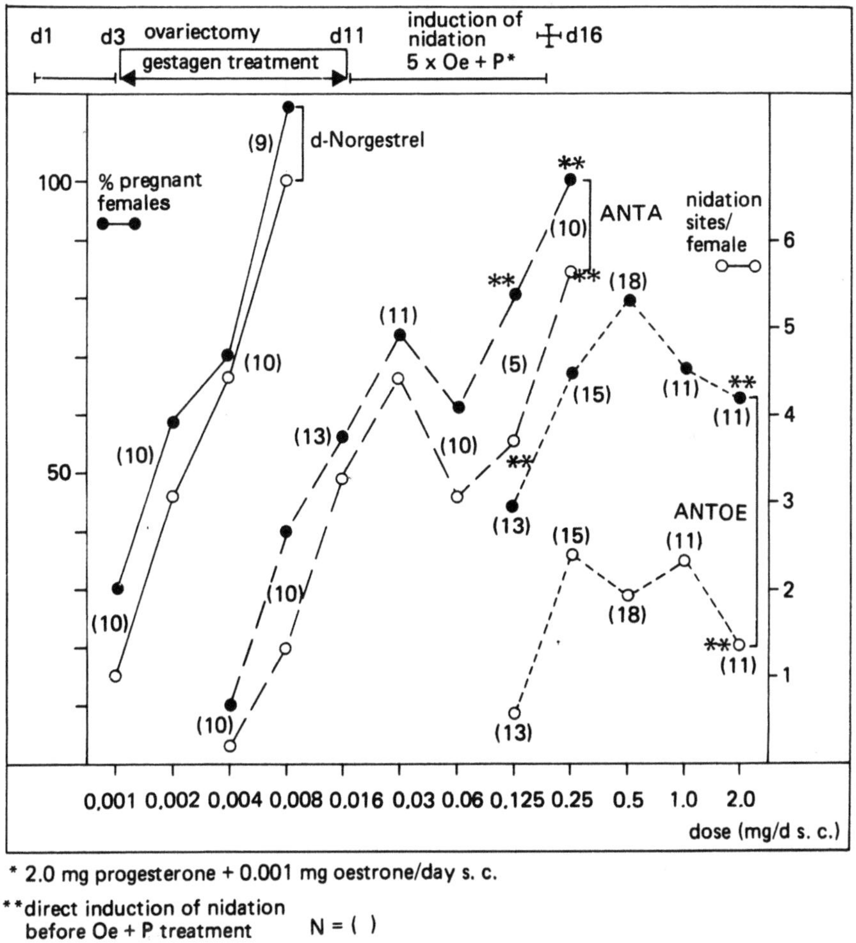

Fig. 7. The blastocyst-maintaining effect of the artificial gestagens d-norgestrel, norethisterone-acetate (ANTA) and -oenanthate (ANTOE) after ovariectomy on day 3 of gestation in rats. (Oe + P*) 2.0 mg progesterone + 0.001 mg oestrone/day s.c. (**) direct induction of nidation before Oe+P treatment

become estrogens before the last progesterone injection was given during the 5-day treatment). To avoid such influences, we used long-term ovariectomized rabbits (interval between ovariectomy and treatment, 17 months!) and also a progestogen that cannot be metabolized quickly into any estrogen compound but is itself a potent progesterone equivalent (17α-hydroxyprogesterone caproate). As demonstrated in Figs. 9 and 10, uteroglobin synthesis and release is stimulated by progesterone and also by 17α-hydroxyprogesterone caproate; the latter, however, shows reduced power in the long-term ovariectomized rabbits.

As was shown by Nishino (this vol. pp. 83-87), the uteroglobin concentration of uterine secretion increases in proportion to the dose of progesterone when intact infantile rabbits are primed with 17β-estradiol and treated with progesterone (or with chlormadinone

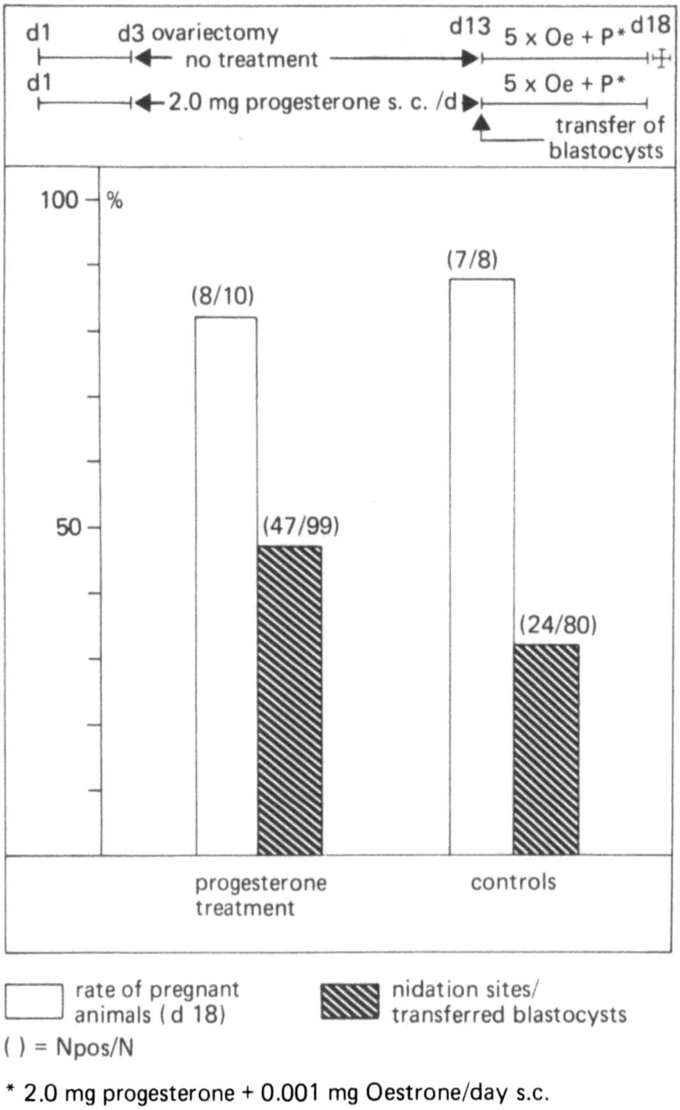

Fig. 8. The receptivity of the rat uterus for transferred day 5 blastocysts after ovariectomy and tubectomy on day 3 of gestation depending on prior progesterone treatment. (Oe + P*) 2.0 mg progesterone + 0.001 mg oestrone/day s.c.

acetate or norethisterone acetate). These results are in good agreement with our findings when progestogen effects are compared in ovariectomized rabbits with and without estrogen priming.

The comparison of the biochemical effects of progesterone under various experimental conditions may provide a means of clarifying the apparent discrepancy between earlier morphological and functional parameters. Another point which needs reevaluation in this connection is the already mentioned statement by Arthur and Chang (1974) that

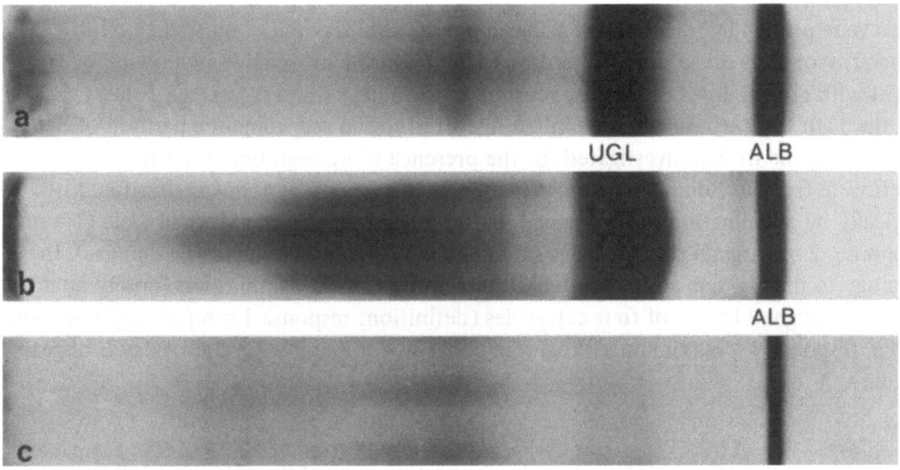

Fig. 9.a-c. Disc electrophoretic patterns of rabbit uterine secretion after different progestogen treatments of ovariectomized animals. Pattern (**a**) represents the released uterine proteins 6 days after one single injection of a long-acting progesterone (125mg 17α-hydroxyprogesterone caproate), which followed an ovariectomy interval of 9 days. Pattern (**b**) and (**c**) represent the protein secretion after an ovariectomy interval of 17 months. In the case (**b**), progesterone treatment was performed during 7 days (10 mg each) and in the other case (**c**) a single progesterone depot injection was given: 125 mg 17α-hydroxyprogesterone caproate. There was no estrogen priming. *UGL* Uteroglobin, *ALB* Albumin

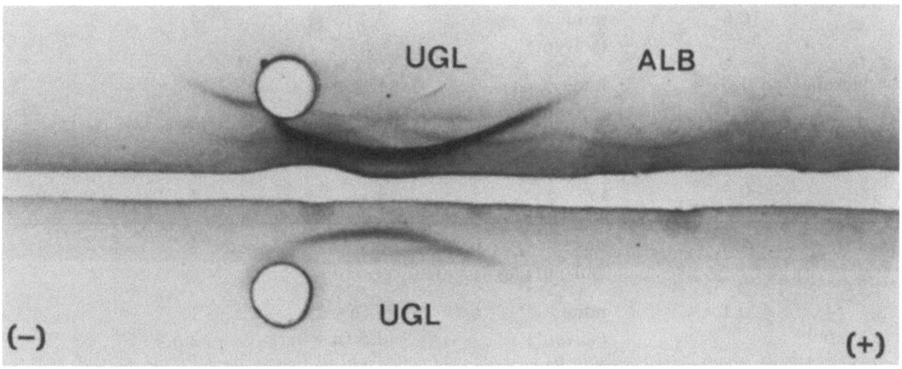

Fig. 10. Immunoelectrophoresis of uterine secretions from the rabbits, as shown in Fig. 9 (b,c). The precipitates were obtained with an anti-rabbit uterine fluid-serum from the sheep. *UGL* Uteroglobin, *ALB* Albumin

potent progestogens from the 19-nortestosterone derivative group, whose strong effect on the endometrium is undisputed, do not stimulate uteroglobin secretion.

In the following, experiments are described in which the effects of progesterone, *d*-norgestrel and chlormadinone acetate on morphological and biochemical parameters are directly to be compared in various experimental conditions.

Adult female rabbits weighing 2.5-3.0 kg were ovariectomized. After 16 days the animals were divided into groups (see Table 1) and treatment was commenced. The animals were primed by injections of either 5 μg estradiol/day in 0.5 ml sesame oil subcutaneously or the vehicle only, for 5 days (days 1-5). The priming was following by treatment with graded doses of the test substances for 6 days (days 6-11) (see Table 1). Late on the 11th day the animals were killed, the uteri rinsed according the method of Beier (1967) and the rinsing investigated for the presence of uteroglobin. After rinsing, the uteri were fixed in Bouin's solution. The degree of gestagen action was classified histologically by examination of HE-stained sections, according to the McPhail scale (1 = no response, 2 = marginal response, 3 = submaximal response, 4 = maximal response). In analogy to the McPhail scale the uteroglobin content of the samples was loosely quantified and assigned to one of four categories (definition: response 1 = no uteroglobin detectable, response 4 = maximum uteroglobin content, comparable to that of day 6 of gestation).

Table 1. Comparison of morphological and biochemical responses of the rabbit uterus after ovariectomy depending on the progestogen dose and on estrogen priming. Uteroglobin determinations were performed in pooled material as indicated

Test compound	Dose mg/day/animal s.c. (days 6-11)	Priming 5 μg E_2 (days 1-5)	Number of animals (n)	McPhail Index (1 = negative, 4 = maximal response) [a]	Uteroglobin response (1 = negative, 4 = maximal response) [a]
	10.0	none (solvent)	6	3.4 (3-4)	4
Progesterone	1.0	none (solvent)	6	1 (n = 5) 3 (n = 1)	1.5 2
	1.0	yes	5	4	4
	0.3	yes	4	4	4
	1.0	none (solvent)	6	3.9	4
d-Norgestrel	0.1	none (solvent)	6	1 (n = 2) 2.5 (n = 4)	1 2.5
	0.1	yes	5	4	4
	0.03	yes	5	4	4
	0.3	none (solvent)	6	3.3	3-4
Chlormadinone	0.03	none (solvent)	6	4 (n = 1) 1.2 (n = 5)	3 1
	0.03	yes	5	4	4
	0.01	yes	5	3.3	4
Controls	0.0 (solvent)	none (solvent)	6	1	1

[a] Arbitrary uteroglobin and McPhail scale responses: negative (1), marginal (2), submaximal (3), maximal response, comparable to that in day 6 of pseudopregnancy (4)

The morphological data (Table 1) show that the endometrium reacts more readily after a preceding estrogen priming with each of the progestogens investigated. On the other hand, each of the artificial progestogens investigated and progesterone were capable of inducing transformation of the endometrium even without estrogen priming, when extremely large doses were given. This indicates a permissive but not essential action of estrogen priming for uteroglobin secretion and morphological responses.

Investigation of the uteroglobin content of the uterine secretion in the same animals always revealed low or negative uteroglobin scores when the McPhail scale scores were low, and high uteroglobin scores when the McPhail scale scores were high. No differences could be found between the various progestogens in this respect. The 19-nortestosterone derivative *d*-norgestrel induced uteroglobin secretion as well as the hydroxyprogesterone derivative dchlormadinone acetate and progesterone itself.

The results demonstrated in Table 1 confirm the close connection between the morphological and biochemical parameters of the progestogens actions. For this reason speculations based on the surmise that there are two different categories of progestogen in this respect cannot be supported (Arthur and Chang 1974). A critical survey of the doses used in the experiments by Arthur and Chang may perhaps explain why no uteroglobin secretion was induced by 19-nortestosterone derivatives in their study. Doses for Arthur and Chang's experiments were calculated on the basis of human doses. This procedure does not take into account the fact that progestogens of the 19-nortestosterone series are significantly less potent in rabbits than might be concluded from their potency in humans. This is particularly true of oral administration (Neumann and Elger 1971; Neumann et al. 1977; Elger and Haase 1974). Thus the possibility cannot be excludes that Arthur and Chang only tested doses of 19-nortestosterone which were far below the effectivity threshold in rabbits.

More recent studies on uteroglobin synthesis have been performed with the aim of shedding more light on the molecular biological level of progesterone-uteroglobin interactions (Jänne et al., 1980, 1981). The determination of the actual uteroglobin content was performed by radioimmunoassay either from uterine fluids or from the soluble cytosol fractions of the endometrial tissue homogenates. Using long-term treatment schedules, mainly 5 days, on intact estrous rabbits, progesterone administration of 1.0 mg/kg body weight per day increased uterine fluid uteroglobin approximately 3000-fold and uteroglobin mRNA activity about 40-fold. In comparative studies Jänne et al. (1981) found that 5α-dihydrotestosterone at a dosage of 5.0 mg/kg body weight per day increased uteroglobin content of uterine fluid 500-fold and its mRNA content in endometrial tissue 15-fold. By contrast, 17β-estradiol treatment (0.03-0.10 mg/day) increased only 10-fold both uteroglobin in uterine fluid and total uteroglobin mRNA activity. The authors conclude that in thus appears that progesterone and androgens, using the progesterone receptor mechanism of the endometrial cell, facilitate more uteroglobin mRNA translation in vivo than 17β-estradiol, suggesting that posttranscriptional control may also be exerted in uteroglobin synthesis. Uteroglobin clearly provides a suitable biochemical marker protein for studies of progesterone action in the mammalian uterus, not only from the point of view of pharmacology and pharmacological endocrinology but also for the basic research demands of molecular biochemists who are interested in studies on the interaction of various steroids in the control of a specific gene product.

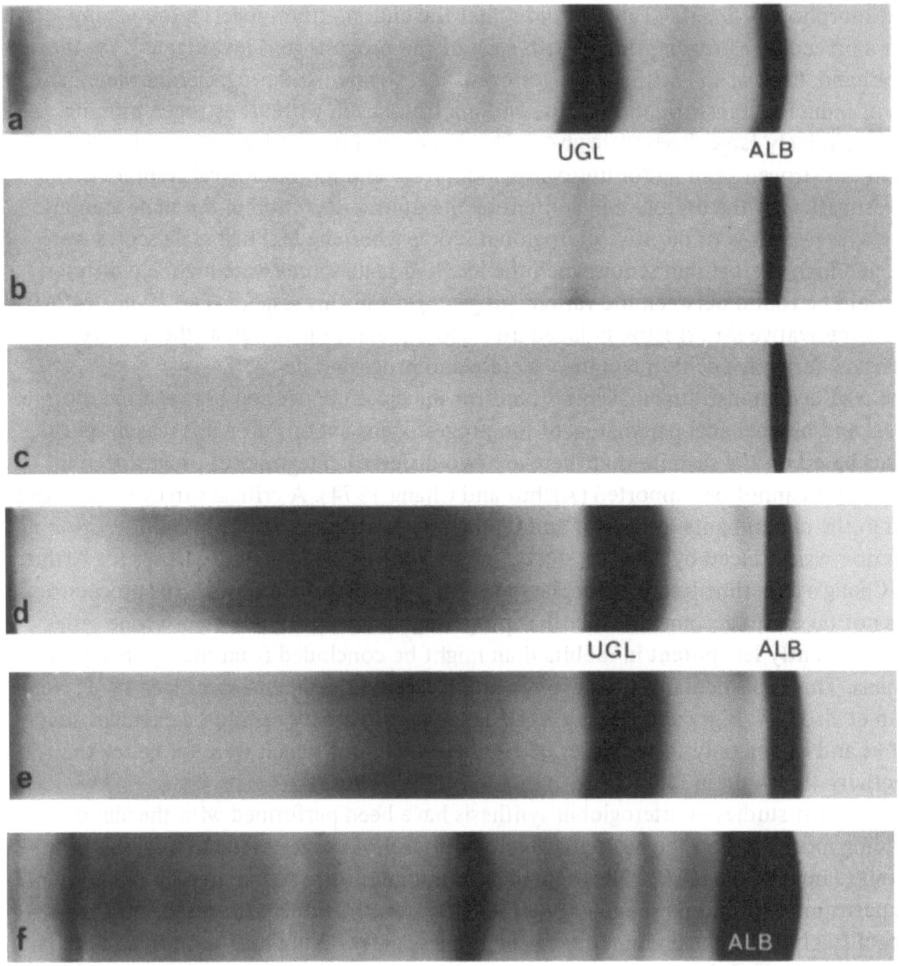

Fig. 11 a-f. Disc electrophoretic patterns of rabbit uterine secretion after different progesterone treatments. The rabbits were ovariectomized before treatment. Patterns (**a**) and (**b**) represent uterine protein secretion (= release) after progesterone injections (**a** = 6 days 10.0 mg progesterone each, **b** = 6 days 1.0 mg progesterone each) without preceding estrogen priming. Pattern (**c**) is the control result after injections with vehicle solutions only, not containing any estrogen or gestagen. Patterns (**d**) and (**e**) represent the protein release after progesterone injections (**d** = 6 days 1.0 mg progesterone each, **e** = 6 days 0.3 mg progesterone each) with preceding estrogen priming (5 days 5 μg estradiol each). The pattern (**f**) shows the blood serum proteins for comparison. ULG = Uteroglobin, ALB = Albumin

Acknowledgment. Parts of this work have been supported by the Deutsche Forschungsgemeinschaft (Special Research Programme "Biologie und Klinik der Reproduktion").

References

Arthur AT, Chang MC (1974) Induction of blastokinin by oral contraceptive steroids: Implications for fertility control. Fertil Steril 25:217-221

Bacic M, Wesselius de Casparis A, Diczfalusy E (1970) Failure of large doses of ehtinyl estradiol to interfer with early embryonic development in the human species. Am J Obstet Gynecol 4:531-534

Beier HM (1966) Das Proteinmilieu in Serum, Uterus und Blastocysten des Kaninchens vor der Nidation. In: Beermann W (Chairman) Biochemie der Morphogenese. Deutsche Forschungsgemeinschaft, Konstanz, pp 1-10

Beier HM (1967) Biochemisch-entwicklungsphysiologische Untersuchungen am Proteinmilieu für die Blastocystenentwicklung des Kaninchens (Oryctolagus cuniculus). Dissertation Naturwiss Fakultät, Marburg/Lahn

Beier HM (1970) Protein patterns of endometrial secretion in the rabbit. In: Hubinont PO, Leroy F, Robyn C, Leleux P (eds) Ovo-Implantation, Human-Gonadotropins and Prolactin. Karger, Basel München New York, pp 157-163

Beier HM (1973) Die hormonelle Steuerung der Uterussekretion und frühen Embryonalentwicklung des Kaninchens. Habilitationsschrift, Medizinische Fakultät, Kiel

Beier HM (1976) Uteroglobin and related biochemical changes in the reproductive tract during early pregnancy in the rabbit. J Reprod Fertil [Suppl] 25:53-69

Beier HM (1980) The role of uterine proteins in the establishment of receptivity of the uterus. In: Leroy F, Finn CA, Psychoyos A, Hubinont PO (eds) Prog reprod Biol vol 7 Blastocyst-Endometrium Relationships. Karger, Basel München New York pp 158-172

Beier HM, Mootz U. Fischer B (1980) New aspects on the physiology of uteroglobin. In: Beato M (ed) Steroid induced Uterine Proteins. Elsevier/North-Holland Biomedical, Amsterdam New York Oxford, pp 47-67

Beling CG, Marcus SL, Markham SM (1970) Functional activity of the corpus luteum following hysterectomy. J Clin Endocrinol 30:30-39

Blye RP (1970) The effect of estrogens and related substances on embryonic viability.Pergamon, Oxford London Edinburgh New York Toronto Sidney Paris; Vieweg, Braunschweig, Advances in the Biosciences, Vol 4, pp 323-343

Blye RP (1973) The use of estrogens as postcoital contraceptive agents. Am J Obstet Gynecol 116: 1044-1050

Böving BG (1959) Implantation. Ann NY Acad Sci 75:700-725

Challis JRG (1974) Physiology and pharmacology of prostaglandins in parturition. In: Population Information Programs, The John Hopkins University (ed) Population Reports, Series G, No. 5, pp 1-12

Chambon Y (1949a) Besoins progestéro-folliculiniques quantitatifs du déciduome traumatique chez la lapine. CR Soc Biol (Paris) 143:701-703

Chambon Y (1949b) Besoins endocriniens qualitatifs et quantitatifs de l'ovoimplantation chez la lapine. CR Soc Biol (Paris) 143:1172-1175

Cooper MJ, Furr BJA (1974) The role of prostaglandins in animal breeding. Vet Rec 94:161-

Coudert SP, Winter JSD, Faiman C (1974) Transient decline in serum progesterone levels during prostaglandins $F_{2\alpha}$ infusion in the midluteal phase of the normal human menstrual cycle. Am J Obstet Gynecol 119:755-761

Csapo AI, Pulkkinen MO, Kaihola HL (1974) The relationship between the timing of luteectomy and the incidence of complete abortions. Am J Obstet Gynecol 118:985-989

Döring GK (1971) Morning-after-pill: Die postkoitale hormonelle Schwangerschaftsverhütung. Dtsch Ärztebl 68:2395-2396

Doyle LL, Barcley DL, Duncan GW, Kirton KT (1971) Human luteal function following hysterectomy as assessed by plasma progestin. Am J Obstet Gynecol 110:92-97

Elger W, Haase F (1974) A comparative endocrinological and pharmacological view of the maintenance or interruption of pregnancy. Pergamon, Oxofrd New York Toronto Sidney Paris; Vieweg, Braunschweig Advances in the biosciences Vol 13, pp 163-180

Elger W, Kosub B, Skuballa W, Korte R (1976) Studies on prostaglandin-induced abortion in guinea pigs. In: Samuelsson B, Paoletti R (eds) Advances in prostaglandin and thromboxane research vol 2, Raven, New York, pp 673-677
Elger W, Korte R, Hasan SH (1977) Changes in serum-progesterone (SP) and their significance in prostaglandin (PG)-induced abortions in early pregnancy of rhesus monkeys. Acta Endocrinol (Copenhagen) 84 [Suppl, 208]: 94 (abstract)
Emmens CW (1970) Postcoital contraception. Br Med Bull 26:45-51
Fraenkel L (1903) Die Function des Corpus luteum. Arch Gynaekol 68:438-545
Fraenkel L, Cohn F (1901) Experimentelle Untersuchung über den Einfluß des Corpus luteum auf die Insertion des Eies. Anat Anz 20:294-300
Frantz AG (1978) Prolactin. N Engl J Med 298:201-207
Froewis J (1963) Wann erfolgt die ausreichende Umschaltung der schwangerschaftserhaltenden Hormonproduktion des Corpus luteum graviditatis auf die Placenta? Wien Klin Wochenschr 75: 368-370
Garmendia F, Kesserü E, Urdanivia E, Valencia M (1976) Luteinizing hormone and progesterone in women under postcoital contraception with d-norgestrel. Fertil Steril 27:1250-1255
Giannina T, Steinetz BG, Rassaert CL, McDougall EA, Meli A (1969) Biological profile of quingestanol acetate (33977). Proc Soc Exp Biol Med 131:781-789
Goldman AS (1970) Animal models of inborn errors of steroidgenesis and steroid action. In: Gibian H, Plotz EJ (eds) Mammalian reproduction. Springer, Berlin Heidelberg New York pp 389-436
Hagenfeldt K (1976) The modes of action of medicated intrauterine devices. J Reprod Fertil [Suppl] 25:117-132
Haspels AA (1976) Interception: Post-coital estrogens in 3016 women. Contraception 14:375-381
Haspels AA, Andriesse R (1973) The effect of large doses of estrogens post coitum in 2000 women. Eur J Obstet Gynecol Reprod Biol 3:113-117
Haspels AA, Linthorst GA, Kicovic PM (1977) Effect of postovulatory administration of 'morning-after' injection on corpus luteum function and endometrium. Contraception 15:105-112
Henzl MR, Segre EJ, Nakamura RM (1973) The influence of oxymetholone on the HCG stimulated corpus luteum. Contraception 8:515-520
Hobson B, Wide L (1972) A comparison between chorionic gonadotrophins extracted from human, rhesus monkey and marmoset placentae. J Endocrinol 55:363-368
Jacob D, McLean Morriss J (1969) The estrogenic activity of postcoital antifertility compounds. Fertil Steril 20:211-222
Jänne O, Hemminki S, Isomaa V, Isotalo H, Kopu H, Orava M, Torkkeli T (1980) Uteroglobin synthesis and its relationship to changes in progesterone receptors, RNA polymerases and poly(A) polymerases in the rabbit uterus. In: Beato M (ed) Steroid induced uterine proteins. Elsevier/North-Holland Biomedical Press, Amsterdam, New York Oxford, pp 319-340
Jänne O, Isomaa V, Isotalo H, Kopu H, Torkello T (1981) Hormonal regulation of uteroglobin synthesis in the rabbit uterus. In: Baulieu EE, Gurpide E, Bardin CW, Mauvais-Jarvis P, Milgrom E (eds) Progesterone and progestins (in Press)
Johansson EDB (1971) Depression of the progesterone levels in women treated with synthetic gestagens after ovulation Acta Endocrinol (Kbh) 68:779-792
Johansson EDB, Gemzell C (1971) Plasma levels of progesterone during the luteal phase in normal women treated with synthetic oestrogens. Acta Endocrinol (Kbh) 68:751-760
Kanojia RM, Ostrowski P, Hahn DW, Allen G, McGuire JL (1975) 17α-Ethyl-20α- and 20β-dihydroprogesterones and other 17α-ethyl-substituted pregnancies as potential contragestational agents. J Med Chem 18:1143-1145
Karim SMM, Hillier K (1974) Prostaglandins: Pharmacology and clinical application. Drugs 8:176-207
Karnaky KJ (1947) Estrogenic tolerance in pregnant women. Am J Obstet Gynecol 53:312-316
Korda AR, Shutt DA, Smith ID, Shearman RP, Lyneham RC (1975) Assessment of possible luteolytic effect of intraovarian injection of prostaglandin $F_{2\alpha}$ in the human. Prostaglandins 9:443-449
Kuchera LK (1974) Postcoital contraception with diethylstilbestrol-updated. Contraception 10: 47-54

Larranaga A, Winterhaltel M, Sartoretto JN (1975) Evaluation of *d*-norgestrel 1.0 mg as a post coital contraceptive. Int J Fertil 20:156-160

Lehmann F, Just-Nastansky I, Czygan PJ, Bettendorf G (1975) Effects of 17α-ethinyl-19-nor-testosterone (ENT) on corpus luteum function. Acta Endocrinol [Suppl 199] (Kbh) 80:160 (abstract)

Loeb L (1923) The effect of exstirpation of the uterus on the life and function of the corpus luteum of the guinea-pig. Proc Soc Exp Biol (Med) 20:441

Loeb L (1927) The effects of hysterectomy on the system of sex organs and on the periodicity of the sexual cycle in the guinea-pig. Am J Physiol 83:202

Lyneham RC, Korda AR, Shutt DA, Smith ID, Shearman RP (1975) The effect of intra-uterine prostaglandin $F_{2\alpha}$ on corpus luteum function in the human. Prostaglandins 9:431-442

Matthies DL, Lyons WR (1971) Luteotropic and luteolytic effects of rat chorionic mammotropin. Proc Soc Exp Biol Med 136:520-529

McLean Morris J (1973) Mechanisms involved in progesterone contraception and estrogen interception. Am J Obstet Gynecol 117:167-176

McLean Morris J, Van Wagnenen G (1966) Compounds interfering with ovum implantation and development. III. The role of estrogens. Am J Obstet Gynecol 96:804-815

McLean Morris J, Van Wagenen G (1967) Post-coital oral contraception. In: Hankinson RKB, Romero H (eds) Proc 8th Intern Conf of the Intern Planned Parenthood Federation, Santiago, Chile, pp 256

Meites J (1977) Evaluation of research on control of prolactin secretion. In: Dellmann H-D, Johnson JA, Klachko DM (eds) Comparative Endocrinology of prolactin. Plenum, New York London, pp 1-215

Mischler TW, Berman E, Rubio B, Larranaga A, Guiloff E, Moggia AV (1974) Further experience with quingestanol acetate as a postcoital oral contraceptive. Contraception 9:221-225

Moawad AH, Kim MH, Zuspan FP, Chagrasulis R, Pishotta FT, Zuspan KJ (1977) Effects of progesterone on the adrenergic mechanisms of the genital tract. Ann NY Acad Sci 286:287-303

Mora G, Faundes A, Pastore U (1974) Clinical evaluation of an oral progestin contraceptive, R-2323, 5 mg, administered at weekly intervals, Contraception 10:145-157

Mora G, Faundes A, Johansson EDB (1975) Lack of clinical contraceptive efficacy of large doses of R 2323 given before implantation or after a missed period. Contraception 12:211-220

Mougdal NR, MacDonald GJ, Greep RO (1971) Effect of HCG antiserum on ovulation and corpus luteum formation in the monkey *(Macaca fascicularis).* J Clin Endocrinol 32:579-581

Myatt L, Bray MA, Gordon D, Morley J (1975) Macrophages on intrauterine contraceptive devices produce prostaglandins. Nature 257:227-228

Neumann F, Elger W (1971) Kritische Überlegungen zu den biologischen Grundlagen von Toxizitätsstudien mit Steroid-(Sexual-)hormone. In: Plotz EJ, Haller J (eds) Methodik der Steroidtoxikologie. Thieme, Stuttgart, pp 6-48

Neumann F, Elger W, Nishino Y, Steinbeck H (1977) Probleme der Dosisfindung: Sexualhormone. Arzneim Forsch 27:296-318

Nygren K-G, Johansson EDB, Wide L (1972) Postovulatory contraception in women with large doses of norethindrone. Contraception 5:445-456

Nygren K-G, Johansson EDB, Wide L (1974) The contraceptive efficacy of 10 mg or 25 mg of norethindrone given orally on day 15-22 of the human menstrual cycle. Contraception 9:249-256

Pala A, ERmini M, Carenza L, Benagiano G (1976) Immunization with hapten-coupled hCG-β subunit and its effect on the menstrual cycle. Contraception 14:579-593

Pauerstein CJ, Fremming BD, Martin JE (1970) Estrogen-indiced tubal arrest of ovum. Antagonism by alpha adrenergic blockade. Obstet Gynecol 35:671-675

Pharriss BB (1971) Prostaglandins in Fertility. Proc R Soc Med 64:10 (abstract)

Pozo E del, Varga L, Wyss H, Tolis G, Friesen H, Wenner R, Vetter L, Uettwiler A (1974) Clinical and hormonal response to bromocriptin (CB-154) in the galactorrhea syndromes. J Clin Endocrinol Metab 39:18-26

Prahalada S, Venkatramaiah M, Jagannadha Rao A, Moudgal NR (1975) Termination of pregnancy in macaques (macaca radiata) using monkey antiserum to ovine luteinizing hormone. Contraception 12:137-147

Prasad MRN, Kaira SP (1967) Mechanisms of anti-implantation action of anti-oesfrogens. In: Hankinson RKB, Romero H (eds) Proc 8th Intern Conf of the Intern Planned Parenthood Federation, Santiago, Chile, pp 413

Rabii J, Kragt CL (1972) Plasma levels of prolactin, FSH and LH in the pseudopregnant rat. Proc Soc Exp Biol Med 141:359-362
Raynaud JP, Bonne C, Bouton MM, Moguilewsky M, Philibert D, Azadian-Boulanger G (1975) Screening for anti-hormones by receptor studies. J Steroid Biochem 6:615-622
Rubio B, Berman E, Larranaga A, Guiloff E, Aquirre JJ (1970) A new postcoital oral contraceptive. Contraception 1:303-313
Sakiz E, Azadian-Boulanger G (1971) R 2323 – an original contraceptive compound. In: James VHT, Martini L (eds) Hormonal steroids. Excerpta Medica, Amsterdam, pp 865
Sakiz E, Azadian-Boulanger G, Laraque F, Raynaud JP (1974) A new approach to estrogen-free contraception based on progesterone receptor blockade by mid-cycle administration of ethyl norgestrienone (R 2323). Contraception 10:467;474
Salahuddin M, Ramakrishnan S, Dubey SK, Talwar GP (1976) Immunological reactivity of antibodies produced by Pr-β-HCG-TT with different hormones. Contraception 13:163-170
Shearman RP (1973) Post-coital contraception. A review. Contraception 7:459-476
Spilman CH, Bergström KK, Beuving DC (1980) Effects of prostaglandins on peripheral progesterone and uterine diamine oxidase in the pregnant hamster. Prostaglandins 20:1061-1074
Spilman CH, Beuving DC, Bergstrom KK (1981a) Uterine diamine oxidase: A marker for progestin action. In: McKerns KW (ed) Reproductive Processes and Contraception. Plenum, New York, pp 509-520
Spilman CH, Wilks JW, Campbell JA (1981b) A progestin bioassay: Uterine diamine oxidase. In: Baulieu EE, Gurpide E, Bardin CW, Mauvais-Jarvis P, Milgrom E (eds) Progesterone and Progestins (in press)
Stevens VC, Crystle CD (1973) Effects of Immunization with hapten-cloupled HCG on the human menstrual cycle. Obstet Gynecol 42: 485-495
Sundaram K (1976) The effectiveness in rhesus monkeys of an antifertility vaccine based on neutralization of chorionic gonadotropin. Contraception 14:639-653
Szontagh FE, Kovacs L (1969) Post-coital contraception with dienoestrol. Med Gynecol Androl Sociol 4:36-37
Takayama M, Greenwald GS (1973) Hormonal requirements for the maintenance of luteal function in hypophysectomized, pseudopregnant rats. J Endocrinol 56:421-429
Talwar GP, Dubey SK, Salahuddin M, Das C (1976a) Antibody response to Pr-β-HCG-TT vaccine in human subjects. Contraception 13:237-243
Talwar GP, Dubey SK, Salahuddin M, Shastri N (1976b) Kinetics of antibody response in animals injected with processed beta-HCG conjugated to tetanus toxoid (Pr-β-HCG-TT). Contraception 13: 153-161
Tredway DR, Mishell DR (1974) Effect of a one weekly oral contraceptive upon gonadotrophin and gonadal steroid levels. Contraception 10:159-169
Van Santen MR, Haspels AA (1980) Interfering with implantation by postcoital estrogen administration. I. Endometrium histology. In: Leroy F, Finn CA, Psychoyos A, Hubinont PO (eds) Prog Reprod Biol vol 7 Blastocyst-Endometrium Relationships. Karger, Basel München New York, pp 310-323
Victor A, Johansson EDB (1975) The effect of weekly administration of an oral progesteron – R 2323 – on ovarian function. Contraception 12:89-94
Yuzpe AA, Lancee WJ (1977) Ehinylestradiol and dl-norgestrel as a postcoital contraceptive. Fertil Steril 28:932-936
Yuzpe AA, Thurlow HJ, Ramzy I, Leyshon JI (1974) Post coital contraception – A pilot study. J Reprod Med 13:53-58

Antifertility Effect of a Long-Acting Progestin (3-Cyclopentyl Propionate of Megestrol Acetate): Prematurity of the Endometrium and Accompanying Changes of Uteroglobin and Progesterone in Uterine Fluid

Yi-chien Yang, Xi-gen Gu, and Shu-xiang Li

Department of Reproductive Physiology, Shanghai Institute of Planned Parenthood Research, Shanghai, Peoples Republic of China

1 Antifertility Effects of a Chinese Progestin in Women

A long-acting progestin (3-cyclopentyl propionate of megestrol acetate, Fig. 1) synthesized in China and called "Progestin No. 1" is a new antifertility agent for women. It is given orally once a month in combination with small amounts of quinestrol (100 mg Progestin No. 1/0.5 mg quinestrol). It has only a 60% antiovulation effect, according to pregnanediol measurements, but possesses marked anti-implantation effects. One of our previous investigations (Yang et al. 1979) demonstrated that Progestin No. 1 could induce premature transformation of the endometrium in women. In general, the endometrium of the control group converted from the proliferative phase to the early secretory phase on day 16 of the menstrual cycle; it then developed to the late secretory phase on day 21 and to the premenstrual phase on day 26. However, the endometrium of the Progestin No. 1-treated group developed to the corresponding phase on days 12, 17, and 20 of the cycle, respectively.

Similar results have also been obtained in rabbits. The prematurity of endometrium could be induced by a single dose of Progestin No. 1 (1.5 mg/kg, without quinestrol) given orally on day -3 before pregnancy. With reference to the McPhail Index (McPhail 1934), the light microscopic study showed that the endometrial glandular development on day 1 in the treated group corresponded approximately to that normally seen on day 3. Under this experimental condition, Progestin No. 1 had 100% anti-implantation effects in rabbits.

The purpose of the present study is to explore whether the compound could induce the premature changes of the ultrastructure of the endometrial surface, and secondly, to investigate the effect of the 3-cyclopentyl propionate of megestrol acetate on uteroglobin secretion and the contents of progesterone and estradiol-17β in uterine fluids.

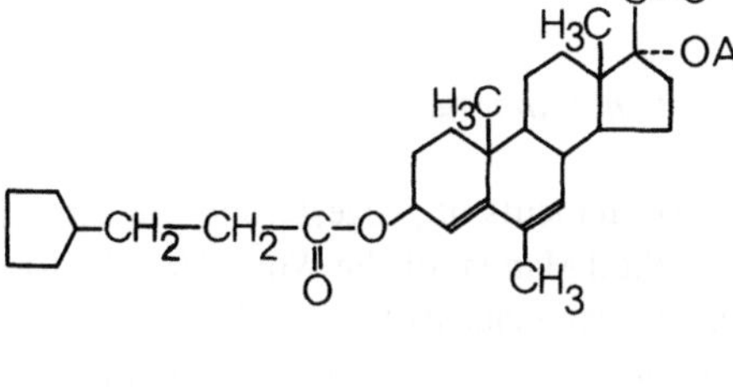

Fig. 1. Chemical structure of 3-cyclopentyl propionate of megestrol acetate, called "Progestin No. 1"

2 Investigations in Rabbits

2.1 Animals

Fifty-one sexually mature female Shanghai Mixed-Breed rabbits (3.6-5.0 kg) were divided into two groups for this study. Pseudopregnancy was induced by injection of 100 I.U. human chorionic gonadotrophin (hCG). Animals of the treated group were fed a single dose of Progestin No. 1 (1.5 mg/ml/kg) on day -3. Only the vehicle (1 ml/kg) was fed to animals in the control group. All animals were killed by cervical dislocation on days 1, 3, 5, 6, 7, or 9 of pseudopregnancy (day 0 being the day of hCG injection). Plasma of most animals was collected daily after hCG injection until the animal was killed. On the day of autopsy, the uterus was dissected free of mesometrial tissue and the cervix. Each uterine horn was flushed with either 1 or 5 ml 0.9% (w/v) sodium chloride. The uterine fluid of one horn flushed with 1 ml saline was centrifuged at 3000 r.p.m. for 10 min and the supernatant fluid was stored at 4°C for disc gel electrophoresis. The 5 ml flushing from another horn was centrifuged at 12 000 g (4°C) for 10 min, the sample was frozen at about -70°C (dry ice and acetone) and lyophylized in a vacuum evaporator. The supernatant was stored at -40°C until steroid analysis. Immediately after flushing, the uterus specimen (about 1 cm in length) was dissected and washed in 0.2 *M* cacodylate buffer (pH 7.4) for SEM study.

2.2 Scanning Electron Microscopy

The uterine specimens pinned on cork were immersed in the chilled fixative (2.5% glutaraldehyde) for 24 h at 4°C. Two pieces of approximately 3 x 3 mm were cut from the specimen, rinsed with buffer and fixed in 1% osmic acid (in buffer). After conductive staining with 1% osmic acid (in water) and then with 2% tannic acid, the tissues were washed with redistilled water, dehydrated in ascending concentrations of ethylalcohol, and placed in amyl acetate for at least 30 min. The specimens were then critical point dried in a Hitachi HCP-2 critical point dryer, coated with gold in a Hitachi HUS-5GB high vacuum evaporator, and finally examined with a Hitachi H-500 scanning electron microscope at 20 kV.

2.3 Disc Gel Electrophoresis

For Analysis by the polyacrylamide gel electrophoresis 60 μl of untreated but centrifuged uterine flushing was used. The concentration of gel was 7.5%, the pH value of Tris-buffer system was 8.9. The amido black stained gel was analyzed in each case by a Shimadzu CS-900 densitometer (560 nm).

2.4 Radioimmunoassay

The measurements of progesterone and estradiol-17β were carried out according to the RIA method Manual of the World Health Organization programme for the provision of matched assay reagents for the RIA of hormones in Reproductive Physiology. The facilities used were kindly supplied and equipped by WHO (Geneva).

3 Effect of Progestin No. 1 on Surface Ultrastructure of the Rabbit Endometrium

During early pseudopregnancy in rabbits the endometrial surface characteristically changes day by day. On day 1, it is occupied predominantly by ciliated cells (Fig. 2a-b). A remarkable transformation is demonstrated on the surface of the endometrial folds on day 3. The epithelial cells are regularly covered by microvilli, only some cells are ciliated. On day 5, there is a further reduction in the number of ciliated cells. The nonciliated cells of the endometrium have very few microvilli, mainly seen at the border of the cell apices. Compared to the above changes normally seen in the control group, we found that premature changes of the endometrial surface occurred in the treated group. As shown in Fig. 2c-d, the ciliated cells are seldom encountered on day 1; a great number of nonciliated cells cover the surface instead. On day 3, the appearance of the luminal surface is similar to that of day 5 in the control group. Some animals even show that epithelial degeneration and loss of epithelial cells occur on day 5 in the treated group.

4 Effect of Progestin No. 1 on the Secretion of Uteroglobin

As shown in Fig. 3, the main fractions of the uterine protein during early pseudopregnancy are albumin and uteroglobin. The released uteroglobin usually is measurable on day 3, then increases to peak values on days 5-9 and declines to 0 on day 10. Table 1 shows the relative percentage of uteroglobin during the period of day 1-9 of pseudopregnancy in the control and treated group. The results indicate that Progestin No. 1 stimulates the secretion of uteroglobin into the uterine lumen. The uteroglobin level on day 1 in the treated group is significantly higher than that in the control group. This finding agrees well with the results of de Visser (see this volume, pp. 89-98), and shows that the used long-acting progestin, the 3-cyclopentyl propionate of megestrol acetate, is able to evoke by a single application the same effects on uteroglobin release as the natural progesterone after daily injections.

5 Effect of Progestin No. 1 on Progesterone Levels in Uterine Fluid

Table 2 shows that after taking Progestin No. 1 the progesterone concentration in uterine fluid increases to a high level as early as on day 3 (15.8 ± 5.3 ng/uterine horn), reaches the peak value (68.2 ± 21.2) on day 5, and then declines abruptly after day 6. These values are statistically different from those of the control group. By contrast, however, neither the progesterone levels in blood plasma, nor the estradiol-17β levels in plasma and uterine flushings seem to be significantly different in the two compared groups (Table 3).

The levels of progesterone in uterine flushings are in a similar range to that previously reported by Fowler et al. (1976), who determined intrauterine progesterone and protein recovered from rabbits rendered pseudopregnant by injection of 100 I.U. hCG.

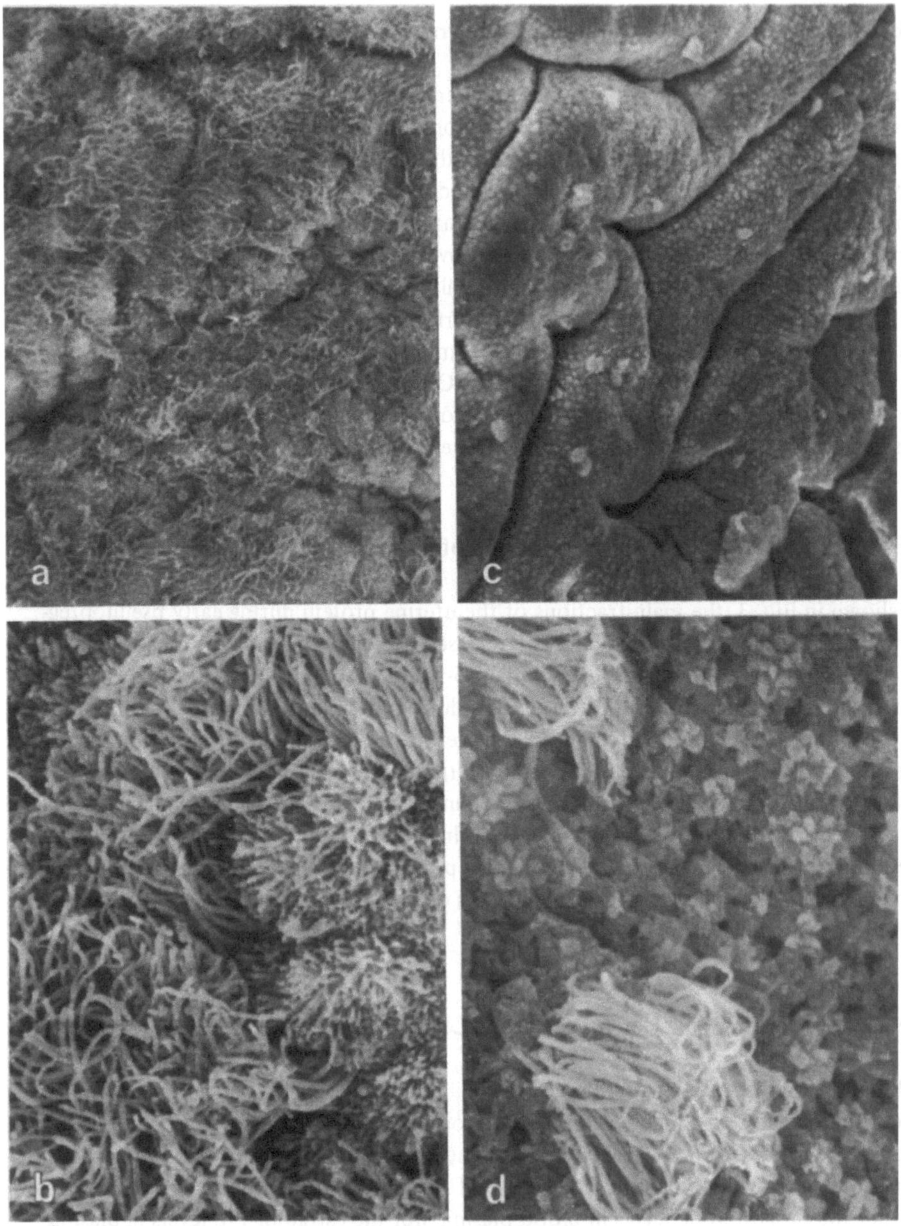

Fig. 2a-d. Scanning electron micrographs of the endometrial surface of the rabbit. Tissue specimens were obtained from the control group of pseudopregnant rabbits on the first day (p.i.), epithelial surfaces are depicted at lower (**a**=x680) and at higher magnification (**b**=x2800). Specimens of treated animals are also taken from the first day of pseudopregnancy (p.i.) and show the desynchronized epithelial transformation at lower (**c**=x280) and higher magnification (**d**=x2800)

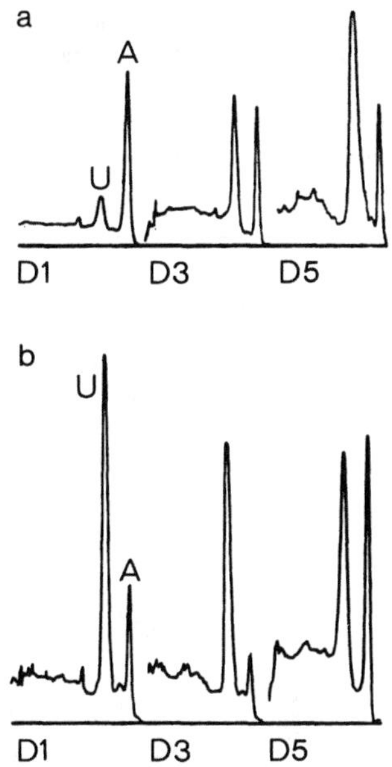

Fig. 3a, b. Electropherogram of the uterine flushings from pseudopregnant rabbits. **(a)** Specimens of control animals (untreated) are obtained at Day 1 (D1), Day 3 (D3), and Day 5 (D5) after hCG injection. **(b)** Animals treated with 3-cyclopentyl propionate of megestrol acetate (Progestin No. 1) reveal different electropherograms at Day 1 (D1), Day 3 (D3), and Day 5 (D5). Polyacrylamide disc electrophoresis, Tris-glycine-buffer pH 8.9, Amido black staining, migration of proteins from left to the right. *A*= Albumin, *U*= Uteroglobin

Table 1. Relative percentage of uteroglobin out of the total amount of uteroglobin plus albumin in the two investigated groups of rabbits: The control group is pseudopregnant; the treated group received 3-cyclopentyl propionate of megestrol acetate (Progestin No. 1)

Days (pseudopregnancy)	1	3	5	6	7	9
Control	17.8 (3)	61.8 (4)	72.4 (4)	81.1 (3)	83.9 (3)	80.6 (4)
Treated	78.2 (4)	77.8 (3)	69.6 (4)	83.1 (2)	79.4 (3)	36.1 (4)

Values are mean values; in parenthesis is the number of animals

6 Conclusions

As demonstrated in the light microscopic study before, the present experimental results show that Progestin No. 1 could also induce the premature changes of the luminal surface of endometrium depicted here on the ultrastructural level. On the other hand, high levels of progesterone and uteroglobin are also reached much earlier than in pseudopregnant animals. It has been well documented that uteroglobin is a specific protein released by the epithelial cells of endometrium when stimulated by progesterone and progestogen contraceptives (Beier 1968, 1974, 1976; Arthur and Daniel 1972; Bullock and Willen 1974). It is also known that uteroglobin can bind steroids, particularly progesterone, and may play an important role as a carrier protein in early embryonic development in rabbits

Table 2. Changes in the concentrations of progesterone in plasma and uterine flushings after administration of Progestin No. 1 in rabbits

Days (pseudo-pregnancy)	Plasma (ng/ml)		Uterine flushings (ng/uterine horn)	
	Control	Treated	Control	Treated
1	2.7±1.2 (12)	2.8±1.2 (8)	3.5±3.9 (4)	3.2±1.8 (4)
2	2.5±0.5 (10)	2.6±0.6 (10)		
3	4.6±1.1 (10)	4.4±0.6 (7)	4.4±4.7 (4)	15.8±5.3 [a] (5)
4	7.1±1.2 (8)	8.1±2.6 (5)		
5	9.3±3.5 (8)	10.9±3.7 (5)	24.5±12.3 (4)	68.2±21.2 [a] (4)
6	9.4±3.8 (6)	12.3±3.8 (3)	24.4±12.6 (4)	17.2±0.5 (2)
7	13.1±3.6 (6)	13.4±3.8 (5)	37.9±9.6 (3)	7.0±2.8 [b] (3)
8	15.1±3.2 (4)	14.3±2.4 (4)		
9	12.9±3.4 (4)	17.7±3.3 (4)	41.9±17.6 (4)	8.0±11.8 (4)

Values are means ± SD; in parenthesis is the number of animals.

[a] $P < 0.025$ compared with control group

[b] $P < 0.01$ compared with control group

(see Beato, this volume pp. 127-140). But one of the questions we are facing is the function of uteroglobin in blastocyst development and implantation. The purpose of this study is partly to throw light on this question. Using the experimental model mentioned above, we have gained interesting results concerning the premature changes of the endometrium and the accompanying changes of uteroglobin and progesterone in the uterine fluid. The results show clearly that it is the change of uterine fluid progesterone level, rather than that of the plasma progesterone, which parallels the changes of uteroglobin. This implies that progesterone in the uterine environment may play an important role in implantation. Finally, it is suggested that the premature changes of the endometrium and the uterine environmental factors caused by Progestin No. 1 may interfere with the synchronization between the development of the embryo and endometrium, and consequently prevent blastocyst implantation.

Table 3. Changes in the concentrations of estradiol-17β in plasma and uterine flushings after administration of Progestin No. 1 in rabbits

Days (pseudo-pregnancy)	Plasma (pg/ml)		Uterine flushings (pg/uterine horn)	
	Control	Treated	Control	Treated
1	17.8±13.2 (12)	17.8±12.2 (9)	34.8±17.3 (4)	23.4±8.8 (4)
2	21.9±19.5 (10)	22.0±20.5 (7)		
3	14.3±9.3 (10)	11.1±8.9 (7)	21.9±2.2 (4)	19.2±7.1 (3)
4	14.4±8.9 (8)	7.0±5.6 (5)		
5	14.7±9.8 (8)	13.4±6.0 (5)	29.0±8.6 (4)	22.4±8.2 (4)
6	12.8±11.1 (6)	7.1±2.9 (3)	17.9±11.0 (4)	33.3±2.9 (2)
7	6.1±8.7 (6)	8.0±3.2 (5)	23.3±1.3 (3)	13.2±5.8 (3)
8	29.8±41.1 (4)	11.8±6.2 (4)		
9	4.4±3.8 (4)	11.8±7.5 (4)	20.4±3.8 (4)	44.6±25.4 (4)

Values are means ± SD; in parenthesis is the number of animals. No values here are significantly different from those of the control group

Acknowledgments. The authors are grateful to Tieh-feng Zhang for his assistance in the SEM study and Xiao-yun Wu and Liu Hong for their technical assistance. Progestin No. 1 was kindly supplied by the Shanghai 12th Pharmaceutical Factory. Financial support for this investigation was gratefully received from the World Health Organization (Geneva). The cooperation with Prof. Henning Beier, Dr. Ursula Mootz, and Maria Petuelli from the Dept. of Anatomy and Reproductive Biology, University of Aachen (W. Germany) is gratefully acknowledged.

References

Arthur AT, Daniel JC Jr (1972) Progesterone regulation of blastokinin production and maintenance of rabbit blastocysts transferred into uteri of castrate recipients. Fertil Steril 23:115-122

Beier HM (1968) Uteroglobin: A hormone-sensitive endometrial protein involved in blastocyst development. Biochim Biophys Acta 160:289-291

Beier HM (1974) Oviducal and uterine fluids. J Reprod Fertil 37:221-237

Beier HM (1976) Uteroglobin and related biochemical changes in the reproductive tract during early pregnancy in the rabbit. J Reprod Fertil (Suppl) 25:53-69

Bullock DW, Willen GF (1974) Regulation of a specific uterine protein by estrogen and progesterone in ovariectomized rabbits. Proc Soc Exp Biol Med 146:294-298

Fowler RE, Johnson MH, Walters DE, Pratt HPM (1976) The progesterone and protein composition of rabbit uterine flushings. J Reprod Fertil 46:427-430
McPhail MK (1934) The assay of progestin. J Physiol (Lond) 83:145-156
Yang YC, Hong MX, Chen HQ (1979) Prematurity of human endometrium after oral administration of Progestin No. 1 compound tablet. Acta Physiol Sinica 31:283-294 (In Chinese with English abstract)

Subject Index

The underlined numbers mean that the entry appears on these pages and the following ones.

R. Baur

Morphometry of the Placental Exchange Area

1977. 37 figures. 65 pages
(Advances in Anatomy 53/1)
ISBN 3-540-08159-3

G. Dallenbach-Hellweg

Histopathology of the Endometrium

English Translation by F. D. Dallenbach

3rd revised and updated edition. 1981.
147 figures, 2 colored plates.
Approx. 330 pages
ISBN 3-540-10658-8
Distribution rights for Japan:
Igaku Shoin Ltd., Tokyo

Functional Morphologic Changes in Female Sex Organs Induced by Exogenous Hormones

Editor: G. Dallenbach-Hellweg

1980. 139 figures, 42 tables. XV, 234 pages
ISBN 3-540-09885-2

H.-W. Denker

Implantation

The Role of Proteinases, and Blockage of Implantation by Proteinase Inhibitors
Translated from the German

1977. 35 figures, 123 pages
(Advances in Anatomy, Embryology and Cell Biology, Volume 53/5)
ISBN 3-540-08479-7

H. Ludwig, H. Metzger

The Human Female Reproductive Tract

A Scanning Electron Microscopic Atlas

1976. 546 micrographs. XI, 247 pages
ISBN 3-540-07675-1

Pathology of the Female Genital Tract

Editor: A. Blaustein

1977. 1206 figures, 39 four-color figures.
XX, 897 pages
ISBN 3-540-90180-9

Plancental Proteins

Editors: A. Klopper, T. Chard

1979. 65 figures, 36 tables. X, 171 pages
ISBN 3-540-09406-7

Springer-Verlag
Berlin
Heidelberg
New York

Monographs on Endocrinology

Editors: F. Gross, M. M. Grumbach, A. Labhart, M. B. Lipsett, T. Mann, L. T. Samuels †, J. Zander

Volume 4
U. Westphal
Steroid – Protein Interactions
1971. 144 figures. XIII, 567 pages
ISBN 3-540-05312-3

Volume 7
E. W. Horton
Prostaglandins
1972. 97 figures. XI, 197 pages
ISBN 3-540-05571-1

Volume 8
E. Gurpide
Tracer Methods in Hormone Research
1975. 35 figures. XI, 188 pages
ISBN 3-540-07039-7

Volume 9
R. E. Mancini
Immunologic Aspects of Testicular Function
1976. 36 figures, 8 tables. IX, 114 pages
ISBN 3-540-07496-1

Volume 10
W. I. P. Mainwaring
The Mechanism of Action of Androgens
1977. 12 figures, 17 tables. XI, 178 pages
ISBN 3-540-07941-6

Volume 12
Glucocorticoid Hormone Action
Editors: J. D. Baxter, G. G. Rousseau
1979. 176 figures, 58 tables.
XIX, 638 pages
ISBN 3-540-08973-X

Volume 14
J. H. Clark, E. J. Peck, Jr.
Female Sex Steroids
Receptors and Function
1979. 116 figures, 18 tables. XII, 245 pages
ISBN 3-540-09375-3

Volume 15
A. T. Cowie, I. A. Forsyth, I. C. Hart
Hormonal Control of Lactation
1980. 64 figures, 7 tables. XIV, 275 pages
ISBN 3-540-09680-9

Volume 19
P. Mauvais-Jarvis, F. Kuttenn, I. Mowszowicz
Hirsutism
1981. 32 figures, 10 tables. XI, 110 pages
ISBN 3-540-10509-3

Springer-Verlag
Berlin
Heidelberg
New York

MIX
Papier aus verantwortungsvollen Quellen
Paper from responsible sources
FSC® C105338

If you have any concerns about our products,
you can contact us on
ProductSafety@springernature.com

In case Publisher is established outside the EU,
the EU authorized representative is:
Springer Nature Customer Service Center GmbH
Europaplatz 3, 69115 Heidelberg, Germany

Printed by Libri Plureos GmbH
in Hamburg, Germany